Contemporary Archipelagic Thinking

RETHINKING THE ISLAND

The Rethinking the Island series seeks to unsettle assumptions by comprehensively investigating the range of topological and topographical characteristics that lie at the heart of the idea of "islandness."

Series Editors

Elaine Stratford, Professor in the Institute for the Study of Social Change, University of Tasmania, Australia

Godfrey Baldacchino, Professor of Sociology and Pro-Rector at the University of Malta, UNESCO Co-Chair in Island Studies and Sustainability

Elizabeth McMahon, Associate Professor in the School of the Arts and Media, University of New South Wales, Australia

Titles in the Series

Theorizing Literary Islands: The Island Trope in Contemporary Robinsonade Narratives, Ian Kinane
Island Genres, Genre Islands: Conceptualization and Representation in Popular Fiction, Ralph Crane and Lisa Fletcher
Postcolonial Nations, Islands, and Tourism: Reading Real and Imagined Spaces, Helen Kapstein
Caribbean Island Movements: Culebra's Trans-insularities, Carlo A. Cubero
Poetry and Islands: Materiality and the Creative Imagination, Rajeev S. Patke
Contemporary Archipelagic Thinking: Toward New Comparative Methodologies and Disciplinary Formations, edited by Michelle Stephens and Yolanda Martínez-San Miguel
Atolls of the Maldives: Nissology and Geography, edited by Stefano Malatesta, Marcella Schmidt di Friedberg, Shahida Zubair, David Bowen, and Mizna Mohamed
Ecocriticism and the Island: Readings from the British-Irish Archipelago, Pippa Marland
The Notion of Near Islands: The Croatian Archipelago, edited by Nenad Starc
An Introduction to Island Studies, James Randall

Contemporary Archipelagic Thinking

Toward New Comparative Methodologies and Disciplinary Formations

Edited by Michelle Stephens and
Yolanda Martínez-San Miguel

ROWMAN & LITTLEFIELD
Lanham • Boulder • New York • London

Published by Rowman & Littlefield
An imprint of The Rowman & Littlefield Publishing Group, Inc.
4501 Forbes Boulevard, Suite 200, Lanham, Maryland 20706, USA
www.rowman.com

86-90 Paul Street, London EC2A 4NE, United Kingdom

First paperback edition 2024

British Library Cataloguing in Publication Information Available

Library of Congress Cataloging-in-Publication Data

The hardback edition of this book was previously cataloged as follows:

Names: Stephens, Michelle Ann, 1969– editor. | Martínez-San Miguel, Yolanda, editor.
Title: Contemporary archipelagic thinking : toward new comparative methodologies and disciplinary formations / edited by Michelle Stephens and Yolanda Martínez-San Miguel.
Description: London ; Lanham, Maryland : Rowman & Littlefield International Ltd., 2020. | Series: Rethinking the island | Includes bibliographical references and index. | Summary: "This anthology explores the archipelagic as both a specific and a generalizable geo-historical and cultural formation, occurring across various planetary spaces including: the Mediterranean and Aegean Seas, the Caribbean basin, the Malay archipelago, Oceania, and the creole islands of the Indian Ocean"— Provided by publisher.
Identifiers: LCCN 2020003472 (print) | LCCN 2020003473 (ebook)
Subjects: LCSH: Archipelagoes—Cross-cultural studies. | Island people—Cross-cultural studies.
Classification: LCC GB471 .C66 2020 (print) | LCC GB471 (ebook) | DDC 304.2/3—dc23
LC record available at https://lccn.loc.gov/2020003472
LC ebook record available at https://lccn.loc.gov/2020003473

ISBN 9781538149911 (paperback)
ISBN 9781786612762 (cloth)
ISBN 9781786612779 (electronic)

To all the graduate students whose engagements with
the archipelagic have inspired us.

To the Rutgers Advanced Institute for Critical Caribbean
Studies and the Center for Cultural Analysis, unique
interdisciplinary spaces for intellectual co-creation.

There's a network of us, strung out about the world.
The Archipelago, we're called.
Isolated above, but connected below.
I ask someone I know, who asks someone they know, who asks . . .
well, you get the gist.
Sapiens invented Google in the 1990s.
We've had it since the Neolithic.

—*Sense8*, "Isolated Above, Connected Below," Season 2, Episode 6

Contents

Acknowledgments

This collection of essays is the culmination of many conversations, conferences, panels, readings, shared thoughts, and journeys. So many colleagues, friends, and family members participated in the different iterations of this project. Their enthusiastic support and thinking informs this volume. The idea of the archipelagic emerged in divergent places and coalesced at Rutgers University, as part of the dialogues and collaborations fostered by the Rutgers Advanced Institute for Critical Caribbean Studies. Thanks to the vision and support of former executive dean of the School of Arts and Sciences Douglas Greenberg and former Rutgers president Richard Levis McCormick, a cluster hire in Critical Caribbean Studies made it possible for the two editors to cross paths with a stellar group of Caribbeanists with whom we discussed preliminary ideas on archipelagic and Caribbean studies. Our dialogues in the research cluster on Archipelagic Studies and Creolization with Yarimar Bonilla, Tatiana Flores, Renée Larrier, Tania López Marrero, Nelson Maldonado Torres, Carter Mathes, and Anjali Nerlekar were crucial in the beginning stages of this project.

During fall 2015 and spring 2016, the Center for Cultural Analysis (CCA) at Rutgers hosted a year-long seminar on "Archipelagoes" that served as the main incubator for the interdisciplinary dialogues at the foundation of this volume. Our colleagues William Galperin and Henry S. Turner, directors of the CCA at the time, Curtis Dunn and John D. Thomas, program coordinators, and Melissa Parrish and Miranda McLeod assisted us in putting together a spectacular group of seminar participants and guest lecturers. We would like to thank all the invited guests to the yearlong seminar on Archipelagoes for sharing with us their works in progress: Godfrey Baldacchino (Sociology, University of Malta); Françoise Lionnett (French and Francophone Studies, UCLA); Fidalis Buehler (Studio Art, Brigham Young University);

Juana Valdés (Printmaking, Florida Atlantic University); Mary Eyring (English, Brigham Young University); Brian Russell Roberts (English, Brigham Young University); Elizabeth DeLoughrey (English and Institute for the Environment and Sustainability, UCLA); Koichi Hagimoto (Spanish, Wellesley College); Craig Santos Pérez (English and Creative Writing, University of Hawai'i); Sherma Roberts (Management Studies, University of the West Indies at Cave Hill, Barbados); and Elaine Stratford (Geography and Environmental Studies, University of Tasmania, Australia). We would also like to thank the seminar participants: Sarah Demott (NYU), Jeremy DeAngelo (Postdoctoral Fellow, CCA), Elena Lahr-Vivaz (Spanish and Portuguese, Rutgers–Newark), Thomas Leppard (Postdoctoral fellow, CCA), Laura Lomas (English and American Studies, Rutgers–Newark), Kathleen Lopez (Latino and Caribbean Studies, Rutgers–New Brunswick), Kyle McAuley (English, Rutgers–New Brunswick), Anjali Nerlekar (African, Middle Eastern and South Asian Languages and Literatures, Rutgers–New Brunswick), Haruki Eda (Sociology, Rutgers–New Brunswick), Jenny Isaacs (Geography, Rutgers–New Brunswick), Enmanuel Martínez (Comparative Literature, Rutgers–New Brunswick), and Tashima Thomas (Art History, Rutgers–New Brunswick). Many of them became contributors to this volume, and Brian Russell Roberts (Brigham Young), Mary Eyring (Brigham Young), Susan Stanford Friedman (University of Wisconsin–Madison), and Lisa Swanstrom (University of Utah) continued and expanded the conversation at the *Archipelagoes/Oceans/Americas Symposium* organized by Brian Russell Roberts and his colleagues at Brigham Young University on October 5–6, 2016. Finally, through the call for papers for this anthology, we added our last group of collaborators: Christopher Lee (Lafayette College), Pippa Marland (University of Leeds), Lizabeth Paravisini-Gebert (Vassar College), Jonathan Pugh (Newcastle University), Gitanjali Pyndiah (University of London), Mimi Sheller (Drexel University), and Lanny Thompson (University of Puerto Rico). We would like to thank this wonderful network of scholars for allowing us to experience the richness, challenges, and transformative power of establishing and maintaining truly inter-, multi-, and transdisciplinary collaborations.

The coeditors of the Rethinking the Island book series, Elaine Stratford, Godfrey Baldacchino, and Elizabeth McMahon, enthusiastically embraced our proposal for an edited volume. Gurdeep Mattu and Natalie Bolderston, carefully shepherded the manuscript through its various editorial stages. The anonymous outside evaluator also provided very useful and detailed suggestions to revise this anthology and make it a better project. We would also like to thank the scholars who generously supported this project through the blurbs included on the back cover. Thanks to Rafael Burgos Mirabal (University of

Massachusetts Amherst) for his crucial editorial assistance with the introduction, and Lisa Rivero for her excellent work preparing the index for this volume. Finally, we would like to thank the various mentors who have guided the coeditors through our undergraduate and graduate trainings, showcasing for us how scholarly work is in many respects an archipelagic act of co-creation.

Yolanda Martínez-San Miguel wants to thank Michelle Stephens, her coeditor, for the many years of reflection about how archipelagic thinking could transform her work in Caribbean studies. It was precisely Michelle's use of the notion of the archipelagic in her first book *Black Empire: The Masculine Global Imaginary of Caribbean Intellectuals* (2005) that made her think about this theoretical framework as a productive point of departure for her future interventions in Caribbean studies. Chica, we have literally traversed many paths together, on Livingston campus, out of the king's office on Rutgers' College Avenue campus, in Provo, Utah, New York, El Yunque National Rainforest in Puerto Rico, Miami, and soon in Malta! She also wants to thank her students and colleagues at Rutgers–New Brunswick and the University of Miami for participating in the long meditation process that informs this volume. Carlos Decena, Nelson Maldonado Torres, Lillian Manzor, Camilla Stevens, and George Yúdice all heard some of the different versions of the questions posed in the introduction, as well as in many of the essays of this anthology. The research fellows participating at the David Rockefeller Center for Latin American Studies, as well as Lorgia García Peña, Gabriela Soto Laveaga, Brian D. Farrell, Andrew W. Elrich, Adrián Enmanuel Hernández-Acosta, Adri Rodríguez-Ríos, Edwin Ortiz, and Katerina González Seligmann (Emerson College) were all important interlocutors during her research semester at Harvard University. She would also like to thank her family and friends for supporting her work. Rígel Lugo and colleagues in *80 grados* read and published some preliminary versions of her work on archipelagic studies and the feedback provided in those readings encouraged her to refine her thinking. Carmen Yolanda San Miguel, María Mercedes (Mereche) Martínez-San Miguel, Alexandra Rodríguez, Eugenio Frías-Pardo, Jossianna Arroyo, Gloria Dámaris Prosper-Sánchez, Gina Ismalia Gutiérrez-Galang, Ben Sifuentes-Júaregui, Mark Trautman, Celinés Villalba-Rosado, José Quiroga, Arnaldo Cruz-Malavé, Marisa Belausteguigoitia, Luis Alvarez Icaza, Juan Carlos Quintero Herencia, Ivette Rodríguez, and Lourdes Martínez-Echazábal have all made possible her thinking and work. Finally, nothing in her professional and academic career would have been possible without the initial support she received from the Ford Foundation when she was a graduate student and a young junior scholar.

Michelle Ann Stephens wants to express her gratitude to Yolanda—who was one of the first two interlocutors to make her aware of the conceptual potential of the archipelagic in her work *Black Empire* (Brian Russell Roberts being the other). Yol, you've been inspiring and collaborating with me every step of the way at Rutgers—in co-creating the cluster on archipelagic studies for Critical Caribbean Studies; inviting me to join you in co-mentoring an undergraduate research group on the archipelagic; and, overall, offering your encouragement, respect and mentorship that has become the foundation for an enduring friendship and intellectual collaboration. Michelle also extends thanks to Executive Deans Doug Greenberg and Peter March, and the faculty and graduate students of Rutgers Advanced Institute for Critical Caribbean Studies (RAICCS), for providing her with another intellectual home in Caribbean studies after her journey through African diaspora studies. She also thanks the graduate students in and visitors to the spring 2013 Archipelagic American Studies seminar, who were so open to thinking with her about the deployment of this heuristic in diverse periods of English literature, and in other fields such as Comparative Literature and Women's & Gender Studies. Brian Russell Roberts, who visited that seminar and also continues to walk this path with her from their very first encounter over ten years ago, also serves as a crucial friend, mentor, and collaborator. Much love and appreciation to Sandra Stephens, Allie Tyre, Prudence Cumberbatch, and Enet Somers-Dehaney for the enthusiastic pep talks over the years; Louis Prisock for his friendship; and Alexandria Stephens Prisock, a continued source of sustenance and pride with each passing year.

This anthology would not have been possible without the generous support of several institutions. Rutgers University funded the CCA seminar and several lecture series on different aspects of archipelagic studies. Rutgers University and the University of Miami provided the necessary funds to cover the subvention of this publication, the copyright permissions to reproduce images and quotes in this anthology, as well as the indexing of this volume. The Wilbur Marvin Visiting Scholarship from the David Rockefeller Center for Latin American Studies at Harvard University and the University of Miami funded sabbatical time in which the introduction and editing of this volume were completed.

This volume is dedicated to our graduate students, who shared our enthusiasm for archipelagic thinking and took the idea in new directions. It is also dedicated to the Rutgers Advanced Institute for Critical Caribbean Studies and the Center for Cultural Analysis, two of the prestigious interdisciplinary centers at Rutgers University, models of intellectual co-creation in which the archipelagic framework found fertile ground, to become an ideal space and source of intellectual inspiration for many colleagues and students.

Archipelagic Poetics

A Foreword

Craig Santos Perez

I am from Guam, which is a small island (212 square miles) located in Micronesia, a region of the Pacific Ocean composed of thousands of small islands. However, Guam never felt "small" because when I was a child, it comprised my entire universe. In my imagination, the island was vast with histories, legends, and storied places, as Pacific scholar Epeli Hau'ofa has highlighted. Every village had its own character and depth, every beach had its own shapes and tides, and every family had its own genealogy and horizon.

My imagination expanded when I first saw Guam on a map and learned that the island is part of the Marianas archipelago. This imaginary came to life when I traveled to the other islands of the archipelago to visit relatives. Later, I would learn more about the other archipelagoes of Micronesia as many people from these islands migrated to Guam. When I was fifteen years old, my family migrated to California, and I experienced the relationships between and among island, archipelago, continent, and ocean.

The different spatial scales that I have lived in have shaped my thinking and poetics. When I write about my home island, I try to capture its "islandness," which is to say, I try to write about the unique roots of the island, its positive insularity. Moving beyond the island, I write archipelagic poems that make connections and comparisons across islands. While being a real space, the archipelago also becomes figurative for other global connections. I write poems that explore planetary archipelagoes through global themes, including food, water, migration, and more. Additionally, an archipelagic poetics also opens space to express solidarity with other peoples, places, and species who have similar struggles and experiences. Furthermore, the archipelagic links the island to the continent, thus making visible profound differences and power dynamics. Lastly, the ocean and oceanic represent the fluid connections and "tidalectics," to quote Kamau Brathwaite, between all spaces and beings.

The archipelagic has also shaped how I conceptualize poetic form. I see each word as an island with its own deep history and meanings. I understand phrases, sentences, and stanzas as textual archipelagoes. I view the blank spaces of the page as watery spaces of silence and contemplation. The reader, then, is invited to navigate the island, archipelagic, and oceanic spaces of the poem.

The poems included in this anthology move from the island of Guam to the continental United States. The poems map and chant the cartography of real and imagined archipelagoes, naming relations and solidarities. The poems praise water and the profound capacity of the ocean. These poems were inspired by a range of Pacific, Caribbean, and American scholars and poets, including Derek Walcott, Elizabeth DeLoughrey, Rob Wilson, Peter Neill, Sylvia Earle, Édouard Glissant, and Albert Wendt. Ultimately, the poems teach me how to navigate the complex coordinates and crosscurrents of my life as an indigenous, migratory, and global islander.

Chapter One

Introduction

"Isolated Above, but Connected Below": Toward New, Global, Archipelagic Linkages

Yolanda Martínez-San Miguel
and Michelle Stephens

Figure 1.1. Kathryn L. Chan, 2002, Limestone plaster, pigments/Fresco. Acknowledgement: Radcliffe Institute for Advanced Study

"ARCHIPELAGO"

We begin with a meditation on the cover art for this anthology, the piece *Archipelago* by Kathryn Chan.[1] We take this artwork as our point of departure for several reasons. The work is composed of pieces of limestone plaster that seem to come from a broken object. There is a certain randomness to the distribution, size, and shape of these pieces, giving an impression of the work overall as emerging from a planned accident. Yet the photographic intervention fixes a structure on that randomness, giving it a shape that, *all of a sudden*, signifies. Geological and human-made archipelagoes share an analogous origin or fate. Created by nature or by sociopolitical and historical processes, archipelagoes are accidents that can become significant, suddenly or over a long period of time.

The colors of Chan's work mix together in shades of blue and green, colors commonly used in maps to signify land and water. And yet, blue and green are unevenly distributed in the different fragments that compose the amorphous arrangement invoked as an archipelago in the title. Land and sea are therefore enmeshed, interlapping (Roberts 2016, 2019) throughout the smaller pieces that configure a network of fragments that are unified on the white paper that serves as surface and background of the work.[2]

In the context of Chan's larger body of work, there are additional layers that complicate our initial reading of her art. In an interview with Patricia Mohammed (2014), the artist reveals that one of the sources inspiring *Archipelago* is Derek Walcott's Nobel Prize lecture "The Antilles: Fragments of Epic Memory." In his speech, Walcott (1992) refers to the complex histories of African and Asian diasporas in the Caribbean as a process of assembling broken pieces:

> Break a vase, and the love that reassembles the fragments is stronger than that love which took its symmetry for granted when it was whole. The glue that fits the pieces is the sealing of its original shape. It is such a love that reassembles our African and Asiatic fragments, the cracked heirlooms whose restoration shows its white scars. This gathering of broken pieces is the care and pain of the Antilles, and if the pieces are disparate, ill-fitting, they contain more pain than their original sculpture, those icons and sacred vessels taken for granted in their ancestral places. Antillean art is this restoration of our shattered histories, our shards of vocabulary, our archipelago becoming a synonym for pieces broken off from the original continent.

Chan therefore connects the apparently disparate pieces of the multiple, translocal human displacements that define the Caribbean as a complex and apparently random process that signifies. Yet both Chan and Walcott also suggest that the process that defines the Caribbean's multilayered ancestry is all but accidental. Coerced diasporas and voluntary displacements, colonialism and imperialism, global capitalism and modernity are all implicated as the forces scattering these painful, ill-fitting, scarred fragments from their original continents and restoring them into a different imaginary and narrative in the "islands of the sea" (Martí 1891). Chan's and Walcott's imaginaries of the Caribbean go beyond *criollismo*, creoleness, *mestizaje*, and *mulataje* to include the presence of Asian populations that inevitably complicate current definitions of Caribbean, Latin American, and American identities.[3]

Another of *Archipelago*'s relevant dimensions derives from the process Chan used to make the work, which Paul Ramírez Jonas describes as "a field of broken frescoes and paintings on terra-cotta. Some of the fragments are conceived as such, and are constructed and painted as parts; others are shards of larger objects. As a reconciliation of fragments, both those that came from a whole and those that will be forever incomplete, the piece seems to encompass her entire project" (Ramirez Jonas 2003). *Archipelago*'s "field" of objects combines broken frescoes on terracotta with pieces painted to blend in with the broken pieces. In this context, the archipelago is both an accident and a designed arrangement of complete and incomplete pieces. Some pieces in this field of objects are broken; others emerge or are fabricated anew and become native, organic to this context. In Chan's articulation of the archipelago, the dispersed network of elements articulating the spatio-visual experience includes both shards and fabricated objects.

Contemporary Archipelagic Thinking takes as its point of departure a more flexible definition of the archipelago, exploring it as a lens that may allow us to engage in interdisciplinary conversations about the ways in which space and time are resignified. These resignifications occur and recur in a complex set of human, object, and (natural or built) surface relations that can congeal into a particular meaning, which can then also become permanent or remain ephemeral. In our understanding, the archipelago calls for a meaning-making and rearticulation that responds to human experiences traversing space and time. Archipelagoes happen, congeal, take place. They are not immanent or natural categories existing independently of interpretation. Yet they can also become an episteme, an imaginary, a way of thinking, a poetic, a hermeneutic, a method of inquiry, a system of relations. They are painful and generative, implicated in native cosmologies or cosmo-visions, or assembled as part of imperial/colonial undertakings. They can refer to multidimensional, focal, spatial forms of thinking that emerge from concrete relationships with inhabited spaces. But, as Walcott suggests, they also require a loving reassembling that signifies beyond the dehumanizing centripetal forces of globalization.

The point of departure for our framing of archipelagic thinking is the Caribbean, and more specifically, works by Édouard Glissant and Derek Walcott. In his *Traité du tout-monde*, Glissant (1997) attempts to conceive notions of global displacement and miscegenation in the Caribbean. *Métissage*, multilingualism, and creolization allow Glissant to question the limitations of the notion of system to grasp the complexity of the human, cultural, and social experiences taking place in the Caribbean region (24). One of the main questions he explores is this: How can one explain the historical and cultural experiences of societies formed as a result of the juxtaposition of several diasporic and translocal populations, hailing from Africa, Asia, and Europe, which severely decimated the indigenous communities of the region?

Notions such as the world-chaos, the trace, relation, and the rhizome are central to how identity is conceptualized as a process in Glissant's work,[4] a process in which there can be more than one point of origin or more than one root and in which totality and globalization are conceived as diversality—or a universal diversity, as the *créolité* group would identify it (Bernabé,

Chamoiseau, and Confiant 1990, 114). Antonio Cornejo Polar (1994, 17) defines this diversality as heterogeneity when referring to cultural formations in Latin America. According to Glissant (1997, 28–29), the concept of relation is lived through creolization and performed through the multilingual contexts in which Creole languages emerge in the Caribbean. In that context of intense difference in continuous contact, opacity and translation work like the two sides of the same coin: opacity indexes a difference that cannot be rendered comprehensible, and translation becomes that art of approaching differences that does not negate the irreducible alterity of the other:

> L'art de traduire nous apprend la pensée de l'esquive, la practique de la trace qui, contre les pensées de système, nous indique l'incertain, le menacé, lesquels convergent et nous renforcent. Oui, la traduction, art de l'approche et de l'effleurement, est une fréquentation de la trace. . . . Traduire ne revient pas à reduire à une transparence, ni bien entendu à conjoindre deux systèmes de transparence. (28–29)
>
> The art of translation teaches us to think about evasive thinking, the practice of the trace that, against the notions of the system, leads us to the uncertain, the threatened, which converge and strengthen us. Yes, translation, the art of approximation and the gentle touch, is a way to get close to the trace. . . . To translate is not to reduce to transparency, nor does it combine two systems of transparency. (Our translation, 28–29)[5]

It is precisely in the context of this negotiation of differences that, according to Glissant, we should characterize the kind of work done within the complexly heterogeneous societies of the Caribbean. The archipelagic is proposed as an alternative framework, epistemology, or way of thinking:

> La pensée archipélique convient à l'allure de nos mondes. Elle en emprunte l'ambigu, le fragile, le dérivé. Elle consent à la pratique du détour, qui n'est pas fuite ni renoncement. Elle reconnaît la portée des imaginaires de la Trace, qu'elle ratifie. Est-ce là renoncer à se gouverner ? Non, c'est s'accorder à ce qui du monde s'est diffusé en archipels précisément, ces sortes de diversités dans l'étendue, qui pourtant rallient des rives et marient des horizons. Nous nous apercevons de ce qu'il y avait de continental, d'épais et qui pesait sur nous, dans les somptueuses pensées de système qui jusqu'à ce jour ont régi l'Histoire des humanités, et qui ne sont plus adéquates à nos éclatements, à nos histoires ni à nos non moins somptueuses errances. La pensée de l'archipel, des archipels, nous ouvre ces mers. (31)
>
> The archipelagic mode of thought suits the pace of our worlds. It borrows from them its ambiguity, its fragility, its derivative drift. It yields to the practice of rerouting, which is neither flight nor renunciation. It acknowledges the scope of the imaginary constructions of the Trace, which it ratifies. Does this mean

> to abrogate self-[governance]? No, it means, precisely, to be in harmony with whatever in the world is scattered through archipelagoes, [the kind of diversity stretched out across the world that joins shores and horizons together]. We become aware of what was continental, thick and weighing heavy on us, in lavish reflection on that system which until our times has governed the History of humanity and which is no longer appropriate to the [ruptures] we experience, neither to our histories nor to our less lavish acts of errantry. [Archipelagic thinking,] of archipelagoes, opens these oceans to us. (31, 237–38)[6]

Glissant criticizes linear, teleological notions of human history that are based on theories of systems to favor conceptualizations that can grasp *éclatements* (ruptures), ambiguity, and detours. He questions continental frameworks that assume synthesis, unity, and coherence as organic to human social experiences, proposing instead the archipelagic as a spatial and symbolic allegory of the theory of relation conceived as differences that interact, collide, and coexist through time and space. The archipelagic functions, then, beyond the geographical or geopolitical. Instead, it is another framework to conceive human social formations and historical experiences in which irreducible differences become the norm instead of an exception that needs to be processed into synthesis, harmony, and consensus.

In "The Antilles: Fragments of Epic Memory," Walcott (1992) also reflects on the difficulties of conceiving an epic memory for the Caribbean in the context of the complex coexistence of Afro-Asian populations in Trinidad. The epic, a foundational literary genre intimately linked to the formation of national or collective identities in Western literature (Lukács 1971; Bakhtin 1981), is confronted in Walcott's text with the notion of epic mobilized in the "Hindu epic the *Ramayana*."[7] Walcott interrogates the epic from its inception as a culturally and historically specific genre and argues that discussions and understandings of the genre need a more complex engagement to be able to offer an account of communities that are not imagined from the vantage point of the sovereign national state. Walcott reflects explicitly on his deep ignorance when confronted with East Asian cultural practices and traditions in Trinidad and his utter dismissal of the immediacy and reality of the Ramleela,[8] which he originally conceived as a theatrical performance instead of as a tradition encompassing forms of being and conceiving of the world that were also deeply Caribbean. Epic memory becomes in his essay "the celebration of a real presence" instead of "something left behind" (Stephens 2015, 283). For Walcott, creolization is a challenge still to be accomplished. His locus of enunciation is that of the writer and thinker who identifies the fragments and not that of the philosopher who attempts to make sense of internal differences by proposing a master narrative that still signifies in a global manner.

The archipelagic is used by Walcott to refer to the historical and cultural formations that take place away from continents and that are transformed through links of solidarity established between the diverse members of the multiply diasporic societies of the Caribbean. For Walcott, the role of art is to create narratives that put together the ill-fitting pieces of diasporic and colonial fragmentation to account for the social and cultural formations that coexist and interact in the Caribbean. Artists build imaginary narratives that could provide a heroic past and collective present in which the African and Asian dimensions of Caribbean histories and identities become visible against the Eurocentric frameworks of Western history:

> All of the Antilles, every island, is an effort of memory; every mind, every racial biography culminating in amnesia and fog. Pieces of sunlight through the fog and sudden rainbows, *arcs-en-ciel*. That is the effort, the labour of the Antillean imagination, rebuilding its gods from bamboo frames, phrase by phrase.
>
> . . .
>
> It is not that History is obliterated by this sunrise. It is there in Antillean geography, in the vegetation itself. The sea sighs with the drowned from the Middle Passage, the butchery of its aborigines, Carib and Aruac and Taino, bleeds in the scarlet of the immortelle, and even the actions of surf on sand cannot erase the African memory, or the lances of cane as a green prison where indentured Asians, the ancestors of Felicity, are still serving time. (Walcott 1992)

For Glissant and Walcott, the archipelagic is conceived of as a framework to theorize multiplicity and difference, while undoing central tenets in historical (root, teleology) and literary (the epic) studies and embracing the multifocal, multisited articulation of human experience.

Contemporary Archipelagic Thinking takes this question we have explored in our individual scholarship about the Caribbean and expands it to a global framework without erasing its local dimensions. In other words, this anthology explores the archipelagic as both a specific and a generalizable geohistorical and cultural formation, occurring across various planetary spaces, including the Mediterranean and Aegean Seas, the Caribbean Basin, the Malay Archipelago, Oceania, and the creole islands of the Indian Ocean. As an alternative geoformal unit, archipelagoes can interrogate epistemologies, ways of reading and thinking, and methodologies informed implicitly or explicitly by more continental paradigms and perspectives. Keeping in mind the structuring tension between land and water and between island and mainland relations, the archipelagic focuses on the types of relations that emerge, island to island, when island groups are seen not so much as sites of exploration, identity, sociopolitical formation, and economic and cultural circulation but also, and rather, as models.

In this moment, archipelagic studies has become a framework with a robust intellectual genealogy (Batongbacal 1988; Baldacchino 2015; DeLoughrey 2001, 2010, 2019; McCall 1994; Roberts and Stephens 2017; Stratford et al. 2011; Thompson 2010, 2017, among others). The strength of our anthology is the diversity of fields and theoretical approaches in the humanities, social sciences, and natural sciences that the included chapters engage. We invited authors to focus more directly on the contributions, challenges, interventions, and productive intersections between contemporary archipelagic thinking and the disciplinary engagements of their work.

We also take as our focus a notion of the archipelagic as a global formation that forces comparative models and methodologies. However, archipelagic comparisons take as their subject a field of objects with multiple, triangulated, networked relations rather than the more typical modes of one-to-one comparison between two paired objects of analysis. Archipelagic thinking, grounded as it is in assemblages of island, continent, and sea, requires a conceptualization of the global that is forced to do more, geographically, geohistorically, and geopolitically, to differentiate islands from each other while theorizing their connectivities and commonalities. It requires micro as well as macro analyses, a focus on materiality as much as metaphor, a turn to the glocal, to find the different valences of the global and the local that are operating simultaneously.

With this glocal organizing principle in mind, we heed Brian Russell Roberts's call in his intervention in this volume to explore a multilingual and multiregional genealogy for archipelagoes, one that emerges from space, time, and humanity as experienced in archipelagic contexts. This genealogy will help to contextualize the contributions and challenges of using archipelagic frameworks in a whole array of research projects taking place in anthropology, archaeology, cartography, cultural and literary studies, disability studies, ecological and environmental studies, ethnomusicology, migration studies, philosophy, sociology, and spatial studies. Aiming to foster multi-, inter-, and transdisciplinary dialogues, *Contemporary Archipelagic Thinking* offers a set of exploratory meditations on the new frameworks that become possible when focusing on the built, systemic interconnections underlying human understandings of space and time.

PLURIVERSAL ETYMOLOGIES: TOWARD MULTILINGUAL GENEALOGIES FOR ARCHIPELAGOES

The basic definition of an archipelago is a cluster or group of islands (Baldacchino 2015, 6), but the scholarship on the subject has explored other angles on this term that focus instead on island-to-island relationships (Stratford et al. 2011, 116). Furthermore, an even wider comparative lens, one that focuses on archipelago-to-archipelago relations, is made possible when, for example, one reconnects the Caribbean to a whole series of archipelagic regions that share some similar kinds of historical, political, social, and environmental features. In the Caribbean more specifically, important artistic meditations linking "poetics and the relationality of the archipelago" (Stratford et al. 2011, 118) have been made possible by the archipelagic framework.

While many meditations refer to the Greek etymology of the word "archipelago,"[9] we would like to displace our point of departure to trace an alternative intellectual genealogy through a more multilingual and multiregional etymology of cognate words. Although still firmly based in the Caribbean, our thinking departs from Benítez-Rojo's foundational formulation of the "repeating island" to focus instead on spatial articulations that are based not necessarily on historico-political similarities but on the creative *rearticulations* promoted by connectivity and interactions. In this regard, archipelagic frameworks privilege spatial and embodied theorizations over synthetic conceptualizations based on biological, racial, ethnic, or cultural mixing (as is the case with tropes of the melting pot, syncretism, *mestizaje*, *mulataje*, creolization, hybridity, or transculturation). In "Island Movements: Thinking with the Archipelago," Jon Pugh (2013, 11) defines the archipelagic as an ontology based on islands as "sites of abstract and material relations of movement and rest, dependent upon changing conditions of articulation or connection." Central in these ontological and theoretical articulations are concrete spatial references as well as significant population displacements, which redefine the ways in which these places are conceived as creative, generative, and transformative locations of belonging and identification. Pugh refers to metamorphosis in his essay to protect the dynamic nature of the archipelagic as a praxis that goes beyond imitation:

> It might be tempting at this point to make a connection to Benítez-Rojo's (1996) notion of the "repeating island," but I feel that this idea is problematic. Too easily it suggests shared island experiences repeating across the chain. While the Caribbean has shared many experiences, indeed is bound by one overarching and archipelago experience called slavery, Walcott's way of conceptualizing inheritance accents the contingency of the form more effectively than does Benítez-Rojo. The real force of the Caribbean island archipelago movement is a metamorphosis that emphasizes invention and creation. (17)

In this context of resisting narratives with a focus on continuous or consistent genealogies, lineages, causalities, or teleologies, notions such as Brathwaite's *tidalectics* (34) and Glissant's rearticulation of Deleuze and Guattari's *rhizome* ("Introduction," *Poetique du divers*) seem to be more productive. In both cases, the movement of the waves of the sea or the intertwined root patterns of the rhizome serve as material and concrete phenomena that are used to theorize the interconnected nature of certain experiences. The archipelagic functions as a mode of thinking or existing that allows us to conceptualize multiple locations or spaces that are simultaneously autonomous but "act in concert" (Baldacchino 2015, 6).

As part of our effort to decentralize single narratives to include multiple locations that act in some sort of coordination, we would like to propose a "pluriversal etymology" of the archipelagic. One can trace these strands of meanings through words that, while not cognates in the truest sense (as in having a common etymological origin), are alike in sharing a related purpose or associated set of meanings. Words and phrases such as "Antilles," the "Antillean confederation," *antillanité* (Glissant), *tuvalu* (eight together), *nusantara* (outer islands), and "sea of islands" (Hau'ofa 1993) represent specific cultural and linguistic articulations of the archipelagic in the Caribbean and the Pacific. As with some of the terms discussed before (rhizome or tidalectics), these cultural references to the archipelagic are based on concrete social, historical, political, and natural spaces resignified by and in the interaction with human beings.

In the case of the Caribbean, there are two concrete references to the region as a conglomerate of islands that are conceived as part of the same unit. First, there is the notion of the Antilles, used since the fifteenth century to refer to the islands of the Caribbean. The name was originally a reference to a legendary or imaginary island located in the Atlantic Ocean, to which Iberian settlers fled during the Muslim conquest of Hispania, thereby linking this region with the preservation of Hispanism (Babcock 1920, 109–10):

> It is from Christian Iberia that the legend of *Antillia* emerged. According to the legend, in c. 714, during the Muslim conquest of Hispania, seven Christian bishops of Visigothic Hispania, led by the Bishop of Porto, embarked with their parishioners on ships and set sail westward into the Atlantic Ocean to escape the Arab conquerors. They stumbled upon an island and decided to settle there, burning their ships to permanently sever their link to their now Muslim-dominated former homeland. (Fritzinger 2016, 25)

One of the most common etymologies of the term proposes that the word refers to the Portuguese *ante-ilha*, "meaning apparently the 'nearer island' or 'an island lying off the coast of a larger,'" and is believed to represent "'pre-Columbian' knowledge of the American continent" (Crone 1938, 260). Others propose that the term may be a deformed transcription of Plato's Atlantis (Fritzinger 2016, 27), or a reference to Antullia (which means "before the shores of Atullia") (Crone 1938, 260). Humboldt proposes a highly unlikely Arabic etymology that means "dragon" (*Examen critique de l'histoire de la géographie du Nuveau Continent*, p. 211, in Babcock 1920, 112). The term "Antilles" is commonly used to refer to the West Indies, or the overseas colonial insular possessions that are the counterpart of the East Indies (or the lands of Southeast Asia).[10] As such, and regardless of the correct etymology, "Antilles" refers to a group of islands located in the Caribbean Sea.

In the history of the region, the Antilles have been invoked as a collective in the Antillean League and the Antillean Confederation, political projects mostly influential in the Spanish Caribbean of the nineteenth century that aspired to create a confederation of several Caribbean islands with the purpose of articulating a multiple-island political state, taking the black republic of Haiti as a point of departure (Buscaglia-Salgado 2003; Chaar-Pérez 2013; Firmin 1910; Jossianna Arroyo 2011; Reyes-Santos 2013). In the mid-twentieth century, another iteration of a similar project in the Anglophone Caribbean would receive the name of the West Indies Federation (1958–1962) (James 1984; Stephens 2015). Both projects had important foundations in the common denominator of the black diasporas that informed the cultural articulation of Afro-Caribbean identities. In both cases the project of creating a regional political organization to preserve the sovereignty of the individual islands was forged in a subsection of the Caribbean (the Hispano or Anglophone Antilles) and culminated in political movements that led to the independence of several islands in the respective subregions. Therefore, from its inception, the notion of the federation or confederation played out the centripetal and centrifugal tensions between nationalism and transnationalism in the Caribbean. In both cases as well, the archipelagic functions as a political articulation that is built or congeals around an imaginary or historico-political project, but one that does not become a definitive or organic articulation for the region.

In the French Caribbean, on the other hand, a notion of the Antilles has been incorporated through Glissant's (1981) theorization of *antillanité*. Originally conceived as a response to the "Négritude movement" developed in the 1930s and 1940s by Léopold Sédar Senghor (Senegal), Aimé Césaire (Martinique), and Léon Damas (French Guyana), this term refers to the diverse cultural articulations of the Caribbean that resist a single origin in Africa and that showcase the multifocal, multiracial, multilingual, and transcontinental identities being configured by the multiple diasporas and the extended colonial contexts in the Caribbean. *Antillanité* then insists on the lack of a single origin, and the discontinuous territory is seen as central to comprehending the collective formations in the Caribbean. Crucial in the articulation of this concept is Glissant's concept of relation, which allows for a more capacious understanding of the cultural and historical developments in the region: "Associated with *antillanité* is the notion of relation, which expresses a preexisting reality, made of muzzled desires and ambitions, a state that allows the establishment of a connection between the other islands of the region that would counterbalance any national sentiment—although a necessity in itself—should such sentiment become overly nationalistic in actions" (Priam 2010).

In the Pacific, *tuvalu* refers to an island nation located between Hawai'i and Australia, and literally means "eight standing together."[11] Stratford et al. offer a summary of the meaning of the term:

> Tuvalu . . . is inherently archipelagic. Its origins are uncertain, but it seems probable that the word "Tuvalu" gained significance only after contact with non-Oceanic peoples. Tu means "to stand" and valu means "eight." The name Tuvalu draws on an identity based on shared senses of competition and cooperation between and among the communities of eight of nine islands in the archipelago—all except one being traditionally inhabited. . . . ["Tuvalu"] is a word that reflects the importance of the eight *fenua* that comprise this archipelagic state: a term denoting an island, its communities, and how community life is enacted in place and made mobile across places. (2011, 123)

A nation composed of islands and atolls, the region conceives of itself as a national unit and has produced a creation myth for the islands (or what Walcott would conceive of as an epic) that accounts for the collective identity contained in a region composed of a discontinuous geographical territory. The idea of a nation that is defined by the shared and explicit contiguity of several insular regions is central in their articulation of space, history, and political systems. Although the declaration of independence in 1978 seems to confirm the legality and popularity of the term, Stratford et al. also recognize the importance of a lived experience in which acting in concert is central as the articulating force behind the local definition and understanding of the term.

In the Indonesian case, the Javanese term *nusantara* would be the equivalent to "archipelago," but it literally means "outer islands." According to Roberts and Stephens (2017, 30), the term indexes an "island-centric mode of envisioning the world beyond Java." Yet what draws our attention now is the articulation of space as a conjunction of several insular territories that conceive of themselves as part of the same unity. In the sixteenth century the term was used to refer to regions that were outside Java's nuclear culture but functioned as colonial possessions (that is, outer islands) (Butcher and Elson 2017, 136). Between 1967 and 1973 the concept was redefined "Wawasan Nusantara" as part of a national guideline. It literally means "archipelagic outlook" and proposed that all the islands relate to Indonesia as a single unit "to mold Indonesian society into an integrated whole in which the population would focus on economic development rather than politics and class, ethnic or regional interests" (205). Today the term means "Indonesian archipelago," and the original colonial referent of the word has disappeared in favor of national affiliation and unification. The archipelagic is therefore activated as an act of interpretation, identification, and human will that in this case is also channeled through the sovereign nation-state, which has also experienced several moments of crisis.

Epeli Hau'ofa is one of the central archipelagic thinkers to engage the topic of creating frameworks and interpretations to conceptualize the Pacific islands in the region of Oceania.[12] In his foundational essay "Our Sea of Islands," Hau'ofa questions the representation of islands as small places that are condemned to be dependent on continental metropolitan centers. Instead, he engages in the process of rearticulating insular lived spaces that do not rely on macroeconomic and global historical and sociopolitical frameworks but dwell instead in the perspectives of the people who inhabit them. Embodied experiences, myths, and imaginaries offer a different perspective in which islands become infinite, interconnected spaces projected to universal and global frameworks. Hau'ofa proposes what could seem to be a minimal linguistic intervention that translates into a complete transformation of the visualization of insular spaces:

> Continental men, namely Europeans, on entering the Pacific after crossing huge expanses of ocean, introduced the view of "islands in a far sea." From this perspective the islands are tiny, isolated dots in a vast ocean. Later on, continental men—Europeans and Americans—drew imaginary lines across the sea, making the colonial boundaries that confined ocean peoples to tiny spaces for the first time. These boundaries today define the island states and territories of the Pacific. I have just used the term "ocean peoples" because our ancestors, who had lived in the Pacific for over two thousand years, viewed their world as "a sea of islands" rather than as "islands in the sea." This may be seen in a common categorization of people, as exemplified in Tonga by the inhabitants of the main, capital island, who used to refer to their compatriots from the rest of the archipelago not so much as "people from outer islands," as social scientists would say, but as *kakai mei tahi* or just *tahi* "people from the sea." This characterization reveals the underlying assumption that the sea is home to such people. (1993, 152–53)

In this passage Hau'ofa recovers a particular idiomatic expression that contains a different worldview. By opening the islands to the ocean, and by proposing islands as a network of discontinuous landmasses connected by human collective experiences, the small isolated imperial conceptualization of the overseas possessions is transformed into a self-contained autonomous and expansive social and collective experience. It is interesting that in this passage Hau'ofa refers to the "outer islands" already mentioned in the definition of *nusantara* as a sign of imperial/colonial exclusion that does not correspond to the embodied experiences of the communities that inhabit these regions. This intervention is precisely the kind that we hope to accomplish by recovering the archipelagic as a lens, framework, epistemology, heuristic, or approach from which to rethink the inhabited experiences of many communities throughout the world. In all the cases we have discussed, the reference is to the lived space as a network of islands in an explicit or concerted effort to signify together.

Although our understanding of the term is derived from and based on the geological and political materiality of archipelagoes as groups of islands, in this anthology we are proposing the archipelagic as a prism or method with which to reengage central disciplinary questions in area, interdisciplinary, and transdisciplinary studies. As a working definition of how archipelagoes and the archipelagic are conceived in this anthology, here they are defined not only as a system of islands but also as a set of humanly constructed relations between individual locations (islands, ports, cities, forts, metropoles, communities). The archipelagic is conceived, therefore, as a set of relations that articulates cultural and political formations (collectivities, communities, societies), modes of interpreting and inhabiting the world (epistemologies), and symbolic imaginaries (as a poetic but also as habitus). It is important for us that the archipelagic has a material foundation that is geographical, geological, and spatial. It is also significant that the archipelagic defines inhabited spaces through the networks of human relations that make them signify in a specific way. *Contemporary Archipelagic Thinking* takes the insights of lived spaces as a point of departure to think about the modes of thinking that become possible when we engage knowledge production as an embodied experience that is predicated on the powerful implications of Glissant's "poetics of relation," defined as "rhizomatic thought . . . in which each and every identity is extended through a relationship with the Other" (1997, 11).

And still, the question arises, why do we need archipelagic frameworks? What kind of work specifically is done by this term?

WHAT'S IN A WORD? ARCHIPELAGOES IN TRANSDISCIPLINARY CONTEXTS

Sense8, the US science fiction television show written and produced by Lana and Lilly Wachowski and J. Michale Strczynski for Netflix (2015–2018), follows the life stories of eight characters of different cultures, ethnicities, and sexual orientations, who are located in Berlin, Chicago, London, Mexico City, Mumbai, Nairobi, San Francisco, and Seoul. The eight characters are *sensates*, who share a psychic connection. The premise of the show is that certain human beings are born in clusters that share psychic connections, which allow them to share experiences that are perceived by one or more of the human senses. In the second season of the show, the notion of the "archipelago" appears, referring to the informal networks linking these psychic human beings globally: "There is a network of us, strung out about the world. The Archipelago we are called. Isolated above, but connected below. I ask someone I know, who asks someone they know, who asks . . . well, you get the gist. Sapiens invented Google in the 1990s. We've had it since the Neolithic" ("Isolated Above, Connected Below," *Sense8*, Season 2, Episode 6). As one character, Old Man of Hoy, describes this global network to another, Riley Blue, this connectivity among human beings exists long before Google and the globalized technological networks invoked by the internet. Rather, like the archipelago, the sensates' link operates like an organic, informal network, one that precedes the technological iterations of the cybernetic platforms. Archipelagoes, like many islands, are rhizomatically "[i]solated above, but connected below," suggesting a form of concerted articulation that has been conceived though other theoretical models that can serve as useful points of comparison and contrast. What is it that the archipelago invokes, then, that is different from the kinds of connections and interactions that are indexed by keywords such as network theory, assemblages, systems theory, or the study of islands, oceans, and constellations?

These are common referents in discussions about archipelagic theories: systems, network and actor-network theory, assemblages, and constellations. Our work also engages in very productive dialogues with island, ocean(ic), and nissology studies. Together they offer important insights and enrich our understanding of contemporary archipelagic thinking and methods, with theories that have been developed to attempt to understand complex human social and historical experiences. Systems theory assumes a structure in which an organized entity functions as a result of minor parts that are interrelated and interdependent (Becker and Seidl 2007; Fischer-Lescano 2011). A foundational text is *General System Theory: Foundations, Development, Applications*, authored by biologist Karl Ludwig von Bertalanffy in 1968. Many of the principles of systems theory are based on our understanding of the natural world or living organisms, articulated as the mostly harmonic or collaborative organization of several organs or particles, such as atoms, protons, neutrons, and electrons, toward the functioning of a bigger unit. This theory assumes a structure in which the existing parts collaborate toward a common goal.

Network theory offers an alternative to systems theory by conceptualizing relationality from a nonorganic and biological perspective and by considering symmetric and asymmetric relations between objects. Although both theories have applications in many different disciplines, network theory has had a significant impact on scientific fields of inquiry, especially in those closely related to electronic and cybernetic technologies. Interesting applications of network theory have informed the study of social experiences: "*What is a network?* A social network consists of a set of actors ("nodes") and the relations ("ties" or "edges") between these actors. . . . The nodes may be individuals, groups, organizations, or societies. The ties may fall within a level of analysis (e.g., individual-to-individual ties) or may cross levels of analysis (e.g., individual-to-group ties)" (Katz et al. 2004, 308). Scholars who use network theory to study social relations have identified some fundamental principles that provide coherence to this school of thought. Some of these principles are that (1) human behavior is closely linked to the web of relationships in which an individual is inscribed instead of a result of isolated variables such as desire, demographic characteristics, and individual attributes; (2) population sectors exist in relation and not as autonomous units; and (3) social systems are more than the aggregates of single parts and often have fluid boundaries (Barry Wellman in Katz et al. 2004, 311–12).

In the 1980s Bruno Latour, Michel Callon, and John Law developed the actor-network theory (ANT), which further complicates the conceptualization of complex social systems. Moving beyond the natural and human-centric models of systems and network theory, ANT considers nonhuman agents and incorporates discursive analysis and the production and dissemination of meaning as a crucial dimension in the articulation of multifocal relationality:

> ANT advances a *relational materiality*, the material extension of semiotics, which presupposes that all entities achieve significance in relation to others. Science, then, is a network of heterogeneous elements realized within a set of diverse practices. . . . Taking seriously the agency of nonhumans (machines, animals, texts, and hybrids, among others), the ANT network is conceived as a heterogeneous amalgamation of textual, conceptual, social, and technical actors. (Crawford 2005, 1)

This theory adds the notion of the actor as things, identities, discursive, and social practices that produce meaning by originating action. Yet, in its efforts to be more inclusive of a wider variety of variables in the conceptualization of relationality and its centrality in the production of meaning, ANT has been criticized for its lack of sensitivity to the effects of asymmetrical power relations, racism, colonialism, and social class in the configuration of the possible relations for a particular individual, social group, or discursive practice.

Bruno Latour has been one of the more widely read thinkers of ANT among humanists and social scientists. In "On Actor-Network Theory: A Few Clarifications Plus More Than a Few Complications," Latour reviews some of the main areas of misunderstanding related to ANT. He describes the main contribution of ANT: "As a first approximation, the A[N]T claims that modern societies cannot be described without recognizing them as having a fibrous, thread-like, wiry, stringy, ropy, capillary character that is never captured by the notions of levels, layers, territories, spheres, categories, structure, systems" (Latour 2017, 176). For Latour, ANT allows for a different mode of thinking and conceptualizing complex relations beyond notions of small/large scale, inside/outside, boundary, close and far, up and down, local and global, inside and outside, among others (5–6). He further clarifies that actors should be conceived of as actants, or generators of action, and as such can include human and nonhuman entities that are not fixed but circulating objects (7–8). Finally, an important consideration for understanding ANT is that while it initially follows the semiotic principle of autonomous signification, it eventually reincorporates material contexts as fundamental in the articulation of meaning: "Building on the semiotic turn, ANT first brackets out society and nature to consider only meaning-productions; then breaking with the limits of semiotics without losing its tool box, it grants activity to the semiotic actors turning them into a new ontological hybrid, world making entities; by doing such a counter-copernican revolution it builds a completely empty frame for describing how any entity builds its world" (187). In Latour's conceptualization of ANT, material and social contexts are significant dimensions in the production of meaning and cannot be completely removed from the analysis of relationality. Together, science and semiotics contribute to create a more comprehensive understanding of action, relationality, and meaning production. Latour even connects ANT with archipelagoes and other natural occurring world relations: "Loci, contingencies or clusters are more like *archipelagos on a sea* than like lakes dotting a solid land" (2017, 177, our emphasis). Likewise, some critics have noted the similarities between this theory and Gilles Deleuze and Felix Guattari's notion of assemblages (Crawford 2005, 3).

The notion of "assemblage," however, allows for a more robust exploration of fluid, ever-changing modes of relationality. Assemblage ("agencement") is a term coined by Deleuze and Guattari to refer to the process of arranging or piecing together heterogeneous elements or objects, modes of expression, social structures, and signs that produce meaning(s) as a result of the "external relations of composition, mixture and aggregation" (Nail 2017, 23). One of the main problems in defining the term is that the translation from French to English misdirects the original meaning of the term in French: "While an assemblage is a gathering of things together into unities, an *agencement* is an arrangement or layout of heterogeneous elements" (22). Many critics and theorists struggle with the lack of specificity of the term in the actual theorization advanced by Deleuze and Guattari:

> What is an assemblage? It is a multiplicity which is made up of many heterogeneous terms and which establishes liaisons, relations between them, across ages, sexes and reigns—different natures. Thus, the assemblage's only unity is that of a co-functioning: it is a symbiosis, a "sympathy." It is never filiations which are important, but alliances, alloys; these are not successions, lines of descent, but contagions, epidemics, the wind. (Deleuze and Parnet, *Dialogues II*, p. 69, from De Landa 2006)

At times, Deleuze and Guattari rely on poetic language to index the complex dimensions of assemblages:

> It is a multiplicity—but we don't know yet what the multiple entails when it is no longer attributed, that is, after it has been elevated to the status of the substantive. One side of a machinic assemblage faces the strata, which doubtless make it a kind of organism, or signifying totality, or determination attributable to a subject; it also has a side facing a *body without organs*, which is continually dismantling the organism, causing asignifying particles or pure intensities to pass or circulate, and attributing to itself subjects that it leaves with nothing more than a name as the trace of an intensity. . . . Literature is an assemblage. It has nothing to do with ideology. There is no ideology and never has been. (2004, 3–4)

Heterogeneity, desire, multiplicity, and contingency are central in the definition of this term. Assemblages are in a constant process of constitution (territorialization), disarticulation (deterritorialization), and rearticulation (reterritorialization) so the meanings of arrangements of elements exist or become in transformation, mutation, redefinition, and rearticulation. The fluidity and heterogeneity of the elements that are conceived as part of an assemblage in Deleuze and Guattari's theory are therefore central for an actual understanding of this term (Parr 2005, 18).[13]

De Landa has attempted to produce a more coherent body of theory from the notion of assemblages. He focuses on the rearticulation of meaning that takes place after a rearrangement of elements: "An ensemble in which components have been correctly matched together possesses properties that its components do not have" (De Landa 2006, 5). He studies the assemblages that belong to the natural world in chapters 5 and 6 of *Assemblage Theory* and culminates his study with the proposal of a virtual diagram of assemblages.

One of the contributions of existing theories on assemblage is that they invite us to consider how meaning and human interpretation are constantly transformed and evolving as a result of the contexts in which different elements interact with each other (among them ideas, objects, cultural manifestations, institutions, practices). Assemblages therefore allow us to conceive the world as a complex network of interactions and relations that produce meanings that are not organic or immanent or cannot be totally explained by breaking them down to their smallest material particles (as claimed by scientific empirical methods) or by assuming a particular teleology of coordination and (re)productivity (as implied in machines, systems, or organs). Yet its foundation is a highly abstract imaginary in which localized, embodied practices and experiences, as well as material (natural organic or inorganic), are conceived without considering the actual limitations of these constant processes of articulation and rearticulation. Although human subjects are included in Deleuze and Guattari's theorization and in De Landa's rearticulation of this theory, some critics note that it is also easy to lose track of the human dimension in the highly abstract articulations of this concept.

Assemblages have been used as a common reference point for theorists working on archipelagic thinking:

> The significance of the assemblage goes beyond that of the simple gathering, collection or composition of things that are believed to fit together. Much like the thinking behind systems theory, *assemblages act in concert*: they actively map out, select, piece together, and allow for the conception and conduct of individual units as members of a group. Just like constellations, assemblages of heavenly bodies that, like Orion the Hunter, take on one (or more) recognizable forms only when their wholeness arises out of a process of articulating multiple elements by establishing connections among them. An archipelago is similar: its framing as "such and such an assemblage" draws our attention to the ways in which practices, representations, experiences and affects articulate to take a particular dynamic form. Archipelagoes are fluid cultural processes, sites of abstract and material relations of movement and rest, dependent on changing conditions of articulation and connection (Stratford et al. 2011) (Baldacchino 2015, 6, our emphasis)

Following from Baldacchino's and Stratford et al.'s insights, as well as Glissant's theorizations about relation, we suggest that there is still some work that the archipelagic does that is not totally evident or present in these other theoretical frameworks on systems, networks, actor-networks, and assemblages.

One of the main differences between archipelagic thinking and the theories discussed so far, and between the theories themselves—of systems, networks, actor-networks, assemblages—is the concrete geographical, historical, social, and political referent that is used as the foundation of each framework. Archipelagoes are concrete and material, spatial, geographical, and geological occurrences that have been resignified by human interaction.

They involve and entail a direct reference to the embodied experience of space that translates into a cognitive process. This experiential dimension can be lost in some of the most abstract articulations of the other theoretical frameworks discussed. The second main difference inheres in our proposal of "thinking"—in place of "studies" or "theory"—in the title of this volume to signal an epistemic formation located beyond institutional disciplines or fields. "Thinking" refers concretely to the actual articulation of knowledge and meaning through human interpretation and mediation. Paul Carter (2019) also conceives of archipelagic thinking as a space "where no discipline enjoys central authority" (12), translating this mode of thinking into a relationally based mode of knowing that promotes inter- and transdisciplinarity. This location of knowledge production abandons any claim of universality, to favor "situated knowledges" (Haraway 1988) and embodied knowledges (Rivera 2016; Anzaldúa 2015; Fanon 1967) as a crucial point of departure in the redefinition of a methodology.

In this regard, *Contemporary Archipelagic Thinking* has important connections with other fields of study that take embodied experiences of spaces, or the relation between social formations and spaces, as their point of departure. This is the case with island and nissology studies, ocean(ic) studies, littoral studies, and the study of constellations. The last framework is perhaps the most similar or closest to archipelagoes, since constellations signal networks of stars that are endowed with particular meanings for navigational, medicinal, (pseudo-)ontological, epic and myth-making, and spatial purposes.[14] Also based on a material foundation that is made up of dynamic, mutable, and contingent articulations of different elements, constellations are the referents for a projection of human interpretation that is based on the constructed relationships between celestial bodies. Yet, although deeply mediated by human interaction and interpretation, traditional constellations are bodies that are signified and inhabited by human knowledge as a distant, somewhat abstract or imaginary category. Even astronomers, who directly interact with these bodies as very concrete and material formations, are constantly revising existing knowledge about the different parts of a constellation, since "stars" are born and die, and our understanding of these distant bodies is constantly transformed as our technologies of observation change and evolve. Although in principle our knowledge and conceptualizations of archipelagoes can experience the same vicissitudes as "distant stars," there is an immediacy in these inhabited spaces that makes archipelagic thinking perhaps easier to conceive of as more concrete or material than the models advanced by constellations. Finally, planetary thinking is another framework that explores the inherent interconnectedness of human existence and its impact on the land, water, climate, and resources available for humanity on a global scale (Elias and Moraru 2015; Pratt 1992; Spivak 2011). Some scholars have explored the productive intersections between archipelagic and planetary thinking, by focusing on climate change as well as the impact of imperialism and colonization on the livelihoods of insular and coastal communities (DeLoughrey 2019; Friedman in the afterword to this volume).

The archipelagic strives to do a different kind of work than network theory, assemblages, systems theory, or the study of islands, oceans, constellations, or the planetary because it takes the geographical and political as a point of departure, as a physical context in which colonial/imperial meanings, structures, practices, and resistances are inscribed or take place. The archipelagic lens allows us to focus on comparative studies to make visible discontinuous and multifocal experiences that are sometimes historically connected but have been disconnected by national, international, transnational, hemispheric, and global frameworks. In other cases, contexts that are not historically linked can be used to analyze structures and practices of imperialism/colonialism that are similar or completely different and elucidate a whole array of articulations of power. These articulations ultimately question the humanity and the knowledges produced and contained by communities that are displaced to favor the economic and political needs of a distant or not-so-distant supremacist center of power. Furthermore, according to Pugh, "Thinking with the archipelago and the spatial turn both seek to denaturalize space so that it is more than a mere backcloth for political or ethical debate. Instead, highlighting ontologies such as assemblages, networks and mobilities, draws out the importance of spatial nuances, differences and connectivities, rather than adhering to that absurd cry 'History,' associated with such grand narratives as progress, development and colonialism" (2013, 20). Moving beyond the teleological mandates of history, the social sciences, and even epic and mythic imaginations, the archipelagic makes possible a whole new series of research questions that interrogate the primacy of continental frameworks in several disciplines of study (such as anthropology, philosophy, literature, or history) to privilege "multiplicities as elemental in the production of knowledge" (Stratford et al. 2011, 124). The archipelagic encourages us to think about how multisited, multifocal experiences contribute to the temporary articulation of historical, political, and symbolical imaginaries, as well as to reenvision environmental and biodiversity questions from frameworks that transcend North-South, South-North, and South-South connections.[15] Furthermore, aside from the questions that become possible with this new theoretical framework, scholars, thinkers, and artists have also advanced some new terms that take the notion of the archipelago as their foundation, thereby offering key contributions of this framework to theoretical and scholarly thinking.

SEMANTIC DISPERSIONS: SOME ARCHIPELAGIC NEOLOGISMS

Two key terms, "archipelagraphy" and "aquapelago," focus on the "interlapping" (Roberts 2016) of land and sea. In 2001, DeLoughrey proposed the notion of "archipelagraphy" as "a historiography that considers chains of islands in fluctuating relationship to their surrounding seas, islands and continents" (23). Aside from connecting cartography, writing, and the symbolic conceptualization of spaces, DeLoughrey's concept reconnects islands with their surrounding bodies of water, to decenter land as the most important referent in conceptualizations of space. Elaine Stratford has subsequently drawn our attention to the visual component of "graphy" to introduce "the idea of archipelagraphs—drawings or maps or other such renderings of islands that may be mobilized as part of a geographical imaginary" (2017, 80). Philip Hayward continues the focus on the interrelationship between water and land through his notion of the aquapelago: "The neologism aquapelago was coined in opposition to the term archipelago, which essentially refers to discrete parcels of land dotted within marine environments. In contrast to the latter concept, various writers argued for a greater recognition and analysis of the integrated terrestrial and marine environments of island aggregates and of human engagement with these" (2012, 84). Helen Dawson identifies the important contribution of this term in archaeological, environmental, geographical, and sociological studies by underlining "both the land and the sea in equal ways" (2012, 20). Both of these terms—"archipelagraphy" and "aquapelago"—invite us to reconsider the ways in which we approach various forms of representation of systems or clusters of locations, as in the case of the portolan charts or navigational maps in which regions are represented as a network of distances, wind and oceanic currents, and other physical accidents that were based on the collected observations of pilots at sea.[16]

Other terms focus on noncontiguous conceptions of political and economic networks in their articulation of the archipelagic. Thinking about precolonial times, in the mid-1970s anthropologist John Murra (1981) coined the term "vertical archipelago" to refer to the Andean indigenous communities' discontinuous conceptualization of the territory in order to benefit from ecological diversity to access food and other resources. Lanny Thompson and Javier Morillo Alicea have worked extensively with the notion of the "imperial archipelago." Morillo Alicea (2005) focuses on the importance of the Philippines in the context of the "Spanish archipelago" during the nineteenth century, calling for a reconceptualization of Latin American colonial studies that recovers the important political connections between insular regions of overseas possessions in the Atlantic and the Pacific. Thompson defines the "imperial archipelago" as the "island territories under U.S. military and political dominion after 1898, namely Cuba, Guam, Hawai'i, the Philippines and Puerto Rico" (2010, 1). In these two cases, the colonial context and the condition of "extended colonialism" (Martínez-San Miguel 2014, 6) invite us to consider the specific geopolitical contexts surrounding insular regions that have functioned as "colonial archipelagoes" (Martínez-San Miguel 2017, 155–57). Craig Santos Pérez expands on this idea to meditate on the particular meaning and trajectory of the notion of "territory" in the United States: "The word *territory* derives from the Latin *territorium* (from *terra*, 'land,' and *-orium*, 'place'). *Territorium* may derive from *terrere*, meaning 'to frighten'; thus, another meaning for *territorium* is 'a place from which people are warned off.' With this in mind, I propose a new term, *terripelago* (which combines territorium and pélago, signifying sea), to foreground territoriality as it conjoins land and sea, islands and continents" (2015, 619–20).

Not all archipelagoes are imagined or conceived as colonial overseas archipelagoes. A foundational example that comes to mind is Rome, conceived as an "archipelago of insulated communities," or a series of communities separated by cultural difference and history and articulated together by the reach of the Roman Empire (Charles Merivale, *History of the Roman's under the Empire*, 1865, in Stratford et al. 2011, 120). Japan, often referred to as the Home Islands, is another archipelago with significant imperial influence in the Pacific. Its archipelago consists of 6,852 islands, 430 of which are inhabited. The "British Isles" (referring, problematically, to England, Scotland, Ireland, and Wales and more than six thousand smaller islands) also comprise an imperial archipelago with presence in the Atlantic and beyond. Since the term "British Isles" is controversial in Ireland, alternative descriptions have been suggested and used, such as Britain and Ireland and, tellingly, the Atlantic Archipelago (Stratford et al. 2011, 117; Norquay and Smyth 2002; Schwyzer and Mealor 2004; Kumar 2003; Armitage and Braddick 2002). In their foundational essay "Envisioning the Archipelago," Stratford et al. (2011, 116–20) also discuss New Zealand, the United States, Canada, and Australia as regions that are also archipelagic although they are usually not imagined as such. Some scholars even suggest that the entire planet can be conceived of as an archipelago of landmasses and bodies of water of different sizes and scales.

Yet what all of these neologisms, based on the concept of the archipelago, share is the desire to point toward the many layers or dimensions of geological, social, historical, or cultural experiences that complicate the meaning of the word and go beyond the superficial or literal insular referent. *Contemporary Archipelagic Thinking* takes this common goal a step further, proposing a multi- and interdisciplinary dialogue that allows us to interrogate the epistemic and institutional limits that inform our study of many regions of the world. This is the invitation we extended to our collaborators, and the present volume collects some of their insights and interventions.

BEYOND DISCIPLINARY FORMATIONS

The works included in this publication dialogue with the research of a group of geographers, historians, cultural critics, and social scientists who focus on islands, oceans, and archipelagoes to propose new research questions. Recently, historians have revived the definition of the Caribbean as a system of islands in studies that refer to the region as the "imperial archipelago" (Thompson 2010; Morillo Alicea 2005) or the "inter-Atlantic paradigm" of colonization (Stevens-Arroyo 1993). Furthermore, several scholars have proposed insular, littoral, oceanic, and nissology approaches for the study of groups, networks, or systems of islands (Hau'ofa 1993; Baldacchino 2015; Stratford 2013, 2017; Pugh 2013; Lionnet 2008; Steinberg 2005; McCall 1994). Notions such as the "aquapelago" and "aquapelagic assemblages" (Hayward 2012; Dawson 2012), "terripelago" (Santos-Perez 2015), and "archipelagraphy" (DeLoughrey 2001) explore the relationship between oceans, islands, nation-states, commonwealths, and overseas possessions in the articulation of other forms of identity formation beyond the framework of a national language or a sovereign state. The archipelagic framework has also been explored through the intersection of comparative island studies and planetary frameworks in recent works on governance and postcolonial politics (Carter 2019) and a critical examination of global climate change from the insular experiences based in the Caribbean and the Pacific (DeLoughrey 2019). Most of this scholarship has been disseminated through articles and chapters in several books and journals, such as *Shima: The International Journal of Research into Island Cultures* and *Island Studies Journal*, and in an "Island Studies" email list recently created by one of the contributors to this volume, Jonathan Pugh. In Pacific studies there is a field of oceanic studies with which we have been in important dialogue. The Instituto de Estudios Avanzados del Litoral in Santa Fe, Argentina, and the Centro Interdisciplinario de Estudios del Litoral in Mayagüez, Puerto Rico, are two interdisciplinary centers that focus on coastal studies, adding another dimension to the study of islands and continents in archipelagic relationships. Our work also dialogues with the Rethinking the Island book series (2014–), coedited by Elaine Stratford, Godfrey Baldacchino, and Elizabeth McMahon. To our knowledge, our anthology is the only edited volume in English that invites contributors to assess how archipelagic thinking transforms existing research methods and disciplinary formations, including collaborations from a wide variety of disciplines in the humanities, social sciences, and natural sciences.

Contemporary Archipelagic Thinking is organized into five sections, intentionally encompassing various areas of the globe, but also pushing against area studies frameworks with analyses that strive to reinvigorate questions that go beyond national and transnational articulations of historical, political, and symbolical regions in order to favor embodied and practical interactions with space. We invited our collaborators to think specifically about how an archipelagic framework, theory, or lens would allow them to transform their research questions beyond the confines of existing disciplinary formations. Collaborators were also invited to send academic essays and shorter reflections or critical pieces engaging issues of methodology or transdisciplinary thinking. The resulting collection is a combination of interventions designed and discussed as part of the annual programming of the Center for Cultural Analysis seminar on archipelagoes held at Rutgers University, New Brunswick, New Jersey, during the 2015–2016 academic year, as well as individual contributions emerging from dialogues at academic conferences and institutional collaborations.

As a mode of thinking that moves swiftly beyond disciplinary and generic boundaries, award-winning poet Craig Santos Perez's contributions to *Contemporary Archipelagic Thinking* demonstrate how archipelagic poetry theorizes and how archipelagic theory can become a poetics. As he expresses in this collection's foreword and in his opening poem, "The Fifth Map," Guam itself has become in his work both a micro-space and a macro-system, in other words, a signifier for many of the movements—between the large and the small, the land and the sea, the colonial, postcolonial, and decolonial, the island, the nation, and the empire—that archipelagic thinking can inspire. With his permission, his folio of poems has been disarticulated from their original collected form and scattered throughout the anthology so that they can both speak with, and in counterpoint to, the viewpoints raised in the chapters. They are inspired, as he describes, by key archipelagic thinkers (Hau'ofa 1993; Brathwaite 1999; Walcott 1992; DeLoughrey 2010; Rob Wilson 2000; Peter Neill 2016; Sylvia Earle 1996; Édouard Glissant 1981, 1990; Albert Wendt 1981). They also represent, like Kathryn Chan's artwork, the many different forms that archipelagic thinking can take, and has taken, in the hands and minds of those who both inhabit and articulate their insularity from an embodied, experiential place.

The four theoretical chapters in the first section, "Space, Scale, Language, and Time: Foundational Epistemological Contributions of Archipelagic Thought," seek to step back and assess where archipelagic thinking as an epistemological field of knowledge has taken us so far and where archipelagic modes of inquiry and methodological approaches may point us in the future. Specifically, we asked authors to reflect on the ways in which archipelagoes can or have become the conceptual foundation of certain cultural notions of space and time, geography and topography. The section begins with a reflection by Elaine Stratford on some of the key contributions of the foundational and generative essay "Envisioning the Archipelago," published in *Island Studies Journal* in 2011 and cowritten with Godfrey Baldacchino, Elizabeth McMahon, Carol Farbotko, and Andrew Harwood. In "Disciplinary Formations, Creative Tensions, and Certain Logics in Archipelagic Studies," Stratford charts the emergence and development of both island and archipelago studies as modes of inquiry whose maturity is evident in the debates each have generated since 2011. The first debate centers on the idea of the archipelago as a "model," an idea that emerges across the humanities and social sciences and is problematized by Peter Hay's caution against abstraction and "concern for the phenomenological ground of island particularity" (2013, 212). Arguing for a more complicated understanding of the merits of metaphor and speaking from the perspective of her own disciplinary field of geography, Stratford points to the productive tensions between the abstract and the empirical that require that they be "neither mutually exclusive, nor combative." A second debate centers around Philip Hayward's (2012) introduction of the "aquapelago" as a challenge to the terrestrial grounding of archipelagic formations. In response, Stratford showcases their strong emphasis in 2011 on the shifting, fractal boundary between land and water in archipelagic territories. The authors' efforts to articulate in "Envisioning the Archipelago" "relational and assemblage thinking" as a central feature of archipelagic epistemologies becomes the basis for Stratford's contemporary thinking on "archipe-logics," a neologism she borrows from Lanny Thompson. For Stratford, contemporary archipelagic thinking involves a "thinking with" the archipelago as a spur to speculative creativity, one that engages with the world through "figures of thought" that highlight connection and move beyond impasse.

Jonathan Pugh's chapter, "The Affirmational Turn to Ontology in the Anthropocene: A Critique," begins with the striking observation that islands have become "exemplary figures of the Anthropocene" across a number of disciplinary fields. As such, they provide a platform from which to evaluate some of the central stakes of the discourse of the Anthropocene. One of those stakes involves what Pugh describes as the "affirmational turn," that is, arguments that assert an ontological shift toward "affirming . . . more-than-human relations." Pugh finds this turn exemplified in the work of Timothy Morton (2016) but also challenged by the work of Frantz Fanon, whom Pugh reads into the canon of archipelagic thinkers. In Fanon, Pugh finds a critique of an approach to the Anthropocene that would ultimately "decenter islanders' own critical capacities and normative political horizons." The question of scale is key for both Morton's and Pugh's deliberations. If the Anthropocene represents the first era in which humans have had a profound and everlasting impact on climate and the environment, it is also the era in which "hyperobjects," such as global warming and other dramatic climatological phenomena, work on islands "through scalar dimensions" that are incomprehensible to humans. These "objects" have become inaccessible to human thought, involving "nebulous, multidimensional forces" well beyond the certitudes of assemblage and network thinking.

For Pugh, however, the cost of such an understanding is the removal of the human from the island itself. Pugh highlights Morton's notion of relationality as useful for island and archipelagic studies, one in which, quoting Morton, "relating isn't some wondering way to fasten islands into chains to make them more exciting. *Relating* is how a thing is, all by itself." However, he faults Morton for arguing for a notion of the ontological totality of the island that is somehow inaccessible to its human inhabitants. Deploying Fanon's 1967 critique of affirmational ontologies writ large—affirming indigenous ways of life or vitalist Negritude ontologies, for example—Pugh draws our attention back to politics, rather than ontology, as the arena for securing island futures in the Anthropocene. In his methodological call for a "critical island studies in the Anthropocene," Pugh seeks to foreground, once again, human action in the creation of epistemological, as opposed to ontological, understandings of the Anthropocene, ones that can combat a "deeply impoverished understanding of human agency."

Earlier in this introduction, we call for alternative linguistic genealogies for the archipelagic. In "What Is an Archipelago? On Bandung Praxis, Lingua Franca, and Archipelagic Interlapping," Brian Russell Roberts extends this call, returning to some of the linguistic genealogies of the word "archipelago" to argue for the documentation of the "multilanguage historical processes that undergird the archipelagic narrative." His innovation is to suggest that English might be the very language, or lingua franca, of archipelagic thinking and that accepting it as such could facilitate myriad projects, as it did in the Indonesian context of the Bandung Conference, the site of Roberts's case study. Here, a "Bandung praxis" comes to represent not merely a fleeting political project of the past with archipelagic tendencies. It stands as a "structure of feeling" with English as an imperfect but necessary lingua franca for communicating both shared and divergent sensibilities across linguistic lines. For Roberts, future archipelagic thinking needs to develop a platform or scaffolding upon which scholars, artists, and thinkers can engage with each other—a "rough and improvisational realm of translation in which multiple linguistic traditions meet each other within the field of a lingua franca, stretching and twisting the lingua franca to their own ends, remaking it in accidental and purposeful ways." It is our hope that the anthology as a whole contributes to his desire for archipelagic thinking to be structured by and as "a sprawling comparative project—multilingual, long durational, transregional, and involving multiple participants." Methodologically, Bandung praxis also represents our adopting a more self-conscious relationship to English, "as an uneven, messy, and deferring mediator" among the multiple language traditions that constitute archipelagic spaces, intellectual traditions, and patterns of thought.

For Lanny Thompson, in "The Chronotopes of Archipelagic Thinking: Glissant and the Narrative of Philosophy," the archipelago is a "geosocial location" that produces knowledge in the form of archipelagic thinking that Glissant named *antillanité*. In viewing *antillanité* as a "philosophical project," Thompson argues for an expansion of contemporary archipelagic thinking from core disciplinary fields such as geography and history to the production of theories of space, time, and the subject, more broadly conceived. Topology, temporality, and ontology (or the concept of being) are foundational in Glissant's archipelagic thought. Thinking with Glissant, Thompson articulates possible conceptual transitions, "from geography to topology," from history to "temporality" or "emplotment," and from identity to "ontology, understood as situated subjectivity" (and therefore very much centering on the human, unlike the affirmational ontologies Pugh describes). Furthermore, these undergirding philosophical premises in Glissant's archipelagic thinking work through key narrative tropes, the chronotopes of the abyss, the plantation, and the island. These very specific literary devices work to configure place, time, and the subject "as mutually interdependent and constitutive." Ultimately, Thompson is able to draw from Bakhtinian theory to show how these three chronotopes interact with each other to serve as the anchors of one of Glissant's most complex formulations, that of "rooted errantry."

Thompson's reflections occurred in the context of the forced errantry foisted on many Puerto Ricans during and in the wake of Hurricanes Irma and Maria. Aware of this context of his chapter, as itself an enactment of the embodied and situated location from which the archipelagic subject writes and thinks, Thompson was moved to contribute a reflection on "the chronotope of the hurricane" as a postscript on the practical and methodological implications of archipelagic thinking. Here he asks the difficult question of whether the chronotype, or in Morton's language, the hyperobject of the hurricane, defeats our human efforts to think our present, diminishing "the relevance of the abstract themes treated" in his chapter. His answers weigh historical relevance against the "violent materiality" of the present and speak in chorus with poet Craig Santos Perez's poetic reflection "Storm Tracking, 2016," written for Fiji after the devastation wrought by Cyclone Winston, "the most powerful storm ever recorded in the Southern Hemisphere." Thompson's chapter, written after two successive, devastating hurricanes in Puerto Rico, ends with his poignant realization that, far from being inaccessible to and withdrawn from human thought, current archipelagic realities require an "expanded vocabulary," that is, even more active theorizing that can take us well beyond the work of historicizing and re-creating genealogies for archipelagic thinking undertaken in previous scholarship.

The chapters in the second section of the collection, "Beyond the Sea as Metaphor: Comparative Maritime Epistemologies," are unified around the common theme of articulating a maritime epistemology from within an archipelagic framework. Inspired by Hester Blum's opening caution, "The sea is not a metaphor" (2010, 670), in her field-defining study "The Prospect of Oceanic Studies," these chapters call for a more literal and material relation to the sea in literary and cultural studies. As Blum continues, "I advocate a practice of oceanic studies that is attentive to the material conditions and praxis of the maritime world, one that draws from the epistemological structures provided by the lives and writings of those for whom the sea was simultaneously workplace, home, passage, penitentiary, and promise" (670). With this in mind, in "An Early Medieval 'Sea of Islands': Area Studies, Medieval Studies, and Traditions of Wayfinding," Jeremy DeAngelo offers a close reading of a literary account of a sea journey that convincingly demonstrates how Anglo-Saxon seafarers of the North Atlantic archipelago could have used a navigational method more commonly associated with the premodern Pacific seafarers of Oceania. This method, one that relies on mapping movement according to the location of stars in the sky and the revolving motion of surrounding islands, extends the maritime beyond the oceanic to the archipelagic and back and forth between them. Methodologically, DeAngelo also demonstrates the merits of grounding medieval textual analysis in a comparative, archipelagic, hermeneutic model.

Moving us to the maritime Mediterranean space between Europe and Africa, in "Archipelago of the Maghreb: Mapping Mediterranean Movement from Transnational Migration to Transregional Mobility," Sarah DeMott pairs archival research with a novel cartographical methodology to argue, visually, for an understanding of migration across the Mediterranean, between Italy and Tunis most specifically, as expressive of an "archipelago logic." Using an archive of baptismal records to track the multiple routes of parents and godparents of babies baptized in Tunisia between 1862 and 1863, DeMott provides four "close mappings" of a shifting Mediterranean space that ultimately emerges into view as an archipelagic network rather than a mere continental bridge between Europe and Africa. This framework, borrowing theoretically from archipelagic thinking and methodologically from digital humanities, reveals the central role of the islands between Tunisia and Italy in producing glocal social worlds. For Mary Eyring, the sea is a site of deformation and disability in "Archipelagic Deformations and Decontinental Disability Studies." In the nineteenth-century maritime world of Herman Melville's fictional sailor Billy Budd, terraqueous spaces provide the possibility for a positive understanding of disability as a common condition the brotherhood of sailors share. Ships provide an important context for Eyring's efforts to think disability and the archipelagic together through the notion of deformation. "Like islands in an archipelago," she asserts, "individual ships were spatially and culturally distinct and yet intimately connected with other ships in and outside of their fleet." It is this maritime culture, what Peter Linebaugh and Martin Rediker (2000, 144) have termed "hydrarchy," that Eyring argues allowed for "identity formations and deformations unavailable in continental settings." Building on a core premise of disability studies, that "environments contribute to the construction of an individual or group's ostensible disability," Eyring argues that the framework of an archipelagic relationality can help us to understand the different meanings of mariners' common condition of disability.

Both the Mediterranean and the stars reappear in a more playful form in Lisa Swanstrom's chapter, "Digital Currents, Oceanic Drift, and the Evolving Ecology of the Temporary Autonomous Zone," which begins by focusing on the 2013 work *Poseidon's Pull* by contemporary artist Reza Safavi. The ironic art installation, which aimed "to provide a communication interface between human beings and the ancient god" of the Ionian Sea, is a more serious instance of digital art resisting the "totalizing authority of nations" by mobilizing the power of "underwater currents and oceanic drift." Tired of old dichotomies pitting nature against technology, Swanstrom, as a scholar in digital studies, focuses on what Jonathan Pugh would describe as "thinking with the archipelago" in order to demonstrate "the importance of digital aesthetics in reframing environmental discourse," in highlighting the agency of natural spaces, in translating natural patterns of action into human readable terms, and in making visible the environmental forces that shape our lives. Taken together, DeAngelo's, DeMott's, Eyring's, and Swanstrom's chapters display how land, sea, and air can work in tandem with each other as an assemblage of coordinates and reference points, moving in relation to each other to guide and map the travels of early Pacific wayfarers, Anglo-Saxon seafarers, migrants across the Maghrebin archipelago, nineteenth-century Atlantic mariners, and twenty-first-century Mediterranean refugees.

The third section, "Archipelagic Environments: Evolving Political Ecologies," assesses the mutually determining impact of ecological interpretive frameworks and the conceptual tools of archipelagic thought. In "Literary

Archipelagraphies: Readings from the British-Irish Archipelago," Pippa Marland traces the "archipelagic turn in British and Irish literary studies," revealing both the resistance to, and the utility of, reimagining an Anglo-centric geographic imagination. Describing earlier, failed efforts to reframe and countermap such formulations as "the British Isles" as an "Atlantic archipelago," Marland identifies more recent efforts by "environmentally oriented literary critics and creative writers" to resituate England and English writing within a broader land- and seascape, denoting "ecological entanglement at a range of different spatial scales." Foregrounding the 2007 launch of the creative writing journal *Archipelago*, followed by a range of recent publications in literary and cultural criticism, she describes a new and emerging "British-Irish archipelagraphy" that has the power to disrupt more traditional geopolitical and cultural understandings of the region. Marland argues overall that an "archipelagic ecocriticism" could address some of the more challenging issues in ecocriticism more broadly—such as the tension between the local and the global and the conceptual challenge of "the vast spatiotemporal scales of the Anthropocene."

Like the gannets that appear on Julian Bell's cover image for the journal *Archipelago* discussed by Marland, shorebirds are one of the primary protagonists in Jenny Isaacs's chapter, "Conservation Archipelago: Protecting Long-Distance Migratory Shorebirds along the Atlantic Flyway." Isaacs follows both the *rufa* Red Knots in their aerial travels along the coastal "Atlantic Flyway" and the island-hopping journeys of the biologists who study them. Along the way, she demonstrates how archipelagic thinking works in praxis as biologists struggle to build and maintain a "'shorebird conservation archipelago'—a network of reserves and conservation organizations that share knowledge and resources across different sites and scales," amid the vagaries and contingencies of local geographic, environmental, cultural, and political conditions. Like Marland, however, Isaacs has a larger goal: to demonstrate how "archipelagraphy . . . may serve as a generative, interdisciplinary mode of environmental study." For Isaacs, applying more-than-human methodologies in geography and political ecology offers opportunities for studying archipelagoes as "multiscalar, multispecies, multidimensional assemblages and liminal environments infused with politics." Far from limiting human agency, such perspectives allow for different approaches to "tracing the transnational conservation apparatus that operates across time, space, and scale" of minority avian species.

For Lizabeth Paravisini-Gebert, on the other hand, "plastic" is the lens through which she works to understand the evolution of the Caribbean "wastescape," from the onset of colonialism and slavery to the archipelago's contemporary, suffocating embrace by global plastic debris. As Paravisini-

Gebert eloquently states, the very "indestructibility" of plastic echoes "the futile nature of attempting to clean up the mess of colonialism through submersion, the thankless task of trying to tidy up the debris of history, the toxicity of the slave trade." In this debris field, visual artists struggle to represent the wealth of meanings held in the empty, negative spaces of plastic bags and bottles washing up on the beaches and shores of Caribbean coasts. Nevertheless, Caribbean artists are responding in their own agentic ways to "forces that have become erratic and unknowable in the Anthropocene." Working through the use of plastic as both medium and message by such Caribbean and Central American artists as Jean-François Boclé, Tomás Sánchez, Tony Capellán, and Alejandro Durán, Paravisini-Gebert's goal is to shift our methodological focus from "examining the environmental crises facing vulnerable island nations as strictly local," small-island concerns. Rather, she argues, "archipelagic approaches invite us to inscribe the threats to islands and archipelagoes brought by climate change . . . materially and metaphorically, as belonging to newly acknowledged multidirectional flows of oceanic forces."

The chapters in the fourth section, "Relational Archipelagics: Redefining Imperial and Postcolonial Studies," address the impact of archipelagic thinking on both the study of empire and the study of the postcolonial nation-state. For Mimi Sheller, precisely because archipelagoes often serve as "offshore" spaces "where both the romance of landscape and the violence of war can be waged," they can also frequently become "pivotal points of politicization and countermobilization." In "Archipelagoes as the Fractal Fringe of Coloniality: Demilitarizing Caribbean and Pacific Islands," Sheller tries to look at both sides of this equation—archipelagoes as "liquid spaces of mutating imperialism" defined by the suspension of "sovereignty, citizenship, and rights," but also "as spaces for counterimperial alternatives, subaltern trajectories of belonging, and subversive itineraries." Taking Vieques Island in Puerto Rico and Tinian and Pågan islands in the Commonwealth of the Northern Mariana Islands as case studies, she demonstrates how simply the resistance to the militarization of archipelagic spaces "can rearticulate their meanings," making islands potential sites of a "fractal fringe of political willfulness."

In "Sardinia 'Lost between Europe and Africa': Archaeology and Archipelagic Theory," Thomas P. Leppard, Elizabeth A. Murphy, and Andrea Roppa add, from their perspective as archaeologists, further problematizations of the space between Europe and Africa by revising previous conceptions of the island of Sardinia. Choosing to study "human action through materials and archaeological landscapes rather than texts," they argue that it is archipelagic thinking that has allowed them to "approach the material records and human pasts of Sardinia in highly local terms," most pertinent for understanding how premodern islanders "subvert and disperse mainland strategies of domination and cultural hegemony" from the European continent. Demonstrating the ways in which, in Sardinia, "the distribution of social power in a disintegrating landscape has been a recurrent strategy to avoid, circumvent, and disappear, in the face of imperial and colonial incorporation," Leppard, Murphy, and Roppa organize these distributive strategies under three themes—archipelagoes, fractality, and palimpsests. Their analysis combines the more materialist methodologies and temporal foci of archaeologic study with a social critique provided by the archipelagic framework. This allows them to demonstrate how such a framework can be applied both beyond the text to forms of material culture and beyond the present to the premodern and pre-Columbian Old World past.

Christopher J. Lee returns us to Bandung and the archipelagic state of Indonesia in his identification of a particular political "format" that emerged in the era of decolonization: "political communities comprised of noncontiguous nation-states." In "Sovereignty between Empire and Nation-State: The Archipelago as Postcolonial Format," Lee argues that these "political archipelagoes" represented provisional attempts on the part of new states "to create new assemblages of group empowerment against the remaining vestiges of Western imperialism," as well as emerging threats from the United States and the Soviet Union. The archipelago thus becomes a model for thinking through "the institutional and spatial possibilities of postcolonial sovereignty." Lee grounds his account in Indonesia, site of the famous Bandung Conference of Asian and African leaders in 1955. Indonesia's geologic status as an archipelago undergirds its importance as the site for the enactment of the archipelago as a new political metaphor, reflecting "the transformation of global anticolonialism from a technique of insurgent grassroots politics to a method of diplomatic statecraft." For Haruki Eda in "Archipelagic Feeling: Counter-Mapping Indigeneity and Diaspora in the Trans-Pacific," a cartographic imaginary helps to shift commonsense understandings of Japan, an archipelagic nation-state, from the hegemonic image of a singular, consolidated, geopolitical form to that of an archipelagic site for the imperially organized management of minority populations. Using his own close-mapping methodology, Eda traces the history and presence of ethnic and indigenous minority populations within Japan—the Ainu, the Ryukyuans, and the Zainichi Koreans—the latter a stateless diasporic Korean community that is, in the literal meaning of *zainichi*, "residing in Japan." Each of these group's minoritized status links back to the obscured imperial relationship between Japan's main islands and majority ethnic group, the Yamato, and the other indigenous and diasporic islanders who constitute ethnic minorities of the nation-state. Critically rereading maps of Japan from the late seventeenth century through the present, Eda reminds us in a very visual way of the Japanese nation-state's geographic, archipelagic form. His "counter-mapping" of Japan's "archipelagraphs" and "cartographs" demonstrates the imperial dynamics that complicate and fracture interinsular relations in Japan and become embedded in the ethnicization and minoritization of indigenous and diasporic populations.

As is evident in Japan's internal imperial history, many of the dynamics operative in empire-island relations or continental-insular relations can appear within archipelagoes themselves. The chapters in the fifth section, "Interisland Dynamisms: Small Islands/Big Worlds," address the issue of scale as it impacts intra-archipelagic relations, where larger or more prominent islands can function as mainlands holding a continental mind-set in relation to their smaller, insular counterparts. Questions of scale also impact how islanders understand their own inter-island identities in relation to their place in both the "small" island and a larger world. In "'Together, but Not Together, Together': The Politics of Identity in Island Archipelagoes," St. Kitts-Nevis serves as one case study among others in Godfrey Baldacchino's brief retracing of the history that resulted in Anguilla's secession. Baldacchino pans out more broadly to explore similar big island/small island dynamics in a number of comparative archipelagic contexts. In a substantive example of a comparativist archipelagic methodology, Baldacchino is able to survey a global, archipelagic landscape of "multiple islands in turbid relationship." Methodologically and philosophically critical of the tendency to conflate island chains with the largest island, "as if they were single islands," Baldacchino points to other frameworks, of both knowledge and consumption (science, geopolitics, tourism), in which archipelagoes can be seen, conceptually, in all their complexity as a feature of insularity that juxtaposes "variation and similarity." Jessica Baker takes up this challenge explicitly in addressing the status of the smaller Caribbean islands as ethnomusicological research sites. In "Small Islands, Large Radio: Archipelagic Listening in the Caribbean," she argues that, far from unengaged with and inconsequential to the regional music scene, small islands such as St. Kitts-Nevis have been instrumental in creating an archipelagic musical consciousness through such mechanisms as "an online streaming radio station and chatroom, largeradio.com, that has created a space and a mode of listening to and engaging with Caribbean music that counters the singularized generality of Caribbean music scholarship by broadening the horizons of the Caribbean imagination to include different kinds of *being from*." The result is a "multicentric conception of Caribbean cultural production through archipelagic listening." Situated within a brief history of ethnomusicological research in the Caribbean, Baker finds Nelson Maldonado-Torres's discussion of "continentality" helpful as a lens for thinking through the unconscious epistemological frameworks that have guided research on musical production in the Caribbean.

Cartography returns as an explicit theme and tool for Anjali Nerlekar in her literary study "The Insular and the Transnational Archipelagoes: The Indo-Caribbean in Samuel Selvon and Harold Sonny Ladoo." Writing more explicitly about inter-island Indo-Caribbean identity than intra-island archipelagic relations, nevertheless Nerlekar finds an archipelagic approach helpful in reading the seemingly opposing strategies of these two authors' efforts to emplace and emplot the Indo-Caribbean subject in a broader world. Nerlekar tracks two Indo-Caribbean authors' struggles to document, in their fiction, the "multiple affiliations of home and away" that structure Indo-Caribbean identities. Selvon and Ladoo offer very different, even opposing "cartographic visualizations" of the layered place of the "Indian" in Trinidad and the place of Trinidad in the wider world. As captured in the title of Selvon's novel, *An Island Is a World*, both Selvon and Ladoo provide complex reconfigurings of Indo-Caribbean belonging as both "local and translocal," insular and transnational. In the very different context of the Indian Ocean islands and archipelagoes, these questions regarding the nature of identity and belonging shape Gitanjali Pyndiah's critical interrogation of the nature of archipelagic being. In this section's final chapter, "On Archipelagic Beings," she revisits the progression of Glissant's work toward the final image of "the world as an archipelago witnessing a global process of creolization." This universalizing trajectory leads Pyndiah to a strong caution: that contemporary archipelagic thinking à la Glissant "runs the risk of omitting to reveal the colonialities of power that inform the human histories of archipelagoes." In a dynamic inter-island and intra-archipelagic comparison, Pyndiah places the theories of two Caribbean philosophers—Martiniquan Glissant and Jamaican Sylvia Wynter—in direct dialogue as their theories have been, or can be, wielded in relation to the history of Black/Creole people in the Mauritian islands. Comparing and contrasting Glissant's notions of creolization with Wynter's mapping of archipelagoes as "first and foremost islands of social exclusion that are connected to a historical process of coloniality," Pyndiah shows how Wynter's paired concepts, of archipelagic being and indigenization, in contrast to Glissantian archipelagic thinking (at least as the latter has been interpreted in the Mauritian context), offer possibilities for theorizing a form of being that derives from the "creative practices" of Black/Creole people. In her conclusion Pyndiah foregrounds "tropological," as opposed to topological and topographical, archipelagic relations, that is, urging contemporary archipelagic thinking to remain attentive to "the 'ongoing' human story mapped by the history as well as geography of the space."

The afterword by Susan Stanford Friedman offers opportunities for reflecting on the foundational epistemological contributions of archipelagic thought, as well as for anticipating and suggesting future directions, developments, and expansions. In "Planetary Constellations: An Afterword," Stanford Friedman proposes a dialogue between the poetic works of Craig Santos Perez and Selina Tusitala Marsh to offer a more expansive meditation on the interdisciplinary implications of the thinking advanced in the chapters collected in this volume. She revisits notions of modernity, agency, and intersectionality to explore how they are transformed by the archipelagic frameworks designed and advanced by the interdisciplinary contributions of the collection. She closes her generous intervention by posing new questions that emerge from what she denominates as an "archipelagic epistemology." Planetary frameworks, environmental studies, the relationship between islands and continents, the role of diaspora—and the centrality of islands, oceans, and bodies of water in all of those interdisciplinary conversations—are some of the current debates Friedman revisits to explore the methodological implications of adding an archipelagic prism. For Friedman, the archipelagic is a mode of thinking that is transforming the study of human and nonhuman phenomena.

* * *

> "There is a network of us, strung out about the world."
>
> —*Sense8*, "Isolated Above, Connected Below," Season 2, Episode 6

This introduction's title—"Isolated Above, but Connected Below"—is taken from the title of an episode of *Sense8* that allows us to offer a final meditation on the ways in which archipelagic thinking transforms multi-, inter-, and transdisciplinary work. The phrase refers first to the archipelago as traditionally constituted by islands that seem to be isolated from each other on the surface of the sea but moves on to the second, deeper actuality that they are connected under water. This conceptualization of the archipelago invites us precisely to enact that complex balance between the singularity of the individual, a community, or a nation and the multiple, often invisible interconnectivities that are constitutive of human, social, historical, political, and symbolic experiences and praxes.

A second source of inspiration that bears mentioning is a shared space for graphic designers in Madrid named "Archipiélago," described as an *espacio de co-creación* (space of co-creation), which our colleague George Yúdice accidentally found during a trip to Spain in July 2019. The idea behind this alternative model of production is that, instead of promoting teamwork toward a common goal, "Archipiélago" provides a space for artists to share the energy of a creativity that is channeled toward their different respective projects but also experienced in concert. In some respects, this model diverges from the definition of the archipelago as a network of locations and subjects acting in concert (or acting through a flexible and extended form of consensus) that we have been using in some sections of this introduction. Instead it points toward one of the possible futures of contemporary archipelagic thinking, focusing on the contiguity of shared creativity as a source of synergy that is meaningful in itself, even when directed toward multiple goals. Designers who share a space enact isolation also in connectivity, while in reverse, connectivity serves as an underlying leitmotif to their isolated works.

This sensibility reflects the type of creative, collaborative, intellectual space this collection has sought to facilitate. *Contemporary Archipelagic Thinking* is an invitation to think beyond the apparent isolation of disciplines, institutional formations, methods, regions, languages, and forms of knowledge and being and to explore the many connections that exist and become possible beyond the surfaces or the bounded realities as conceived by disciplinary formations or by historical legacies ingrained in the constitution of departments, colleges, and schools in academic institutions. The collaborators included in this volume come from a wide array of disciplines, institutions, and regions of the world; yet they have accepted the invitation to deploy the archipelagic as a geopolitical and spatial condition that can be translated into an epistemology, a method, and a common framework to promote connectivities contained below the surface. They accepted the provocation of invoking a community of scholars willing to engage in conversations that exceed the boundaries of islands and nations, departments, institutions, disciplines, and cultural formations. In this sense, archipelagic thinking models a form of collaborative research in the humanities, social sciences, and sciences that we would like to conceive of as co-creativity. The conversations that come to fruition in this volume in many respects parallel the isolation in connectivity and the connectivity in isolation invoked in the notion of archipelago that informs the plot structure of *Sense8* and the shared space for graphic designers in Madrid.

In the process of doing our work for this collection, a third source of archipelagic inspiration comes to mind: Caribbean intellectual C. L. R. James's hope that the project of West Indian independence could borrow from the flexibility and movement he felt epitomized the ancient, archipelagic Greek city-states, which "moved so far and fast . . . were so small that everybody had a grasp of what was going on. Nobody was backward; nobody was remote; nobody was far in the country; and people [now] have methods of transport that bring us very rapidly together" (2013, 105). We invited our colleagues to cross paths and come together with us to imagine how archipelagic thinking could enrich, transform, and challenge their disciplinary trainings and research methods. As we all drift back into our multiple academic, professional, and personal lives, what is left behind is a network of ideas and several moments in time in which we all tried to co-create, to act in concert, and to imagine what would happen if we activated our own modes of thinking otherwise in theory, in practice, and beyond.

NOTES

1. See Kathryn L. Chan Art (https://www.kathrynchan.com).
2. Brian Russell Roberts's notion of "interlapping" modes of archipelagacity as a variation and intensification of overlapping was presented at the "Archipelagoes/Oceans/Americas Symposium" held at Brigham Young University, Provo, Utah, October 6–7, 2016. He further expands on this notion in his chapter included in this book.
3. See work by Kathleen Lopez, Ignacio Lopez, Aisha Khan, and A. Dominique Curtius on Asian migration to Latin America and the concept of "coolitude" as a supplement to existing notions of *mestizaje* and *mulataje*. Khal Torabully coins the term "coolitude" to refer to indentured laborers of Asian descent.
4. Glissant's thinking echoes Néstor Perlongher's notion of *devenir*, which also conceives of identity as a process that does not culminate in a hegemonic conflation with otherness, although Perlongher's focus is on gender and sexuality ("Los devenires minoritarios"). Both Glissant and Perlongher use theorizations by Deleuze and/or Guattari about processes of identity formation, and in his essay "La desaparición de la homosexualidad," Perlongher takes the archipelago and the constellation as points of departure to conceptualize male gay identity (105).
5. The translation of this quote is ours, but we would like to thank Katerina González Seligmann for her advice on how to translate this passage by Glissant.
6. We are using the translation included in Hiepko's chapter, with revisions included in brackets for the translation provided for the word *éclatements* as well as other phrases and expressions. We would like to thank Katerina González Seligmann for her advice on how to translate this passage by Glissant.
7. The *Ramayana* is a Sanskrit epic, traditionally attributed to Valmiki, that concerns the banishment of Rama from his kingdom, the abduction of his wife Sita by a demon and her rescue, and Rama's eventual restoration to the throne (http://www.yourdictionary.com/ramayana).
8. Ramleela (Ramlila) is the reenactment of the life of Rama, the hero in the Hindu epic story of the *Ramayana* (first millennium BC), considered one of the oldest genres of Indian literature. It also refers to the ritual performance of a series of festivities linked to Hinduism and celebrated in the fall in many towns and villages in India, as well as in other regions with a significant Indian diasporic community, as is the case in Trinidad.
9. The term "archipelago" is believed to have originated in Italian, as a reference to "the Aegean Sea," and comes from the Latin *archi-* (main, chief, principal) and

-pelago (gulf, abyss, sea) and/or from the Greek *pelagos* (high, open, or main sea) (https://www.etymonline.com/word/archipelago).

10. "The term 'indies' apparently only came into use in Europe in the 1400s. Its precise context is lost in antiquity. Traders of the time probably grouped all of South and East Asia into a geographical region linked to India. In other words, they referred to the 'indies' as the larger region including India and its trading nations to the east and northeast" (Lineback and Lineback Grizner 2010).

11. The concrete region is a nation that became independent from the United Kingdom on October 1, 1978, and became a member of the United Nations in 2000.

12. Oceania has three major subregions: Melanesia, Micronesia, and Polynesia; it includes roughly fifteen countries (Australia, Fiji, Indonesia, New Zealand, Tonga, and Tuvalu are some of them) and twenty-five dependencies and is inhabited by 40 million people.

13. Deleuze and Guattari's notion of assemblage is also closely related to their theorization of the rhizome, which refers to the complex process of entanglement that characterizes many social formations. Based on the subterranean, entangled root system of certain kinds of plants, the rhizome is conceived as an alternative model to think about complex collective and social articulations. Glissant expands the theorization of rhizomes as part of his "poetics of relations" and links it with archipelagic thinking, as we have discussed.

14. By medicinal and ontological we are referring to medieval theories of humors and the effect of stars on human behavior, a form of thinking that is still the foundation of zodiac signs and horoscopes in contemporary societies.

15. An example of an archipelagic approach to biodiversity is found in Nieto-Blázquez, Antonelli, and Roncal 2017, a study that questions continental colonization to favor the existence of flora originating in the Caribbean islands that eventually colonizes continental regions.

16. In many of these maps the representation of "windrose networks" (or navigational lines traced in relation to the magnetic North Pole) produces space as a network of routes and locations that can be conceived as the epitome of an archipelagic visuality. Although meant for practical use, these charts did not take into account the curvature of the Earth and were found to be useless for navigation in open sea (Campbell 1987).

REFERENCES

Anzaldúa, Gloria. 2015. *Light in the Dark/Luz en lo oscuro: Rewriting Identity, Spirituality, Reality*. Durham, NC: Duke University Press.

"Archipelago." Etymology Online. https://www.etymonline.com/word/archipelago.

Armitage, David, and Michael Braddick. 2002. *The British Atlantic World, 1500–1800*. New York: Palgrave.

Arroyo, Jossianna. 2011. "Revolution in the Caribbean: Betances, Haití and the Antillean Confederation." *La Habana Elegante*, Segunda Época 49 (Primavera-Verano). http://www.habanaelegante.com/Spring_Summer_2011/Invitation_Arroyo.html.

Babcock, William H. 1920. "Antillia and the Antilles." *American Geographical Society* 9, no. 2 (February): 109–24.

Bakhtin, M. 1981. *The Dialogic Imagination*. Austin: University of Texas Press.

Baldacchino, Godfrey. 2015. "More Than Island Tourism: Branding, Marketing and Logistic in Archipelago Tourist Destinations." In *Archipelago Tourism: Policies and Practices*, 1–18. London: Routledge.

Batongbacal, Jay L. 1988. "Defining Archipelagic Studies." In *Archipelagic Studies: Charting New Waters*, edited by Jay L. Batongbacal, 183–94. Diliman, Quezon City: UP Systemwide Network on Archipelagic and Ocean Studies in cooperation with the UP Center for Integrative and Development Studies.

Becker, Kai Helge, and David Seidl. 2007. "Different Kinds of Openings to Luhmann's Systems Theory: A Reply to la Cour et al." *Organization* 14, no. 6: 939–44.

Benítez-Rojo, Antonio. 1989. *La isla que se repite*. Hanover, NH: Ediciones del Norte.

Bernabé, Jean, Patrick Chamoiseau, and Raphaël Confiant. 1990. *Éloge de la Créolité/In Praise of Creoleness*. Translated by Mohamed B. Taleb-Khyar. Baltimore: Johns Hopkins University.

Blum, Hester. 2010. "The Prospect of Oceanic Studies." *PMLA* 125, no. 3 (May): 670–77.

Brathwaite, Kamau. 1999. *ConVERSations with Nathaniel Mackey*. New York: We Press.

Buscaglia-Salgado, José. 2003. *Undoing Empire: Race and Nation in the Mulatto Caribbean*. Minneapolis: University of Minnesota Press.

Butcher, John G., and Robert Edward Elson. 2017. *Sovereignty and the Sea: How Indonesia Became an Archipelagic State*. Singapore: National University of Singapore Press.

Campbell, Tony. 1987. "Portolan Charts from the Late Thirteenth Century to 1500." In *The History of Cartography*, 1:371–463. Chicago: University of Chicago Press.

Carter, Paul. 2019. *Decolonising Governance: Archipelagic Thinking*. New York: Routledge.

Centro Interdisciplinario de Estudios del Liroral. Universidad de Puerto Rico—Mayagüez. http://www.uprm.edu/p/ciso/centro_interdisciplinario_de_estudios_del_litoral_.

Chaar-Pérez, Kahlila. 2013. "'A Revolution of Love': Ramón Emeterio Betances, Anténor Firmin, and Affective Communities in the Caribbean." *Global South* 7, no. 2 (fall): 11–36.

Chan, Kathryn. 2002. *Archipelago*. Limestone plaster, terracotta and paint.

Cornejo Polar, Antonio. 1994. *Escribir en el aire: Ensayo sobre la heterogeneidad socio-cultural de las literaturas andinas*. Lima: Horizonte.

Crawford, Cassandra S. 2005. "Actor Network Theory." In *Encyclopedia of Social Theory*, edited by George Ritzer, 1–3. Thousand Oaks, CA: Sage.

Crone, G. R. 1938. "The Origin of the Name Antillia." *Geographical Journal* 91, no. 3 (March): 260–62.

Curtius, A. Dominique. 2010. "Gandhi et Au-Béro, ou comment inscrire les traces d'une mémoire indienne dans une négritude martiniquaise." *L'Esprit Créateur* 50, no. 2 (summer): 109–23.
Dawson, Helen. 2012. "Archaeology, Aquapelagos and Island Studies." *Shima: The International Journal of Research into Island Cultures* 6, no. 1: 17–21.
De Landa, Manuel. 2006. *A New Philosophy of Society: Assemblage Theory and Social Complexity*. London: Continuum.
Deleuze, Gilles, and Félix Guattari. 1986. *Kafka: Towards a Minor Literature*. Minneapolis: University of Minnesota Press.
———. *A Thousand Plateaus*. 2004. Translated by Brian Massumi. London: Continuum, 2004 [trans. of *Mille Plateaux*. Paris: Les Éditions de Minuit, 1980].
Deleuze, Gilles, and Claire Parnet. 1987. *Dialogues II*. New York: Columbia University Press.
DeLoughrey, Elizabeth. 2001. "'The Litany of Islands, the Rosary of Archipelagoes': Caribbean and Pacific Archipelagraphy." *ARIEL: A Review of International English Literature* 32, no. 1: 21–51.
———. 2010. *Roots and Routes: Navigating Caribbean and Pacific Island Literatures*. Honolulu: University of Hawai'i Press.
———. 2019. *Allegories of the Anthropocene*. Durham, NC: Duke University Press.
Earle, Sylvia. 1996. *Sea Change: A Message of the Oceans*. New York: Ballantine Books.
Edmunds, George F., and George S. Boutwell. 1901. *Insular Cases*. Boston: New England Anti-imperialist League.
Elias, Amy J., and Christian Moraru, eds. 2015. *The Planetary Turn: Relationality and Geoaesthetics in the Twenty-First Century*. Evanston, IL: Northwestern University Press.
Fanon, Frantz. 1967. *Black Skin, White Masks*. Translated by Charles Lam Markmann. New York: Grove Press [French orig. 1952].
Firmin, Anténor. 1910. "Haiti et la Confédération Antilliene." Lettres de Saint Thomas. Paris: V. Giard et E. Brière.
Fischer-Lescano, Andreas. 2011. "Critical Systems Theory." *Philosophy and Social Criticism* 38, no. 1: 3–23.
Fritzinger, Jerald. 2016. *Pre-Columbian Trans-Oceanic Contact*. N.p.: Lulu.com.
Glissant, Édouard. 1981. *Le discours antillais*. Paris: Editions du Seuil.
———. 1990. *Poétique de la relation*. Paris: Gallimard.
———. 1997. *Traité du tout-monde*. Paris: Gallimard.
Haggard, H. Rider. 1885. *King Solomon's Mines*. London: Cassell and Co.
Haraway, Donna. 1988. "Situated Knowledges: The Science Question in Feminism and the Privilege of Partial Perspective." *Feminist Studies* 14, no. 3 (autumn): 575–99.
Hau'ofa, Epeli. 1993. "Our Sea of Islands." *A New Oceania: Rediscovering Our Sea of Islands*. Suva, Fiji: School of Social and Economic Development.
Hay, P. 2013. "What the Sea Portends: A Reconsideration of Contested Island Tropes." *Island Studies Journal* 8, no. 2: 209–32.
Hayward, Philip. 2012. "Aquapelagos and Aquapelagic Assemblages." *Shima: The International Journal of Research into Island Cultures* 6, no. 1: 1–10.

———. 2015. "The Aquapelago and the Estuarine City: Reflections on Manhattan." *Urban Island Studies* 1, no. 1: 81–95.

———. 2012. "The Constitution of Assemblages and the Aquapelagality of Haida Gwaii." *Shima: The International Journal of Research into Island Cultures* 6, no. 2: 1–14.

Hiepko, Andrea Schwieger. 2003. "Creolization as a Poetics of Culture: Edouard Glissant's 'Archipelagic Thinking." In *A Pepper-Pot of Cultures: Aspects of Creolization in the Caribbean*, edited by Gordon Collier and Ulrich Fleischmann, 237–59. Amsterdam: Editions Rodopi.

Instituto Estudios Avanzados del Litoral, Universidad Nacional del Litoral, Santa Fe, Argentina. https://iealitoral.com.ar/en.

Island Studies Journal. 2006–.

James, C. L. R. 1984. *At the Rendezvous of Victory: Selected Writings*. London: Allison and Busby.

———. 2013. *Modern Politics*. Oakland, CA: PM Press.

Katz, Nancy, David Lazer, Holly Arrow, and Noshir Contractor. 2004. "Network Theory and Small Groups." *Small Group Research* 35, no. 3: 307–32.

Khan, Aisha. 2004. *Callaloo Nation: Metaphors of Race and Religious Identity among South Asians in Trinidad*. Durham, NC: Duke University Press.

Kumar, Krishan. 2003. *The Making of English National Identity*. New York: Cambridge University Press.

Ladoo, Harold Sonny. 1974a. *No Pain like This Body*. Toronto: House of Anansi.

———. 1974b. *Yesterdays*. Toronto: House of Anansi.

Latour, Bruno. 2017. "On Actor-Network Theory: A Few Clarifications Plus More Than a Few Complications." *Philosophical Literary Journal Logos* 27: 173–97 (accessed at www.bruno-latour.fr/sites/default/files/P-67 ACTOR-NETWORK.pdf).

Lineback, Neal, and Mandy Lineback Grizner. 2010. "The Indies: West and East." *Geography in the News*.

Linebaugh, Peter, and Marcus Rediker. 2000. *The Many-Headed Hydra: Sailors, Slaves, Commoners, and the Hidden History of the Revolutionary Atlantic*. Boston: Beacon Press.

Lionnet, Françoise. 2008. "Continents and Archipelagoes: From E Pluribus Unum to Creolized Solidarities." *PMLA* 123, no. 5: 1503–15.

López, Kathleen. 2013. *Chinese Cubans: A Transnational History*. Chapel Hill: University of North Carolina Press.

López-Calvo, Ignacio. 2008. *Imaging the Chinese in Cuban Literature and Culture*. Gainesville: University Press of Florida.

Lukács, Georg. 1971. *The Theory of the Novel: A Historico-philosophical Essay on the Forms of Great Epic Literature*. Translated by Anna Bostock. Cambridge, MA: MIT Press/London: Merlin Press. [German orig. 1914–1915].

Maldonado-Torres, Nelson. 2006. "Post-Continental Philosophy: Its Definition, Contours and Fundamental Sources." *Worlds and Knowledges Otherwise* 1–3: 1–29 (accessed at https://globalstudies.trinity.duke.edu/sites/globalstudies.trinity.duke.edu/files/file-attachments/v1d3_NMaldonado-Torres.pdf).

Martí, José. 1891. "Nuestra América." *Revista Ilustrada*. January 1.

Martínez-San Miguel, Yolanda. 2014. *Coloniality of Diasporas: Rethinking Intra-colonial Migrations in a Pan-Caribbean Context*. New York: Palgrave Macmillan.
———. 2017. "Colonial and Mexican Archipelagoes: Reimagining Colonial Caribbean Studies." In *Archipelagic American Studies*, edited by Brian Russell Roberts and Michelle Ann Stephens, 155–73. Durham, NC: Duke University Press.
McCall, Grant. 1994. "Nissology: A Proposal for Consideration." *Journal of the Pacific Society* 17, nos. 2–3: 93–106.
Melville, Herman. 1851. *Moby Dick*. New York: Harper and Bros.
Mignolo, Walter D. "On Pluriversality." *Convivialism Transnational*. http://convivialism.org/?p=199.
Mohammed, Patricia. 2014. "Chinese Art: A Little-Known Heritage." *Trinidad and Tobago Guardian*. November 9. http://www4.guardian.co.tt/arts/2014-11-09/chinese-art-little-known-heritage.
Morillo Alicea, Javier. 2005. "Uncharted Landscapes of 'Latin America': The Philippines in the Spanish Imperial Archipelago." In *Interpreting Colonialism: Empires, Nations and Legends*, 25–53. Albuquerque: University of New Mexico Press.
Morton, Timothy. 2016. "Molten Entities." In *New Geographies 08: Island*, edited by D. Daou and P. Pérez-Ramos, 72–76. Cambridge, MA: Universal Wilde.
Murra, John V. 1981. "Los límites y las limitaciones del 'archipiélago vertical' en los Andes." Segundo congreso peruano del hombre y la cultura andina. Trujillo, October 1974. *Maguaré* 10, no. 1: 93–98.
Nail, Thomas. 2017. "What Is an Assemblage?" *SubStance* 6, no. 1 (142): 21–37.
Neill, Peter. 2016. *The Once and Future Ocean: Notes toward a New Hydraulic Society*. Sedgwick, ME: Leete's Island Books.
Nieto-Blázquez, M. E., A. Antonelli, and J. Roncal. 2017. "Historical Biogeography of Endemic Seed Plant Genera in the Caribbean: Did GAARlandia play a role?" *Ecology and Evolution* 7, no. 23: 10158–74. https://doi.org/10.1002/ece3.3521.
Norquay, Glenda, and Gerry Smyth. 2002. *Across the Margins: Cultural Identity and Change in the Atlantic Archipelago*. Manchester, UK: Manchester University Press.
Parr, Adrian, ed. 2005. *The Deleuze Dictionary*. New York: Columbia University Press.
Perlongher, Néstor. 1997. "La desaparición de la homosexualidad." *Prosa Plebeya. Ensayos 1980–1992*, 85–90. Buenos Aires: Colihue.
———.1997. "Los devenires minoritarios." *Prosa Plebeya. Ensayos 1980–1992*, 65–75. Buenos Aires: Colihue.
Pratt, Mary Louise. 1992. *Imperial Eyes: Travel Writing and Transculturation*. London: Routledge.
Priam, Mylène. 2010. "Antillanité." In *The Oxford Encyclopedia of African Thought*, edited by Abiola Irele and Biodun Jeyifo. Oxford: Oxford University Press. http://www.oxfordreference.com/view/10.1093/acref/9780195334739.001.0001/acref-9780195334739-e-033.
Pugh, Jonathan. 2013. "Island Movements: Thinking with the Archipelago." *Island Studies Journal* 8, no. 1: 9–24.
Ramirez Jonas, Paul. 2003. "Kathryn Chan." *BOMB* 82 (winter): https://bombmagazine.org/articles/kathryn-chan.

Reyes-Santos, Irmary. 2013. "On Pan-Antillean Politics: Ramón Emeterio Betances and Gregorio Luperón Speak to the Present." *Callaloo* 36, no. 1: 142–57.

Ritzer, George, ed. 2004. *Encyclopedia of Social Theory*. Thousand Oaks, CA: Sage Publications.

Rivera, Mayra. 2016. *Poetics of the Flesh*. Durham, NC: Duke University Press.

Roberts, Brian. 2016. "Archipelagoes/Oceans/Americas Symposium." Brigham Young University, Provo, Utah, October 6–7.

———. 2020. "What Is an Archipelago? On Bandung Praxis, Lingua Franca, and Archipelagic Interlapping." In *Contemporary Archipelagic Thinking: Toward New Comparative Methodologies and Disciplinary Formations*, edited by Michelle Stephens and Yolanda Martínez-San Miguel. London: Rowman & Littlefield International.

Roberts, Brian Russell, and Michelle Ann Stephens. 2017. "Introduction. Archipelagic American Studies: Decontinentalizing the Study of American Culture." In *Archipelagic American Studies*, 1–54. Durham, NC: Duke University Press.

Santos Perez, Craig. 2015. "Transterritorial Currents and the Imperial Terripelago." *American Quarterly* 67, no. 3: 619–24.

Schwyzer, Philip, and Simon Mealor. 2004. *Archipelagic Identities: Literature and Identity in the Atlantic Archipelago*. Farnham, UK: Ashgate Publishing.

Selvon, Sam. 1955. *An Island Is a World*. London: Wingate.

Shima: The International Journal of Research into Island Cultures. 2007–. (ISSN: 1834-6057) (www.shimajournal.org).

Spivak, Gayatri. 2011. *Death of a Discipline*. Cambridge, MA: Harvard University Press.

Steinberg, Phillip. 2005. "Insularity, Sovereignty and Statehood: The Representation of Islands on Portolan Charts and the Construction of the Territorial State." *Geografiska Annaler* 87, no. 4: 253–65.

Stephens, Michelle. 2015. "Federated Ocean States: Archipelagic Visions of the Third World at Mid-century." In *Beyond Windrush: Rethinking Postwar Anglophone Caribbean Literature*, edited by J. Dillon Brown and Leah Rosenberg, 222–37. Jackson: University Press of Mississippi.

Stevens-Arroyo, A. 1993. "The Inter-Atlantic Paradigm: The Failure of Spanish Medieval Colonization of the Canary and Caribbean Islands." *Comparative Studies in Society & History* 35, no. 3: 515–43.

Stratford, E., G. Baldacchino, E. McMahon, C. Farbotko, and A. Harwood. 2011. "Envisioning the Archipelago." *Island Studies Journal* 6, no. 2: 113–30.

Stratford, Elaine. 2013. "The Idea of the Archipelago: Contemplating Island Relations." *Island Studies Journal* 8, no. 1: 3–8.

———. 2017. "Imagining the Archipelago." In *Archipelagic American Studies*, edited by Brian Russell Roberts and Michelle Stephens, 74–94. Durham, NC: Duke University Press.

Stratford, Elaine, Godfrey Baldacchino, and Elizabeth McMahon, eds. 2014–. *Rethinking the Island* book series. London: Rowman & Littlefield International (accessed at http://www.rowmaninternational.com/our-publishing/series/rethinking-the-island).

Tancons, Claire. 2007. "Lighting the Shadow." *Third Text* 21, no. 3: 327–39. http://dx.doi.org/10.1080/09528820701362464.

Thompson, Lanny. 2010. *Imperial Archipelago: Representation and Rule in the Insular Territories under U.S. Dominion after 1898*. Honolulu: University of Hawai'i Press.

———. 2017. "Heuristic Geographies: Territories and Areas, Islands and Archipelagoes." In *Archipelagic American Studies*, edited by Brian Russell Roberts and Michelle Ann Stephens, 57–73. Durham, NC: Duke University Press.

Torabully, Khal. 1992. *Cale d'étoiles—Coolitude*. Sainte Marie, La Réunion: Éditions Azalées.

———. 1996. "The Coolies' Odyssey." *UNESCO Courier* 49, no. 10: 13–16.

Venuti, Lawrence. 2006. *The Scandals of Translation: Towards an Ethics of Difference*. London: Routledge.

von Bertalanffy, Karl Ludwig. 1968. *General System Theory: Foundations, Development, Applications*. New York: George Braziller.

Wachowski, Lana, dir. 2017. "Isolated above, Connected below." *Sense8*. Episode 6, Season 2. Los Angeles, CA: Elizabeth Bay Productions.

Walcott, Derek. 1992. "The Antilles: Fragments of Epic Memory." The Nobel Prize. December 7. https://www.nobelprize.org/nobel_prizes/literature/laureates/1992/walcott-lecture.html.

Wendt, Albert. 1981. "Towards a New Oceania." In *Writers in East-West Encounter: New Cultural Beginnings*, edited by Guy Amirthanayagam, 202–15. London: Palgrave Macmillan.

Wilson, Rob. 2000. *Reimagining the American Pacific: From South Pacific to Bamboo Ridge and Beyond*. Durham, NC: Duke University Press.

Part I

SPACE, SCALE, LANGUAGE, AND TIME

Foundational Epistemological Contributions of Archipelagic Thought

The Fifth Map

Craig Santos Perez

For Hsinya

1

The first map my dad hangs in the hallway
is an aerial view of Guam.
The island nearly fills the entire space. "Where
is our village?" I ask him. "In the center,"
he says. "Here: Mongmong." I whisper
the names of other villages:
"Hagatna, Barrigada, Tumon, Dededo, Agat . . ."
I once imagined these places as completely separate,
but now I see how they're only different parts
of the same tropical body.

2

The second map my dad hangs in the hallway
is an aerial view of the Marianas archipelago.
I count 15 islands extending in a vertical
crescent. I recognize the shape of Guam,
the largest and southernmost in the linked
chain. My dad pronounces the names
of the northern islands: "Rota, Aguijan,
Tinian, Saipan, Farallon de Medinilla,
Anatahan, Sarigan, Guguan, Alamagan,
Pagan, Agrihan, Asuncion, Maug,
and Farallon de Pajaros." I tell him:
"They look like the beads of a rosary."

3

The third map my dad hangs in the hallway
is an aerial view of Micronesia.
"'Micro' means 'tiny,'" he says. "And 'nesia'
means 'islands.'" Two thousand dots scattered
across the Western Pacific.
My dad points: "Here's the Marianas,
and here's Palau, Yap, Chuuk, Pohnpei, Kosrae,
the Marshalls, Nauru, and Kiribati."
The archipelagoes resemble constellations.
"Are the people there Chamorro too?"
I ask him. "No," he answers.
"But they're our cousins."

4

The fourth map my dad hangs in the hallway
is an aerial view of the Pacific Ocean,
rimmed by the Americas and Asia.
Countless archipelagoes divided into three regions:
Micronesia, Melanesia, and Polynesia.
My dad traces a triangle between Hawai'i,
Easter Island (Rapa Nui), and New Zealand.
"This is Polynesia," he says. "'Poly' means many."
Then he draws an imaginary circle around
Papua New Guinea, the Solomon Islands,
Fiji, Vanuatu, and New Caledonia.
"This is Melanesia," he says. "'Mela'means black."
The ocean, connecting these vast distances,
looks like a blue continent. "Remember,"
he says. "We are all relatives."

5

My dad never hung a fifth map in the hallway.
I first see it when I travel, as an adult, to Taiwan.
The tour guide shows me an aerial view
of Austronesia. "'Austro' means 'south,'" she says.
A highlighted area, the shape of a full sail, stretches
from Madagascar to the Malay peninsula
and Indonesia, north to the Philippines and Taiwan,
then traversing Micronesia and Polynesia.
"Austronesians migrated to escape war,
famine, disease, and rising seas," she says.

It's difficult to imagine that 400 million people
alive today, who speak over 1000 different languages,
all descend from the same mother tongue,
the same genetic family. I examine the map closely,
navigating beyond the violent divisions
of national and maritime borders, beyond
the scarred latitudes and longitudes
of empire, to navigate the cartography
of our most expansive legends
and deepest routes.

Chapter Two

Disciplinary Formations, Creative Tensions, and Certain Logics in Archipelagic Studies

Elaine Stratford

In the context of writing and reading, the phrase "gather round" is a prompt to focus in and settle. It conjures spaces and times to come together—like pulling material into folds that are then stitched, or collecting the fruits of a harvest, or centering around a person or thing for some purpose. For me at least, such ideas are reminiscent of both islands and archipelagoes. So, in this chapter, I use "gather round" as an invitation to revisit and augment some of the recent foundational debates that have helped produce the study of archipelagoes and inclinations to archipelagic thinking. The focus of my story is the essay "Envisioning the Archipelago" (Stratford et al. 2011).

The chapter is in two parts. The longer, first part is historical and necessarily partial, and in it I revisit certain ideas about islands and archipelagoes that inform this collection and hinge on different ontologies, epistemologies, and values; I bolster those ideas by reference to work by Tariq Jazeel (2016). Then, in the shorter second part I respond to an invitation from the editors to think about the methodological implications of engaging with archipelagic frameworks by reference to work by Lanny Thompson (2017, 66). That work presents several ideas about the possibility that "archipelagic studies might have its own 'methodo-logic': an 'archipe-logic,'" as do observations about thinking advanced by Marjorie Garber (2012), on whom I also draw.

CREATIVE TENSIONS IN DEBATES ABOUT A RELATIONAL TURN IN ISLAND AND ARCHIPELAGIC STUDIES

Godfrey Baldacchino once reflected on the meanings of a term that, by virtue of its suffix, signifies the state(s) or condition(s) of islands. Island*ness*, he wrote, is "an intervening variable that does not determine, but contours and conditions physical and social events in distinct, and distinctly relevant, ways" (Baldacchino 2004, 278). Two years later, framing the advent of *Island Studies Journal*, Baldacchino (2006, 6) suggested that island studies is a powerful "focus of inquiry, straddling as well as going beyond conventional disciplines" in order to contribute to knowledge at many scales. Baldacchino also pointed to nascent agreement that island studies need not be constituted as a discipline—although he did not dismiss its consolidation as such. And he understood that this threefold capacity to focus, straddle, and exceed is a key strength of island studies. On that basis, Baldacchino proposed that the field's mandate could involve "sharing, advancing and challenging existing theorization on islands and island studies; while avoiding, delimiting or debunking false or partial interpretations of the island condition" (10).

One reason for these predictive and programmatic statements seems clear: it was fitting to make them at the launch of an endeavor such as a journal—whether or not the motivation was to advance claims for the legitimacy of a discipline or a focus of inquiry. Another reason becomes clear in Baldacchino's observation that those in the island studies community were then experiencing a sense of "urgency to develop and nurture" an audience, scholarship, and respectable platform for the field he described (10). And in ways that were and remain tantalizing, he argued that the field need not remain focused on islands alone—indeed, perhaps it should not. Instead, work should also interrogate the relations between mainlands and islands and between islands and islands. These relations he referred to in terms of "the beckoning study of archipelagoes" (10), which has also been prefigured by, for example, Benítez-Rojo (1996), Glissant (1997), and DeLoughrey (2001, 2007).

Island and archipelagic studies are, in fact, now constituted by a range of disciplinary formations, a matter elaborated upon again later in this volume. As a form of intellectual devotion, the idea of the discipline has been described by Mark Edwards (2006) as a communal pursuit in which one's work speaks to and circulates in enduring communities of inquiry and practice. Edwards also delineates the several hallmarks of disciplinary activity. Among them are journal publications, stand-alone monographs, edited collections, books in series, and reviews. There are conferences, funding bodies, and standardized norms to guide learning, certification, teaching, and research. There are professional and collegial associations typified by a sense of professional responsibility and the desire to leave a legacy connected to knowledge formation and dissemination. And there are systems and processes to enshrine authority and protect self-regulation.

A hallmark of disciplinary formations is the presence of debates both within and across their more-or-less porous boundaries. Doubtless, there are debates aplenty in island and archipelagic studies as currently configured, and that is what one might hope for. In relation to archipelagic studies at least, it would be odd were such not the case, for, as Michelle Stephens and Yolanda Martínez-San Miguel note in their introduction to this volume, such studies reframe how we conceive social and cultural formations and historical and contemporary experiences to acknowledge and allow for irreducible differences. And while Stephens and Martínez-San Miguel refer specifically to the ways in which archipelagic thinking might unsettle continental paradigms and perspectives, equally such thinking has the capacity to keep open, relational, and unsettled a range of diverse paradigms and perspectives within the field of inquiry (the discipline?) itself.

The *archipelago*, then, may be conceived of as a model—as well as being "real" in the terms defined by the United Nations Convention on the Law of the Sea: "a group of islands, including parts of islands, interconnecting waters and other natural features which are so closely interrelated that such islands, waters and other natural features form an intrinsic geographical, economic and political entity, or which historically have been regarded as such" (United Nations General Assembly 1982, Article 46). The potential that rests in thinking of the archipelago as a model was certainly advanced in the social sciences by Baldacchino (2006, 10) in relation to development. It was also aired in the humanities by Edmond and Smith (2003, 7), who drew on Dening's scholarship as one deploying the island as "a model, rather than simply a site" of investigation (see, for example, Dening 1980, 1994, 2004). Edmond and Smith then referred to the heterogeneity of human cultures. They unsettled what they saw as a prevailing and counterintuitive idea that islands are homogeneous and therefore offer "the possibility of the complete description of something static, unmixed and singular" (12). They underscored the point that islands and islandness are profoundly mobile/mobilizing/mobilized and offered "a model of how to live complexly rather than through the simplifications and essentialisms that have characteristically been projected" onto them (12).

The idea that the archipelago could serve as a model was also explored by Stratford et al. (2011, 114), and an invitation to engage was extended to others when we asked, So "what model/s [might apply] and how can island studies scholars attend to the range, specificity and dynamism of their interrelation?" And this idea of the archipelago-as-model has been revisited here by Stephens and Martínez-San Miguel in their introduction, in which they write that "the archipelagic focuses on the types of relations that emerge, island to island, when island groups are seen not so much as sites of exploration, identity, sociopolitical formation, and economic and cultural circulation but also, and rather, as models."

The notion that islands (and archipelagoes) may serve as models has been subject to criticism. Most notable is this from Peter Hay (2013, 212): "I wish to resist this call for abstraction, arguing instead for that concern for the phenomenological ground of island particularity that I deem paramount. I made that call, back in [2006] . . . because I did not want 'island' . . . to be relegated (yes, 'relegated') to the status of idea, concept, metaphor" (see also Hay 2006). And yet, metaphor—and thus abstraction—underpin the networks of meanings that we constitute and are enmeshed in. As my colleague Elizabeth McMahon (personal communication, 2018) has noted of reality, *abstraction* is "the invisible interconnection of the material, the historical, and the cultural. I think it is one of the great benefits of the interdisciplinary reach of island studies that it harnesses expertise across this reach—thereby arriving at understanding that is genuinely COMPLEX (itself an archipelagic term)." Or as Lakoff and Johnson (2008, 1) have proposed,

> Metaphor is pervasive in everyday life, not just in language but in thought and action. Our ordinary conceptual system, in terms of which we both think and act, is fundamentally metaphorical in nature. The concepts that govern our thought are not just matters of the intellect. They also govern our everyday functioning, down to the most mundane details. Our concepts structure what we perceive, how we get around in the world, and how we relate to other people.

My own "home" discipline of geography is fully-fledged and structured by concepts and replete with all the formations Edwards described above. It is also one in which metaphors help and co-constitute real experiences of, for example, space, scale, place, region, environs, boundaries, flows, and so forth. Less so now than in the past, geography has been strongly linked to area studies, under which island studies is sometimes listed as a discipline. Perhaps it is of more than passing interest that work in *Island Studies Journal* over its first decade was written by those tending to identify predominantly as geographers, then as anthropologists (Stratford 2015, 147, Figure 3), and then as other discipline experts.

Either way, the Greek scholar Eratosthenes (276–194 BC) described geography as "earth-drawing" (*γεωγραφία*), and the discipline has a multifaceted history comprising theoretical musings about the nature of life on earth and empirical practices in the field. In fact, it is replete with debates about the relative merits or drawbacks of abstraction and particularity, deduction, and induction. Tariq Jazeel (2016) has thoughtfully grappled with these debates, initially by reference to what he calls Geography's geography (see also Jazeel 2009). He has pinpointed and critiqued tendencies in both area and discipline studies to collapse into the terms "world leading" and "international" a set of knowledge claims that are overwhelmingly Anglo-American. He has noted other tendencies to dismiss work oriented to "local political or community contexts" (653), to downplay the empirical in favor of grand abstractions, and to see the two as hierarchical and binary. It is this last point that I want to emphasize here in relation to ideas about the archipelagic as a relational model. I suggest that theoretical and empirical and abstract and concrete scholarly activities *need be neither mutually exclusive nor combative*, although ideological and disciplinary practices may render them so.

Let me elaborate on this assertion by way of example. My own work often starts inductively: I experience particular phenomena or events—walking on an island shore, for instance, looking out to sea to another island shore nearby, or imagining other coast(line)s thousands of kilometers away. I make meanings from those experiences on the basis of others now past, including those mediated by reading or hearing about others' experiences or theorizations—for example, of gazing across horizons. I seek to deepen, question, shift, add to, change, or otherwise transform those meanings by further emotional, psychological, social, physical, or intellectual labors. I might ask locals about "that island over there" and its relationship to "here," then search for opportunities to test the reach and impact of those meanings by reference to others' ideas from elsewhere (see, for example, Stratford 2013). Abstraction and empirical work are both present and co-constitutive.

Interestingly enough, "Envisioning the Archipelago," which Hay found dissatisfying and unsettling, began life in a discussion I had with Godfrey Baldacchino during an eight-week residency he had at the University of Tasmania in 2011. In more recent conversations, we recalled that he thought then that we might begin to think about the notion of the archipelago as topos, field, and assemblage and consider how those labors could profit from an island epistemology, and vice versa (Baldacchino, personal communication, 2018). That paper emerged from a collective venture born of a subsequent three-day conversation in my office. There, my coauthors and I asked ourselves, What about real islands so compelled our attention? What experiences had we—of Tuvalu, Tasmania, and the Torres Strait, or of Malta, Canada, the Pacific, or the Caribbean? How had we made sense of those experiences? And then, slowly, how had we made sense of our experiences in terms of our scholarly engagements with island studies from our disciplines—geography, sociology, literary studies, and environmental studies? How was it that each of us had come to be curious about the ways in which our experiences of island studies discourses and practices suggested the need to rethink them in order to more comprehensively consider archipelagic relations? Note that by these terms I mean, unambiguously, islands themselves, the waters between them (and not necessarily defined in relation to them alone), the water column and benthic environment below these, the air above them, and all manner of actors and actants—human, nonhuman, and more-than-human. That is the gist of relational and assemblage thinking: it is what was meant then and what is intended now.

In the intervening period, we have been intrigued and delighted by the range and amount of work that our seedbed paper has informed. Among the most challenging of those works is that by Philip Hayward (2012, 2), who coined the term "aquapelago" on the grounds that our focus had been on the "archipelago as a *terrestrial* aggregate" (emphasis added)—a view that we dispute, given how we delineated our terms, as reflected in the paragraph above. There is also the aforementioned paper by Hay (2013), who justifiably argues that those engaged in island and archipelagic studies must never forget and must always honor island peoples, places, conditions, relations, and issues.

And indeed, our point was that the centrality of metaphor and abstraction to human experience and thought means that these can be powerful tools for just such labors to honor real experiences (for example, see McMahon 2013, 2016). Indeed, my recollection is that we wanted precisely to engage in metacognitive and metanarratological work. At no time did we think it legitimate (or assert the need) to reject any particular approach to studying islands; tolerance and openness are, we think, a hallmark of archipelagic thinking. Rather, we invited expansion and intensification and sought to do so generously and generatively. Had Jazeel's words been written in 2011, we would have paid heed to his point that in the "face of the teleology of disciplinary progress it can become all too easy to forget any obligation toward place and the responsibilities it demands [as well as to the] . . . finely grained texture of area, the politics of particular places, communities and grounded yet distant contexts" (654). We were fully motivated by such obligations and responsibilities.

At the same time, I appreciate Jazeel's decision to emphasize an important point about representation not standing "in a mimetic relation to place, to be judged merely on the basis of accuracy. It stands also in a constitutive relationship to place; as we well know, imaginative geographies are produced through representation" (654). Hence, too, the importance we placed in 2011 upon philosophical and literary engagements with island and archipelagic studies—engagements that demand abstraction and yet still deeply and profoundly enrich one's personal phenomenal experiences. Crucially for someone producing a geographical narrative, Jazeel first suggests that the labors of representation of place also serve to produce place, then underscores how this effect "redoubles the responsibility toward field-sites incumbent on all of us who work on particular places, even if those places are only accessed textually or archivally" (654).

Thus, theorizing and reification are not synonyms after all. Abstraction need not be divorced from particularity and can be used harmoniously in relation to the particular. And a model refers equally to a likeness made to scale, something fashioned (given form) in order to be shared, an exemplar, and a frame of reference. On such understanding, I cannot see how modeling necessitates the elision of particularity. I wonder, too, if disciplinary formations are a kind of model in the sense of something formed to be shared, and to exemplify then are they not also simultaneously made up of multitudes of particularities, like archipelagos? And if that is the case, island and archipelagic studies cannot help but be oriented to abstraction and to the specificities that Hay values.

I do understand that disciplinary formations may risk being attended by orthodoxies or forms of authoritarian thinking. As Marilena Chauí (2011, 145) notes of such tendencies, this is knowledge that "supports itself in the already seen (an exemplary fact), in the already thought (previous theory), in the already enunciated (authorized discourse)." It is mischievous knowledge, too: as when facts are rendered mere examples, or tests for theories are reduced to formal schemas or models—that is, empty frameworks rather than newly formed insights. In relation to our work on island and archipelagic studies, Hay (2013) thought he discerned the emergence of a "party line," but at no time was the work from 2011 a declaration of authority-over. Rather, it was and remains an invitation to think and act relationally, and it called for innovations in methodology. In that respect, our agenda was then, and I think it remains, to suggest the efficacy of theory as strategy. Again, in words provided by Jazeel (2016, 658), we sought to "enable differentially situated problem-spaces to speak to one another, to relationally constitute one another."

Because of Baldacchino's numerous speculations about the possibilities that disciplinary formations might coalesce around the field of island studies, and in light of the foregoing discussion about conceivable risks attending those formations, I want briefly here to revisit and clarify some of the points related to "Envisioning the Archipelago," not least because the advent of the present volume will have significant and particular effects on and emotional resonance for those working in the field and others who are interested in it but position their work outside it.

We first described our intention: to "understand archipelagos: to ask how those who inhabit them or contemplate their spatialities and topological forms might view, represent, talk and write about, or otherwise experience disjuncture, connection and entanglement between and among islands" (Stratford et al. 2011, 114). The word "inhabit" is important here, and it refers to archipelagoes and not to islands. So, too, is the word "between": nothing in our statement precluded the sea between islands in archipelagoes—for example, by dwelling in an ephemeral way during voyages or more permanently by constructing stilt houses or living on vessels at the near shore. Indeed, though perhaps naively, we presumed that the etymology of the term "archipelago" (chief sea, with reference to the Aegean) would be sufficient to flag our understanding that island-to-island relations could never preclude the maritime and maritimity.

We then made clear a foundational assumption of the paper—that archipelagic relations exist. We proposed that they can be mapped (traced from direct and indirect means ranging from experience, observation, and conversation to consideration of archives, prior maps, or other texts). And we suggested that once (or as) perceived and interpreted, these archipelagic relations could be understood in terms of their implications and their affects. Crucially, for me at least, we underscored the point that archipelagic relations and those things that derive from them are found in states of in-between-ness, and we also suggested that in those states are found individual and collective capacities both to act and be acted upon by what Seigworth and Gregg (2010, 1) called "forces or intensities . . . that pass body to body (human, nonhuman, part-body, and otherwise)"—which is to say, *experiences*.

In the ensuing discussion, we sought to exemplify our speculations. We posited that there has been a measure of work on "islands' singularity, unique histories and cultures, crafted and inscribed by the border between land and sea . . . and on [islands'] differences from . . . larger settings" (Stratford et al. 2011, 114). It is perhaps these studies that Hay (2013) wished to protect in what he called their radical particularity, but I note that their worth was not being questioned by us. Rather, the suggestion was that island-to-island relations were worthy of more attention—theoretically, methodologically, empirically. In like vein, Ralph Crane and Lisa Fletcher (2017, 106–7) have also responded to Hay's call for radical particularity, emphasizing the "value in appreciating the non-phenomenological ways in which meanings are attached to islands [and one might add, to archipelagoes]." Their aim was "not to deny or discount the meanings produced through direct engagement with islands but rather to show that the conceptualisations and representations of islands do not need to be restricted to 'real islands.'" While Hay did concede that most islands are archipelagic and that island-to-island relationships exist, he held that they are so specific as to render comparison futile, commonalities elusive. On this matter we continue to differ if for no other reason than thinking—as doing and experience—is never futile per se. Granted, it would have been inconsistent for Hay (2013, 210) not to have valorized "island specificity and the construction of island meanings in the unique terms of emotional dialogue between the hard biophysicality of each island and the people who live, or regularly interact, with it (and each other)." But equally, such identity politics in place really does depend on abstractions: the idea of specificity itself is one; so, too, are emotion, interaction, identity, politics, and place. They are also always and already material—and that was never at issue for us.

In 2011, we also referred to the boundary of land and water, but, following Dening (2004), we emphasized that this "line" is shifting, fractal, and paradoxical—a point that does not obviate another: namely, the encirclement of land by water constitutes spaces of insidedness and outsideness that are more or less ephemeral or durable. And while we noted the tendency to see those spatialities as further constitutive of ideas of isolation and insularity, we neither added pejorative judgements about them nor suggested that the ideas to which we had referred were ones to which we subscribed. Hence our reference to work on the Aegean Sea by Christy Constantakopoulou (2007, 2), who suggested of her case study that its "islands were understood as distinct 'closed' worlds, ideal locations for the extraordinary and the bizarre, but at the same time they were also perceived as parts of a complex reality of interaction." These ideas about shifting, fractal, paradoxical, and interactive spaces and/or relations led us to Deleuze and his idea that the archipelago offers a model of a world in process. This particular phrasing resonated then, and still does, because it *constitutes the "model" as formative* rather than punitive or restrictive. And we further qualified the point that we understood archipelagic relations as one might a complex ecology and as ontogenic—as generative. In ontogenesis we saw the possibility for island peoples to continue to determine for themselves/ourselves how they/we engage with each other and with others.

Our work also sought to shape an agenda by which "to understand how those who inhabit the archipelago, or how those who contemplate its spatialities, view, represent, talk and write about, or otherwise experience disjuncture, connection and entanglement between and among islands" (Stratford et al. 2011, 124). Again, the terms "inhabit" and "between" are crucial here, and neither was intended to preclude either the maritime or experience (at least as I have sought to delineate them above). In the process, we thought about the possible effects of ideas such as *reterritorialization*. By referring to it, we implied (and perhaps might have made clearer the point) that we saw thinking as a form of doing, remapping as a form of experiencing, and relating itself as a crucial mechanism by which to bridge (and sometimes to breach). And relating is, of course, rhetorical and performative, none of which implies duplicity, cant, illusion, flimsiness, or an abandonment of the real, which is always already partially constituted by the imaginary and symbolic—by theorizations and a priori frames, or models.

My present engagements with the respectful and helpful critiques that followed the publication of "Envisioning the Archipelago" have taken a long time to coalesce and have been prompted by the development of the present collection. In my estimation, Hayward, Hay, and my coauthors and I have written at cross purposes, and yet we have all been striving to think about how to think about these aquatic and terrestrial regions (and see also Peters, Steinberg, and Stratford 2018). In hindsight, then, "Envisioning the Archipelago" was not the invention of some disciplin*ing* party line but an invocation of a set of ideas about radical relationality in which the water was not backgrounded but integral—or why summon the archipelago in the first place? Certainly, I made this point again in 2013 when I suggested that archipelagoes are geographical forms that invite those in the field/discipline to think about island relations. At that time, I qualified that statement with a prefatory description of the term "archipelago"—one that placed front and center the sea: nominally "a group of islands, more properly a sea studded with islands" (Stratford 2013, 3). And, by reference to conversations with my colleague Elizabeth McMahon, I noted that thinking with the archipelago may "enable island scholars and others to radically recentre positive, mobile, nomadic geopolitical and cultural orderings between and among island(er)s" (4). It is a possibility exhibited in this present volume and one that merits further development.

SPECULATIONS ON METHODOLOGY

In this second and shorter part of the chapter, I want to consider the idea of an "archipe-logic" advanced by Lanny Thompson (2017, 66). I come to Thompson because I think his work is a lucid exposition of the benefits of archipelagic thinking. It also invites further reflections on the implications of archipelagic thinking for methodology, a focus both of this collection and the work colleagues and I produced in 2011.

Thompson argues that the idea of the archipelago compels one first to study places in terms of the complex connections and processes that characterize them and then to traverse the spaces between them. Doubtless, one *could* apply the term "archipelago" metaphorically to consider various urban clusters such as were deployed in Walter Christaller's (1966) central place theory in geography, best represented by his early network diagrams of central Germany, dotted with large and small "islands"—settlements—in seas of rural or industrial lands (and see also Barnes and Minca 2013). But that is not, I think, Thompson's intent, for his own examples are literal islands and archipelagoes. By extrapolation, the archipe-logic to which he refers is meant to be applied to places that are islands (and associated entities, including airs and waters, flora and fauna, political systems, and cultural practices). In this sense, the archipe-logic to which he refers *is* a specific methodology to think—in his terms, *about*, *with*, and *from* archipelagoes.

Thompson submits that "thinking *about* archipelagoes means to understand them, not simply as natural formations but rather as constituted places, islands brought together in complex historical and geographical configurations" (66–67). Methodologically, "thinking about" is a complex task dealing with connection that can be done by reference to any number of perspectives—positivist, interpretivist, constructivist, or critical—and any number of associated quantitative methods. Yet this phrase "thinking about" seems first to compartmentalize that which is being considered before it connects. I might think about a pending event. Someone else might think about what to have for dinner. Another person might think about what it might mean to replace one slow ferry for two fast ones that double the number of vehicles that might be on-island at any given time. In the first instance, "I" am one entity; the event I am going to is another. In the second, the hungry person is one entity, and the food to be consumed constitutes another, a meal. In the third instance, each ferry is an entity, the island another, each vehicle too. Each implication is an entity. And in all three cases I might think not just about the entities but about the relations between them. But those relations also seem derivative (though not necessarily secondary) considerations. There is, for example, the sense of connectedness or distancing I may feel in relation to others or to the event itself by attending it, but that stems from my thinking about these matters. There is the anticipation of being satiated or disappointed by the union of flavor and taste bud when someone eats. And there are varied relational derivatives from ferry "upgrades" to an island that might touch on tourists and residents, invasives and indigenous species, pollution and clean air, and so on.

Thinking *with* the archipelago, for Thompson, "highlights the connections among material, cultural, and political practices that are spread out across islands and continents and are generated through the complex movements of capital and commerce, media and technology, ideas and ideology, and inventive, resourceful people" (67). Etymologically, "with" denotes association, combination, and union. Methodologically, "thinking with" is another complex task, and one on which Marjorie Garber (2012) also provides helpful insights. She traces the origins of the phrase "good to think [with]" to Claude Lévi-Strauss (1963), who first deployed the phrase in *Totemism*, published in French in 1962 and in English the following year—the square brackets will be explained shortly.

Garber describes the phrase as ubiquitous in, for instance, "feminism, science, architecture, taxes, the body, food, hypertext, networks, the liberal tradition, capitalism, and the brain" (96). Even so, it is vitally important for at least three reasons. First, it has "validating power" for scholarship in the humanities, which too often are now tested against measures that ill suit them. Second, the phrase has explanatory power in relation to apparent binaries. Having decided it was important to consider "how to make opposition, instead of being an obstacle to integration, serve rather to produce it"—this in response to fellow anthropologist Radcliffe Brown's work on totemic traditions—Lévi-Strauss (1963, 89) proposed that, being both real and symbolic, totemic animals were "good to eat" and "good to think." (There is, in fact, no "with" in the translation of his work. Animals are good to think [with].) Either way, for Garber (2012, 97), as for so many of us, the phrase is "not about the referent, the thing in the world; it is a celebration and a validation of thinking." Third, the phrase has imaginative power. According to Garber's reading, the act of thinking [with] may well be spurred by useful empirical observations such as derive from work in, for example, the social, physical, or life sciences. Yet, "its payoff is speculation, which is then reattached to, embodied in, or reembodied in the objects, concepts, or beings that gave rise to it" (97). And speculation—and the imagination that it implies—enables facts to "reemerge as figures of speech or, more precisely, figures of thought: metaphors, metonyms, personifications, allegories, categories, oppositions, analogies" (97). Such, I recall, were the concerns of our 2011 essay and our invitation to "gather round" an emerging debate and disciplinary formation.

What is particularly exciting about Garber's ideas is that they map onto those advanced by Thompson when he emphasizes the point that thinking with is an act that highlights connections. Thus, for Garber, the "next move, which has proven equally important to theoretical work in the humanities, is to question the boundary between the terms of the opposition, and to use that questioning as a way of thinking beyond an impasse" (97). It is for such reasons that abstractions and models need not be detached from particularity and can be used agreeably in relation to the particular—ideas captured in figures of speech, fashioned and formed to be shared as exemplars and frames of reference.

And finally, Thompson (2017, 68) writes, there is thinking *from* archipelagoes because they are "geosocial locations for the production of knowledge" in multiple and relational ways. The logical conclusion of Thompson's labors—and I hope this insight applies to others of us working in this space—is that those engaged in island and archipelagic studies need to "go beyond the strict adherence to the limits of academic disciplines and move explicitly towards comparative transdisciplinary inquiry, interpretation, and research. . . . The logic of archipelagic studies should allow us to trace relationships and map their dispersions in space and time . . . discontinuous connections . . . fluid movements . . . complex spatial networks . . . in sum, moving islands" (69–70).

Methodologically, it seems to me that those working in the field are being asked both to see archipelagoes as geosocial and to reflect on how the geosocial embraces different social locations and geographical standpoints, their interactions, differences, entanglements, and co-constitutive dynamics. Thus, thinking *from* islands and archipelagoes means not just standing on the shores or the margins (or indeed in the littoral zone, or up to one's knees, or from below the waves) "speaking back" to others elsewhere. It means deploying what is implicit in the term "from"—movement, departure, forward motion, and accomplishment; this is speaking to and up and into, and it seems ontogenic to me. In short, we are, I think, entering a new period of fertile thinking that is founded on the maturation of the field and in which we can continue to refine and deepen the ways in which we think about, from, and in relation to islands and archipelagoes and revitalize methodological and other insights about them. This volume is an important part of that process.

REFERENCES

Baldacchino, G. 2004. "The Coming of Age of Island Studies." *Tijdschrift voor Economische en Sociale Geografie*, 95, no. 3: 272–83.

———. 2006. "Islands, Island Studies, Island Studies Journal." *Island Studies Journal* 1, no. 1: 3–18.

Barnes, T. J., and C. Minca. 2013. "Nazi Spatial Theory: The Dark Geographies of Carl Schmitt and Walter Christaller." *Annals of the American Association of Geographers* 103, no. 3: 669–87.

Chauí, M. 2011. *Between Conformity and Resistance: Essays on Politics, Culture, and the State*. Dordrecht: Springer.

Christaller, W. 1966. *Central Places in Southern Germany*. Translated by Carlisle W. Baskin. Englewood, NJ: Prentice Hall.

Constantakopoulou, C. 2007. *The Dance of the Islands: Insularity, Networks, the Athenian Empire, and the Aegean World*. Oxford: Oxford University Press.

Crane, R., and L. Fletcher. 2017. *Island Genres, Genre Islands*. London: Rowman & Littlefield International.

DeLoughrey, E. 2001. "'The Litany of Islands, the Rosary of Archipelagoes': Caribbean and Pacific Archipelagraphy." *ARIEL: A Review of International English Literature* 32, no. 1: 21–51.

———. 2007. *Routes and Roots: Navigating Caribbean and Pacific Island Literatures*. Honolulu: University of Hawai'i Press.

Dening, G. 1980. *Islands and Beaches: Discourse on a Silent Land, Marquesas: 1774–1880*. Honolulu: University of Hawai'i Press.

———. 1994. *Mr. Bligh's Bad Language: Passion, Power and Theatre on the Bounty*. Cambridge: Cambridge University Press.

———. 2004. *Beach Crossings: Voyaging across Times, Cultures, and Self*. Philadelphia: University of Pennsylvania Press.

Edmond, R., and V. Smith. 2003. "Editors' Introduction." In *Islands in History and Representation*. London: Routledge, 1–31.

Edwards, M. U. 2006. *Religion on Our Campuses: A Professor's Guide to Communities, Conflicts, and Promising Conversations*. New York: Palgrave Macmillan US.

Garber, M. 2012. *Loaded Words*. Oxford: Oxford University Press.

Glissant, E. 1997. *Poetics of Relation*. Translated by Betsy Wing. Ann Arbor: University of Michigan Press.

Hay, P. 2006. "A Phenomenology of Islands." *Island Studies Journal* 1, no. 1: 19–42.

———. 2013. "What the Sea Portends: A Reconsideration of Contested Island Tropes." *Island Studies Journal* 8, no. 2: 209–32.

Hayward, P. 2012. "Aquapelagos and Aquapelagic Assemblages." *Shima: The International Journal of Research into Island Cultures* 6, no. 1: 1–11.

Jazeel, T. 2009. "Reading the Geography of Sri Lankan Island-ness: Colonial Repetitions, Postcolonial Possibilities." *Contemporary South Asia* 17, no. 4: 399–414.

———. 2016. "Between Area and Discipline: Progress, Knowledge Production and the Geographies of Geography." *Progress in Human Geography* 40, no. 5: 649–67.

Lakoff, G., and M. Johnson. 2008. *Metaphors We Live By*. Chicago: University of Chicago Press.

Lévi-Strauss, C. 1963. *Totemism*. Translated by R. Needham. Boston: Beacon Press.

McMahon, E. 2013. "Reading the Planetary Archipelago of the Torres Strait." *Island Studies Journal* 8, no. 1: 55–66.

———. 2016. *Islands, Identity and the Literary Imagination*. London: Anthem Press.

Peters, K., P. Steinberg, and E. Stratford, eds. 2018. *Territory beyond Terra*. London: Rowman & Littlefield International.

Seigworth, G. J., and M. Gregg. 2010. "An Inventory of Shimmers." In *The Affect Theory Reader*, edited by G. J. Seigworth and M. Gregg, 1–25. Durham, NC: Duke University Press.

Stratford, E. 2013. "Guest Editorial Introduction: The Idea of the Archipelago: Contemplating Island Relations." *Island Studies Journal* 8, no. 1: 3–8.

———. 2015. "A Critical Analysis of the Impact of Island Studies Journal: Retrospect and Prospect." *Island Studies Journal* 10, no. 2: 139–62.

Stratford, E., G. Baldacchino, E. McMahon, C. Farbotko, and A. Harwood. 2011. "Envisioning the Archipelago." *Island Studies Journal* 6, no. 2: 113–30.

Thompson, Lanny. 2017. "Heuristic Geographies: Territories and Areas, Islands and Archipelagoes." In *Archipelagic American Studies*, edited by Brian Russell Roberts and Michelle Ann Stephens, 57–73. Durham, NC: Duke University Press.

United Nations General Assembly. 1982. UN Convention on the Law of the Sea. United Nations. http://www.un.org/Depts/los/convention_agreements/texts/unclos/part8.htm, last accessed 13 December 2012.

Chapter Three

The Affirmational Turn to Ontology in the Anthropocene

A Critique

Jonathan Pugh

ISLANDS AS THE EXEMPLARY FIGURES OF THE ANTHROPOCENE

Previously often on the periphery of academic debate, today the island has become the emblematic figure of the Anthropocene across the sciences, social sciences, humanities, and popular press (Chandler and Pugh 2018; Hessler 2018; Pugh 2018). How islands are understood in these wider debates therefore not only tells us new things about islands, it also reveals how the fundamental stakes of the Anthropocene itself are being framed and engaged. The focus of this chapter is the recent and prolific rise of what has come to be known as the *affirmational turn* in broader Anthropocene scholarship (see Lorimer 2012; Ulmer 2017; Chandler, 2018a, 2018b; Chandler and Pugh 2018; Schmidt and Koddenbrock 2018; Bargués-Pedreny and Finkenbusch 2019). The affirmational turn is defined in this chapter as *the proliferation of novel ontologies in debates about the Anthropocene, which seek to decenter islanders much more fundamentally than relational and archipelagic thinking*[1] *to date*. Thus, it is explicitly about *affirming the humbling powers of more-than-human relations in the Anthropocene*. The chapter takes the island research of Timothy Morton, the most high-profile Anthropocene philosopher writing more generally today, as exemplary of the affirmational turn. The key argument is that the recent shift to affirm ontology more widely in the Anthropocene and island studies runs up against a salient challenge raised by the Martinican scholar Frantz Fanon, who famously drew attention to how merely affirming ontology "rapidly becomes unproductive" (1967, 190) and leads "up a blind alley" (172). For Fanon, human thought, understanding, and claim making can only be understood as dynamic processes of work, struggle, and labor. Seen in this light, I raise the follow-up questions, Why does the work, labor, and struggle associated with producing affirmational ontologies seem to pay off so often for scholars working in the social sciences and humanities today, where the figure of the island features prominently in these debates? What does the lure and appeal of affirmational ontologies and thinking with islands tell us about what it means to be an Anthropocene and island scholar today? The answers to these questions, I argue, are central both to understandings of islands in the Anthropocene and to what it means to work in contemporary islands and Anthropocene scholarship.

Structurally, the next section explores both the key characteristics of the broader affirmational turn in contemporary Anthropocene scholarship and Morton's more general position within this. The section following turns to Morton's engagement with island studies and his own novel island ontology. The chapter then explores what Fanon can usefully bring to a critique of Morton's affirmational ontology and, indeed, to the increasingly prolific affirmation of ontology more generally in islands and Anthropocene scholarship. The final section turns to the broader critical methodological implications of these key arguments for islands and Anthropocene studies and calls for the initiation of *critical island studies in the Anthropocene*.

AFFIRMING THE POWER OF THE MORE-THAN-HUMAN IN THE ANTHROPOCENE

The growing argument from a broad range of scientists is that human actions have transformed planet Earth to such a degree that we have now left the last 11,500 years of the Holocene and live in a new geological epoch called the Anthropocene—a term coined by Eugene Stoermer in the 1980s and popularized by Paul Crutzen in the 2000s (Crutzen and Stoermer 2000; Crutzen 2002). Fueled by carbon dioxide emissions and biosphere degradation, the Anthropocene is a new epoch of climate volatility and instability, characterized by such forces as accelerating global warming, ice-sheet melting, sea-level rises, and carbon sinks weakening. The salient concern for this chapter is that for most leading contemporary scholars, the Anthropocene profoundly challenges and disrupts previous ways of thinking about humans and their relation to the world (Morton 2013; Stengers 2015; Tsing 2015; Latour 2017).

Scholarship in the social sciences and humanities has been marked by the proliferation of novel and alternative ontologies to modern frameworks of reasoning that seek to affirm the humbling powers of more-than-human relations (Morton 2013; Stengers 2015; Danowski and Viveiros de Castro 2016; Clark and Yusoff 2017; Latour 2017). For such authors, it is variously the power of the Anthropocene (Danowski and Viveiros de Castro 2016),

Gaia (Stengers 2015; Latour 2017), the lithosphere (Clark and Yusoff 2017), or hyperobjects (Morton 2013), such as global warming, that is simply too great for the human intellect to grasp, control, or hold back. There has been this broader affirmational turn to decenter humans within the vastly complex characteristics of elemental, atmospheric, oceanic, geological, and other more-than-human relations. For leading scholars, although anthropos may have caused the Anthropocene, today the tables are now turned; our transforming planet is revealing how humbled humans are by the overwhelming power and forces of the world. Humans are reduced to having this more adaptive and responsive role—to thinking about how we can better learn to sense, attune, and become aware of what the transforming planet is telling us (Chandler 2018a; Chandler and Pugh 2018). This new tendency is what I call the "affirmational turn." As Danowski and Viveiros de Castro illustrate, the Anthropocene is "a *passive* present, the inert bearer of a geophysical karma which it is entirely beyond our reach to cancel" (2016, 5). Similarly, for Clark, "the impression that deep-seated forces of the earth can leave on social worlds is out of all proportion to the power of social actors to legislate over the lithosphere" (2010, 220–21). For Clark and Yusoff, the Anthropocene foregrounds "a before, a beneath, a beyond to the human presence that draws our attention to other modes of relating" (2017, 16).

Timothy Morton famously characterizes the Anthropocene as this new "age of asymmetry" (2013, 159), where what he calls massive hyperobjects, such as global warming, nuclear plumes, or other boundary-disrespecting pollutants, unfold through trillions of overlapping phases that are simply too vast to be understood or controlled by humans. Global warming unfolds through vastly interlinking spatiotemporalities—from the more immediate and intensified violence of hurricanes on a small island to large-scale sea-level rises, complex changes in atmospheric conditions, and the hundred thousand years or so it takes for increased levels of carbon to dissolve in the surrounding oceans. Thus, for example, the massively overwhelming hyperobject of global warming works on an island through scalar dimensions that are simply beyond human comprehension:

> Infinity is far easier to cope with. Infinity brings to mind our cognitive powers, which is why for Kant the mathematical sublime is the realization that infinity is an unaccountably vast magnitude beyond magnitude. But hyperobjects are not forever. What they offer instead is *very large finitude*. I can think infinity. But I can't count up to one hundred thousand. I have written one hundred thousand words, in fits and starts. But one hundred thousand years? It's unimaginably vast. Yet, there it is, staring me in the face, as the hyperobject global warming. (Morton 2013, 60, emphasis in original)

As Morton (2013) writes, it is simply impossible for the human intellect to stand back and coherently grasp how an island exists within the *totality* of global warming; such hyperobjects extend vastly in space and time to define both the present and the future. For Morton then, modernity's human/nature divide completely collapses through the overwhelmingly disruptive power of global warming: the "more maps we make, the more real things tear through them. Nonhuman entities emerge through our mapping, then they destroy them" (130). The very existence of global warming brings "about the end of modernity" (94).

As for most related commentators writing on the Anthropocene today, for Morton the end of modern frameworks of reasoning and collapse of the human/nature divide "is precisely not an instant vaporization, but rather a lingering coexistence with strange strangers" (2013, 95). A hyperobject, such as a nuclear explosion, does "not bring about an apocalypse in the sense of a total dissolution of things, but rather it brings about the end of the world as a horizon or limit that exists 'over there,' over yonder, like Nature, or indeed, like God" (94). Indeed, for Morton, the fact that all islands in the world *already exist within vastly multidimensional hyperobjects* such as global warming—in the inescapable present and in the inescapable future—makes hyperobjects themselves "the last god" (50). Like all gods, these are gods that we cannot know in their overwhelming power and mystery. For Morton, hyperobjects are all pervasive; yet they remain essentially withdrawn from the human intellect in their totality.

Morton is well known for his style of writing (Wark 2017; Schmidt and Koddenbrock 2018). He is particularly good at generating the affective sensation of hyperobjects such as global warming being withdrawn from human knowledge in their totality, even as they are everywhere penetrating, surrounding, and binding everything on the planet together. Morton is fond of employing analogies to heighten this affective sensation of these weirdly disorienting forces of the Anthropocene, regularly comparing global warming to "the Force" in *Star Wars*, where he says that global warming "surrounds me and penetrates me" but, in its totality, remains a mystery (2013, 28; see also 80, 85, 141). Much as Luke Skywalker senses "the Force" while sitting on his island retreat in *The Last Jedi*, for Morton global warming is similarly sensed in the trees, oceans, rocks, surfaces, and surrounding oceans of an island, but there is always this essentially ungraspable mystery of the humbling ontological "rift" between the vast multidimensions of global warming and the humans who live on the island (18). Thus, if Latour is the philosopher who has influenced many islands scholars to focus upon mapping and tracking relations, and if Glissant more cautiously argues the impossibility of grasping Relation (that is, focus on the immanence of movement), then, for

Morton, it is hyperobjects themselves that have finally "done what two and a half decades of postmodernism failed to do, remove humans from the center of their conceptual world" (181). All islands in the world today already exist *within* global warming. No island or islander exists somewhere, "just over there," outside this last god of the Anthropocene.

Unsurprisingly, given how the island has become the central figure of the Anthropocene, today's leading scholars, such as Morton (2013, 2016), often write about islands explicitly in order to illustrate their novel ontologies. What will be particularly interesting here for islands scholarship is that, in doing so, Morton (2016) takes time to directly challenge and seeks to radically move on from the dominant themes and conceptual frameworks of the relational and archipelagic turns in island studies to date. For Morton (2016), given that all islands already exist *within* hyperobjects such as global warming, even as we embed ourselves down in such concerns as island networks, mobilities, archipelagoes, constellations, actor-networks, and assemblages (see, for example, DeLoughrey 2001; Hau'ofa 2008; Sheller 2009; Hayward 2012; Stratford et al. 2011; Pugh 2013; Ronström 2018; Evans and Harris 2018), other, more nebulous, multidimensional forces increasingly find their way into our consciousness too. For Morton (2016) then, the dominant approaches of the relational and archipelagic turns feel too sure, too modern, in their mapping of island and islander relations and the immanence of island movements. As will be elaborated in the next section, Morton's island ontology profoundly switches the register of debate away from a concern with island interconnections, networks, mobilities, actor-networks, assemblages, and so forth, to instead consider how *islands themselves have a relational ontology, all the way down; in particular, how islanders are humbled within the vast, multidimensional, but essentially withdrawn, strange, and alien forces of the Anthropocene.*

MORTON'S AFFIRMATIONAL ISLAND ONTOLOGY

Morton (2016) has recently elaborated his affirmational ontology in the island essay "Molten Entities." This labor begins by drawing upon the work of the widely acclaimed artist Olafur Eliasson, who displayed a large block of ice from Greenland at the Paris Climate Change Conference (COP21). Using this work, called *Ice Watch*, "Eliasson was hoping to show how the ice invites us humans into something like a dialogue or dance. The ice is not simply an unformatted surface waiting for us humans to make it significant. The molten edges of the ice block, displaced in a Paris square, become a way to think about how beings are *intrinsically in motion* because they are intrinsically melting, fragile" (71, my emphasis). For Morton (2016, 71), Eliasson's high-profile artwork reveals how we can approach the "ambiguous edges" of island relations very differently from the concerns for island connections, networks, and mobilities that have dominated the relational and archipelagic turns in island studies to date (Hau'ofa 2008; Sheller 2009; Stratford et al. 2011; Hayward 2012; Pugh et al. 2009; Pugh 2013; Roberts and Stephens 2017; Martínez-San Miguel 2014; Grydehøj 2019). In her seminal work, DeLoughrey writes, "No island is an isolated isle[, so instead] a system of archipelagraphy—that is, a historiography that considers chains of islands in fluctuating relationship to their surrounding seas, islands and continents—provides a more appropriate metaphor for reading island cultures" (2001, 23). While Morton agrees that the "idea that 'No man is an island' is obviously very popular right now as a progressive concept" (2016, 71), his own island ontology is very different, because for Morton, as noted, *relationality is ontological to being itself, all the way down*:

> If you look at the coastline of an island from space, you will see something fairly regular—perhaps it's rather triangular. When you look close up, say from a hang glider, you will see all kinds of curves and folds that you didn't see from space. And when you crawl around the surface of the coastline as an ant about three millimeters long, you will find something very different again—not just impressionistically different, but extensionally different: the circumference will be a different length. Indeed there may be circumstances—ways of measuring that island—that cause its circumference to be infinite. This is rather like what happens when you examine something like a Koch Curve, the fractal shape in which triangles are populated with smaller versions of themselves to infinity. One ends up with a shape that is bounded yet infinite. The Koch Curve is strangely "more than itself" at every point. An island is a cornucopia, or TARDIS, that contains more of itself on the "inside" than it appears to have on the outside. This is because they always exceed how they appear, even to themselves. They melt out of themselves, without moving *in* space or time and without being pushed by anything. (71–73, emphasis in original)

Thus, for Morton, "relating isn't some wondering way to fasten islands into chains to make them more exciting. *Relating* is how a thing is, all by itself" (73). His key point is that islands are relational not only because we can explore island connections, mobilities, and networks but because of the ontological withdrawnness of the island itself. Morton profoundly switches the register of debate by *affirming the humbling power of that impassable rift that opens up between island relations in their ontological reality and the capacity of humans to experience this totality*. Indeed, as we have seen in the previous section, Morton consistently affirms how acclimatizing to, becoming attuned to, and developing an awareness of objects such as islands in the Anthropocene "means to approach, then diminish, from a certain fullness" whose total reality is fundamentally inaccessible to humans (2013, 74). This reconfigured island ontology works much more explicitly and harder, at a deep and fundamental level, to decenter humans—foregrounding human impotence—when compared to relational and archipelagic thinking to date.

Such an approach is different from what has been called "thinking *with* the archipelago" (Pugh 2013, 9; Glissant 1997; DeLoughrey 2007; Stratford et al. 2011), but also from the more recent turns to vitalism and new materialism in island studies as well (Hayward 2012). For Morton, these more recent turns too readily insist "on regressing to fantasies of embeddedness" (2013, 18) These "so-called posthuman games are *nowhere near posthuman enough* to cope with the time of hyperobjects" (201, emphasis in original). For Morton, much like the earlier relational and archipelagic turns with only a few minor modifications, the recent posthuman turn is too certain in its abilities to be able to read and understand material affects, to enter into a "multispecies dialogue" (Evans and Harris 2018, 1), when what instead needs to be foregrounded are the richly disorienting and pervasive forces of the Anthropocene that are essentially withdrawn from the human intellect.

While Morton turns away from relational and archipelagic thinking and from new materialism and vitalism in island studies, for him hyperobjects such as global warming are particularly good illustrations of what his favorite philosophical tradition, Graham Harman's object-oriented ontology (OOO), reveals about all objects—cups, houses, capitalism, global warming, islands—being withdrawn from the human intellect in their essence and totality. For Harman, Latour's actor-network theory (influential in relational thinking in island studies) had previously gone some way in pushing the notion that no "metalanguage" exists through which humans can grasp meaning (that is, meaning is always implicated in relational more-than-human entanglements). But we now need to push this a whole lot further and in new directions. As Harman illustrates, the

> praiseworthy aim [of Latour's actor-network theory] was to free us from an older tradition in which society was viewed as a self-contained realm where humans did all the acting and objects were passive receptacles for human mental or social categories. . . . To say that objects mediate relations is to make the crucial point that unlike herds of animals, human society is massively stabilized by such nonhuman objects as brick walls, barbed wire, wedding rings, ranks, titles, coins, clothing, tattoos, medallions, and diplomas. . . . What this still misses is that the vast majority of relations in the universe do not involve human beings, those obscure inhabitants of an average-sized planet near a middling sun, one of 100 billion stars near the fringe of an undistinguished galaxy among at least 100 billion others. If we forget that objects interact among themselves even when

> humans are not present, we have arrogated 50 percent of the cosmos for human settlement, no matter how loudly we boast about overcoming the subject-object divide. A truly pro-object theory needs to be aware of relations between objects that have no direct involvement with people. . . . [Moreover, to] treat objects solely as actors forgets that a thing acts because it exists rather than existing because it acts. Objects are sleeping giants holding their forces in reserve, and do not unleash all their energies at once. (2016, 5–7)

For Morton, contemporary phenomena such as global warming—and islands existing within the massively multidimensional forces of the Anthropocene more generally—are sound exemplars of these sleeping giants: as noted, global warming is a massive *hyper*object that unfolds not all at once on islands but through vastly interweaving spatiotemporalities that stretch from the immediate to the hundreds of thousands of years. All of this is beyond the grasp of the mere human intellect. This profound switch in attention to *affirming hyperobjects* suggests even more clearly than actor-network theory, relational and archipelagic thinking, and new materialism that there is no "metalanguage"—nowhere to stand "outside"—of global warming through which we could somehow look down and grasp it from above. In such understandings island relationality in the Anthropocene becomes too rich, too intense, too withdrawn to be grasped by the older frameworks of reasoning associated with the relational and archipelagic turns and the more recent posthuman and more-than-human ontologies.

For Morton then, we should invest hope neither in modernity nor in the more radical framings of relations that many commentators seek to engender by way of relational thinking, actor-networks, assemblages, vitalist ontologies, and others. Morton is more aligned with the increasingly influential work of Ray Brassier (2007), Quentin Meillassoux (2010), and Claire Colebrook (2016), where it is today argued that it is not possible to re-enchant our engagement with the world in the ways these previous frameworks suggested. Morton switches to what could be called engaging a world "beyond hope," where humans are *much more* humbled because human meaning is inextricably withdrawn from the world. As noted, for Morton (2013), hyperobjects, not humans, have *already* defined our present *and* our future. It is these hyperobjects, such as global warming, and not humans that are "the last god" (50). As he concludes his influential book *Hyperobjects: Philosophy and Ecology after the End of the World*,

> Just to be clear, let me restate it in stronger terms. Nonhuman beings are responsible for the next moment of human history and thinking. It is not simply that humans became aware of nonhumans, or that they decided to ennoble some of them by granting them a higher status—or cut themselves down by taking away the status of the human. These so-called posthuman games are *nowhere*

> *near posthuman enough* to cope with the time of hyperobjects. They are more like one of the last gasps of the modern era, its final pirouette at the edge of the abyss. The reality is that hyperobjects were already here, and slowly but surely we understood what they were already saying. They contacted us. (Morton 2013, 201, emphasis in original)

AFFIRMATIONAL ONTOLOGY LEADS US "UP A BLIND ALLEY"

Morton's work aligns well with the *affirmational turn* that runs through many recent strands of Anthropocene philosophy, marked by the prevalence of approaches that seek to "shock and awe" readers about just how much humans are humbled by the overwhelming power of more-than-human relations in the Anthropocene (Schmidt and Koddenbrock 2018, 1; Chandler 2018a, 2018b). For Chandler (2018a, 2018b), the affirmational turn is about affirming the end of modern frameworks of reasoning and a telos of progress, while at the same time leaving us with impoverished ways of understanding the possibilities for change—for example, reducing humans to merely sensing, correlating, and attuning to hyperobjects, such as global warming. For other critical commentators, such as Schmidt and Koddenbrock, the affirmational turn comes in the form of a "'presentist persuasion' [that] relies upon the existence of revelators," such as Morton, who "set themselves to work and persuade others by pointing out undeniable signs and effects of" the Anthropocene (2018, 2). Having also discussed Morton's work with an increasing number of islands scholars in recent years, I sense that their worry is about how such affirmational approaches draw upon and think with islands and islanders. The affirmational turn humbles the human and the islander themselves within vast relational forces, foregrounding their impotence, lack of agency and ability to grasp relations and direct change. This, I think, is important to consider in more detail. In order to take on and develop these critical concerns, it is useful to turn to a salient challenge posed by Fanon, who famously drew our attention to how affirming ontology more generally leads us "up a blind alley" (1967, 172).

In his monumental book the *Wretched of the Earth*, Fanon (1967) usefully frames his criticism of intellectuals living in countries subjected to colonialism in terms of, on the one hand, those who do not challenge or seek to overthrow dominant powers but rather embrace the practices that support these powers and which these powers often forward and proliferate themselves; and, on the other hand, those who retreat into affirming ontology, as if passionately framing the stakes in terms of alternative ways of being will somehow result in new political futures. To quote Fanon, "The intellectual throws himself in frenzied fashion into the frantic acquisition of the culture of the occupying power and takes every opportunity of unfavorably criticizing his own national culture, or else takes refuge in setting out and substantiating the claims of that culture in a way that is passionate but rapidly becomes unproductive" (190). Fanon was, of course, strongly against both approaches, which today could be updated, on the one hand, to those Anthropocene scholars who educate islanders about how to become more adaptive, adjustive, and resilient in the Anthropocene (rather than engaging and overthrowing oppressive powers and structural inequalities), and, on the other, to those who retreat into affirmational ontologies, like hyperobjects. While it could be argued that, in practice, both logics conflate and reduce the critical stakes for engaging the Anthropocene to merely attuning, correlating, and ontologizing the rich intensification of relationality,[2] Fanon's more general and useful critical point is that affirming ontology and politics are different things.

At his own time of writing, for Fanon those intellectuals who retreated into affirmational ontology were those who sought to "affirm the existence of an African culture" (1967, 172–73). Affirming African ways of being produced "a vigorous style, alive with rhythms, struck through and through with bursting life; it is full of colour, too, bronzed, sun-baked and violent" (177). Although this style "astonished the peoples of the West, [for Fanon it] has nothing racial about it, in spite of frequent statements to the contrary" (177). Fanon famously criticized the affirmational ontologists for their retreat into a "universal standpoint" (176) that left "individuals without an anchor, without a horizon, colourless, stateless, rootless—a race of angels" (175). For Fanon, the challenge is to not reduce our engagement to merely affirming ontology but instead to foreground how human action, thought, and claim making are the more dynamic products of work, struggle, and labor. Thus, Fanon writes that the stakes should be explicitly framed in terms of an *objective and instrumental struggle*—to work hard, transgress, and find new ways to move on from mere static affirmation of "what is" as the normative political horizon for action. As he says to those who retreat into affirmation of "African culture," "When a people undertakes an armed struggle or even a political struggle against a relentless colonialism, the significance of tradition changes" (Fanon 1967, 180).

It is extremely important to note that Fanon's salient and overriding criticism of affirmational ontology came from his deep belief and understanding that human action, thought, and claim making cannot be reductively grasped in terms of ontology but are rather the dynamic products of ongoing struggle, work, and labor: whether "the fight is painful, quick, or inevitable, muscular action must substitute itself for concepts" (177). For Fanon, retreating into affirming ontology—of whatever kind and variety—

is therefore unproductive, and while the intellectual who engages in such approaches wishes to attach to the people,

> instead he [*sic*] only catches hold of their outer garments. And these outer garments are merely the reflection of a hidden life, teeming and perpetually in motion. That extremely obvious objectivity which seems to characterize a people is in fact only the inert, already forsaken result of frequent, and not always very coherent adaptations of a much more fundamental substance which itself is continually being renewed. The man [*sic*] of culture, instead of setting out to find this substance, will let himself be hypnotized by these mummified fragments which because they are static are in fact symbols of negation and outworn contrivances. Culture has never the translucidity of custom; it abhors all simplification. In its essence it is opposed to custom, for custom is always the deterioration of culture. The desire to attach oneself to tradition or bring abandoned traditions to life again does not only mean going against the current of history but also opposing one's own people. When a people undertakes an armed struggle or even a political struggle against a relentless colonialism, the significance of tradition changes. (180)

While Fanon was writing about different circumstances and at a different time, the distinction he makes between retreating into affirmation of ontology, on the one hand, and how human action, thought, and claim making are in practice the work of labor and struggle, on the other, provides a salient lesson for many of today's debates about the Anthropocene. McKenzie Wark has recently written that, interesting and novel as they are, we need to "move on from the *contemplative* thought" of affirmative ontologies such as Morton's (2017, 272, emphasis in original). Likewise, the central and key argument in this chapter is that Fanon's critique offers us a useful way to move on because, although affirmational ontologies can certainly reveal conditions and enable new and interesting questions to be posed, for Fanon the overriding consideration would be that such statements are nevertheless being made by *human beings*. That is, affirmational ontologies are in themselves the product of human work, struggle, and labor and reveal important things about *ourselves*—about how *we* see the stakes of engagement.

Thus, it is Timothy Morton's work that brings into being entities as hyperobjects—but not in the way Morton suggests. Where he errs is in downplaying people's roles in framing hyperobjects and producing the world in thought *as well as* in its ontological materiality. For Fanon, the Anthropocene could only be poorly grasped if understood in ontological terms alone. It would be more useful to approach such scholarship as part of a much broader and prolific contemporary social and academic struggle to *decenter* humans and *their* critically reflective capacities. It is *this* struggle that is immanent to the Anthropocene, in the world, and is a *reflection* of how the critical stakes of the Anthropocene are being worked, framed, and engaged.

For Fanon, whether certain ontologies become more alluring and influential tells us what *we* think it means to engage and struggle with our world. How these ontologies reconfigure island relationalities, what they foreground and downgrade in the process, therefore not only makes us aware of new things about islands in the Anthropocene (as in Morton's novel affirmational ontology). It also tells us important things about *ourselves* as scholars engaging the Anthropocene—about our own preoccupations, how these are changing, opening up, and importantly limiting engagement with islands and the Anthropocene in new ways. Morton is therefore misguided to argue that hyperobjects "have done what two and a half decades of postmodernism failed to do, remove humans from the center of their conceptual world" (2013, 181). For all the emphasis upon the more-than-human in such contemporary ontologies, for Fanon, *we* are the critique; we are expressions of the world in thought as well as being. Thus, affirming hyperobjects, like global warming, as "the last god" (50) would be, for Fanon, a conceit, because *we* are the coproducers of our gods, our islands, and our world—something that, as Fanon recurrently argues, is the product of work, struggle, and labor and therefore reveals things about ourselves and how *we* seek to frame and engage the stakes for action.

CONCLUSION: "PRACTICAL AND METHODOLOGICAL IMPLICATIONS"

What makes Morton's affirmative approach so attractive for Anthropocene scholars, resulting in his becoming the highest-profile Anthropocene philosopher today? How does it lure us? What does it enable? How does it cohere? What is its mode of affect? What are the effects? How does his affirmational ontology show us the world or enable us to enter the world by seeing it as an immanent and "real" product of the world, rather than merely seeing this as a transcendental and "subjective" claim about the world? Such questions are important to ask in debates about what it means to do both island studies and Anthropocene scholarship. In flagging questions about the affective appeal of the affirmational turn, the exclusion of the human subject and the decentering of islanders cannot be assumed as "given" in the world ontologically; rather, they are important in terms of affective presentation and "poetic" style—and to how *we* are lured into, engage, and understand the critical stakes of the Anthropocene (I purposefully do not say "duped" because what we are talking about is a broader social phenomenon and way of thinking). Ontologically speaking, for all the novel complexity, there is little new in Morton's stating that reality is infinitely richer than human categorization. But it is both interesting and telling to consider why such affirmational approaches have risen to the fore and have widespread appeal today.

From this key concern for the affective appeal of affirmational ontologies, it is possible to begin to stake out an intervention, methodology, and conclusion to this chapter that I hope might move the debate on. Although I have been critical of Morton in this chapter, for me, more fundamentally, it is not a matter of determining what affirmational approach is "right" or "wrong," ontologically speaking. Morton's work is novel and interesting for how it makes us think with islands and the Anthropocene in new ways. But, for me, a more interesting, and indeed useful, way of reframing the stakes of debate is in terms of how the widespread appeal or lure of such ontologies reflects how our purposes as researchers, scholars, and activists have changed.[3] Informed by Fanon's insistence that human understanding, thought, and claim making can only be understood as dynamic processes of work, struggle, and labor, what I am fundamentally concerned with then is this: What does it mean today to do work as a scholar—and an islands scholar in particular—engaged with the Anthropocene?

On that basis, I suggest that approaching such scholarship itself as immanent to the world, as in the world, may enable some important critical reflections. Broader disciplinary shifts that position islands as symbolic for many wider debates about the Anthropocene bring island studies in relation to cognate fields and beyond, and here certain views of human (islander) potentiality can be seen as being enabled or excluded. How are islanders' critical capacities enabled when reframed in terms appropriate to the recent affirmational turn in broader Anthropocene scholarship today? What limitations on islander potential emerge when we affirm the existence of today's more-than-human, object-oriented, or vitalist ontologies? By foregrounding how and why certain ontologies become more alluring and affect *our* thinking and that done by islanders, scholarship becomes a way *into* the world—into understanding how the critical stakes of the Anthropocene are being understood and engaged—rather than a way of abstracting and separating ourselves from them.

Much contemporary Anthropocene scholarship—exemplified by Morton—celebrates how humans are humbled within the vastly complex and disorienting relations of the Anthropocene and points to our inability to grasp and take control of these. This scholarship plays an important role in constantly affirming the lack of critical agency and controlling capacities of the human subject. We live in challenging times. As suggested by Adelene Buckland, "Apprehending the scale of the crisis is as much the *problem* as it is the solution, and this problem is a frequent refrain in Anthropocene discourse" (2018,

220, emphasis in original). However, the problem, as I see it, is that impotence itself has become the central and running mantra of much of this Anthropocene scholarship more generally.

From the perspective of someone working in a Western university, it seems to me particularly telling that Morton's shift in logic synchronizes well with the longer antihumanist march of critical theory. Here the humbling powers of more-than-human relations in the Anthropocene have conveniently handed such critical theorists the final stake that they had long wanted to drive into the heart of modernity (as just some of many examples, see Danowski and Viveiros de Castro 2016; Morton 2017; de la Cadena, Marisol, and Blaser 2018; Latour 2018; for critique, see Chandler and Pugh 2018; Chandler and Reid 2018). While I am not personally a supporter of modernity, today's dominant Anthropocene logics consistently decenter the human, adopting an increasingly antihumanist rhetoric and stance. We have shifted our approach and the stakes of engaging the world a long way from Fanon's distinctively humanist approach to challenging structural inequalities and relations.

So, what to do? I do not think we can approach the stakes until some underlying and fundamental concerns have been addressed. Political struggle cannot be theorized in the abstract, we no longer live in Fanon's world, and the stakes are always best approached by examining the existing social conditions and critically excavating frameworks of reasoning that bind, justify, and hold these together. Thus, we can presently say that for large numbers of academics, policymakers, and activists getting to grips with and engaging the conditions of the Anthropocene, political struggle is understood in terms less of subjugation to a widespread, collective political project than of correlating to and/or speculating upon the world in new ways. Here it is vitally important to engage how the Anthropocene is increasingly held to humble humans within vast, complex, and disorienting relations. Island relationality is understood as too rich, too intense, for human critical capacities to grasp and take control of, and the human subject is understood as more decentered. As Morton illustrates well,

> No longer are my intimate impressions "personal" in the sense that they are "merely mine" or "subjective only": they are footprints of hyperobjects, distorted as they always must be by the entity in which they make their mark—that is, me. I become (and so do you) a litmus test of the time of hyperobjects. I am scooped out from the inside. My situatedness and the rhetoric of situatedness in this case is not a place of defensive self-certainty but precisely the opposite. That is, situatedness is now a very uncanny place to be . . . (2013, 5)

David Chandler hits the nail on the head when he says that in such contemporary debates the "Anthropocene is not merely the recognition of the importance of climate change or global warming; but neither is it merely a critique of modernity: for a growing number of theorists, it is affirmed as a new framework for understanding and acting in a world, which can never be considered a 'home'" (2018b, 3). In a world withdrawn from the human intellect, we are forced to speculate upon meaning; hence, the rise of the "speculative turn," which Morton exemplifies, gains increasing prominence.

Yet, as Fanon explains, it is always useful to be reminded that *people* frame and understand their worlds as they do. Today human flourishing is understood not in terms of widespread subjugation to a coherent, alternative political project, but increasingly in terms of speculating upon and attuning to the richness of the Anthropocene—much as a Buddhist gardener attunes to relations rather than seeking to direct and control them. The critical stakes, I suggest, are both different and more urgent than this; but locked within a deeply impoverished understanding of human and political agency, in a way that Fanon would surely abhor, are reflections of a profound shift away from the active capacities of islanders (and all of us) to develop higher political aspirations. Therefore, I call for a more critical island studies in the Anthropocene.

NOTES

Thanks, in particular, to David Chandler (whom I regularly work with when thinking about islands and the Anthropocene). Also thanks to Julian Reid, Stephanie Wakefield, Godfrey Baldacchino, Elizabeth DeLoughrey, Elaine Stratford, Yolanda, Michelle, and the anonymous reviewers for some really useful comments on earlier drafts of this paper.

1. Both relational and archipelagic thinking explore how islands are composed of relational movements, networks, and connections, but the relational turn does not always employ the "archipelagic" trope to do this. Thus, while often closely related, archipelagic thinking can be seen as a subset of the broader "relational turn" in islands scholarship (see Pugh, 2016).
2. Thanks to Stephanie Wakefield for this important observation.
3. Here my argument is strongly influenced by my ongoing work with David Chandler and how we are both really interested in what it means to think with islands in the Anthropocene.

REFERENCES

Baldacchino, Godfrey. 2006. "Islands, Island Studies, Island Studies Journal." *Island Studies Journal* 1: 3–18.

Bargués-Pedreny, Pol, and P. Finkenbusch. 2019. "From Critique to Affirmation in International Relations." *Global Society* 33, no. 1.

Brassier, Ray. 2007. *Nihil unbound*. Houndsmills, UK: Palgrave Macmillan.

Brathwaite, Kamau. 1999. *Conversations with Nathaniel Mackey*. Staten Island, NY: We Press.

Buckland, Adelene. 2018. "'Inhabitants of the Same World': The Colonial History of Geological Time." *Philological Quarterly* 97, no. 2: 219–40.

Chandler, David. 2018a. "The Death of Hope? Affirmation in the Anthropocene." *Globalizations* 16. no. 5: 695–706. doi: 10.1080/14747731.2018.1534466.

———. 2018b. *Ontopolitics in the Anthropocene: An Introduction to Mapping, Sensing, and Hacking*. Abingdon, UK: Routledge.

Chandler, David, and Jonathan Pugh. 2018. "Islands of Relationality and Resilience: The Shifting Stakes of the Anthropocene." *Area*. June 11. doi.org/10.1111/area.12459.

———. Forthcoming. "Islands and the Rise of Correlational Epistemology in the Anthropocene: Rethinking the Trope of the 'Canary in the Coalmine." *Island Studies Journal*.

Chandler, David, and Julian Reid. 2018. "'Being in Being': Contesting the Ontopolitics of Indigeneity." *European Legacy* 23, no. 3: 251–68.

Clark, Nigel. 2010. *Inhuman Nature: Sociable Life on a Dynamic Planet*. Kindle ed. London: SAGE.

Clark, Nigel, and Katherine Yusoff. 2017. "Geosocial Formations and the Anthropocene." *Theory, Culture & Society* 34, no. 23: 3–23.

Colebrook, Claire. 2016. "What Is the Anthropo-political?" In *Twilight of the Anthropocene Idols*, by Tom Cohen, J. Hillis Miller, and Claire Colebrook, 81–117. N.p.: Open Humanities Press.

Crutzen, Paul J. 2002. "Geology of Mankind." *Nature* 415, no. 6867: 23.

Crutzen, Paul J., and Eugene F. Stoermer. 2000. "The 'Anthropocene.'" *Global Change Newsletter* 41: 17–18.

Danowski, Déborah, and Eduardo Batalha Viveiros de Castro. 2016. *The Ends of the World*. Cambridge, UK: Polity.

de la Cadena, Marisol, and Mario Blaser, eds. 2018. *A World of Many Worlds*. Durham, NC: Duke University Press.

DeLoughrey, Elizabeth. 2001. "The Litany of Islands, the Rosary of Archipelagos: Caribbean and Pacific Archipelagraphy." *ARIEL: A Review of International English Literature* 32, no. 1: 21–51.

———. 2007. *Routes and Roots: Navigating Caribbean and Pacific Island literatures*. Honolulu: University of Hawai'i Press.

Evans, Mike, and Lindsay Harris. 2018. "Salmon as Symbol, Salmon as Guide: What Anadromous Fish Can Do for Thinking about Islands, Ecosystems and the Globe." *Shima: The International Journal of Research into Island Cultures* 12, no. 1: 1–14.

Fanon, Frantz. 1967. *The Wretched of the Earth*. London: Penguin Books.

Glissant, Édouard. 1997. *Poetics of Relation*. Ann Arbor: University of Michigan Press.

Grydehøj, Adam. 2017. "A Future of Island Studies." *Island Studies Journal* 12, no. 1: 3–16.

———. 2019. "Critical Approaches to Island Geography." *Area*. March 10. https://doi.org/10.1111/area.12546.

Grydehøj, Adam, and Ilan Kelman. 2017. "The Eco-island Trap: Climate Change Mitigation and Conspicuous Sustainability." *Area* 49, no. 1: 106–13.

Grydehøj, Adam, Xavier Barceló Pinya, Gordon Cooke, Naciye Doratlı, Ahmed Elewa, Ilan Kelman, Jonathan Pugh, Lea Schick, and R. Swaminathan. 2015. "Returning from the Horizon: Introducing Urban Island Studies." *Urban Island Studies* 1, no. 1: 1–19.

Harman, Graham. 2016. *Immaterialism*. Cambridge, UK: Polity.

Hau'ofa, Epeli. 2008. *We Are the Ocean*. Honolulu: University of Hawai'i Press.

Hayward, Philip. 2012. "Aquapelagos and Aquapelagic Assemblages." *Shima: The International Journal of Research into Island Cultures* 6, no. 1: 1–11.

Hessler, Stefanie, ed. 2018. *Tidalectics: Imagining an Oceanic Worldview through Art and Science*. London: TBA21-Academy.

Latour, Bruno. 2017. *Facing Gaia: Eight Lectures on the New Climatic Regime*. Hoboken, NJ: John Wiley & Sons.

———. 2018. "Down to Earth Social Movements: An Interview with Bruno Latour." *Social Movement Studies* 17, no. 3: 353–61.

Lorimer, Jamie. 2012. "Multinatural Geographies for the Anthropocene." *Progress in Human Geography* 36, no. 5: 593–612.

Martínez-San Miguel, Yolanda. 2014. *Coloniality of Diasporas: Rethinking Intra-colonial Migrations in a Pan-Caribbean Context*. New York: Palgrave Macmillan.

Meillassoux, Quentin. 2010. *After Finitude: An Essay on the Necessity of Contingency*. London: Bloomsbury Publishing.

Morton, Timothy. 2013. *Hyperobjects: Philosophy and Ecology after the End of the World*. Minneapolis: University of Minnesota Press.

———. 2016. "Molten Entities." In *New Geographies 08: Island*, edited by D. Daou and P. Pérez-Ramos, 72–76. Cambridge, MA: Universal Wilde.

———. 2017. *Humankind: Solidarity with Nonhuman People*. London: Verso.

Nakashima, Douglas, Kirsty Galloway McLean, Hans Thulstrup, Romas Castillo, and Jennifer Rubis. 2012. *Weathering Uncertainty: Traditional Knowledge for Climate Change Assessment and Adaptation*. Paris: UNESCO.

Pugh, Jonathan. 2013. "Island Movements: Thinking with the Archipelago." *Island Studies Journal* 8, no. 1: 9–24.

———. 2016. "The Relational Turn in Island Geographies: Bringing Together Island, Sea and Ship Relations and the Case of the Landship." *Social & Cultural Geography* 17, no. 8: 1040–59.

———. 2018. "Relationality and Island Studies in the Anthropocene." *Island Studies Journal* 13, no. 1.

Pugh, Jonathan, Deborah Thien, Noortje Marres, David Featherstone, Liza Griffin, Swapna Banerjee-Guha, Benedikt Korf, et al. 2009. "What Are the Consequences of the 'Spatial Turn' for How We Understand Politics Today? A Proposed Research Agenda." *Progress in Human Geography* 33, no. 5: 579–86.

Quammen, David. 2018. *The Tangled Tree: A Radical New History of Life*. New York: Simon and Schuster.

Roberts, Brian Russell, and Michelle Ann Stephens, eds. 2017. *Archipelagic American Studies*. Durham, NC: Duke University Press.

Ronström, Owe. 2018. "On Seriality." *Shima: The International Journal of Research into Island Cultures* 12, no. 1: 15–20.

Schmidt, Mario, and Kai Koddenbrock. 2018. "Against Understanding: The Techniques of Shock and Awe in Jesuit Theology, Neoliberal Thought and Timothy Morton's Philosophy of Hyperobjects." *Global Society* 33, no. 1: 66–81.

Sheller, Mimi. 2009. "Infrastructures of the Imagined Island: Software, Mobilities, and the Architecture of Caribbean Paradise." *Environment and Planning A* 41, no. 6: 1386–403.

Stengers, Isabelle. 2015. *In Catastrophic Times: Resisting the Coming Barbarism*. London: Open Humanities Press.

Stratford, Elaine, Godfrey Baldacchino, Elizabeth McMahon, Carol Farbotko, and Andrew Harwood. 2011. "Envisioning the Archipelago." *Island Studies Journal* 6, no. 2: 113–30.

Tsing, Anna Lowenhaupt. 2015. *The Mushroom at the End of the World: On the Possibility of Life in Capitalist Ruins*. Princeton, NJ: Princeton University Press.

Ulmer, Jasmine B. 2017. "Posthumanism as Research Methodology: Inquiry in the Anthropocene." *International Journal of Qualitative Studies in Education* 30, no. 9: 832–48.

Walcott, Derek. 1998. *What the Twilight Says*. Boston: Faber & Faber.

Wark, Mackenzie. 2017. "Timothy Morton: From OOO to P(OO)." In *General Intellects: Twenty-Five Thinkers for the Twenty-First Century*. London: Verso Books.

Chapter Four

What Is an Archipelago?

On Bandung Praxis, Lingua Franca, and Archipelagic Interlapping

Brian Russell Roberts

In its entry for the term "archipelago," the *Oxford English Dictionary* (*OED*), in both the second edition of 1989 and the current online edition, notes that the word derives from the Italian *arcipelago*, with *arci-* signifying "chief, principal" and *-pélago* signifying "deep, abyss, gulf, pool." The entry clarifies that the term, while drawing on Greek roots, does not derive from "ancient or mediaeval Greek" but was developed first in early modern "western languages" such as Italian, Spanish, Portuguese, and French. From there, it was imported into Modern Greek. The *OED*'s first definition for "archipelago" appears as follows: "The Aegean Sea, between Greece and Asia Minor," with the first such usage listed as occurring in 1268 in a treaty "between the Venetians and the emperor Michael Palaeologus." The term's second definition gestures toward the first definition's planetary deployment as a metaphor: "Hence (as [the Aegean] is studded with many isles): Any sea, or sheet of water, in which there are numerous islands; and [in a transferred sense] a group of islands." The dictionary's examples for this latter usage date from 1529 to about 1860, with the term "archipelago," this erstwhile synonym for the Aegean, now applied to island groups situated in what we now think of as Indonesia, the Arctic Ocean, the Pacific, and the North Atlantic ("Archipelago," 1989; "Archipelago," 2017).[1] The definitions and examples provided by the *OED*'s second and online editions remain unchanged from those included in the dictionary's first edition (1933) and, looking further back, from those included in the *OED*'s late-nineteenth-century forerunner, *A New English Dictionary on Historical Principles* (*NED*). Indeed, as of the present writing, the online *OED* provides a verbatim repetition of the *NED*'s original definition and usage examples for "archipelago" as first offered in the 1880s ("Archipelago," 1888; "Archipelago," 1933).

Within discussions of archipelagic thinking (whether in the humanities, social sciences, law, or diplomacy), these 130-year-old definitions and etymologies (or their derivations) are widely cited.[2] Indeed, we might consider the *NED*/*OED* as having provided a baseline narrative within archipelagic thought—namely, the term "archipelago" arose as a name for the island-studded Aegean Sea and was subsequently applied metaphorically to other island-studded seas and island groups throughout the world. This narrative is at once compact (contained in a single sentence) and astonishingly sweeping, unfolding over the course of centuries (from the thirteenth century to the twenty-first century) and comprehending, though only obliquely acknowledging, the planet-spanning conflicts (material and epistemic) that have been attendant to continuums of Indigenous knowledges; varied approaches and responses to colonization and empire; myriad processes of creolization and mestizaje; anticolonial and postcolonial thought and practice; developments in human perception of terraqueous materialities and objects, including oceans, islands, and continents; efforts in linguistic and cultural translation; and schemes of active and passive miscommunication, genocide, enslavement, liberation, trans-Indigenous solidarities, and cultural reconstitutions. These vectors, together with uncounted others, constitute the backdrop that informed, and was reciprocally informed by, the processes by which (as cryptically outlined by the *OED*) a term for the Aegean Sea came to be applied metaphorically to groups of islands that span the planet.

In fleshing out these processes' imbrications with the archipelago concept, English is far from adequate. If this were not clear already (based on the sprawling tangles of history adumbrated in the paragraph above), it becomes apparent in the fact that nine of the *OED*'s eleven examples of the historical usage of "archipelago" are taken from the English language, and none of its examples seems to clearly illustrate the major pivot upon which the *OED*'s narrative hinges, namely, a moment in which the term is being deployed by a user in a self-consciously metaphorical way, during which the user is clearly making an overt analogy between the Aegean Sea and a sea or island group elsewhere in the world. If the *OED*'s English-language examples cannot provide readers with concrete instances of the term's planetary pivot into a metaphor, is it fair to assume that the metaphorical superimposition occurred more prominently in European languages besides English? Richard Scholar, drawing on French and Italian sources in his essay "The Archipelago Goes Global: Late Glissant and the Early Modern *Isolario*," has gestured in this direction, arguing that "the island-studded Aegean is . . . cast in the role of mother and sovereign to all the other seas of the globe," with the European-derived archipelago concept emerging as "ready for export as a structure of thought to other parts of the globe in the age of European discovery and conquest" (2015, 39–40).[3]

Scholar's chapter is a compelling gesture, but I am nonetheless unaware of any scholar who has undertaken, to any substantial degree, a project that has sought to document the multilanguage historical processes that undergird the archipelagic narrative implied in the *NED*/*OED*. Such a project would require not only a facility with English, Italian, Spanish, Portuguese, French, Dutch, and other colonial languages; it would also require a facility with the host of Indigenous languages spoken by the many groups whose members in various ways resisted and acquiesced to invasion and colonization by speakers of the colonizing languages. How might Taino epistemologies regarding island interconnection have interacted with Spanish projections of Caribbean islands as a New World archipelago analogous to the Aegean? How have Polynesian methods of island grouping interacted with traditions of French archipelagic thought? What Indigenous notions of island linkages and separations have been lost and persisted in the face of the Spanish and subsequently US-American application of the archipelago concept to island groups such as the Philippines and the Marianas? How did the Dutch term *archipel* interface with fourteenth-century Javanese and later twentieth-century Indonesian notions of *Nusantara*?

Various facets of the answers to such questions must be acknowledged as, to borrow a phrase from George B. Handley, having "inevitably and irrevocably fallen into historical oblivion" (2007, 42). But for many of these questions, the oblivion is revocable, with its revocation contingent on linguistic facility by scholars. Still, the very question of linguistic facility presents a conundrum: Imagine a Venn diagram illustrating the overlap between scholars with a research agenda in archipelagic thought and scholars who have facility with at least ten of the languages germane to planetary archipelagic research. I suspect that such a diagram would showcase a very limited degree of overlap. (I, for one, would not be present in the overlap.) And yet, to my mind, a sprawling comparative project—multilingual, long durational, transregional, and involving multiple participants—would be one of the most urgent of all projects that could be undertaken in the realm of archipelagic thinking. Far beyond the *NED*/*OED*'s concise historical definitions and sparsely documented usage examples, such a project would move toward asking and answering, in deeply material and culturally aware ways, a question that must lie at the very foundation of research into archipelagic thinking: What is an archipelago?

Derek Walcott once contemplated an analogous question: What is the nature of the island? He concluded that he was not ready to answer it:

"Except by hints. Contradictions. Terrors. The opposite method to the explorer's" (Walcott 2005, 52). Neither does this chapter undertake anything close to the project that would be required to begin to answer the question posed in its title: What is an archipelago? Rather, the present chapter points toward a hint and a contradiction, which is that even as English is far from adequate to the sprawling archipelagic project, it may be that English, as a historically contingent lingua franca and in a way analogous to what Edward Kamau Brathwaite has called "English in a different way" (1984, 5), currently offers this seemingly impossible project one of its highly significant avenues toward realization. In a way that refuses what Aamir R. Mufti has recently critiqued as English's frequent default treatment "as [a] neutral or transparent medium" within realms of world literary language and global neoliberalism (2016, 16), I want to frame English not as a centralizing linguistic tradition for archipelagic thought but as a type of eccentric linguistic architecture that permits dozens and even hundreds of languages to attain a type of noncentered relationality that I refer to as *interlapping*. For this model of interlapping, as showcased in the present chapter's conclusion, I look toward Brathwaite's work on palimpsestous Caribbean English. And for a model of English as a noncentering lingua franca—one that operates not in the positivist mode of what Walcott calls the explorer's method but rather as a mode of "disorientation at odds with the temporal rhythms" often associated with "progress and civilization" (Rafael 2009, 18)—I look toward the watershed anticolonial Asian-African Conference, held in the Indonesian city of Bandung in 1955. In looking toward the Bandung Conference, I am pursuing something more material than what has been critiqued as a rarified and dehistoricized postcolonial "Bandung spirit."[4] Rather, I am pursuing the archipelagic in what I term "Bandung praxis."

In 1955, the postcolonial country of Indonesia hosted the Asian-African Conference, popularly known as the Bandung Conference, which was a gathering of representatives from twenty-nine newly and increasingly independent nations. In its name and animating spirit, the conference was continental and even transcontinental in scope—that is, it professed to represent a postcolonial conjunction of the continents of Asia and Africa.[5] But in its locale it was archipelagic—it was held in Indonesia, which is often regarded today as the world's largest archipelagic state, constituted by some eighteen thousand islands spanning from continental Asia in the north to Australia in the south and sprawling from the Indian Ocean in the west to Melanesia and the Pacific in the east (see Figure 4.1).[6]

The conference was also archipelagic in its praxis, in a way that comes into focus via commentary by the black US writer Richard Wright, who attended and reported on the meeting. In his 1957 book *White Man, Listen!*, Wright

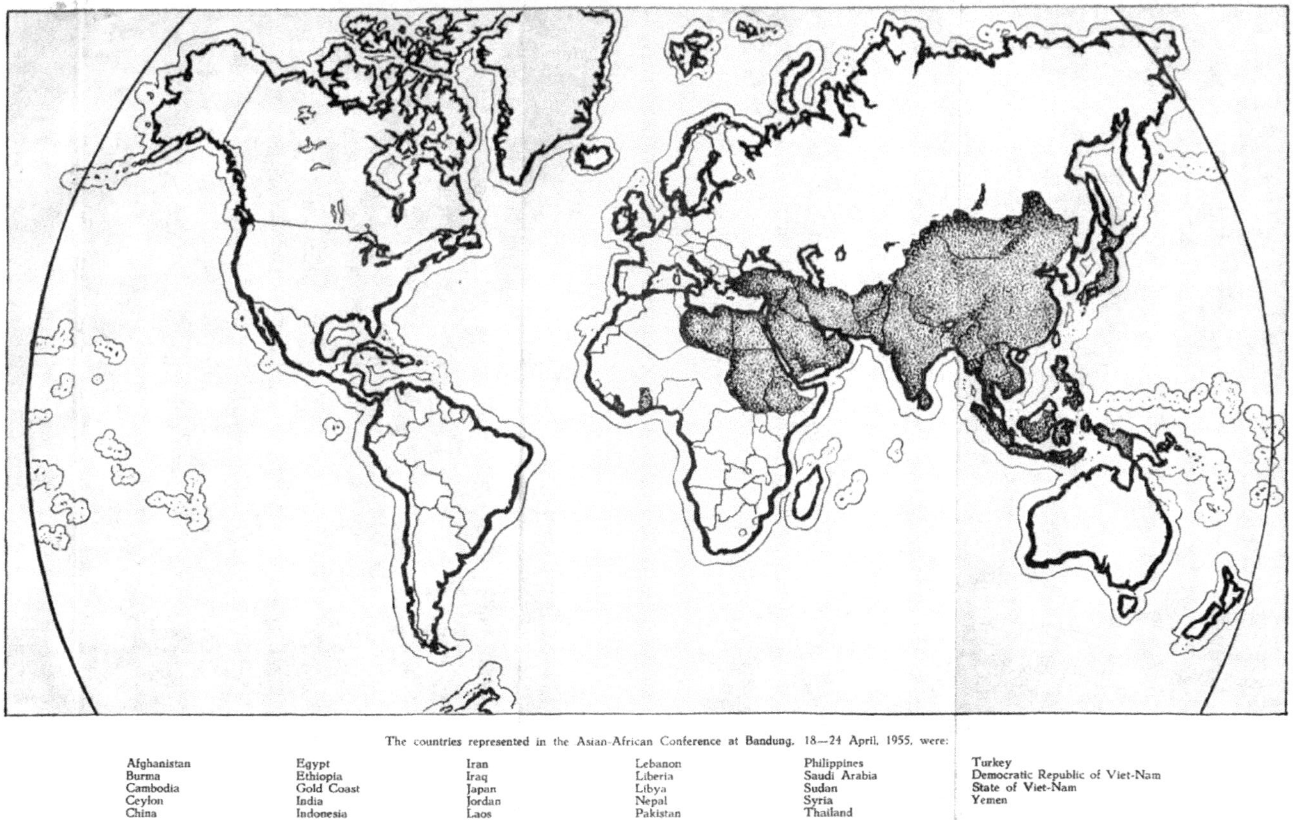

Figure 4.1. Asian-African Map. Appearing as an end-paper in the Indonesian Ministry of Foreign Affairs' book *Asia-Africa Speaks from Bandung* (1955), this map represents (in black) the countries whose official delegates participated in the Bandung Conference. The map's use of ocean-based lines around large and small landmasses, an example of what has been called coastal vignetting, gives heightened visibility to the world's islands and archipelagoes, many of which would otherwise be virtually invisible.

narrates the postcolonial elites who attended the conference not in continental terms but in terms of an archipelago. At one point, he states, "As the waters of Western imperialism recede from the land masses of Asia and Africa, and when we begin to study the residue left behind, we shall find some strange formations indeed" (2008b, 683). Later he narrates one of the strange formations that has become evident as the waters are receding: the "elite in Asia and Africa constitutes islands of free men" (722). According to Wright's model, then, the Bandung Conference, based in the Indonesian archipelago, functioned to bring conference-goers (these islands of free men) into an archipelagic relationality with one another, archipelagizing them in a way allied with what Narendran Kumarakulasingam (2016) has recently described as the Bandung Conference's effect of "de-islanding" in the wake of colonialism.

In practice, the Bandung Conference functioned to archipelagize the islands of free men by means of English, which was indeed the conference's lingua franca, even if, as we shall see, it was assuredly not what Vicente L. Rafael has discussed as "English as a kind of universal lingua franca" that assumes "all other languages ought to be reducible to its terms and thereby assimilable" to its Euro-American structures of power (2009, 4). As stated in a speech during the Bandung Conference's opening session by the Indonesian prime minister and conference chair Ali Sastroamidjojo, "The whole world is indeed watching us with hopes and expectations. Let us therefore all speak the same language, however much our tongues and our ideologies may differ, the language which not only the peoples of Asia and Africa will understand, but which will be understood by the whole world" (1955, 38).[7] This was a language, as Sastroamidjojo indicated, that ought to be trained against the Euro-American "illusion that stockpiling of atom and hydrogen bombs can bring about peace," against "the chains of colonialism," and with the intent that "the rest of the world take due notice that it is the voice of nearly two-thirds of the world's population which will be heard from this Conference-hall" (34, 35, 38).

Indeed, Wright listened to this "voice" in the conference hall and found a version of English as a lingua franca that was far from a reinforcement to the colonially minded "rest of the world" to which Asia-Africa was speaking. Wright stated in his 1956 book *The Color Curtain: A Report on the Bandung Conference* that "English . . . was the dominant language of the Conference," and he detailed his vision of this unsettling lingua franca:

> I felt while at Bandung that the English language was about to undergo one of the most severe tests in its . . . history. Not only was English becoming the common, dominant tongue of the globe, but it was evident that soon there would be more people speaking English than there were people whose native tongue was English. . . . What will happen when millions upon millions

> of new people in the tropics begin to speak English? Alien pressures and structures of thought and feeling will be brought to bear upon this our mother tongue and we shall be hearing some strange and twisted expressions. . . . But this is all to the good. (2008a, 591, 592)[8]

Taking English as the official language of the conference—and thereby bringing islands of free men into a strange and twisted archipelagic relationality—was a critical component of what I am discussing as Bandung praxis, or the concrete and pragmatic practices attendant to the organizational aspects of an Asian-African Conference that is too often taken (within the realm of anticolonial thought and postcolonial studies) as disembodied myth and spirit. The notion of Bandung praxis seeks to clear a space for looking back, historically, at the Bandung Conference and identifying practices that, by analogy, may be adapted and redeployed in other circumstances. Here, I am specifically interested in Wright's intimations regarding the conference's English as "strange" and "twisted," two terms that I repeat throughout this chapter, self-consciously rejecting their pejorative sense in standard US-American English and following Wright in his evident admiration for the destabilizing epistemological promise so often concomitant with that which is strange and twisted.

Notably, this linguistic aspect of Bandung praxis, which used a lingua franca to forge an international archipelago of postcolonial nation-states, was homologous with another, contemporaneous Indonesian project that was operating at the national rather than international scale, namely, postcolonial Indonesia's long-running aspirations to use a lingua franca to unite a disparate archipelago of ethnic groups sprawling across a disparate set of islands. The lingua franca was the Indonesian language, or *Bahasa Indonesia.* A rare case among postcolonial countries, Indonesia adopted neither the colonial language (Dutch) nor the majority ethnic language (Javanese) as its national language. Rather, Indonesia's nationalist politicians arrived at Malay as an Indigenous language that had for centuries functioned among many of the islands as well as the Malay Peninsula as an interethnic and interisland language of trade and governmental administration. By the early twentieth century, Malay had become an Indigenous language of modernity, an alternative to Dutch.[9] With the formation of the postcolonial Indonesian state, this language—now in a standardized form as the Indonesian language—became the state-sponsored lingua franca, which Indonesia's residents (80 million in 1955) were to learn as a common language after their myriad and usually mutually incomprehensible regional languages.[10] In 1954, independent Indonesia's first president, Sukarno, celebrated the postcolonial Indonesian tradition of advocating for the Indonesian language as a unifying force. He presided over a meeting at which it was declared that Indonesian was "a language of unity 'from Sabang to

Merauke,'" or from the northern tip of the western island of Sumatra to one of Indonesia's most easterly outposts on the island of New Guinea. And then in 1955, the same year as the Bandung Conference, during a presidential visit to the central Javanese city of Solo, the public was directed to look back to Indonesia's mythic *Sumpah Pemuda* (Youth Pledge) of October 1928, which was a set of three principles adopted in the city of Batavia (renamed Djakarta in the 1940s) by high-status and generally young Indonesian postcolonial nationalists hailing from several regions and islands of the Dutch East Indies. As recalled during Sukarno's 1955 visit to Solo, the *Sumpah Pemuda* had showcased "djandji berbahasa satu, bertanah air satu dan berbangsa satu!" (the promise to have one language, one homeland, and one nation).[11] To put it in terms of Antonio Benítez-Rojo's description of "the character of an archipelago" as exhibiting "discontinuous conjunction" (1996, 2), the several hundred distinct linguistic traditions on thousands of inhabited islands constituted the archipelago's discontinuity, while the national language was set forth as the archipelago's overlay of conjunction.

Previously, the islands of what we now think of as Indonesia were so disparate (culturally, linguistically, ethnically, spatially) that one of their major material points of overlap rested in a common history of Dutch colonization, but even as Indonesian leaders were required to establish their postcolonial nation-state within the confines of former Dutch colonial borders (Butcher and Elson 2017, 46–47), they rejected the colonial language and selected, adapted, and forged the Indonesian language as a lingua franca for the country's thousands of inhabited islands, advancing an ideology of neo-Indigenous linguistic unity that offered something in excess of a common history of oppression under the Dutch. In 1951, while the future Bandung Conference chair Ali Sastroamidjojo was Indonesian ambassador to the United States, a pamphlet published by the Embassy of Indonesia in Washington explained the Indonesian language this way: even before Indonesian independence, "the Malayan language, which is flexible and adaptable, came to be used widely . . . throughout Indonesia; thus it gradually became the *lingua franca* of Indonesia," and "in 1945, after the declaration of independence, the Indonesian Government appointed a Permanent Commission to guide the growth and expansion of the new language" (1951, 47–48).

Analogous to the islands of Indonesia, the new nation-states of Asia-Africa were so far-flung (culturally, linguistically, ethnically, spatially) that their major point of commonality resided in their shared histories of colonialism and decolonization. And analogous to the Indonesian language within the Indonesian archipelago, English became, in practice at Indonesia's Bandung Conference, a lingua franca and language of unity across the gathering's archipelagic assortment of "islands of free men." In fact, at the Bandung

Conference's opening session, President Sukarno placed emphasis on Indonesia as both analogous to and a mise en abyme of Asia-Africa: "Sisters and Brothers, Indonesia is Asia-Africa in small. It is a country with many religions and many faiths. . . . Moreover, we have many ethnic units. . . . But thank God, we have our will to unity." Sukarno proposed archipelagic Indonesia's motto—"Unity in Diversity"—as a model for Asia-Africa's embrace of a "unifying force which brings us all together" (1955, 27–28). If the conference's uniting spirits were anticolonialism, antiracism, and peace, then Bandung's facilitating praxis, as indicated by Ali Sastroamidjojo and Wright, was the lingua franca of English, one of the major colonial languages.[12] But the major English-speaking colonial powers (Britain and the United States) were not invited to contribute to the conference discussions that this lingua franca was facilitating in Bandung. Especially in the absence of US and British speakers, the conference's lingua franca was a decentered English, one that, as Wright said, introduced "strange and twisted expressions" and, consequently, proffered "alien pressures and structures of thought and feeling."

Writing in the mid-1950s, Wright was using the phrase "structure of feeling" before it attained wide purchase in the study of culture, as famously advanced by Raymond Williams. Although Williams's initial description of "structure of feeling" was offered in 1954, the year before Wright attended the Bandung Conference, it is all but certain that Wright and Williams coined the phrase independently of one another. And yet juxtaposing Wright's usage with Williams's contemporaneous definition offers crucial insight into Wright's commentary and the mechanisms by which a structure of feeling may be accessed. Such insights consequently shed light on the interlocking questions of translation and archipelagic frameworks that surround Indonesia's archipelagic self-perception and Bandung praxis's role in forging a twentieth- and twenty-first-century definition and usage for the term "archipelagic."

As critics looking into the past, Williams explains, "we examine each element [of cultural life] as a precipitate, but in the living experience of the time every element was in solution, an inseparable part of a complex whole" (1954, 21). To experience the elements in solution (as lived intuitively rather than as precipitated and analyzed) is to experience the "structure of feeling" (21). For Williams in 1954, the structure of feeling was "only realizable through experience of the work of art itself, as a whole" (22). But Wright's usage of the phrase plots another mechanism by which a structure of feeling may be accessed: the rough and improvisational realm of translation in which multiple linguistic traditions meet each other within the field of a lingua franca, stretching and twisting the lingua franca to their own ends, remaking it in accidental and purposeful ways by introducing expressions whose "strange and twisted" qualities mark the entrance of "alien . . . structures of

. . . feeling" into and across the lingua franca.[13] Hence, in Wright's model we find a mechanism for accessing the structure of feeling aside from the work of art; there are moments of translation—sometimes intentional and also those that are impromptu, unpolished, accidentally nonidiomatic, and otherwise "strange and twisted" within the lingua franca—that afford access to a circulating miscellany of affective structures. As these structures circulate within the lingua franca, the structures—together with their destabilizing epistemological promise—may be noticed and consequently accessed precisely because of their strange and twisted qualities. Here, the lingua franca becomes a basin of structures of feeling brought in from other languages, with the lingua franca undergoing deformations as these structures arrive and thrive. At the same time, the deformed lingua franca serves to communicate the structures not only from the non–lingua franca cultures to the native lingua franca culture but also, and perhaps more importantly, between and among the non–lingua franca cultures that converge within the lingua franca.

How to access these processes? One method would be to listen to spoken words for strange and twisted expressions, which is the approach that no doubt prompted Wright's commentary on the conference as a window into nonnative English's future. Words that are specifically spoken within a lingua franca, though more ephemeral than written words, would tend to be improvised translations and hence would most likely harbor a greater number of phrases that, according to Wright's model, would offer access to structures of feeling imported from non–lingua franca cultures. But spoken words tend to be more difficult to access across time, so I want to turn toward written words, which have the benefit of being less ephemeral, even if they also tend to have the disadvantage of having gone through editorial processes that have placed them in a more standardized (hence less twisted) form of the lingua franca. Consider, for instance, the following passage from President Sukarno's opening speech at the Bandung Conference: "Yes, some parts of our nations are not yet free. That is why all of us cannot yet feel that journey's end has been reached. No people can feel themselves free, so long as part of their motherland is unfree" (1955, 23). Here, Sukarno's English bears certain impresses of the Indonesian language: an Indonesian inclination toward the passive voice ("has been reached"), the mild eccentricity of a missing definite article or possessive pronoun (rather than "journey's end," it may have been more standardly idiomatic to say "the journey's end" or "our journey's end"), and a usage of "themselves" that sounds strange in context. These eccentricities are hardly revelatory windows into Indonesia-based structures of feeling regarding archipelagic thinking. But Sukarno's word "motherland," which is a standard translation of the Indonesian term *tanah air*, indeed constitutes such a window. In the case of *motherland*/*tanah air* we see that when approaching a standardized translation (such as that found in a prepared and edited speech), gaining access to a non–lingua franca culture's structures of feeling may hinge on comprehending the structures that have been standardized *out* of the translation.

To access the archipelagic structure of feeling that has been standardized out of Sukarno's speech, recall the summary of the *Sumpah Pemuda* as showcased during the Indonesian president's visit to the city of Solo. Again, the summary reads as follows: "djandji berbahasa satu, bertanah air satu dan berbangsa satu!" In the well-researched and insightful article from which I am drawing this quotation, my friend and collaborator (on several projects, including *Indonesian Notebook: A Sourcebook on Richard Wright and the Bandung Conference*) Keith Foulcher translates this Indonesian phrase as "the promise to have one language, one homeland, and one nation." As non–Indonesian speakers may understand from Foulcher's translation, the *Sumpah Pemuda* seeks to use "one language" to forge "one nation" out of many islands and ethnicities. But there is something still more archipelagic that resides in this 1955 summary of the *Sumpah Pemuda*. Foulcher's translation of the *tanah air* component of the phrase *bertanah air satu* is "homeland." Indeed "homeland" (or "fatherland" or, as Sukarno used, "motherland") is the standard domesticating translation of *tanah air*, even if, as Foulcher (and any other Indonesian speaker) knows, this Indonesian term is made up of two components: *tanah* (land) and *air* (water), such that nondomesticating English translations for the term could be "water land" or "watery land" or "land of water."[14] Hence, a nonstandard translation, one that is more literal but has little idiomatic purchase in a more standard English, makes an archipelagic Indonesian structure of feeling visible, namely a historically and geomaterially produced lived experience of patria not as solid ground but as made up of both land and water. This structure of feeling, experienced in various ways by millions of Indonesian speakers across a sprawling archipelago and abroad in a diaspora, converges with Édouard Glissant's terminology in his own work toward shifting structures of feeling, in French, by supplanting the term *terre mere* (motherland) with the term *terre mer* (sea land), or as J. Michael Dash translated and explicated it, "a marine habitat, neither land nor sea, . . . that propels the subject into . . . that roaring global (dis)order in which everything imaginable exists in a glorious cacophony" (2017, 359).

But surely, one might argue contrariwise, *tanah air* and its potential to be eccentrically (albeit literally) translated as "water land" or "watery land" is of little consequence to English as a lingua franca. Many speakers of English will consider Indonesian (if they consider Indonesian at all) to be among the languages that has very little commerce with or relevance to English. What's more, if *tanah air* has "homeland," "fatherland," and "motherland" as its standardized English-language translations, the term will have had virtually no opportunity to introduce an alien structure of feeling (i.e., patria as terraqueous) to English as a lingua franca. Does it, then, constitute merely a bit of archipelagic trivia to know that the Indonesian term for *patria* might more literally be translated as "water land"? And might this bit of trivia rise to the level of actually being interesting merely because of the convergence of *tanah air*/"water land" with some wordplay advanced by the towering archipelagic theorist Édouard Glissant?

But this line of dismissive questions can hardly be entertained if we consider Indonesia's central role in creating, within the United Nations Convention on the Law of the Sea (UNCLOS), the very category of the "archipelagic State," which, as elaborated in UNCLOS (1982), is "a State constituted wholly by one or more archipelagos," with "archipelago" defined as "a group of islands, including parts of islands, interconnecting waters and other natural features which are so closely interrelated that such islands, waters and other natural features form an intrinsic geographical, economic and political entity, or which historically have been regarded as such" (United Nations 1983, 15). UNCLOS further specifies that "an archipelagic State may draw straight archipelagic baselines joining the outermost points of the outermost islands" and that "the sovereignty of an archipelagic State extends to the waters enclosed by the archipelagic baselines . . . described as archipelagic waters" (15–16).

In the mid- to late 1950s, Indonesians understood the terraqueous term *tanah air* to mean the "area over which a state's jurisdiction extends" ("Tanah," 1959),[15] consistent with Indonesia's 1957 announcement that its "Government declares all waters around, between and those connecting the islands as included in the State of Indonesia" (Danusaputro 1974, 86). Subsequent to this declaration, strange and twisted if purposeful translations of *tanah air* went hand in hand with Indonesia's ongoing arguments toward arriving at an internationally agreed-upon definition of an "archipelagic State" that would bear the impress of archipelagic notions aligned with Indonesia's own traditions.[16] Consider, for instance, the arguments advanced in the 1970s by Mochtar Kusumaatmadja, a Yale-trained Indonesian diplomat and the major architect of the country's vision for international archipelagic recognition.[17] In June 1972, at the Seventh Annual Conference of the Law of the Sea Institute at the University of Rhode Island, Mochtar spoke during a session titled "Major Positions, Problems and Viewpoints Regarding the Needs and Interests of Developing States." He explained,

> Here I may perhaps mention one term in our language which expresses [a] perhaps subconscious feeling. In our language, as in many languages, we have a

word for "native country." The French word is "patrie," the German word "das heimat"; in Indonesia, it is "tanah air," which means "land and water."

This, I think, is a very strong argument for the viewpoint that we should adopt a different way of looking at things, because this is a word not coined by lawyers who have made comparative studies, not by geographers, but this is a word that comes from the people who have lived in these islands and these archipelagoes, and they feel it is part of them. The water is part of their everyday lives. They depend on it for their living, and it is a very real thing, and this impresses me more than any argument, especially if it is made by lawyers. If a simple man says "tanah air" (i.e., land and water), then I think he *means* it. . . . This may not be a very good legal argument, but this is a thing we firmly believe in. (Kusumaatmadja 1973, 174)

Further deploying this argument regarding *tanah air*, Mochtar spoke in July 1974 during the Third United Nations Conference on the Law of the Sea. The record summarizes this portion of his remarks as follows:

On 13 December 1957, the Indonesian Government had proclaimed Indonesia an archipelagic State. It had stated, among other things, that all the waters around and between the islands of Indonesia, regardless of their width, were the natural appurtenances of the land territory of the Republic and formed part of the internal or national waters under its absolute sovereignty. That concept emphasized the unity of the land and water territories of Indonesia, as reflected in the word "tanah-air", which in the Indonesian language was the equivalent of "fatherland" and literally meant "land-water." (United Nations 2009, 187)

Consistent with Ali Sastroamidjojo's pragmatic approach to a Bandung praxis involving English as a lingua franca in 1955, Mochtar was in 1974 speaking in English in a diplomatic context—on one hand this was simply so that the world could understand, but it was also with the aim of altering the world's understanding, deforming the lingua franca, stretching it so that the term "archipelago" would not merely signify a string of islands. Rather, the lingua franca's "archipelago" would now be grafted onto new Malay/Indonesian etymological roots, with the effect that when UNCLOS offered its English-language definition of "archipelagic State" in 1982, the notion of *tanah air* (and an accompanying notion of straight-line archipelagic baselines that Mochtar himself had helped innovate) was incorporated into the English-language definition specifying that "an archipelagic State may draw straight archipelagic baselines" to form a perimeter around the state's outer islands, with "the sovereignty of an archipelagic State" extending not simply to the land but to the "archipelagic waters" that are "enclosed by the archipelagic baselines" (United Nations 1983) (see Figure 4.2).[18]

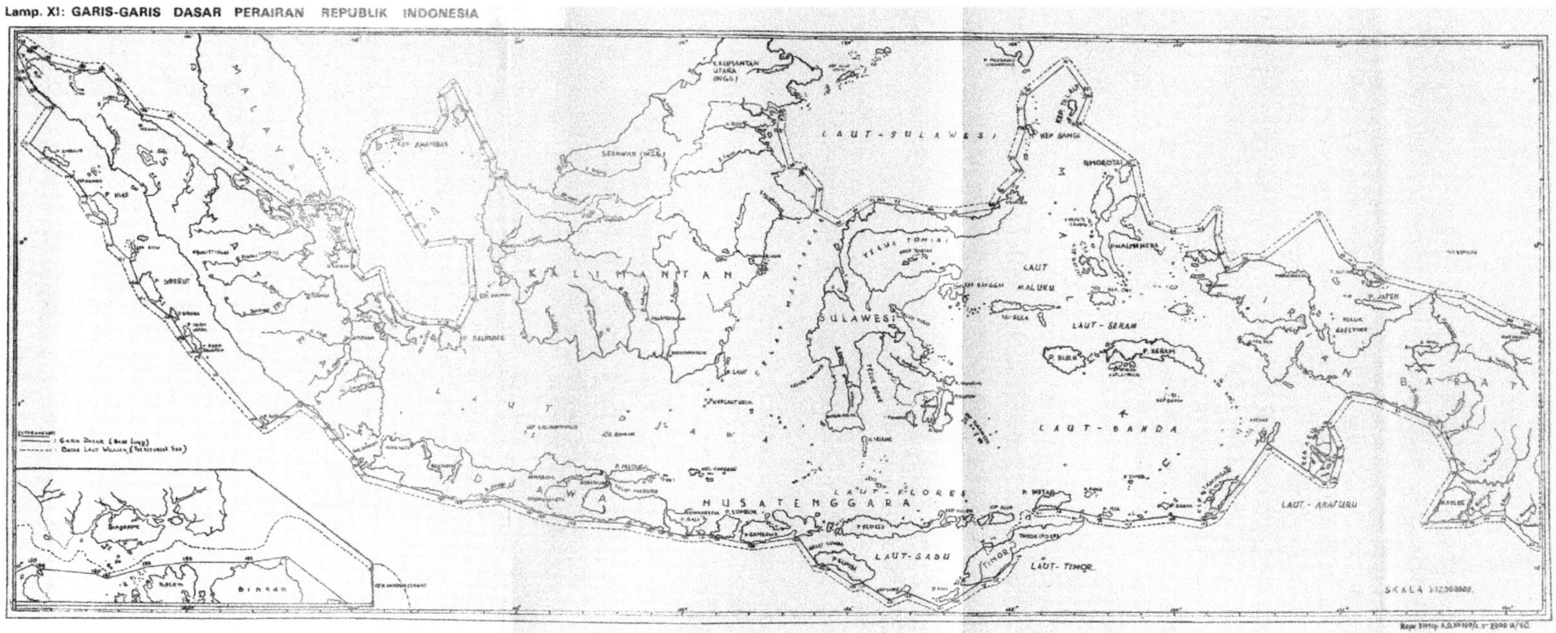

Figure 4.2. Baselines Map. In 1960, Indonesia distributed this map of its archipelagic baselines, drawn around the country's outer islands and occasional land borders; waters enclosed by these baselines were claimed as sovereign territory. Map reproduction from Sumitro Lono Sedewo Danuredjo, *Hukum Internasional Laut Indonesia: Suatu Usaha untuk Mempertahankan Deklarasi 1957*, vol. 2 (Djakarta: Bhratara, 1971).

In this way, and in a way that is equally as etymologically significant as the *OED*'s narrative of the Aegean's deployment as a planetary metaphor, Indonesia self-consciously deformed the lingua franca, introducing to the international world an alien structure of feeling, a deformed version of the term "archipelago," one whose Indonesian etymology, even if presently absent from the *OED*, has silently given *tanah air*—as a structure of feeling—the force of international law vis-à-vis the term "archipelago." Within this international arena, Indonesian structures of feeling have found transit through the lingua franca into the myriad official and unofficial languages of the countries that have become signatories or otherwise affiliated with UNCLOS, as *tanah air*, with "archipelago" as its vehicle, has transformed official and practical perceptions of archipelagic spaces in a multitude of languages. If, as Rebecca L. Walkowitz has persuasively argued, we require a "project of unforgetting" to counteract the tendency of English-language literary works to "'forget' that [they have] benefitted from literary works in other languages" (2015, 23), then archipelagic thinking requires an analogous project of unforgetting, one centered on what I have gestured toward earlier in this chapter as archipelagic thinking's revocable oblivions regarding the agonistic histories of translation and untranslatability that reside within the scholarship's English-language lingua franca.

The interactions between the terms *tanah air* and "archipelago" in the context of Bandung praxis regarding English as a lingua franca constitute one narrative only among the vast number of interactions that remain, due to language barriers, within a type of contingent and revocable historical oblivion. And following Ali Sastroamidjojo's pragmatic approach to English in Bandung, as well as modernism/modernity scholar Susan Stanford Friedman's vision of "scholars working in translation [as] essential" to bringing knowledge of non-Eurocentric modernisms "into the *lingua franca* of the field" (Friedman 2010, 492), I want to suggest that scholars of archipelagic thinking become more self-reflexive about the uneven ways English has already been the lingua franca of the field. When and how has the scholarly English that contemplates archipelagic thinking been eccentric, twisted, and strange with destabilizing structures of feeling and epistemologies? When and how has it, unfortunately, become a "vanishing mediator" (Mufti 2016, 16)? And in what circumstances, more consistent with the mode of Bandung praxis that assailed Wright with its discomfiting epistemological promise, has it been a mode of "reverse assimilation," requiring "readers comfortable with standardized English to acquire" not simply new terms from other languages but "a new standard," the "mak[ing of] English into a foreign language" (Walkowitz 2015, 36–37, 41)? Consider as test cases the ways in which English has functioned as a lingua franca for discussions of archipelagic thinking in touchstone arenas ranging from the *OED* to UNCLOS and in essay collections from *A New Oceania: Rediscovering Our Sea of Islands* (Wadell, Naidu, and Hau'ofa 1993) to *Archipelagic Studies: Charting New Waters* (Batongbacal 1998). Or the way it has functioned as a lingua franca in providing access to watershed archipelagic texts ranging from Antonio Benítez-Rojo's 1996 *The Repeating Island* (cited with more frequency than the Spanish-language source text *La isla que se repite* [Benítez-Rojo 1989]) to Édouard Glissant's 1997 *Poetics of Relation* (cited with more frequency than the French-language source text *Poétique de la Relation* [Glissant 1990]).[19] Or consider the way English has been a lingua franca in venues ranging from *Island Studies Journal* to *Shima: The International Journal of Research into Island Cultures*, the latter of which (even as it accepts only English-language submissions) has chosen a non-English-language title (*shima* means both "island" and "community" in Japanese) to indicate "the publication's international project and its intent to engage with research and scholarship beyond . . . anglophone hegemony" (Editorial Board, 2007, 3).

In many ways, then, as scholars of archipelagic thinking have frequently used English as a lingua franca to work across multiple language traditions, we have already seen the emergence of the project that, in this chapter's introduction, I have called for, namely, a sprawling comparative project—multilingual, long durational, transregional, and involving multiple participants—that would take us far beyond the *NED*/*OED*'s concise historical definitions and sparsely documented usage examples. Yet, due to language barriers, some of the research along these lines that is relevant to English's situation as a lingua franca has fallen outside mainstream archipelagic research, with the Indonesian case standing as a particularly vivid example. Indeed, this is a multilayered story, cutting back and forth among Indonesian and English and several other languages. This is seen in an English-language document circulated by the Indonesian delegation to the United Nations Conference on the Law of the Sea held in Geneva in 1958. In this document (under the headings "What is an archipelago?" and "Is Indonesia an archipelago?"), Indonesia advanced English-, French-, German-, and Dutch-language dictionary definitions and usage examples, including the *NED*'s definition and one of *NED*'s seventeenth-century usage examples that references Indonesian islands as "that great Archipelago" (Danusaputro 1980, 156–61, 160).[20] Continuing this mode of etymological discussion, as Indonesia moved toward the triumph of its archipelago principle within the forum of UNCLOS, Indonesian thinkers in the 1970s referenced the *NED* as a means of accessing the Italian origin of the term "archipelago" and "artinya yang asli . . . SUATU LAUT (atau WILAYAH AIR) DENGAN PULAU2 DIDALAMNYA" (its original meaning . . . A SEA [or REGION OF WATER] WITH ISLANDS IN

IT) (Danusaputro 1972, 53).[21] They matched such etymological discussions with Indonesian perceptions of patria as constituted by both the islands and the sea, which in turn bolstered their finally successful twisting and reshaping of the term "archipelago" on the international stage (Danusaputro 1974, 67; Danusaputro 1975, 43). In the case of these etymological discussions, English facilitated an uneven dialogue among French, Dutch, German, Indonesian, Javanese, thirteenth-century Italian, and Greek roots in a way that, via UNCLOS 1982, changed the worldwide definition of the term "archipelago." In so doing, English as a lingua franca also evoked a "small world" (to borrow a term from network theory and from Irad Malkin's work on archipelagic networks in the ancient Mediterranean) among archipelagic thinkers and traditions, moving archipelagic languages and cultures from simply being connected to their "nearest neighbors" and introducing a link that "drastically reduces the degrees of separation among all the nodes, increasing the connectivity of the entire system" (Malkin 2011, 28). Again we come to the hint and contradiction: English has been inadequate for archipelagic thinking; English as a lingua franca has been foundational to archipelagic thinking.[22]

Contemporary Archipelagic Thinking, the title of the collection within which this chapter appears, admirably reaches beyond what *Shima* refers to as Anglophone hegemony, riffing (at least according to my ear) on a French-language phrase coined by Glissant—that is, *la pensée archipélique*, which appears in his *Introduction à une Poétique du Divers* (1996), *La Cohée du Lamentin* (2005), and *Philosophie de la Relation* (2009), none of which exists properly within Anglophone hegemony because none has yet been published in an English-language version (Glissant 1996, 43, 45; Glissant 2005, 75; Glissant 2009, 45). As a follow-up, though, and as a question spurred by Glissant's own suggestion that "la traduction est . . . une des espèces parmi les plus importantes de cette nouvelle pensée archipélique" (translation is . . . among the elements most essential to this new archipelagic thought),[23] one might ask about what difference it would make to the volume's lingua franca—what difference in terms of making it strange, twisted, foreign—if it drew (instead of upon a language whose close relation to English suggests a commonsense equivalency between *archipélique* and *archipelagic*) upon the Indonesian term *Nusantara* or the Hawaiian-language *pae 'āina*, both of which have "archipelago" as a standard English-language translation, but which both set sail, as will any ocean-island term that is a noncognate with "archipelago," from radically different epistemological shores.[24]

Hence, joining with the editors of *Shima* and with the Francophone leanings of the present collection's title, I have turned toward Bandung praxis as offering an avenue toward twisting English into a useful lingua franca even while seeking to avoid Anglophone hegemony. For me, this twisted

English of Bandung praxis is further opened up by recourse to a term Edward Kamau Brathwaite uses in the opening pages of his 1974 book *Contradictory Omens: Cultural Diversity and Integration in the Caribbean.* Here, Brathwaite describes the Caribbean as a place of "inter-lapping" (1977, 5), a relational state that in Brathwaite's specific commentary plots a Caribbean relationality to North America and Africa that is distinct from conventional *over*lapping. Brathwaite does not elaborate on his definition of the term, but I take it as a scholarly instantiation of what he, in an address given around the same time as *Contradictory Omens* was published, referred to as "English in a different way" (1984, 5).[25] As Brathwaite narrates, the language of the Anglophone Caribbean arose in the context of "ancestral languages still persisting in the Caribbean," including Amerindian, Hindi, "varieties of Chinese," and "survivals of African languages," especially involving "a submergence" of African linguistic forms "adapted to the cultural imperative of the European languages" and in turn "influencing the way in which the English, French, Dutch, and Spaniards spoke their own languages" (6–8). This is "an English which is like a howl, or a shout or a machine-gun or the wind or a wave" (13). This is an English that is "submerged/emerging," a mode of language that is "fluid/tidal" (42, 49).

I take *interlapping*—Brathwaite's scholarly howl and twisted word-image—as fundamental to archipelagic thinking generally as well as to a more immediate project of grasping how, within the arena of archipelagic thinking, we might view archipelagic English not as English-centric but as English-*eccentric*. In terms of interlapping's relation to general archipelagic thought, we may take the ocean-island complex of the archipelago as a figure of interlapping in that it is founded on mutual palimpsest, where islands seem to write over the ocean, and oceans lap up again onto the islands' shores, where various components within the island-ocean complex reciprocally and cyclically inscribe and reinscribe themselves upon each other. The term is evocative of Deleuze's commentary on certain islands as offering a "reminder that the sea is on top of the earth," while other islands remind us that "the earth is . . . under the sea, gathering its strength to punch through to the surface," with earth and sea "in constant strife" (2004, 9). Or, treating the same interlapping dynamic but without Deleuze's antagonistic imaginary, Elizabeth Bishop (1983) offers a less assured version: "Land lies in water . . . / Or does the land lean down to lift the sea from under . . . ?" (3). Within this mode of interlapping, both sea and land are, to borrow from Brathwaite on Caribbean English, "submerged/emerging."

In terms of its utility for figurations of archipelagic English, the notion of interlapping aids in the imagination of archipelagic English as an occasion and mechanism by which multitudinous language traditions may converge in a dispersed way without fully meeting at a centralizing point, analogous perhaps to how Amitav Ghosh describes the Asian-African-European maritime language of Laksari, "that motley tongue, spoken nowhere but on the water, whose words were as varied as the port's traffic, an anarchic medley" (2009, 104). As is the case in Brathwaite's "English in a different way" that has interlapping ancestral languages deferring illusions of full translatability, non–lingua franca languages become the structure's components, interlinked islands (in Wright's terms and Bandung's praxis) that come together to exchange and collaboratively imagine (on purpose and via the vicissitudes of less formal translational processes) traditional and new forms of archipelagic thinking. Their variegated and interlapping modes of figuring and metaphorizing multi-island and island-ocean cohesions across the planet offer a crucial reminder that the English/European/Mediterranean word "archipelago" can only ever constitute a catachresis, or a term used in lieu of any standardized or stable term for what "archipelagic" thinkers are seeking to access.[26] Over the course of decades and centuries and of hiatuses and events, these eccentric conjunctions become sprawling lattices of interlinking and interanimating structures of feeling whose crucial patterns and variations both constitute and trace the shape of an English-eccentric question, a radically multilingual query of hints and contradictions and terrors that seeks a noncentered pileup of non-English epistemologies and structures of feeling: What is an archipelago?

PRACTICAL AND METHODOLOGICAL IMPLICATIONS

Departing from approaches to the Bandung Conference that look toward this 1955 event in search of "the Bandung spirit," this chapter advances an approach that seeks "Bandung praxis," or the concrete and pragmatic practices of the conference which, by analogy, may be adapted and redeployed in other circumstances. Bandung praxis—particularly on the point of the conference's use of English as a lingua franca—becomes a template for archipelagic studies to consider its own relation to English as an uneven, messy, and deferring mediator among the multiple language traditions that have used terms for interisland and water-land connectivity that are not cognates of the term "archipelago." With the Bandung Conference's lingua franca as a background, English comes into view as both inadequate to archipelagic thinking and fundamental to archipelagic thinking, allied with Barbadian theorist Edward Kamau Brathwaite's notion of "English in a different way." This mode of English—an archipelagic English that functions as a lingua franca in persistent flux—is advanced as one way of pursuing radically revised and retextured answers to the question, What is an archipelago? Whereas traditional definitions of the term "archipelago" tend to rely on the 130-year-old etymological and historical account offered by the *Oxford English Dictionary*, this chapter traces the Indonesian term *tanah air* as a little-known but crucial component of the present-day etymology of the word "archipelago" and its significance within the international world. Inasmuch as this case study points to how little we know about the interrelation between the term "archipelago" and its noncognate terms in other languages, this chapter argues that a sprawling project—multilingual, long durational, transregional, and involving multiple participants—would be one of the most urgent of projects that could be undertaken in archipelagic studies, as it would redefine, and make harder to define, the field's founding geography, the archipelago.

NOTES

Thanks to John Butcher, Keith Foulcher, and Marlene Hansen Esplin for useful comments on this chapter.

1. The 1989 and 2017 versions of the *OED* state that while the term was an Italian coinage, "a true Italian compound," the move to make this compound was "suggested probably by the medieval Latin name of the Aegean Sea, *Egeopelagus*."

2. The *NED*'s definition and etymological history are predated by discussions in various editions of *Encyclopedia Britannica*, which, in its first edition in 1771, offered the following definition for "archipelago": "in geography, a general term for a sea interrupted with islands; but more especially denoting that between Greece and Asia" ("Archipelago," 1771). By the mid-nineteenth century, the encyclopedia's entry for the term centered its discussion on describing "that part of the Mediterranean extending from European Turkey and Greece on the west, to Asia Minor on the east, and stretching southward to the island of Candia," while a brief addendum indicates, "The name Archipelago, which was primarily given to the Aegean Sea, is now applied to various other seas which contain numerous islands, as the Eastern Archipelago, Caribbean Archipelago, &c" ("Archipelago," 1853). The encyclopedia's ninth edition in 1875 continues to focus on the Aegean while using language that resembles the *NED*'s discussion of the subsequent decade: "a name which, though it is now applied to any island-studded sea, was formerly the distinctive designation of what, though still known as the Archipelago, is often distinguished as the Grecian Archipelago" ("Archipelago," 1875).

3. For an apparent historical example of a self-consciously metaphorical use of the term "archipelago," see Scholar 2015, 48–49.

4. For an overview of some correctives to critical reliances on a dehistoricized Bandung spirit and myth, see Roberts and Foulcher 2016, 2–4. The present chapter complements narratives and analysis in Stephens 2015.

5. For an overview of the conference and its perceived importance in postcolonial studies, see Ashcroft, Griffiths, and Tiffin 2013, 23–25. For primary sources on the

conference's origins and immediate outcomes, see Ministry of Foreign Affairs, Republic of Indonesia 1955, 11–13, 161–69.

6. For information on the generally international and specifically Indonesian interest in claiming ocean space after World War II (as mentioned in the caption for Figure 4.1), see Butcher and Elson 2017, 46–76.

7. There is perhaps some ambiguity in Ali Sastroamidjojo's quotation here. In this instance, was he speaking literally, directly referencing the English language as the conference's lingua franca? Or was he speaking metaphorically of a vague common language of peace and anticolonialism? The former scenario is likely, given the conference record's emphasis on the fact that speeches delivered in Chinese were "immediately after translated into English by an interpreter speaking before the Conference" (Ministry of Foreign Affairs, Republic of Indonesia 1955, 63 and 180). In either case, however, the quotation is an accurate description of English's place at the Bandung Conference.

8. Wright's observations presage those who study English as a lingua franca (ELF), who in recent decades have noted that "roughly only one out of every four users of English in the world is a native speaker of the language" and that "English is being shaped at least as much by its non-native speakers as by its native speakers" (Seidlhofer 2005, 339).

9. The narrative of Malay as a strictly Indigenous lingua franca needs some qualification, given that for centuries this language, native to only small portions and populations of the archipelago, was promoted as a lingua franca throughout the Dutch East Indies, chiefly for administrative and missionary purposes: "the Indies Malay which up to the beginning of the twentieth century had principally served the colonial power's interest, within two decades was generating an organized self-awareness among the indigenous people of themselves as 'Indonesians'" (Hoffman 1979, 92).

10. On the Indonesian population in 1955, see Sukarno 1955, 29; on Indonesian and Malay, see Roberts and Foulcher 2016, xxiii, 232–34.

11. These 1954 and 1955 narratives, quotations, and translations are drawn from Foulcher 2000, 388.

12. In his opening speech, Sukarno (1955) stated, "We are united . . . by a common detestation of colonialism in whatever form it appears. We are united by a common detestation of racialism. And we are united by a common determination to preserve and stabilise peace in the world" (22).

13. In ways that converge with Wright's description and my riffing on that description, scholars of ELF have discussed "the most important ingredients of a lingua franca: negotiability, variability in terms of speaker proficiency, and openness to an integration of forms of other languages" (House 2003, 557).

14. On domesticating translation practices, see Venuti 2008, 13–15.

15. After Wright's 1955 visit to Indonesia for the Bandung Conference, his Indonesian hosts wrote of Wright's own "tanah air," applying this terraqueous term to the United States ("Synopsis," 1955, 27).

16. Indonesia, of course, was not the only country involved in the fight for recognition of the notion of the category of "archipelagic State." A Czech member of the International Law Committee seems to have coined the term in the mid-1950s

(Butcher and Elson 2017, 60–61), and the other archipelagic states that negotiated for recognition as such in the run-up to UNCLOS 1982 included Fiji, the Philippines, Mauritius, and the Bahamas. For a deeply impressive history on Indonesia and the archipelago concept, see Butcher and Elson 2017.

17. For background on Mochtar Kusumaatmadja, see Sumardjo 1999.

18. On Mochtar's involvement in innovating these baselines, see Butcher and Elson 2017, 66–67, 239. Thanks to John Butcher for help locating the map that appears in Figure 4.2.

19. Citation figures based on Google Scholar as of May 2019.

20. For context on this meeting, see Butcher and Elson 2017, 87.

21. My translation from Indonesian.

22. Even as I am advancing the idea of Bandung praxis's counterhegemonic English as a lingua franca, it is absolutely true that such traditions of decentered English have not existed in the absence of US efforts to recenter the lingua franca. For instance, in 1965 US President Lyndon B. Johnson "signed a statement to the effect that because of the growing world need for English as a *lingua franca*, the United States would sponsor and promote the teaching and use of English abroad as 'a major policy'" (Gordon 1978, 49).

23. For this quotation, see Glissant 1996, 45. Thanks to Daryl Lee for consulting on this translation from French.

24. On the terms *Nusantara* and *pae 'āina*, see Gaynor 2005, 34–40, and McDougall 2017, respectively.

25. I take "inter-lapping" as a scholarly instantiation of Brathwaite's commentary on "nation language," or the language of the Anglophone Caribbean, because at the conclusion of his book on that topic, Brathwaite (1984) discusses scholarly engagement with nation language and strives for a "new world to make new words and we to overstand how modern ancient is" (50). Brathwaite's coinage of "overstand" rather than "understand" switches out the prefix in a way analogous to his coinage of "inter-lapping" rather than "overlapping." The talk on which *History of the Voice* is based was given at the 1976 Carifesta in Jamaica (Brathwaite 1984, 49).

26. On catachresis and notions of multi-island cohesions, see Roberts and Stephens 2017, 30–31.

REFERENCES

"Archipelago." 1771. *Encyclopaedia Britannica; or, A Dictionary of Arts and Sciences, Compiled on a New Plan*. Vol. 1. Edinburgh: Colin Macfarquhar.

"Archipelago." 1853. *The Encyclopaedia Britannica; or, A Dictionary of Arts, Sciences, and General Literature*. 8th ed. Vol. 3. Edinburgh: Adam and Charles Black.

"Archipelago." 1875. *The Encyclopaedia Britannica; or, Dictionary of Arts, Sciences, and General Literature*. 9th ed. Vol. 2. American rpt. Philadelphia: J. M. Stoddard & Co.

"Archipelago." 1888. *A New English Dictionary on Historical Principles*. Vol. 1. Oxford: Clarendon Press.

"Archipelago." 1933. *Oxford English Dictionary* 1st ed. Oxford: Clarendon Press.
"Archipelago." 1989. *Oxford English Dictionary* 2nd ed. Oxford: Clarendon Press.
"Archipelago." 2017. *Oxford English Dictionary*. Online ed. Oxford University Press.
Ashcroft, Bill, Gareth Griffiths, and Helen Tiffin. 2013. *Postcolonial Studies: The Key Concepts*. 3rd ed. London: Routledge.
Batongbacal, Jay L., ed. 1998. *Archipelagic Studies: Charting New Waters*. Quezon City: University of the Philippines Printery.
Benítez-Rojo, Antonio. 1989. *La isla que se repite: El Caribe y la perspectiva posmoderna*. Hanover, NH: Ediciones del Norte.
———. 1996. *The Repeating Island: The Caribbean and the Postmodern Perspective*. Translated by James E. Maraniss. 2nd ed. Durham, NC: Duke University Press.
Bishop, Elizabeth. 1983. "The Map." In *The Complete Poems, 1927–1979*. New York: Noonday Press.
Brathwaite, Edward. 1977. *Contradictory Omens: Cultural Diversity and Integration in the Caribbean*. Mona, Jamaica: Savacou.
Brathwaite, Edward Kamau. 1984. *History of the Voice: The Development of Nation Language in Anglophone Caribbean Poetry*. London: New Beacon Books.
Butcher, John G., and R. E. Elson. 2017. *Sovereignty and the Sea: How Indonesia Became an Archipelagic State*. Singapore: National University of Singapore Press.
Danusaputro, St. Munadjat. 1972. "Nusantara Indonesia." *Ketahanan Nasional* (November): 48–63.
———. 1974. "Wawasan Nusantara and the International Sea System." *Indonesian Quarterly* 2, no. 4 (July): 52–87.
———. 1975. "The International Sea System in Perspective." *Indonesian Quarterly* 3, no. 4 (July): 3–43.
———. 1980. *Tata Lautan Nusantara dalam Hukum dan Sejarahnya*. Bandung: Penerbit Binacipta.
Dash, J. Michael. 2017. "The Stranger by the Shore: The Archipelization of Caliban in Antillean Theatre." In *Archipelagic American Studies*, edited by Brian Russell Roberts and Michelle Ann Stephens, 356–70. Durham, NC: Duke University Press.
Deleuze, Gilles. 2004. "Desert Islands." In *Desert Islands and Other Texts, 1953–1974*, edited by David Lapoujade, 9–14. Translated by Michael Taormina. Los Angeles: Semiotext(e) Foreign Agents Series.
Editorial Board. 2007. "An Introduction to Island Culture Studies." *Shima* 1, no. 1: 1–5.
Embassy of Indonesia. 1951. *The Cultural Life of Indonesia: Religion, the Arts, Education*. Foreword by Ali Sastroamidjojo. Washington, DC: H. K. Press.
Foulcher, Keith. 2000. "*Sumpah Pemuda*: The Making and Meaning of a Symbol of Indonesian Nationhood." *Asian Studies Review* 24, no. 3: 377–410.
Friedman, Susan Stanford. 2010. "Planetarity: Musing Modernist Studies." *Modernism/Modernity* 17, no. 3 (September): 471–99.
Gaynor, Jennifer. 2005. "Liquid Territory: Subordination, Memory and Manuscripts among Sama People of Sulawesi's Southern Littoral." PhD diss., University of Michigan.
Ghosh, Amitav. 2009. *Sea of Poppies*. New Delhi: Penguin.

Glissant, Édouard. 1990. *Poétique de la Relation*. Paris: Gallimard.
———. 1996. *Introduction à une Poétique du Divers*. Paris: Gallimard.
———. 1997. *Poetics of Relation*. Translated by Betsy Wing. Ann Arbor: University of Michigan Press.
———. 2005. *La Cohée du Lamentin: Poétique V*. Paris: Gallimard.
———. 2009. *Philosophie de la Relation: Poésie en étendue*. Paris: Gallimard.
Gordon, David C. 1978. *The French Language and National Identity (1930–1975)*. The Hague: Mouton Publishers.
Handley, George B. 2007. *New World Poetics: Nature and the Adamic Imagination of Whitman, Neruda, and Walcott*. Athens: University of Georgia Press.
Hoffman, John. 1979. "A Foreign Investment: Indies Malay to 1901." *Indonesia* 27 (April): 65–92.
House, Juliane. 2003. "English as a Lingua Franca: A Threat to Multilingualism?" *Journal of Sociolinguistics* 7, no. 4 (November): 556–78.
Kumarakulasingam, Narendran. 2016. "De-islanding." In *Meanings of Bandung: Postcolonial Orders and Decolonial Visions*, edited by Quỳnh N. Phạm and Robbie Shilliam, 51–60. London: Rowman & Littlefield.
Kusumaatmadja, Mochtar. 1973. "Supplementary Remarks." In *The Law of the Sea: Needs and Interests of Developing Countries—Proceedings of the Seventh Annual Conference of the Law of the Sea Institute, June 26–29, 1972, at the University of Rhode Island, Kingston, Rhode Island*, edited by Lewis M. Alexander, 172–77. Kingston: University of Rhode Island.
Malkin, Irad. 2011. *A Small Greek World: Networks in the Ancient Mediterranean*. New York: Oxford University Press.
McDougall, Brandy Nālani. 2017. "'We Are Not American': Competing Rhetorical Archipelagoes in Hawai'i." In *Archipelagic American Studies*, edited by Brian Russell Roberts and Michelle Ann Stephens, 259–78. Durham, NC: Duke University Press.
Ministry of Foreign Affairs, Republic of Indonesia, ed. 1955. *Asia-Africa Speaks from Bandung*. [Jakarta]: Ministry of Foreign Affairs, Republic of Indonesia.
Mufti, Aamir R. 2016. *Forget English! Orientalisms and World Literatures*. Cambridge, MA: Harvard University Press.
Rafael, Vicente L. 2009. "Translation, American English, and the National Insecurities of Empire." *Social Text* 27, no. 4 (101) (winter): 1–23.
Roberts, Brian Russell, and Keith Foulcher. 2016. "Richard Wright on the Bandung Conference, Modern Indonesia on Richard Wright." In *Indonesian Notebook: A Sourcebook on Richard Wright and the Bandung Conference*, edited by Brian Russell Roberts and Keith Foulcher, 1–31. Durham, NC: Duke University Press.
Roberts, Brian Russell, and Michelle Ann Stephens. 2017. "Archipelagic American Studies: Decontinentalizing the Study of American Culture." In *Archipelagic American Studies*, edited by Brian Russell Roberts and Michelle Ann Stephens, 1–54. Durham, NC: Duke University Press.
Sastroamidjojo, Ali. 1955. "Address by Ali Sastroamidjojo, President of the Conference." In *Asia-Africa Speaks from Bandung*, edited by the Ministry of Foreign Affairs, 31–39. [Jakarta]: Ministry of Foreign Affairs, Republic of Indonesia.

Scholar, Richard. 2015. "The Archipelago Goes Global: Late Glissant and the Early Modern *Isolario*." In *Caribbean Globalizations, 1492 to the Present Day*, edited by Eva Sansavior and Richard Scholar, 167–95. Liverpool, UK: Liverpool University Press.

Seidlhofer, Barbara. 2005. "English as a Lingua Franca." *ELT Journal* 59, no. 4 (October): 339–41.

Stephens, Michelle. 2015. "Federated Ocean States: Archipelagic Visions of the Third World at Mid-century." In *Beyond Windrush: Rethinking Postwar Anglophone Caribbean Literature*, edited by J. Dillon Brown and Leah Rosenberg, 222–37. Jackson: University Press of Mississippi.

Sukarno. 1955. "Speech by President Sukarno of Indonesia at the Opening of the Conference." In *Asia-Africa Speaks from Bandung*, edited by the Ministry of Foreign Affairs, 19–29. [Jakarta]: Ministry of Foreign Affairs, Republic of Indonesia.

Sumardjo, Jakob. 1999. "Biografi Prof. Dr. Mochtar Kusumaatmadja, S.H., LL.M." In *Mochtar Kusumaatmadja: Pendidik and Negarawan—Kumpulan Karya Tulis Menghormati 70 Tahun Prof. Dr. Mochtar Kusumaatmadja, S.H., LL.M.*, edited by Mieke Komar, Etty R. Agoes, and Eddy Damian, 3–28. Bandung: Penerbit Alumni.

"Synopsis." 1955. *Konfrontasi* (May–June): 25–28.

"Tanah." 1959. *A Malay-English Dictionary*. Part 2 (L–Z). London: MacMillan & Co.

United Nations. 1983. *Law of the Sea: Official Text of the United Nations Convention on the Law of the Sea with Annexes and Index*. New York: St. Martin's Press.

———. 2009. *Third United Nations Conference on the Law of the Sea, 1973–1982*. Vol. 1: *Summary Records of Plenary Meetings*. Second Session: 42nd Plenary Meeting A/CONF.62/SR.42. http://legal.un.org/diplomaticconferences/1973_los.

Venuti, Lawrence. 2008. *The Translator's Invisibility: A History of Translation*. 2nd ed. London: Routledge.

Wadell, Eric, Vijay Naidu, and Epeli Hau'ofa, eds. 1993. *A New Oceania: Rediscovering Our Sea of Islands*. Suva, Fiji: University of the South Pacific.

Walcott, Derek. 2005. "Isla Incognita." In *Caribbean Literature and the Environment: Between Nature and Culture*, edited by Elizabeth M. DeLoughrey, Renée K. Gosson, and George B. Handley, 51–57. Charlottesville: University of Virginia Press.

Walkowitz, Rebecca. 2015. *Born Translated: The Contemporary Novel in the Age of World Literature*. New York: Columbia University Press.

Williams, Raymond. 1954. "Film and the Dramatic Tradition." In *Preface to Film*, by Raymond Williams and Michael Orrom, 1–55. London: Film and Drama Limited.

Wright, Richard. 2008a. *The Color Curtain: A Report on the Bandung Conference*. In *Black Power: Three Books from Exile*, by Richard Wright, 429–609. Introduction by Cornel West. New York: HarperPerennial.

———. 2008b. *White Man, Listen!* In *Black Power: Three Books from Exile*, by Richard Wright, 631–812. Introduction by Cornel West. New York: HarperPerennial.

Chapter Five

The Chronotopes of Archipelagic Thinking

Glissant and the Narrative of Philosophy

Lanny Thompson

INTRODUCTION: ARCHIPELAGIC THINKING

Archipelagic thinking focuses on the analysis of discontinuous but interconnected insular places rather than the continuous territorial spaces of the state, the unitary regions of area studies, the isolated island, and the deterritorialized flatness of the global. In this view, archipelagoes are not natural phenomena but rather spatial and historical formations, assembled, reconfigured, and shaped largely by imperial, colonial, and postcolonial processes. Archipelagic studies highlight the connections among material, cultural, and political practices, which are spread out across islands, oceans, and continents and which are generated through the complex movements of capital and commerce, media and technology, ideas and ideology, and inventive, resourceful, and mobile people (Thompson 2017). The epistemic premise is that archipelagoes are geosocial locations that provide a vantage point from which to produce knowledge and shift the "geography of reason" (Gordon 2005).

The term "archipelagic thinking" was introduced by Édouard Glissant in one of his later works, *Traité du tout-monde*, but it was previously known as *antillanité* and then also as the "poetics of Relation" (Glissant 1989, 1997a, 1997b). It bears a strong family resemblance to the notion of "tidalectics" (Brathwaite and Mackey 1999) and is more distantly related to the notion of "repeating islands" (Benítez-Rojo 1992). While it is born of the Caribbean, similar ways of thinking may be found also throughout the archipelagoes of the Pacific. These ideas have in common the search for a "methodological tool that foregrounds how a dynamic model of geography can elucidate island history and cultural production, providing the framework for exploring the complex and shifting entanglement between sea and land, diaspora and indigeneity, and routes and roots" (DeLoughrey 2009, 2).

More than method for historical, cultural, and literary analysis, archipelagic thought may be understood also as a philosophical project, encompassing aesthetics, ethics, epistemology, and ontology (Drabinski and Parham 2015, 1–6). Archipelagic thinking, as a process of rethinking geography and history, implicates abstract notions of space and time, of topology and temporality. Furthermore, by emphasizing the formation of identities, archipelagic thought encompasses theories of the subject and ontology. In this chapter, I propose to show how archipelagic thinking might move from geography and history in order to address topology, temporality, and ontology. First, topology, understood as emplacement, makes explicit those geographical contexts where we relate with others, construct places, and engage in spatial thinking. This is a conceptual transition from geography to topology. Second, temporality, understood as emplotment, brings together and relates those same places within historical narratives that suggest notions of time in relation to space. Finally, ontology, understood as situated subjectivity, understands that place, temporality, and being belong together as intricately and intractably connected (cf. Malpas 2012).

This chapter seeks to explore and elaborate, both conceptually and methodologically, these underlying philosophical premises of archipelagic thought as elucidated in the writings of Édouard Glissant. In order to integrate this movement toward topology, temporality, and ontology, I enlist the analytic of the chronotope as proposed by Mikhail Bakhtin for literary studies. The analytic of the chronotope is particularly suited for a study of the philosophy of Glissant. First, he employs a literary style even in his philosophical works. Second, Glissant is quite explicit regarding the dimensions of space and time in his work, especially in the context of Caribbean geography and history. Third, much of his literary analysis has all the attributes of a chronotopic analysis even though he never uses the term (see Glissant 1989, 134–57). In other words, I use the analytic of the chronotope to describe some of the specific chronotopic figures in Glissant's philosophy in order to elucidate his notions of topology, temporality, and ontology.

CHRONOTOPES IN LITERATURE AND PHILOSOPHY

"Chronotope" means literally "time-space." Mikhail Bakhtin proposed this neologism to refer to a "unit of analysis for studying texts" according to their particular configuration of "temporal and spatial categories" (1981, 425). With his approach, he sought to describe the narrative structures that distinguish literary genres, especially the novel and its variants. In his analysis of broad literary genres, Bakhtin also proposed that any number of minor chronotopes (elements or motifs) might be integrated dialogically with the major ones (252). Multiple chronotopes "may be interwoven with, replace or oppose one another, contradict one another or find themselves in ever more complex [dialogical] interrelationships" (252). Bakhtin's work suggests that the chronotope is at once a method and a specific description of genres, elements, and motifs. In other words, it is a method that results in a description of literary texts, emphasizing their temporal and spatial motifs and their dialogical interaction.

Bakhtin's original descriptions of genres and motifs have been influential in literary analysis, and several authors have applied them, without much modification, to Caribbean literature. These chronotopes include the road, idyll, historical inversion, and carnival (Berman 2009; Dash 1995; Hart 2004; Tlostanova 2013). And yet, Bakhtin's original analysis arose out of a European context—its mythology, its novels, its villages and towns—far removed from the Caribbean. For this reason, other authors have sought to define the unique chronotopic configurations associated with Caribbean experience and history. Perhaps the most influential thinker in this regard was Paul Gilroy, who proposed the ship as a chronotope to describe the "black Atlantic" constituted by the slave trade, which initiated the African diaspora. He understood ships to be mobile microcosms that linked places in Europe, Africa, the Caribbean, and beyond. The ship was "a living, micro-cultural, micro-political system in motion" (1993, 4). It was the "living means by which the points within that Atlantic world were joined" (16). Moreover, he uses the chronotope of the ship as a means to "rethink modernity via the history of the black Atlantic and the African diaspora into the western hemisphere" (17). In other words, this chronotopic analysis of the ship highlights the importance of the slave trade to the processes of modernization, industrialization, the histories of European ports, and all of the "interfaces with the wider world" (17). Gilroy also introduced the chronotope of the crossroad to designate places where people, cultures, and materials met and mixed throughout their American landfalls. This chronotope served to refute the "dualistic structure which puts Africa, authenticity, purity, and origin in crude opposition to the Americas, hybridity, creolisation, and rootlessness" (199). Guillermina de Ferrari finds considerable resonance between the works of Gilroy and Glissant, stressing that both posit a "spatiotemporal matrix of pure relationality" (2012, 190). For Gilroy this matrix is constituted through the slave ship; for Glissant it is the plantation. According to DeLoughrey (2009), Gilroy's chronotope of the ship expresses a paradox: the open space of the ocean in contrast to the enclosed structure of the ship. Glissant describes this paradox as characteristic of the plantation. In addition, Glissant introduces a number of topographical referents—in addition to ships—that are important to his unique notion of archipelagic thinking, as we shall see.

The perspectives of Gilroy and Glissant (our particular interest here) have the additional advantage of stressing subjectivity that is situated within narrative formations of collective identities rather than the more categorical or deterministic theories of the social sciences (Brubaker and Cooper 2000). In her reworking of the concept of identity in social theory, Margaret Somers argues that "it is through narrativity that we come to know, understand, and make sense of the social world, and it is through narratives and narrativity that we constitute our social identities" (1994, 606). She defines "narrative" in a chronotopic way, although she does not use the term. She writes that "narratives are constellations of *relationships* (connected parts) embedded in *time and space*, constituted by *causal emplotment*" (616, original emphasis). In a similar fashion, Guillermina de Ferrari argues that Glissant employs "narrative-based identity" rather than "history-based identity" (2012, 196–99). While the latter refers to the formation of distinct groups defined by commonality or sameness, the former appeals to encumbered selves embedded in multiple social relationships.

These literary and narrative analyses have made invaluable contributions and have suggested ways to deepen the integration of the philosophical notions of space, time, and subjectivity. The analytic of the chronotope provides a way to reveal further the philosophical notions implicit in narrative studies. Michael Holquist (2010) has shown that the term "chronotope" is a reworking of the Kantian concepts of time and space. For Kant, time and space were a priori categories that subjects brought to bear on their practical experience. The subject was split between, on the one hand, the universal, transcendent categories of time and space and, on the other hand, the empirical cognitive activities in the real world. In other words, Kant posited a disjuncture between mind, with its transcendental categories, and sensate subjects in the experiential world. The adequate mediation of the a priori categories and empirical observations was a problem of cognition and, ultimately, of knowledge. However, Kant did not address the roles of language and narrative in this process of mediation.

Bakhtin was equally concerned about time, space, and the subject but reworked the Kantian problematic and its categories. The concept of the chronotope transfers a priori time and space to the context of narratives, especially in the novel but more broadly in all cultural representations of lived experience. First, Bakhtin molded the categories of time and space into a single term—chronotope—which expressed their necessary interdependence and inseparability. Second, he argued that language, especially when expressed as narrative, mediated the activities of subjects in the real world. Third, he argued that chronotopes were necessarily a part of all narratives, even though their characteristics varied according to literary genre and, in a more general sense, particular cultural expressions. For Bakhtin, the subject is also split, not between the transcendental and the empirical but rather between narrative chronotopes and the real world. In this view, acting subjects in the world mediate their experience in the world by means of language and narrative (Holquist 2010). Furthermore, the possibilities of ethical human action—subjectivity—are framed by narrative chronotopes that situate actors within a temporal and spatial frame that limits but does not determine their possibilities (Steinby 2013, 122). Although Glissant never makes explicit reference to chronotopes, he also situates the ethical possibilities of subjects within the particularities of Caribbean history and geography. Through narrative tropes he reflects upon the ideas of justice and responsibility, belonging and civility (De Ferrari 2012, 208).

In sum, the analytic of the chronotope need not be confined only to a description of the novel and other literary genres; rather it might be returned to philosophy. My suggestion here is that the analytic of the chronotope provides us with a method that allows us to link narratives and philosophy in a way that intertwines space, time, and subjectivity. In other words, I stress the distinction between the particular motifs that Bakhtin attributed to European literature and the method of chronotopic analysis he used to describe them. I will elaborate upon the idea, only implied by Bakhtin, that the analytic of the chronotope allows a reflection on the interconnectedness of topology, temporality, and ontology in philosophy. More specifically, here I propose that the analytic of the chronotope may be useful in understanding the philosophical underpinnings of archipelagic thinking as expressed by Édouard Glissant.

THE CHRONOTOPE OF THE ABYSS

Glissant's philosophical narratives, when refracted through a chronotopic analysis, reveal a number of major and minor motifs that are dialogically interconnected. The opening to *Poetics of Relation* introduces the motif of the slave ship of the Middle Passage from Africa to the Caribbean. The hold of the slave ship is described as an abyss, and for those who survived, it was a place of social death and bare life. The slave ship crossed another abyss, that of the bottomless, boundless ocean. Yet it opened up to the beach, the entrance to a land that was both new and imposed. The third abyss consists of all that was left behind and remained to a large extent forgotten, except in fragmentary fashion. Any return to this former life was impossible. After crossing the abyss, the slave ship arrived and discharged its human cargo. This arrival is described metaphorically as a birth that constitutes the foundational moment of creolization. Glissant does not write literally from this location; rather, "thought draws the imaginary of the past; a knowledge becoming" (1997a, 2). That is, the narrative calls upon the imaginary of the past to situate its thinking, ground it in the present, and project a future. The text posits the abyss as a threshold, as the passage from a place of no return to absolute exile.

This passage marks, at the same time, the beginning of a long diasporic journey, understood as creolization and, as we shall see, errantry. The abyss constituted a foundation that is antifoundational, without root, without linear filiation or territorial boundaries. The beginning of creolization was based on the creative combination of uprooted and dispersed fragments of language, culture, and memory. The flotsam and jetsam of the passage provided the elements for thinking with fragments. The chronotope of the abyss collapses space and time. They are folded together into the microcosm of the ship where neither historical nor biographical time exists. After crossing the abyss, the subject does not emerge intact but rather is birthed anew, in a way different from the previous. In Glissant, then, the abyss is a meta-threshold of three dimensions: the bare life (Agamben 1998) in the hold of the ship, the unfathomable deepness of the ocean, and the unsurmountable distance from all previous life of the past. The chronotope of the abyss situates the origin of Creole society on the shifting sands of the complete fragmentation of time, space, and subjectivity, both individual and collective. Yet the abyss opens up to a new topography, begins a different narrative, and emplaces the subject on the road to Becoming by the collective processes of creolization. This subject emerges neither from the realm of nothingness (non-Being) nor with a fixed identity (Being), as we shall see in more detail later. The short narrative of the abyss is a dialogic prelude to the chronotopes of the plantation and the island. Rather than a firm foundation, the abyss is a condition or premise, ever present.

John Drabinski (2012, 148) argues that these dislocations characteristic of "exilic life" are marked by the traumatic fragments of past experiences. Relocation requires that memory serve to imagine futures that can lead potentially to spaces of creative self-assertion and self-creation, both individual and collective. In this way, the past, present, and future interact across spatial divides. In a similar fashion, Esther Peeren argues that "dispersed communities connect themselves to each other and to the homeland by forging relationships across space and time through a shared performative (habitual and mnemonic) construction of time-space: a shared chronotope" (2006, 73). These "diasporas" are characterized by a "dwelling-in-displacement" or a "dwelling-in-dischronotopicality," when the spatiotemporal environment of the place of origin is distinct from that of the place of arrival (72). The experience of migrations to and from the Caribbean archipelago suggests that the absolute exile of the abyss was the foundation of a continual process of dislocation and relocation, whether conceptualized as exilic life or as diaspora. As we shall see, Glissant uses the term "rooted errantry" to refer to these complex processes.

CARIBBEAN CHRONOTOPES: PLANTATIONS AND ISLANDS

Throughout *Poetics*, Glissant poses a spatial distinction between closed, structured relationships, represented by the plantation, and the exuberant openness of *Relation*, represented by the Caribbean topography: islands and beaches, the archipelago and the sea. Glissant understands the plantation as an enclosed social hierarchy, partially autarkic but actually highly dependent on the world economy, with a technical mode of production based upon slavery. He describes the plantation as a closed place with certain structural and functional components, an institutional site that organizes and distributes relations of power by means of a delimited, internal grid of relationships (cf. Wolf and Mintz 1957). It was shaped by the cyclical rhythms of production and the places occupied by subjects within the social hierarchy.[1] The plantation combined the concrete temporality of agricultural cycles and the spatial locations of subjects within the social domain, a chronotope of structured racial oppression, economic exploitation, and sexual abuse. In the plantation enclosure, there was a temporal-spatial contiguity; time and space were local and interlocking.

Even though the plantation was an oppressive structure, it was also the origin of linguistic and cultural creolization that arose as forms of resistance (Murdoch 2012, 2013). The creole multilingualism that emerged there was an unpredictable creation arising from a violent clash of cultures: "The place was closed, but the word derived from it remains open" (Glissant 1997a, 75). The plantation was an enclosure that paradoxically opened out to the larger world. Despite its pretension to autarky, the plantation was constituted by both internal structures and external factors; it was a place where these elements met, mixed, and were transformed. Even though the plantation was marked by structured oppression, subjects, in order to survive, resisted through the creation of language, cuisine, and music, all fashioned from the material and cultural fragments available. What took place there anticipated what would eventually come to pass on a global scale, including, in synchronic fashion, an interconnection of time and space far beyond the scale of local places.

In contrast to the plantation, the motifs of islands, beaches, and the sea serve as topographical referents for the mediation of closure and openness. In the final chapter of *Poetics*, Glissant reflects on a place of his own experience, the "burning beach" that manifests an unseen, underlying volcanic turbulence as evidenced by the hot spots in the mangroves that vent steam. In the following quote, the closed plantation, an archetypical Caribbean place, is situated within the currents of the sea and the submarine depths—here invoked as rivers—which connect us to the mountains of the *marrons* and the sea of islands.

> This tie between the beach and the island, which allows us to take off like *marrons* . . . is thus tied to the dis-appearance—a disappearing—in which the depths of the volcano circulate. I have always imagined that these depths navigate a path beneath the sea in the west and the ocean in the east and that, though we are separated each in our Plantation, the now green balls and chains have rolled beneath from one island to the next, weaving shared rivers that we shall open up when it is our time and where we shall take our boats. (Glissant 1997a, 206)

Islands, with their liminal beaches, mediate closed places (like plantations) and the open horizons of the sea. Even the plantation, a highly structured and delimited place, opens up to the movement of errantry. Likewise, from the vantage point of the burning beach, insularity is not confining, not isolated, nor inward-looking only. As Glissant writes, "Ordinarily, insularity is treated as a form of isolation, a neurotic reaction to place. However, in the Caribbean each island embodies openness. The dialectic between inside and outside is reflected in the relationship of land and sea. It is only those who are tied to the European continent who see insularity as confining. A Caribbean imagination liberates us from being smothered" (1989, 139).

The isolated, desert island has been a standard chronotope in European literature, especially in the genre of nautical adventures, the Robinsonades (DeLoughrey 2009, 13–14). In contrast, Glissant contradicts this chronotope of isolation by setting up here, and indeed throughout *Poetics*, analogous and parallel distinctions—closed/open and internal/external—that express the paradox of islands. Furthermore, these distinctions also correspond to the conceptual difference between "relation" and "*Relation*." "To the extent that our consciousness of Relation is total, that is, immediate and focusing directly upon the realizable totality of the world, when we speak of a poetics of Relation, we no longer need to add: relation between what and what? This is why the French word *Relation*, which functions somewhat like an intransitive verb, could not correspond, for example, to the English term *relationship*" (1997a, 27).[2]

Relationships occur between specific things. A plantation (or even a culture) may be defined by its particularities, the internal relationships of its components, as well as the external relationships that affect it from outside. The totality of the internal and external relationships may be called "relation" (in lower case). Sociology, anthropology, and the humanities study these "structural components and dynamic relationships" and create theoretical models to understand them. Structural analysis "helps us to imagine better," but it has its limitations (Glissant 1997a, 170). It closes off rather than opens up; it focuses on fixed relationships rather than unpredictable creative chaos, which is relayed, relativized, and related in the "overall rhythm" of *Relation.* Indeed, these "models claim to base the matter of *Relation* in relationships," that is, they attempt to explain *Relation* by means of the social structures that constitute the entanglements of relation (173). However, *Relation* goes beyond relation: "The idea of relation does not limit Relation, nor does it fit outside it" (185). Conceptually, then, *Relation* is related to, but different from, relation, the latter understood as structured internal and complex external relationships. In contrast, *Relation* partakes of flows and movements in which languages, cultural practices, and identities come in contact and, in doing so, influence and change one another. *Relation* refers to the spontaneous creation through which cultural differences are continuously brought into play with unpredictable results.

These motifs—the abyss of the slave ship, the structure of the plantation, the liminality of islands—refer to dynamics of closure and opening, of internal and external, of birth and becoming, of capture and escape to the hills, of oppressive places and open language, and of slave ships and small boats launched to neighboring islands. On the one hand, we find the structured relationships and systems that make up the worldwide entanglements of relation, slave ships, plantations, societies, cultures, and so forth. These are closed series that constitute places. On the other hand, *Relation,* by means of rooted, rhizomatic errantry, links and relates insular places and creates unpredictable archipelagoes. In contrast to the more static relationships among structural elements, *Relation* signifies a fluid openness and movement arising from the complex and chaotic dynamics of bringing things into relationship on a grand scale. This dialectic of closed/open, internal/external, and relation/*Relation* always occurs through the constitution of places brought into connection. As Glissant writes, "*Relation* exists in being realized, that is, in being completed in a common place" (1997a, 203). In other words, the openness of *Relation* is manifested in concrete, structured places of interaction.

The term "rooted errantry" expresses this dynamic of closed places connected through open spaces. In one of his more famous quotations regarding *créolité* and *Relation,* Glissant sets up a series of spatial contrasts based, once again, on the dynamic of closed and open: "What took place in the Caribbean, which could be summed up in the word *creolization,* approximates the idea of Relation for us as nearly as possible. It is not merely an encounter, a shock . . . , a *métissage,* but a new and original dimension allowing each person to be there and elsewhere, rooted and open, lost in the mountains and free beneath the sea, in harmony and in errantry" (1997a, 34).

The parallel construction here contrasts rooted and open. On the one hand, rooted refers to the "there," "lost in the mountains," and "in harmony." On the other hand, the open is "elsewhere," "free beneath the sea," and "in errantry." In this context, two different notions of "root" are distinguished, somewhat in the fashion of Gilles Deleuze and Félix Guattari. The arboreal root, which is singular, is contrasted to the rhizomatic root. "Rooted errantry" refers to the rhizomatic, not the arboreal, root. The rhizomic topography of islands opens up and connects by means of the openness the ocean provides. That is, Glissant roots local places on islands with the context of movement, as illustrated by the sea. Thus, "root identity" is discarded in favor of "relational identity," which in turn finds its topological ground in rhizomatic movement, in "rooted errantry."

Islands are points of arrival and departure, of the constitution of place and of errantry (Murdoch 2012). Subjects depart from a place in which they have set down roots and set off on a journey that often retraces the chains of imperial dominion even to the very centers of power and wealth. It is no accident that Caribbean migrants often migrate to the countries that, in the past or present, held them as colonial subjects and their lands as colonial possessions. Caribbean migrants must leave their countries and take the routes—by land, sea, or air—that have been forged by empires; in this way they are immersed in the inequalities of imperial archipelagoes that are shaped by cultural racism. In this geographical movement, subjects are transformed: they refashion and remake themselves accordingly and establish roots in new places usually under difficult, discriminatory circumstances (Grosfoguel 2017).

In sum, Glissant expresses the connections between time and space by means of the dialogical relation among the chronotopes of the abyss, of the plantation, and of the island. His concept of "rooted errantry" dialogically condenses these three chronotopes. In other words, rooted errantry forms a major chronotope, a complex dialogic of the abyss (as prelude), the plantation (contiguity of local time and space), and archipelagic islands (synchronic temporal flows that connect dispersed places). The chronotopic ground of archipelagic thinking is at once rooted in island places and errant through synchronic connectedness and movement. These are spatiotemporal coordinates that ground archipelagic thinking. Let us now consider more fully the notions of space and time.

THE TOPOLOGY AND TEMPORALITY OF ROOTED ERRANTRY

Poetics is permeated by the spatial ideas—but not the terminology—of Gilles Deleuze and Félix Guattari: striated, smooth, and holey space (Hantel

2012). In *A Thousand Plateaus*, these authors distinguish the striated space of states, the smooth space of nomads, and the holey spaces that constitute the intermediaries between the two extremes. Striated space is sedentary and territorial; its space is constituted by means of a uniformly measurable and quantifiable grid. Striated space forms the basis of state power and sedentary civilizations (Deleuze and Guattari 2008, 415). Smooth space refers to heterogeneous and open spaces that facilitate the movement and bringing together of "any type of nonmetric, acentered, rhizomic multiplicities" (371). Finally, holey space intersects with smooth and striated spaces; it "connects" with nomadic space while it "conjugates" with striated, sedentary space. "The ambivalent nature of holey space turns on the distinction between connection and conjugation: connections imply an intensification of different deterritorializing flows that reciprocally accelerate; conjugation, on the other hand, 'indicates their relative stoppage' because the flows are brought under the control of a single code" (220). Glissant's particular term for this connection and conjugation is "rooted errantry."

However, he does not merely adopt the nomadology of smooth spaces or the "single code" of the striated spaces. Rather, Glissant evokes these ideas obliquely and transforms them through reframing them historically and spatially, by changing the scale—that is, the geographical context (Hantel 2013, 102–4). He clearly situates his elaboration of these ideas firmly within the Caribbean and shifts the terminology to the topography of an archipelagic region constituted by connected islands composed of rhizomic places. For example, his narrative of conquest, migration, and settlement in the New World does not correspond to either sedentary or nomadic spaces, which are extrapolated from continental spaces theorized by Deleuze and Guattari. In the Caribbean, the nomads do not come from the steppes of Asia. Rather, they were European conquerors who left their settlements and entered into an invading or "arrowlike" nomadism, characterized as a "totalitarian drive for a single, unique root" (Glissant 1997a, 14). Rather than passively accepting "nomadology," Glissant develops his own concept of nomadism in the context of European colonization. Even though he writes favorably of the notion of the rhizome, he opts instead for the term "errantry" to refer to movements that are rooted in places rather than nomadic deterritorialized smooth space. Rooted errantry results in a reterritorialization in which localities are incessantly relinked, relayed, and repeated (Murdoch 2012).

From the Caribbean, Glissant narrates a history that subverts the Hegelian dialectic, on the one hand, and notions of modern linear time, on the other. First, he makes frequent allusions to three-part, pseudo-dialectical sequences but without any suggestion of teleology, fulfillment, or completion. For example, in his description of the migration between core and periphery, he sets out three phases: from core to periphery, from periphery to core, and finally complex movements that neither negate nor supersede this simple dichotomy; rather they confound and complicate it. Second, he undercuts the modern, Eurocentric interpretation of history through frequent references to the centrality of the Caribbean to world history, even though the region is often considered to be without history. Glissant thus presents a sophisticated critique of teleological historical time, both linear and dialectical, as often found in conventional historiography (Wilson 1999).

We have seen previously that the chronotope of the plantation, and of closed places in general, expresses the contiguity of local space and time. On a global scale, this chronotope is multiplied and becomes more complex; rather than the internal contiguity of time and place, the chronotope of islands embedded in archipelagoes expresses a chronotope of various temporal rhythms that are synchronic and spread out over connected places. In his description of the varied topographies of the mountainous north, the central plains, and the littorals of the south, Glissant suggests that Martinique recapitulates the whole geography and history of the New World. He concludes, "So history is spread out beneath this surface, from the mountains to the sea, from north to south, from the forest to the beaches. Maroon resistance and denial, entrenchment and endurance, the world beyond and dream. (Our landscape is its own monument: its meaning can only be traced on the underside. It is all history.)" (Glissant 1989, 11).

In this discussion of landscape as an expression of history, Glissant deploys a notion of synchronous space and time. In addition, Glissant suggests a modification of Derek Walcott's famous assertion that the "sea is history." This is neither a disagreement nor a contradiction but rather a supplement that connects the sea with island landscapes. An even more explicit expression of the synchronicity of space and time is the phrase he used to describe his play, *Monsieur Toussaint*. In the preface to the first edition, Glissant states that his work is a "prophetic vision of the past." According to this preface, the play addresses two types of Caribbean readers. The first experiences an "absence of history," which has been "obscured or obliterated by others." This group, without history, lives superficially in the present. The second group is burdened by the tyranny of history that it finds difficult to escape; it experiences "darkness and despair." The play's "poetic endeavor" is to join the two in a common struggle in order to remedy the ontological lack—the "insecurity of being"—that they all share (Glissant 1981, 17).

Victor Figueroa (2015, 191–92) argues that this prophetic vision of the past is a call to recuperate those histories ignored or marginalized by conventional Eurocentric historiography in order to project a future that seeks to rectify those inequities and injustices that persist even to the present.

This prophetic vision is marked by a synchronicity of space and time. In Nicole Kaplan's words, the play presents a "non-chronological unfolding of events, the continuous interaction of past and present" (2007, 208). In addition, the simultaneity of past and present is framed by two spaces, the insular space where the Haitian revolution unfolds and the carceral space where the imprisoned Toussaint reflects upon the consequences of his actions. According to Glissant, "The simultaneity of the two time frames in which Toussaint lives (that of the insular space and that of the prison) does not stem from technical sophistication. The equivalence of past and present is essential in light of what he has or has not accomplished and of what he expects—or no longer expects" (1981, 17–18).

This short section establishes a synchronicity that is framed in the imaginary by two distant places that exist simultaneously with their own temporalities. In this context, the protagonist is encumbered, but not overdetermined, since he always has the ability to make choices that are unpredictable, especially with respect to their future consequences (Figueroa 2015, 193–94). As we shall see in the next section, this chronotope expresses a complex ontology in which subjects are immersed in structured relationships and yet possess a creative potential for choice and ethical decisions. In this view, coerced, oppressed, and encumbered subjects still assert their capacity to act and to resist whether in subtle or explicit ways.

DIALOGICS OF SUBJECTIVITY: BEING-OF-THE-WORLD

Some years ago, Peter Hallward criticized Glissant for emphasizing the singularity of individuals rather than their specificity. In this argument, singularity refers to the self-creation of unique individuals entirely in opposition to or entirely devoid of the confining limitations of sociohistorical context. Specificity, in contrast, refers to the complete immersion of individuality in social, historical, and cultural conditions. Hallward argues that Glissant's ontology posits subjects as nonrelational singularities rather than specificities located within a complex myriad of social connections. Unlike in the early works, such as the politically infused *Caribbean Discourse*, in the more philosophical *Poetics of Relation* collective subjects recede as historical agents of political movements and social change (Hallward 1998, 2001). This interpretation has provoked considerable debate. In her review of this debate, Guillermina de Ferrari (2012, 194) accepts the distinction between specific and singular individuals but argues, in contrast to the dichotomous thinking of most commentators, that Glissant proposes that Caribbean subjects are *both* specific and particular, both encumbered and free, both socially embedded and errant.

On this view, individuality is not previous to social relationships; rather, it emerges within these social structures. In ontological terms, there is not a prior, essential Being; rather being is a process of becoming in relation to others in structured relationships.

In a similar fashion, Clevis Headley (2012, 76) argues that Glissant rejects the logic of either/or in favor of both/and with regard to particularities and universality. More specifically, Glissant elaborates an "existential ontology" and posits a "creolizing of Being, meaning that he approaches being from the perspective of difference, relation, and immanence but not transcendence" (59). Furthermore, the metaphor of the sea "connotes rhythmic motion and repetition, for the sea is not a substance, not a structurally organized closed system, and it lacks discrete discernible boundaries" (59). Headley locates Glissant's ontology firmly within the orbit of the philosophy of Gilles Deleuze, stressing three central concepts: immanence (autopoiesis, or self-organization and self-creation, as the immanent characteristic of materiality), duration (heterogeneous markers of continual change), and difference (Being as pure difference and multiplicity). This ontological perspective is based on the premise that "identities, namely phenomenal identities, are secondary products of differential processes, which are themselves not dependent on prior identities" (65). In the words of Headley, Glissant's ontology is radical in its attempt to "approach existence as an historically and geographically situated reality" (77).

Headley's exposition, however, does not address the question of Being in relation to space and time, except to note briefly that Glissant is writing from the "novel historical space" of the Caribbean and from the "underside of modernity," that is, from "the perspective of those formerly excluded from the universalist consciousness of European philosophy"(Headley 2012, 59). Likewise, Drabinski (2012, 146) concludes that *Relation* is not simply another name for Being in the conventional philosophical sense. However, in order to address fully the question of the philosophy of archipelagic thinking, we must integrate topology and temporality with ontology.

In order to understand better Glissant's position with respect to specific and singular beings in relation to space and time, it will be useful to analyze his critique of European ontology that is found in a short chapter in *Poetics*. This chapter consists of four dense pages composed entirely of aphorisms (Glissant 1997a, 185–88). Neither the style nor the method is characteristically philosophical in the European sense. Rather than cite authors by explicit reference, he "incites" them by reframing and reworking concepts, treating them as "commonplaces," as well-known and freely available notions (Oakley 2008, 2012). *Poetics* establishes a critical distance with respect to various definitions of Being in order to elaborate its principal concept, *Relation*, and to ground it ontologically. Glissant begins his discussion with the rejection of two commonplace ideas of Being. First, he rejects the notion of a transcendental self, designated as "Being as Being" (Glissant 1997a, 185). This concept refers to a subject that is a "self-important entity that would locate its beginning in itself" (160). Being as Being is "self-sufficient" and so "cannot bear having any interaction attached to it" (161). This formulation apparently makes reference to the Cartesian tradition, which posits an individual subject with a fully autonomous consciousness with respect to an external world. In contrast, *Relation* "does not partake of Being" for it is not grounded in this kind of ontology (160).

Second, Glissant dismisses the relational subject of phenomenology, which he identifies with the statement "Being is relation" (1997a, 185). Again, in the absence of citations in the text, I suggest that Glissant makes reference to an idea that may be found in Emmanuel Levinas, as well as other phenomenologists (Dastur 2017). For Levinas, the fundamental ground of Being is the face-to-face relationships with Others. Human intentionality, language, consciousness, and ethics are founded, not on a mere coexistence of individuals, but rather on the prior existence of Others (Mensch 2015; Cailler 2011). Glissant contrasts this view with his own. He states that we must "renounce the fruitful maxim whereby Being is relation, to consider that Relation alone is relation" (Glissant 1997a, 170). Here, *Relation* substitutes Being as the central ontological category and is contextualized within the complex structures of relationships, designated as "relation."

In order to understand Glissant's aphorism, we must refer to another phenomenological variant, that of Martin Heidegger. According to Heidegger, ontology concerns only those conscious Beings (*Dasein*, or Being-there) who inquire as to the meaning of Being. Being-there (*Dasein*) is a complex phenomenon consisting of Being-in-the-world, of Being-with other human beings, and of Being-with-oneself. In contrast to Beings, the world consists of mere beings, that is, ontic entities that are objects of use and study according to their characteristics and classification. Even though this ontology distinguishes Being and ontic entities, it recognizes that they are existentially connected. Not only are all Beings at once entities, but also Beings exist within the ontic world of entities: Being-in-the-world. However, on this view, Beings, if they are resolute, do not succumb to the past and present structures of the world but rather project an authentic future. Those human beings who do not assert their autochthonous consciousness are implicitly reduced to ontic beings (Schalow and Denker 2010, 11, 61, 70–71, 199–200, 244; Maldonado-Torres 2007).

In contrast to the Heideggerian distinction between ontological Beings and ontic beings, Glissant states that we must "abandon the apposition of Being and beings" (1997a, 170). Moreover, he reverses the ontological priority of Being over ontic beings: *Relation* is not "(of) Being, but (of) beings" (186). This is consistent with the previous aphorism "*Relation* is relation." This is precisely because the entanglements of human beings in their structured relationships—also known as relation—are the very ontic foundations of *Relation.* These ontic foundations underlie the interconnection of all human beings, considered in their existential facticity in the *tout-monde*, that is, the totality of the world. On this view, human beings are a part of—and not merely in—the world. In this way, Glissant's ontology links specific subjects, encumbered in social relationships, with the singularities of opaque, errant, and autopoietic beings. Glissant states, "The being-of-the-world realizes [ontological] Being:—in [ontic] beings" (187).[3] This aphorism turns Heidegger upside down; it suggests that ontological Being is grounded in ontic beings, which are thoroughly of the world. Rather than the aloof authenticity of Being-in-the-world, Glissant emphasizes encumbered beings who are fully of the world but, nevertheless, fully capable of autopoiesis. It signals a move away from ontological notions of Being that stress authenticity and rootedness that are often associated with nationalism (Maldonado-Torres 2004).

Moreover, Glissant, similar to Frantz Fanon, criticizes the coloniality of Being, that is, the philosophical denial, whether implicit or explicit, of the subjectivity of oppressed peoples of the Caribbean and other colonies (Maldonado-Torres 2007). Sylvia Wynter argues that Glissant's shift from *L'Être* (Being) to *l'étant* (beings) corresponds to the replacement of "ontogeny with sociogeny" (1989, 645), as proposed by Fanon. The former term refers to the development of individual subjectivity, while the latter refers to those social processes, understood both as structures and ideology, that are internalized by individuals. Both oppose "Ontological Lack," the exclusion of colonized people from full human status, as suggested by those ontological categories of autonomous, authentic subjects. Furthermore, Glissant rejects the category of nothingness, another European obsession from Heidegger to Sartre. Even considering the horrors of the abyss and the plantation, Glissant concludes, "Non-Being does not precede Relation, which is not expressed on the basis of any break. The non-Being of Relation would be impossible completion" (1997a, 187). Glissant, then, refuses to base his ontology on anything but emplaced and emploted subjects. The issue here is that colonized beings matter.

In sum, *Relation* is both ontic, in that it is emplaced in the worldwide entanglements of relation, and ontological, in that it refers to the realization of being-of-the-world through subjects in their unpredictable singularity and chaotic connectedness. In this way, Glissant unifies human subjects (ontic and specific) that are embedded in structured relationships with creative, singular beings within the same philosophical scheme. Encumbered social beings escape into *Relation* when they take a poetic journey to totality, to *tout-monde*, above and beyond the relationships of specific, embedded subjects. These place-making, journey-taking subjects enact resistance and autopoiesis by means of the *opacité* of language, culture, and aesthetics. These philosophical categories recognize the structured relationships that necessarily envelop ontic beings and simultaneously posit language and culture as creative forms of resistance and change. Aesthetics as much as politics is understood as part of that struggle (Vété-Congolo 2015, 103–7, Drabinski 2012).

THE CHRONOTOPE OF THE HURRICANE: A POSTSCRIPT ON THE PRACTICAL AND METHODOLOGICAL IMPLICATIONS OF ARCHIPELAGIC THINKING

I wrote most of this chapter in San Juan, Puerto Rico, long before the landfalls of Hurricanes Irma on September 4, 2017, and María sixteen days later. The entire island endured a prolonged state of emergency for several months. The devastation was general, but it was not equally distributed. Even though the entire electrical grid collapsed, many large companies, banks, prosperous businesses, and middle- to upper-class homes soon had electricity, either because their sector was a priority or because they had previously invested in their own generators. Since my extended family lives in San Juan in sturdy, concrete homes, we did not suffer the same the catastrophic losses as thousands of families living in rural, mountainous regions, in flooded areas, or in self-constructed wooden houses with zinc or shingled roofs. Any analysis that does not address the social inequalities of the Caribbean cannot capture the fundamental contours of the current disaster and its enduring materiality. For this reason, and after many months of strictly material concerns, I now elucidate further the relevance of the abstract themes treated in this chapter and expand further the chronotopes of my archipelagic thinking.

Following Glissant, I have argued that archipelagic thinking, expressed abstractly as *Relation*, condenses the chronotopes of the abyss, the plantation, and the island. These chronotopes express synchronicities of space and time on three different levels or scales. In the abyss of the slave ship, time and space collapsed around subjects who worked with the fragments of their cultures in order to survive the harsh reality of their circumstances. The abyss opened onto the plantation, a place that was at once oppressively structured and yet open to the wide world of Europe, its colonies, and the consequent circulation of people, material goods, and ideas. The encumbered subjects of these closed places resisted by creating and sharing language, cuisine, and music—in short, culture. The chronotope of the island envisions an even wider horizon, one of the simultaneity of encumbered and autopoietic subjects engaged in worldwide entanglements of relation. As beings-of-the-world, these subjects experience both the repetitive, cyclical rhythms of localities as well as the synchronicity of the different temporalities of connected global places.

Archipelagic thinking works with and assembles fragments from the flotsam and jetsam of history. It produces counternarratives in order to engage the colonial contexts of continental thinking and modern philosophy. It seeks to rethink history from the Caribbean in order to rectify those pernicious inequities and injustices that persist even to the present. This is not a simple, linear emplotment but rather an expression of the polyrhythmic simultaneity of the worldwide synchronicity of places and temporalities. It not a teleology leading to a determinate end; rather it refers to the adventures of rooted errantry. Archipelagic thinking expresses the dialogics of structural relationships and creative *Relation*. It posits place-making and journey-taking subjects. The chronotopes of the abyss, the plantation, and the island all come into dialogic relation through the overarching, major chronotope of rooted errantry, also conceptualized as *Relation*, characterized by the synchronicities of time, of emplaced ontic beings encumbered by structured relationships, and of beings-of-the-world fashioned through autopoiesis.

The chronotope of the current catastrophe in Puerto Rico is not entirely different from those of the abyss and plantation, but I think it requires an expanded vocabulary. The archipelago has long since transitioned from the slave ship and the plantation to other modes of exploitation and dispossession through other means of accumulation of wealth, property, and power. The chronotopes of the abyss and plantation are historically adequate but do not quite resonate in our lived experiences with the same deep frequency today. Given the recurrent manifestation of hurricanes, their abiding threats, and their presence in the collective memory of the Caribbean, it now seems to be obligatory to advance the chronotope of the hurricane (Schwartz 2015). In chronotopic terms, the hurricane collapses space and time; it reduces everything to the immediate, to survival in the here and now. And yet, in its aftermath, the surviving subjects emerge intact, albeit thoroughly changed. Familiar places—dwellings and communities, beaches and landscapes and roads and bridges, businesses and workplaces—if they still exist, are no longer the same. Time—expressed in the rhythms of work, school, recreation, and even day and night—has been disrupted. The errantry provoked by the hurricane was so widespread that almost everyone experienced a change of place, either as those who had to move, whether permanently or temporarily, or as those who provided shelter for the displaced. Everyone seeks to pick up the pieces and return to normality. The paradox of the hurricane is its peculiar threshold: things can never be the same, and yet that is what is most commonly sought.

The most enduring aspect of normality in the Caribbean is its continuously restructured inequalities. In Puerto Rico, the current fiascos—before and after the hurricanes—are too many to enumerate. They have provoked the moral language of corruption, fraud, and illegality, but this is an analytical distraction. This is because Caribbean societies are modern, bureaucratic, legalistic, and very closely connected to their metropoles, which in turn created the very conditions for their mismanagement, which is just a euphemism for crass expropriation by local power holders and legalized accumulation by local and international corporations. The metropolitan centers that originally carved out and divided the Caribbean are still very much present and continue to oppress through a variety of political structures, some openly colonial and others nominally postcolonial. These governments, both metropolitan and local, delimit highly structured places that are not in any way disordered. Rather, they are overwhelmed by laws and legislators, by lawyers and loopholes that create the contexts and connections for global capital accumulation, whether commercial, financial, or productive. There is nothing marginal or backward about the Caribbean. The archipelago is composed of carefully delimited and highly structured islands of extreme inequality. The hurricane has amplified and laid bare the precariousness and vulnerability spread out across the islands at all levels, but to different degrees.

The hurricane reveals structured inequality, but it also provokes action. It urges us to think in terms of subjectivity and ethics. The most immediate solutions to the catastrophe fall within the limited spheres of extended families, neighborhoods, and community organizations and not within the larger context of the inherent injustices of the colonial state. Furthermore, the hurricane has increased the growing diaspora that has followed the paths forged previously by others and that traverse the chains of empire: Florida, Texas, New York, and a multitude of other destinations in the United States. This errantry is firmly rooted in those places of departure that abide in memory and those places of arrival where imagination projects an alternate future. Furthermore, the philanthropic work of various organizations is certainly laudable but at best ameliorative rather than transformative. In addition, the top-down strategies of local and federal governments have proven woefully inadequate. And so arises a second paradox: the hurricane clears a way toward a rethinking of our historical reality with an intent to forge a more just future for Puerto Rico; and yet the road to social justice as a collective project is neither clear nor certain. Archipelagic thinking must confront the paradoxes of surpassing the threshold of catastrophe, of seeking alternatives to the current structures, of forging a prophetic vision of the past. Today, archipelagic thinking must be grounded in the dialogic chronotope of hurricanes and all their violent materiality.

NOTES

I would like to thank my colleagues who have taught me about the Caribbean, geography, and the chronotope: Jorge Giovannetti, Juan Giusti, Carlos Guilbe, and Evelyn Dean. Christie Capetta provided copyediting in addition to Jennifer Kelland at Rowman & Littlefield. I wrote this chapter before the publication of John Drabinski's *Glissant and the Middle Passage: Philosophy, Beginning, Abyss* (Minneapolis: University of Minnesota Press, 2019). I have been influenced by his previous publications, as noted in the bibliography.

1. Glissant, just as Antonio Benítiz-Rojo, raises the (slave) plantation to an ahistorical synecdoche that stands for the whole Caribbean. However, the plantation was only one institution of the social formations in the Caribbean (Giusti-Cordero 2011).
2. The same distinction and capitalization are in the original French edition. In this quotation, the English edition sets off the two concepts with italics, while the French edition uses quotation marks (Glissant 1997a, 27; Glissant 1990, 39–40). Following the English edition, I italicize upper-case *Relation* throughout.
3. In French, the term is *l'être du monde* without hyphens; see Glissant 1990, 201. I have hyphenated the term here to clearly mark it as an ontological category.

REFERENCES

Agamben, Giorgio. 1998. *Homo sacer*. Translated by Daniel Heller-Roazen. Stanford, CA: Stanford University Press.

Bakhtin, Mikhail M. 1981. *The Dialogic Imagination: Four Essays*. Translated by Caryl Emerson and Michael Holquist. Austin: University of Texas Press.

Benítez-Rojo, Antonio. 1992. *The Repeating Island: The Caribbean and the Postmodern Perspective*. Durham, NC: Duke University Press.

Berman, Carolyn Vellenga. 2009. "The Known World in World Literature: Bakhtin, Glissant, and Edward P. Jones." *Novel* 42, no. 2: 231–38.

Brathwaite, Edward Kamau, and Nathaniel Mackey. 1999. *Conversations with Nathaniel Mackey: An Evening with Nate Mackey and Kamau Brathwaite*. Staten Island, NY: We Press.

Brubaker, Rogers, and Frederick Cooper. 2000. "Beyond 'Identity.'" *Theory and Society* 29, no. 1: 1–47.

Cailler, Bernadette. 2011. "Totality and Infinity, Alterity, and Relation: From Levinas to Glissant." *Journal of French and Francophone Philosophy* 19, no. 1: 135–51.

Dash, J. Michael. 1995. *Edouard Glissant*. Cambridge: Cambridge University Press.

Dastur, Francoise. 2017. *Questions of Phenomenology: Language, Alterity, Temporality, Finitude*. New York: Fordham University Press.

De Ferrari, Guillermina. 2012. "The Ship, the Plantation, and the Polis: Reading Gilroy and Glissant as Moral Philosophy." *Comparative Literature Studies* 49, no. 2: 186–209.

Deleuze, Gilles, and Félix Guattari. 2008. *A Thousand Plateaus: Capitalism and Schizophrenia*. Translated by Brian Massumi. London: Continuum.

DeLoughrey, Elizabeth. 2009. *Routes and Roots: Navigating Caribbean and Pacific Island Literatures*. Honolulu: University of Hawai'i Press.

Drabinski, John. 2012. "Aesthetics and the Abyss: Between Césaire and Lamming." *CLR James Journal* 18, no. 1: 126–52.

Drabinski, John E., and Marisa Parham, eds. 2015. *Theorizing Glissant: Sites and Citations*. Lanham, MD: Rowman & Littlefield.

Figueroa, Victor. 2015. *Prophetic Visions of the Past: Pan-Caribbean Representations of the Haitian Revolution*. Columbus: Ohio State University Press.

Gilroy, Paul. 1993. *The Black Atlantic: Modernity and Double Consciousness*. London: Verso.

Giusti-Cordero, Juan A. 2011. "But Where Are 'The People'? Unfinished Agendas in *The People of Puerto Rico*." *Identities* 18, no. 3: 203–17.

Glissant, Édouard. 1981. *Monsieur Toussaint*. Translated by Joseph Foster and Barbara Franklin. 1st Eng. ed. Washington, DC: Three Continents Press.

———. 1989. *Caribbean Discourse: Selected Essays*. Translated by J. Michael Dash. Charlottesville: University Press of Virginia.

———. 1990. *Poétique de la relation*. Paris: Gallimard.

———. 1997a. *Poetics of Relation*. Translated by Betsy Wing. Ann Arbor: University of Michigan Press.

———. 1997b. *Traité du tout-monde*. Paris: Gallimard.

Gordon, Lewis R. 2005. "From the President of the Caribbean Philosophical Association." *Caribbean Studies* 33, no. 2: xv–xxii. doi: 10.2307/25613483.

Grosfoguel, Ramón. 2017. "Las migraciones coloniales del Caribe a Estados Unidos y Europa Occidental: Colonialidades diferenciadas en cuatro centros del sistema-mundo." *Kamchatka* 9 (July): 225–50. doi: 10.7203/KAM. 9.9550.

Hallward, Peter. 1998. "Édouard Glissant between the Singular and the Specific." *Yale Journal of Criticism* 11, no. 2: 441–64.

———. 2001. *Absolutely Postcolonial: Writing between the Singular and the Specific*. Manchester, UK: Manchester University Press.

Hantel, Max. 2012. "Errant Notes on a Caribbean Rhizome." *Rhizomes: Cultural Studies in Emerging Knowledge* 24: 1–39.

———. 2013. "Rhizomes and the Space of Translation: On Édouard Glissant's Spiral Retelling." *Small Axe* 42: 100–112. doi: 10.1215/07990537-2378955.

Hart, David W. 2004. "Caribbean Chronotopes: From Exile to Agency." *Anthurium: A Caribbean Studies Journal* 2, no. 2: 1–22.

Headley, Clevis. 2012. "Glissant's Existential Ontology of Difference." *CLR James Journal* 18, no. 1: 59–101. doi: 10.5840/clrjames20121816.

Holquist, Michael. 2010. "The Fugue of the Chronotope." In *Bakhtin's Theory of the Literary Chronotope: Reflections, Applications, Perspectives*, edited by Nele Bemong, Pieter Borghart, Michel De Dobbeleer, Kristoffel Demoen, Koen De Temmerman, and Bart Keunen, 19–33. Gent, Belgium: Academia Press.

Kaplan, Nicole. 2007. "Edouard Glissant. 2005. Monsieur Toussaint: A Play." *Caribbean Studies* 35, no. 2: 205–9.

Maldonado-Torres, Nelson. 2004. "The Topology of Being and the Geopolitics of Knowledge: Modernity, Empire, Coloniality." *City* 8, no. 1: 29–56.

———. 2007. "On the Coloniality of Being." *Cultural Studies* 21, nos. 2/3: 240–70. doi: 10.1080/09502380601162548.

Malpas, Jeff. 2012. *Heidegger and the Thinking of Place: Explorations in the Topology of Being*. Cambridge, MA: MIT Press.

Mensch, James R. 2015. *Levinas's Existential Analytic: A Commentary on Totality and Infinity*. Evanston, IL: Northwestern University Press.

Murdoch, Adlai. 2012. "Glissant's Opacité and the De-nationalization of Identity." *CLR James Journal* 18, no. 1: 14–33.

Murdoch, H Adlai. 2013. "Édouard Glissant's Creolized World Vision: From Resistance and Relation to Opacité." *Callaloo* 36, no. 4: 875–90.

Oakley, Seanna Sumalee. 2008. "Commonplaces: The Rhetoric of Difference in Heidegger and Glissant." *Philosophy and Rhetoric* 41, no. 1: 1–21. doi: 10.1353/par.2008.0001.

———. 2012. "InCitation to the Chance: Glissant, Citation, Intention, and Interpretation." *CLR James Journal* 18, no. 1: 34–58.

Peeren, Esther. 2006. "Through the Lens of the Chronotope: Suggestions for a Spatiotemporal Perspective on Diaspora." *Thamyris/Intersecting: Place, Sex and Race* 13, no. 1: 67–77.

Schalow, Frank, and Alfred Denker. 2010. *Historical Dictionary of Heidegger's Philosophy*. Lanham, MD: Scarecrow Press.

Schwartz, Stuart B. 2015. *Sea of Storms: A History of Hurricanes in the Greater Caribbean from Columbus to Katrina*. Princeton, NJ: Princeton University Press.

Somers, Margaret R. 1994. "The Narrative Constitution of Identity: A Relational and Network Approach." *Theory and Society* 23, no. 5: 605–49.

Steinby, Liisa. 2013. "Bakhtin's Concept of the Chronotope: The Viewpoint of an Acting Subject." In *Bakhtin and His Others: (Inter)subjectivity, Chronotope, Dialogism*, edited by Liisa Steinby and Tintti Klapuri, 105–25. London: Anthem Press.

Thompson, Lanny. 2017. "Heuristic Geographies: Territories and Areas, Islands and Archipelagoes." In *Archipelagic American Studies*, edited by Brian Russell Roberts and Michelle Ann Stephens, 57–73. Durham, NC: Duke University Press.

Tlostanova, Madina Vladimirovna. 2013. "Transcultural Tricksters beyond Times and Spaces: Decolonial Chronotopes and Border Selves." *Language. Philology. Culture.* 2–3: 9–31.

Vété-Congolo, Hanétha. 2015. "The Ripening's Epic Realism and the Martinican Tragic Unfulfilled Political Emancipation." In *Theorizing Glissant: Sites and Citations*, edited by John Drabinski and Marisa Parham, 103–27. London: Rowman & Littlefield.

Wilson, Norman J. 1999. *History in Crisis? Recent Directions in Historiography*. Upper Saddle River, NJ: Prentice Hall.

Wolf, Eric, and Sidney Mintz. 1957. "Haciendas and Plantations in Middle America and the Antilles." *Social and Economic Studies* 6, no. 3: 380–412.

Wynter, Sylvia. 1989. "Beyond the Word of Man: Glissant and the New Discourse of the Antilles." *World Literature Today* 63, no. 4: 637–48.

Storm Tracking, 2016

Craig Santos Perez

For Fiji, after Cyclone Winston, the most powerful storm ever recorded in the Southern Hemisphere

This is when the warm ocean gives birth to a cyclone—
This is when we give a human name to a thing we can't control—
This is when wind triggers warning system & rain echoes violent refrain,
This is when evacuation is another word for shelter & home is another word for debris
and broken sugar cane—
This is when we hold our children close and whisper: *don't be scared, we won't let go*—
This is when flood waters whisper: *let go*—
This is when survivors become drone footage & social media—
This is when disaster trauma attracts tourism—
This is when the world finally sees us: only after the eye of a storm sees us—
This is when we chant: *we will overcome*
This is when we fundraise and benefit event,
when canned goods and bottled water become gift and debt—
This is when corporations promise *green development* and politicians siphon aid and investment—
This is when disaster justifies another military coup—
This is when we migrate with or without dignity,
and home becomes another word for remittance,
This is when we wish we could afford to give more—
This is when looking for bodies becomes prayer,
when counting bodies becomes prayer,
when counting donations becomes prayer,
when counting days until the next storm becomes prayer
This is when our sea of vulnerable islands
becomes an archipelago of prayer—

Part II

BEYOND THE SEA AS METAPHOR

Comparative Maritime Epistemologies

Chanting the Waters

Craig Santos Perez

In solidarity with the Standing Rock Sioux tribe and all peoples protecting the sacred waters of this Earth

water is life becuz our bodies are 60 percent water
becuz my wife labored for 24 hours through contracting waves
becuz our sweat is mostly water & salt
becuz she breathed & every other breath is birthed from the ocean
water is life becuz our lungs are 80 percent water
becuz water broke forth from her body
becuz amniotic fluid is 90 percent water
becuz our daughter crowned like a new island
water is life becuz our blue planet is 70 percent water
becuz some say water came from asteroids & comets
becuz some say the ocean formed within the earth from the beginning
becuz water broke forth from shifting tectonic plates
water is life becuz the ocean is 99% of the biosphere
becuz we say our gods created water
becuz no human has found a way to safely create water
water is life because we can't drink oil
becuz water is the next oil
becuz 185,000 miles of U.S. oil pipelines leak everyday
becuz we wage war over gods & water & oil
water is life becuz only 3 percent of global water is freshwater
becuz the water footprint of an average american is 2000 gallons a day
becuz it takes 660 gallons of water to make one hamburger
becuz more than a billion people lack access to clean drinking water
becuz in some countries women & children walk 4 miles every day to gather clean
water
& carry it home

becuz we can't desalinate the entire ocean
water is life becuz if you lose 5 percent of your body's water you will become feverish
becuz if you lose 10 percent of your body's water you will become immobile
becuz we can survive a month without food but less than a week without water
water is life becuz we proclaim water a human right
becuz we grant bodies of water rights to personhood
becuz some countries signed the UN Convention on the Law of the Sea
becuz my wife says the Hawaiian word for wealth, waiwai, comes from their word for
  water, wai
water is life becuz corporations steal, privatize, dam, & bottle our waters
becuz sugar, pineapple, corn, soy, & gmo plantations divert our waters
becuz concentrated animal feeding operations consume our waters
becuz pesticides, chemicals, oil, weapons, & waste poison our waters
water is life becuz we say stop, you are hurting our ancestors
becuz they say we thought this was a wasteland
becuz we say stop, keep the oil in the ground
becuz they say we thought these bones were fuel
becuz we say stop, water is sacred
becuz they say we thought water is a commodity
becuz we say we are not leaving
becuz they say we thought you were vanishing
becuz we are water warriors & peaceful protectors
becuz they call us savage & primitive & riot
becuz we bring our feathers & lei & sage & shells & canoes
  & drones & hashtags & totems
becuz they bring their bulldozers & drills & permits & surveillance drones & helicopters
becuz we bring our treaties & the UN Declaration on the Rights of Indigenous Peoples
becuz they bring their banks & politicians & police & private militia
  & national guard & lawyers
becuz we bring our songs & schools & lawyers & prayers & chants & ceremonies
becuz they bring their barking dogs & paychecks & pepper spray & rubber bullets
becuz we bring all our relations & all our generations & all our livestreams
water is life becuz our drumming sounds like rain after drought echoing
  against our taut skin
water is life becuz our blood is 90 percent water
becuz every minute a child dies from water-borne diseases
becuz every day thousands of children die from water-borne diseases
becuz every year millions of children die from water-borne diseases
water is life becuz my daughter loves playing in the ocean
becuz someday she will ask us, "where does the ocean end?"
becuz we will point to the dilating horizon
water is life becuz our eyes are 95 percent water
becuz we will tell her that the ocean has no end
becuz we will tell her that the sky & clouds carry the ocean

becuz we will tell her that the mountains embrace the ocean into a blessing of rain
becuz we will tell her that the ocean-sky-rain fills aquifers & lakes
becuz we will tell her that the ocean-sky-rain-lake flows into the Missouri River
becuz we will tell her that ocean-sky-rain-lake-river returns to the ocean & connects us
   to our cousins at Standing Rock
becuz we will tell her about the sacred stone of a mother holding her child
becuz we will tell her that the Sioux are still there & still breathing
water is life becuz our hearts are 75 percent water
becuz, while my daughter is sleeping, i will chant to her, my people's word for water:
   "hanom, hanom, hanom," so her dreams of water carry us home
becuz water is life, water is life, water is life

Chapter Six

An Early Medieval "Sea of Islands"

Area Studies, Medieval Studies, and Traditions of Wayfinding

Jeremy DeAngelo

ARCHIPELAGIC STUDIES AND THE AFFINITY OF PREMODERN LITERATURE

The study of an archipelago and its cultures is often, from a certain perspective, a focus on surfaces. It is usually interested not so much in the whole of the islands themselves but rather in those portions that emerge from the ocean and the flora, fauna, and human societies they sustain. Moreover, when we discuss the connections between these islands, the intervening sea is almost always reduced to two dimensions, with virtually no attention given to the depths beneath.[1] You will see, in multiple works of geography, ethnography, and cultural commentary included in this chapter, the same assumption made. Yet—as I imagine the geologist may lament—an island is only made possible by the vast bulk of its mass: the rock carved by the currents, pushed upward by tectonic processes, or stacked layer by layer through volcanic activity, all of which lies below the water line. So, too, do archipelagic literary studies typically skim across the temporal surface—in most cases, they have been about contemporary societies, from the colonial or postcolonial periods, taking full advantage of their ability to range across new places and peoples but declining to plumb the depths of history. Yet there is little reason this should be. After all, these islands have always been there, at least in the span of human time; as compelling and urgent as the present era may be, there is more there, under the waves.

The reason that this long history is not so readily seen is connected to the types of literature and culture that are associated with the archipelago and how these assumptions impact their study. The issue, familiar to those who study literature produced outside North America or Europe, is the old postcolonial problem of the center versus the periphery. And while the study of English literature can move laterally to other corners of the globe and vertically to other periods, both of these lines remain as the axes of a graph, oriented toward the vertex of modern British/American literature rather than casting their ideas and queries in multiple fruitful directions. Yet it is precisely because of their similar situations in both world geography and the modern academy that those who study both contemporary and premodern archipelagic literatures are well positioned to help each other. Through the use of a test case demonstrating how Pacific Islander culture can help shed light on a crux in Old English studies, I hope to indicate ways in which both of these disparate island literatures can be brought into conversation with one another.

One need not delve deeply into discussions on the place of Caribbean, Pacific, Indian, or Australasian literatures to find reservations expressed about their place in the larger field of English studies. A particular concern is that the study of these texts reproduces the dynamics of colonial and postcolonial relationships by treating them as secondary to the literature of North America or Europe or by using academic study as another way to commodify or domesticate the products of non-Western societies.[2] They are often segregated from other subjects in categories such as "world literature" or "area studies," becoming, as Vicente L. Rafael puts it, "managed sites of higher education . . . producing a field of exclusions. For within the interdisciplinary optic of the liberal notion of area studies, the area and presumably its populations remained at a safe remove, managed by the operations of the social sciences into stages of comparable development, cultural groupings, or discrete ethnolinguistic realms" (1994, 97). The result is a pervasive feeling that influence and interest do not flow both ways. "Value proceed[s] *'from'* the old terrain of Anglo-America and *'to'* the rest of the English-speaking world," Rita Raley argues, saying elsewhere that this "highlights the extent to which these presumably new disciplinary alignments ultimately legitimate the center" (1999, 66, 68–69).[3] Enforcing that legitimacy is the presumption of influence radiating solely from that center. The literature of the global South is so infrequently permitted to speak universally rather than locally or deemed capable of providing insights into unrelated American or British texts. These assumptions in turn impact debates over the prerequisites for majors, the allocation of resources, and what specialties are necessary to create well-rounded scholars, issues as sensitive as they are important.

What is striking is how closely such anxieties parallel the concerns of premodern scholars. The past, after all, is a foreign country, and some of the same demands in engaging ancient or medieval literature correspond to the learning necessary to come to grips with another, contemporary culture. As a result, the fields have developed similar contours. Rafael, for example, compares area studies to Classics with "their stress on linguistic competence as the key to scholarly research." The same description can be used for medieval studies, additionally as when he diagnoses his own field with "a resistance to theory" (Rafael 1994, 93, 101). And like those studying world literatures, medievalists (and classicists, I imagine, but I will restrict my comments to my own discipline) feel relegated and encroached upon by the scholarly center. Considerations such as the perception of the field by the wider academy, the isolation of many specialists at their institutions, and the dismissal of premodern periods as inapplicable to certain theories or methodologies all speak to the peripheral status of medieval studies.[4] The overall result is two disciplines that are consistently underestimated. Outside perceptions of both medieval studies and world literature often presume that their topics are unrepresentative of the Western or the modern. One is too old and the other too new. The Middle Ages are thought to be too alien, too far from our current age to be able to speak effectively to present-day concerns.[5] The global South, in contrast, is rife with "new" concepts—colonialism, diversity, multilingualism—that are seen as parvenu and subsidiary to the enduring interests of the canon.

Of course, anyone who studies either topic would know this to be false. And in addition to the work we are already doing, proof in this regard can also be furnished through demonstrating the applicability of modern world literature to medieval studies and vice versa, by using one to deepen the knowledge of another. In bypassing the disciplinary center, direct conversation would allow the vast cultural breadth of literary study to engage with its profound chronological depth. And archipelagic and island studies are particularly useful vehicles through which to foster this collaboration. As I mentioned to start, the theorization of islands as spaces with impact upon the form and expression of a culture has been largely restricted to modern postcolonial literatures and societies. Their application here is obvious; yet a focus on them exclusively is too narrow a view. As historians such as J. G. A. Pocock have reminded us, Britain itself, and the larger North Sea with it, is an archipelago, as immersed in the sea and active in it as any other maritime culture (2005, 16). How scholars have attempted to define or delineate that archipelago varies. Some imagine an entire North Atlantic Rim, with a shared historical and geographical experience stretching in an arc from Canada's eastern seaboard to the Maghreb, while others restrict it to Great Britain's political boundaries.[6] For the Old English period (c. 450–1066 CE), it makes particular sense for the archipelago to include Ireland along with Britain, as well as Scandinavia and all of the northern islands settled by the Norse in this era. This encompasses all of the places from which the settlers of Britain in this period arrived; for this reason, too, we should also include the northern shore of Europe from Brittany to Denmark, for this is where the English themselves (and, later, the Normans) originated before coming to Britain and the region to which the Celtic Britons immigrated in the wake of the Saxon migrations. It was also the base of operations for the Frisian people, who dominated the trade routes in the North Sea prior to the advent of the Vikings. As one can tell from this attempt at definition, the medieval history of this region, like that of virtually any other archipelago, is that of people moving across the sea to settle or conquer and of cultural and trade networks connecting parcels of land from across the ocean. The sea is ever present in its earliest literature, and the best-known text from this period, *Beowulf*, takes place in an archipelago and dramatizes its hero's movement among and engagement with the cultures present within it. So many others, representing the entire range of the Middle Ages in Britain—*Wulf and Eadwacer*, *The Battle of Maldon*, *Brut*, *Branwen ferch Llŷr*, *Havelok the Dane*, and *The Man of Law's Tale*, to name a few—feature the archipelago and the interactions within it as the backdrop for their stories.

THE EARLY ENGLISH SEA OF ISLANDS

It should not surprise, then, that comparative work on island literatures brings important insights to medieval literature (and one should expect the reverse to be true as well, though I will restrict myself to my own field here). For Old English literature, examination through the prism of archipelago studies helps to combat a common implication in much of the scholarship that the early English were particularly unengaged with the world outside their kingdom(s) or even outside their local communities. The presumption that the English, specifically, held a particular contempt for or fear of the world outside their own communities is quite widespread.[7] This is in part due to the prominence of certain works in the corpus, such as the elegies *The Seafarer* and *The Wanderer*, which depict sea travel as uncomfortable and unpleasant. Yet these examples do not exhaust the examples of seafaring in Old English literature; in fact, ocean travel is everywhere, depicted in a great variety of ways. This speaks to the English peoples' experience with it. After all, they owed their existence as a culture to their ancestors' decision to sail across the North Sea from the Continent and settle Britain; this oceangoing identity is expressed in their literature again and again (Howe 1989, 8–32). Moreover, once settled, the English did not cease to travel throughout their archipelago. To one Irish writer, they were summed up best as the "Saxon snámhach"[8] ("seafaring Saxons"), an epithet that reflected actual practice (Meyer 1897, 112–13; *Genealogical Tracts I* 1932, 24). The English were extensively engaged with the Continent as well, with the church in Rome and the Carolingian court in France.[9] Trade networks also forged connections. For example, the variety of artifacts found in monuments such as the Sutton Hoo burial (which was enclosed in a ship) indicates a commercial reach that stretched from the Mediterranean to the Baltic (Carver 1990, 117–19; Lebecq 2007, 170–79; Carver 2010, 1–24).

The early English, therefore, demonstrated extensive engagement with the sea, yet hold the opposite reputation. It is a paradox that one can find applied to Pacific Islanders in the modern day as well. Epeli Hau'ofa targeted this same misapprehension in his essay "Our Sea of Islands." Western observers of Oceania, he writes, conceived of its islands as insular and the cultures upon them, consequently, as cut off from opportunity and innovation. This is an attitude the islanders, and even Hau'ofa himself, had internalized. Yet Hau'ofa urges Pacific Islanders, instead of imagining themselves as living upon "islands in the sea," to perceive their world as "a sea of islands." The distinction between the two is dependent upon whether one imagines the ocean as preventing or facilitating connections with the outside world. Those who live lives dominated by land imagine the sea to be an obstacle. But the people of Oceania know better:

> The world of our ancestors was a large sea full of places to explore, to make their homes in, to breed generations of seafarers like themselves. People raised in this environment were at home with the sea. They played in it as soon as they could walk steadily, they worked in it, they fought on it. They developed great skills for navigating their waters, and the spirit to traverse even the few gaps that separated their island groups.
>
> Theirs was a large world in which peoples and cultures moved and mingled unhindered by boundaries of the kind erected much later by imperial powers. From one island to another they sailed to trade and marry, thereby expanding social networks for greater flow of wealth. They travelled to visit relatives in a wide variety of natural and cultural surroundings, to quench their thirst for adventure, and even to fight and dominate. (Hau'ofa 1993, 7–8)[10]

Although it is separated from the postcolonial context in which Hau'ofa offers his remarks, I would suggest that the North Atlantic of the early English would benefit too from being thought of as "a sea of islands," an archipelago whose geography expanded rather than restricted opportunity.[11] What colonists thought of the Pacific Islanders closely parallels scholars' views of the medieval English, and just as these assumptions cut the islanders off from imagining certain narratives for themselves, so too does scholarly misapprehension prevent us from coming to grips with some of the capabilities of early medieval peoples.

COLLABORATION BETWEEN MODERN AND MEDIEVAL ISLAND STUDIES IN PRACTICE

Yet, in some ways, this work in reconceptualization is only the first step. Once we reorient our views to recognize the similarities between the modern and the medieval—both island literatures, despite their numerous differences in time, culture, and history—it is possible to forge productive lines of inquiry between the two. There is, in fact, much in the practical conduct of Pacific people upon the sea that can inform Old English studies. This is especially the case for skills and expertise that survive in the Pacific that must have existed, in some form, in early medieval England.

Sailing in the North Sea in the Middle Ages was not and could not be conducted the same way as it was in the Mediterranean, despite the two regions' proximity. In the premodern Mediterranean, sea travel was conducted within sight of the coast; this was not possible in the North Sea, where sandbars and other hazards made such a practice far too dangerous. Instead, sailors had to navigate via dead reckoning—by using markers available to them out on the open ocean, far from land (Wooding 1996, 7, 16–18). This, too, is the reality of sailing in the Pacific, where the ratio of water to land is of course immense, which means that navigators had to develop a similar expertise as North Sea sailors. The details of early English sailing are largely lost; however, by placing what we do know alongside our vastly more detailed understanding of the traditional forms of seafaring in the Pacific, it is possible to gain greater appreciation for how North Sea sailors may have navigated their own watery environment.[12]

In the Caroline Islands, the system developed for navigation is known as *etak*. An *etak* is a leg of a journey, determined through using the location of an island—known as an *etak* island—as a reference point. The island need not be in sight; rather its relative location is known. The Caroline Islanders, in describing their journeying, visualize their boats as stationary and the ocean and landmarks as moving around them. As one anthropologist has noted, they perceive the world moving past them just as we do when riding in a car or on a train (Gladwin 1970, 183). And just as with the illusion of the world passing by our window, objects closer to the vehicle appear to be moving by more quickly than those further away. The Caroline Islanders use this sensation to divide their journeys into any necessary number of *etaks*. With the location of the reference island known, they observe its movement beneath consecutive constellations in order to determine the *etaks* and their length. So, in Figure 6.1, the journey between *a* and *b* is divided into two *etaks* based on the movement of *etak* island *c* from beneath star sign *x* to beneath star sign *z* over the course of the journey. Its movement beneath star sign *y* is the indication that the first *etak* is over and the second has begun. For a more complex example of the same process, see Figure 6.2.

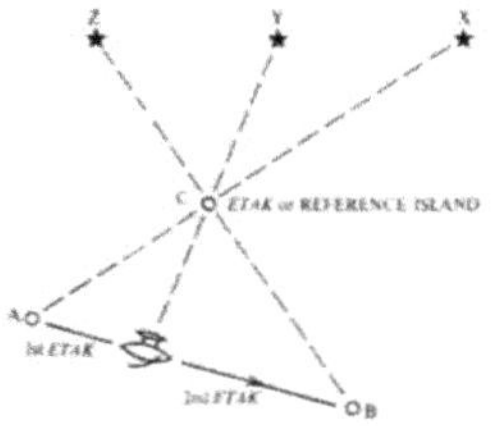

Figure 6.1. A Visual Representation of the Etak System. (Lewis 1994, 133).

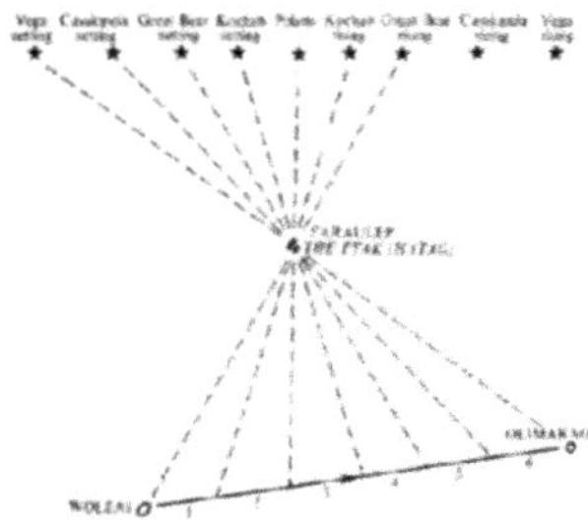

Figure 6.2. A More Elaborate Example. (Lewis 1994, 136).

The *etak* system of navigation, of course, could not have impacted practice in the medieval North Sea; nor could the indigenous sailors of that region have developed the same procedure for dead reckoning. However, there is evidence of a broadly similar process at work, contained in a text known as *The Old English Orosius* (Bately 1980). This is a version of the polemical world history *Historiarum adversam paganos libri septem* (*A History against the Pagans in Seven Books*) by the fifth-century cleric Paulus Orosius. In the late ninth century, King Alfred had it translated into Old English. One of the work's values to English readers was its first book, which was a detailed geography of the known world. Of course, the advance of four centuries and its new northern audience had made the geography both outdated and incomplete in some respects. For this reason, the translation was supplemented with further material, including the testimony of a Norwegian named Ohthere. A retainer of King Alfred and a chieftain who "ealra Norðmonna norþmest bude" ("settled the furthest north of all Norwegians") in *Halgoland* in northern Norway (Bately 1980, 13), he had taken exploratory journeys up the coast in order to see if anyone lived further north than him. He eventually sailed further than whale hunters were known to go and rounded Finnmark and the Kola Peninsula. He explains the denizens and economy of the district, as well as the region's location with respect to other landmarks, and briefly discusses travel toward *Scringes heal*, a trading town in southern Norway.

Ohthere's directions south are what concern us here, as they have proven to be a difficult crux in the text. According to him,

& ealle ða hwile he sceal seglian be lande; & on þæt steorbord him bið ærest Iraland, & þonne ða igland þe synd betux Iralande & þissum lande; þonne is þis

> land oð he cymð to Scirincgesheale, & ealne weg on þæt bæcbord Norðweg. (Bately 1980, 16)
>
> He would sail by the coast the whole way. To starboard is first of all *Iraland* and then those islands which are between *Iraland* and this land, and then this land until he comes to *Scringes heal*, and Norway is on the port side the whole way.

In the most common explanation of this account, Ohthere is first saying that Ireland, then some unnamed islands, and then Britain ("this land") will be on a sailor's right-hand side as he sails south from *Halgoland* to *Scringes heal*. The problem with this description is that, when one looks west from the Norwegian coast, Ireland is always obscured by Britain and therefore useless as a navigational aid (Figure 6.3).

This dilemma has led experts to suggest a number of alternative identities for *Iraland*. William C. Stokoe Jr. sums up the confusion nicely:

> *Iraland* has been glossed by various interpreters of the passage as Ireland, Iceland, Scotland, "Ireland-cum-Scotland," and even as Gotland and Jutland. Moreover, with the exception of Greenland, every island in the North and Irish

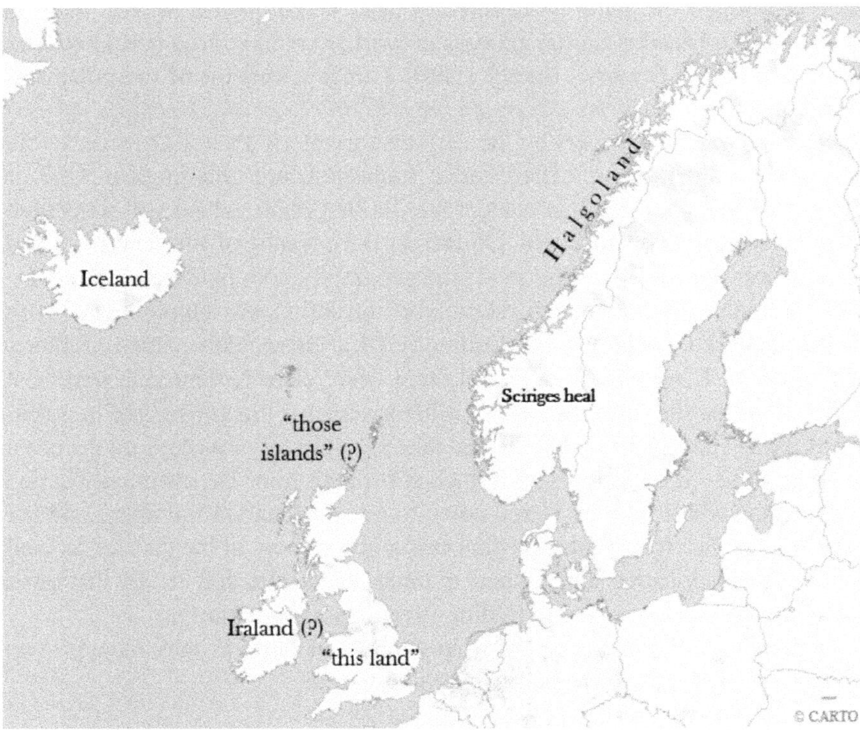

Figure 6.3. The North Sea, according to Ohthere, with the most widely accepted designations for his terminology.

> Seas, as well as some in the Skagerrak and the Kattegat, has at some time been named to render "þā īgland þe sind betux Iraland and þissum lande." Glosses for "this land" are fewer, but still Britain, England, Wessex, and Norway have been offered. (1957, 299)

Iceland was considered due to its position relative to Norway and the tendency in Old Norse for *s* to be replaced by *r*.[13] Scotland and "Ireland-cum-Scotland" also possess largely open water between themselves and Norway, and at the time there was a tendency to confuse the people of both places (Stokoe 1957, 299–301). Yet Iceland had only just been discovered in 874, and given that the *Orosius* refers only to the mythical Arctic island of Thule anywhere else, it is unlikely that the English had yet heard of the island's actual existence or of the Norse coinage *Ísland* by this point. What's more, from *Halgoland*, Ohthere describes nothing residing west of him but "þa widsæ on ðæt bæcbord" ("open sea to port") as he faces north, suggesting that he knew nothing of Iceland at the time either (Bately 1980, 16; Malone 1930, 142–43). As for Scotland, it has the obvious disadvantage of being a part of Britain (i.e., "this land") and is therefore not a separate landmass at all, which makes it unclear what the islands between *Iraland* and *þissum land* are supposed to be (Stokoe 1957, 301).

In attempting to untangle this knot, scholars have made two observations that shed light on the matter. The first was Kemp Malone's discovery that the geographical section of the *Old English Orosius* overlaid two different systems of orientation. One is that of Orosius, which is like our own today, where cardinal north points to true north—toward the North Pole. The second system, however, is a local one, which is skewed forty-five degrees, so that the coast of Norway—the "North Way"—points directly north. In this system, what the people of the North Sea understood as "north" is in fact our northeast, "northeast" is in fact our east, and so on. In the shifted system, moreover, looking "south" from *Halgoland* does in fact grant one an unobstructed view of the Mayo County coast of Ireland, with the "islands between" being the Hebrides (Malone 1930, 142–43) (Figure 6.4).

The other contribution is that of Stokoe, who seeks to understand Ohthere's directions from the point of view of a medieval sailor. After all, Ohthere does not use directional terms in his account, but rather *steorbord* ("starboard") and *bæcbord* ("port"). Stokoe points out that, unlike the cardinal directions, the designations starboard and port have "a moving, not a fixed, point of reference" that does not delineate any one direction toward something but rather "sweeps in an arc of one hundred and eighty degrees," allowing for more flexibility in his directions (1957, 304). Stokoe goes on to suggest that Ohthere is speaking not of the actual landmasses when he writes of *Iraland*, Britain, or the islands between but rather of the sea routes to these places. He

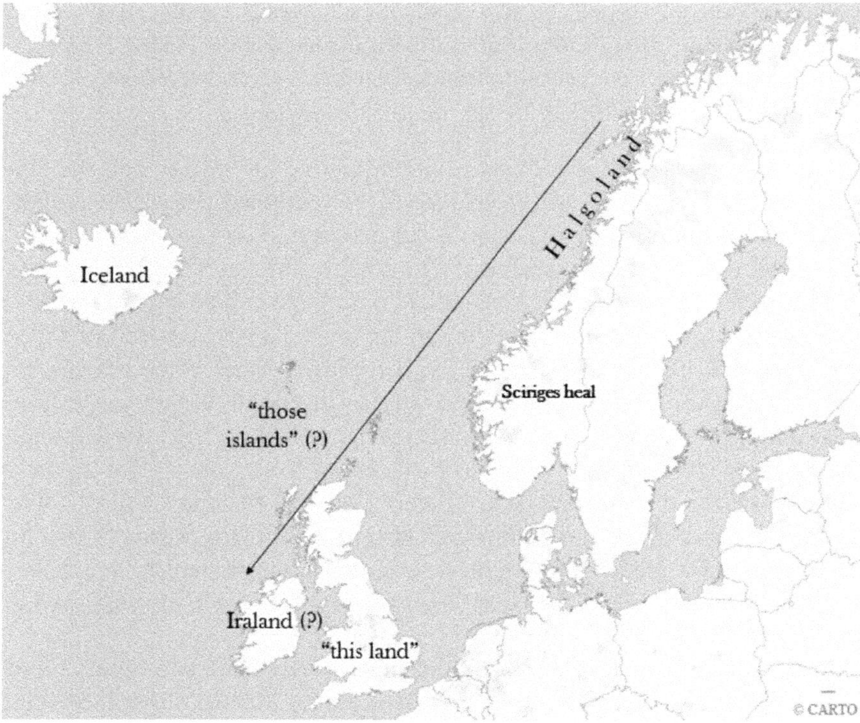

Figure 6.4. Malone's Interpretation

points out how hazardous a journey it would be to sail directly to Ireland as the gull would fly, this route being strewn with "skerries (ship-killing rocks), tide-roosts, whirlpools, and confused winds in some of the most storm-infested and fog-bound waters on the seven seas" (302). The route from Norway to Ireland, therefore, was less direct and more of a "dog-leg," where a sailor tacked west (starboard) from Norway and then veered sharply south after several days (Figure 6.5).

Ireland, then, was "first on the starboard of Ohthere's sailor in one sense: at this early stage in his voyage, had he been bound for Ireland, he would have changed course to starboard" (304). Similarly, Stokoe proposes the Shetlands and the Orkneys as the islands in between, not because they are physically situated between Ireland and Britain but because they were common stopping-off places for northern sailors traveling from one to the other, making them "in between" along the route (305).

Both of these insights have the benefit of seeking to conform scholarship to local practice. However, note how these proposed methods of navigation parallel the *etak* system. All of these navigational strategies involve not

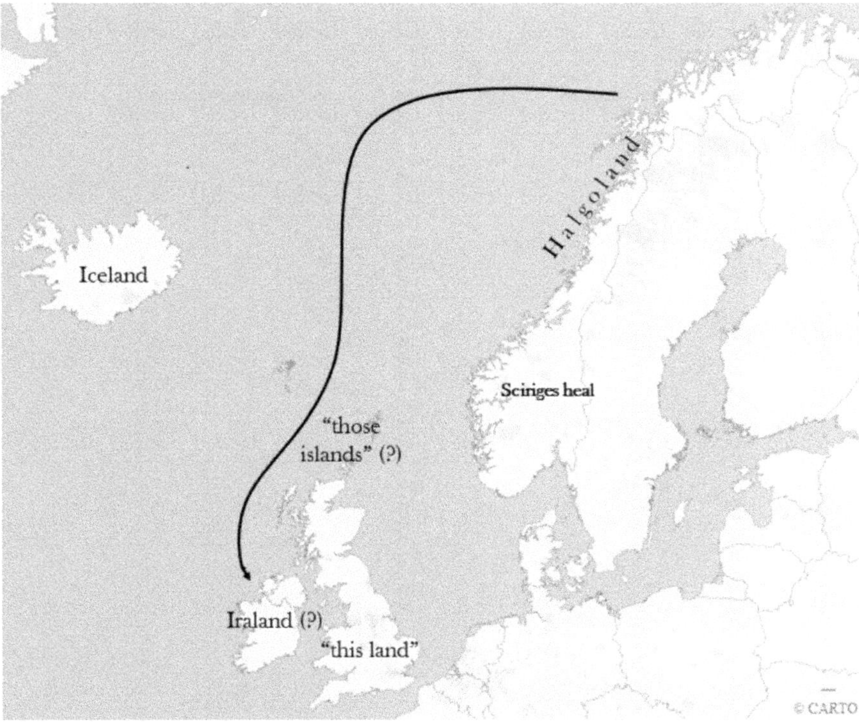

Figure 6.5. Stokoe's Interpretation

landmarks that can be seen from the position of the boat but rather those miles out of sight. Stokoe's reading, especially, also treats the perspective from the ship as the dominating one, just as with the *etak* system: it is the world that moves around the stationary ship, not vice versa. Whether it be a route or an island, what is used for orientation is an understood point rather than a discernable one. Moreover, no matter what the interpretation, given Ohthere's directions, navigation depends upon dividing the journey into legs—*etaks*—determined through triangulation with these distant fixed points. In fact, a similar map can be created using Ohthere's directions, which is quite reminiscent of the *etak* system. The main difference is that the fixed point, the *etak* island, *Iraland* in this case, is the furthest point from the navigator, being essentially the vertex of the angle formed by connecting the points (Figure 6.6). In the Pacific, the *etak* island is between two shifting points, serving as the fulcrum between the position of the boat and the series of constellations that the island moves underneath.

It is also important to note how this navigational system used by Ohthere is apparently mutually intelligible to him, the Norwegian sailor, and his audience,

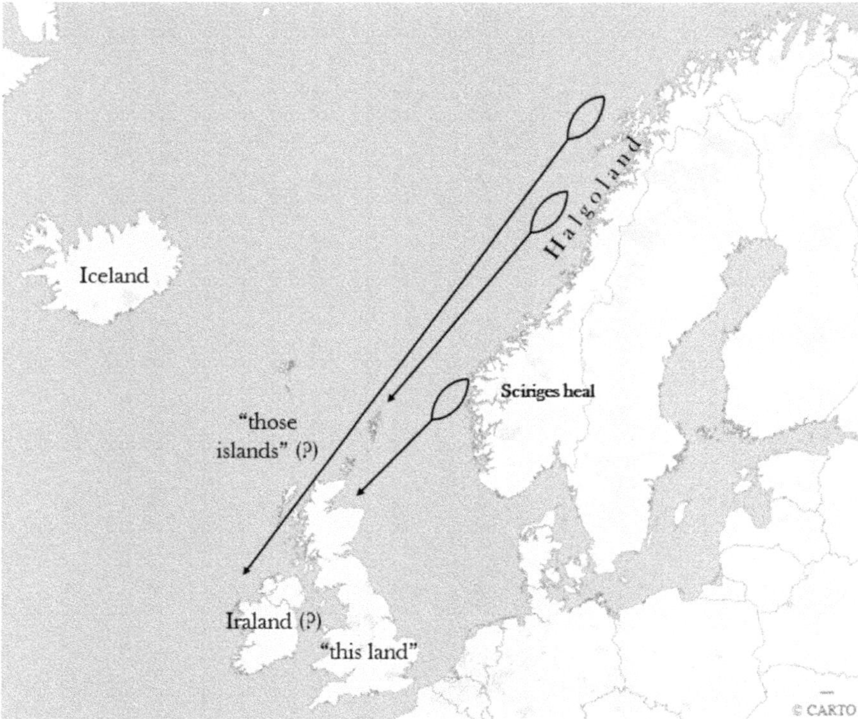

Figure 6.6. Ohthere's directions understood as an "etak" system.

Alfred's court. Elsewhere in his account, he is careful to explain aspects of his northern existence that would not be understood by an English audience, such as walruses and the goods that can be made from their hides and the techniques used in reindeer herding. That he feels no need to elaborate on his directions means that he is working within a system deployed by multiple cultures within the North Atlantic rather than simply the Norse. He is speaking with a technical terminology and an understanding borne out of generations of experience with the sea, just like the practices of Pacific Islanders. And the result is a broadly analogous navigational system, one which, in the medieval case, would not be possible to fully recreate without recourse to the example of the Pacific Islanders.

LITERARY EXPLORATION

It may seem counterintuitive to suggest a relationship between navigation practice and literature, but both Pacific Islanders and medieval writers have made their connections explicit. Writing and traveling were frequently linked in the early literature of the North Atlantic, which borrowed the tradition from Greek and Roman writing.[14] Their statements emphasize the boldness of launching into the unknown, onto the open sea or the blank page, and the uncertainty of the outcome. Navigation, consequently, is not only a practical measure but a statement of purpose and perspective. It is an expression of how a people orient themselves and how they intend to advance through the world. Edward Said describes both travel and scholarly activity as exploratory:

> The image of traveler depends not on power, but on motion, on a willingness to go into different worlds, use different idioms, and understand a variety of disguises, masks, and rhetorics. Travelers must suspend the claim of customary routine in order to live in new rhythms and rituals. Most of all, and most unlike the potentate who must guard only one place and defend its frontiers, the traveler *crosses over*, traverses territory, and abandons fixed positions, all the time. (1994, 17)

Gregory Clark, picking up where Said left off, notes that the works of writers "cross the many boundaries that territorial conceptions of identity and rhetoric prompt them so persistently to draw" (1998, 11–12). So, too, does the connectivity of the waters in the archipelago encourage the same type of exploration and self-examination in defiance of norms (Greene 2000, 140). Pacific Islanders understand the strength in the unique view afforded them by their surroundings. Vincent Diaz encourages indigenous readers to apply *etak*'s unique perceptions of time and space to both literature and the world. The "indigenous subjectivity" as he puts it, is a product of culture and history that should not be so readily discarded, for both cultural and practical reasons (Diaz 2011, 25–27).

We can see how the English peoples' practical experience of the sea may find expression in their literary work if we let the *etak* system guide us again. Among the Pacific Islanders, the first and last two *etaks* of a journey are of an entirely different type than the others. Instead of being determined through reckoning the boat's position relative to an island and a constellation, these *etaks* are established by the immediate surroundings. The first and last *etaks* of a journey are known as the "*etaks* of sight." This means that within these *etaks*, the land—the point of either departure or arrival—can be seen. At the beginning of a journey, once the island disappears over the horizon, the first *etak* has ended; similarly, the moment the destination can be seen with the naked eye, the vessel has entered the final *etak* of the voyage. Abutting these two *etaks*, however, are two known as "the *etak* of birds." Of the varieties of seabirds, some, like albatross and petrels, are truly pelagic; coming across them in the ocean gives no indication of one's position or proximity to land. Most seabirds, however, must return to land to rest and therefore do not fly too very far out. Therefore, depending on the time of day, navigators in the Pacific who see any of these shoreline birds know that land must be nearby—they are within the *etak* of birds. Observation has allowed Pacific navigators even more precision. They know, for example, that terns and noddies never stray farther than twenty-five miles from shore, while boobies and frigatebirds might venture as far as fifty miles out (Lewis 1994, 163). Once these avian markers disappear, however, then the vessel is truly in the open sea, and the *etaks* of the journey must be determined in the usual way.

Pacific Island navigators are not unique in using the behavior of birds to aid in navigation. One prominent European example is Christopher Columbus, who, on his first voyage, tried to learn the habits of tropical species to determine whether the *Niña*, *Pinta*, and *Santa María* were anywhere near land (Morison 1942, 281–84). We similarly see European explorers of the Pacific employ knowledge of bird behavior in navigation, though they are silent on where they received such intelligence (de Queirós 1904, 187; Robertson 1948, 11, 12, 14, 15, 103, 109–10, 112–13, 246–48). One presumes that they brought some basic knowledge with them, but at some times they can seem surprisingly uninformed, such as when navigator George Robertson (1948, 16, 35, 101) is unable to identify terns or oystercatchers, species that exist in Britain as well as in the Pacific. The indigenous people, as we see, could be much more precise. Moreover, while Europeans evidently entered the Pacific and sailed the world with a homegrown expectation that the behaviors of birds could help guide them, reliance on the observations of wild species appears to be particularly prominent in the Pacific. Elsewhere, it is more common that captive birds were depicted as being used (Hornell 1946, 142–49). But this may be due to factors that have nothing to do with the particulars of the region, such as what records just happen to be extant.

The early English, for their part, appear to have been observant of the bird life around them on the sea. If we pay attention to the fauna mentioned in Old English literature, we can discern a broad overlap between the types of seabirds given significance and those integral to the Carolinian *etak* system. In the poem *The Seafarer*, for instance, the birds named are usually identified as the swan, the gannet, the curlew, the tern, and the eagle—all ones that stay close to land, incidentally.[15] But what is particularly helpful is how several of the species useful in navigation in the Pacific are closely related to species in the North Atlantic and have similar habits. Terns, for example, are found in both places, and boobies and gannets are of the same family (*Sulidae*), formerly thought to be the same genus. Each fills roughly the same niche in its respective ecosystem. We have no indication that the medieval English used the habits of birds to aid in navigation like the Pacific Islanders, but given their engagement with the sea, we should expect them to be familiar with the different species and their behaviors. And for that reason, we should be attentive as to whether mentions of these birds indicate something deeper than we may initially suspect.

Take, for example, the gannet (*Morus bassanus*), which appears eight times in the Old English corpus, outside the glosses. It is best known as one of the elements of the kenning *ganotes bæþ* ("gannet's bath"), which is usually taken as a generic synonym for the sea. However, it is useful to note the particular contexts in which gannets are evoked. Beowulf describes diplomacy and commerce between Geatland and Denmark as occurring "ofer ganotes bæð" (Klaeber 1950, line 1861). In the year 975 of the *Anglo-Saxon Chronicle*, manuscript D, King Edgar is eulogized as being famous "ofer ganetes beð" (Cubbin 1996, 46), while in the poem commemorating him, the ealdorman Oslac is described as having been exiled over that expanse as well (Dobbie 1942a, line 26a). In *The Rune Poem*, an oaken ship successfully navigates the gannet's bath (Dobbie 1942b, line 79a). In all these cases, the particular aspect of the ocean invoked by the context in which the bird is raised is its importance in mediating human interaction, either connecting individuals through travel, trade, or communication or dividing them through its expanse. The particular part of the ocean referenced, then, is the part that forms the network represented by the archipelago: the coasts where travelers and merchants meet, the short sea lanes between islands and mainland. The gannet's bath, in other words, is the water close to human activity, just as it is with real gannets. Robert L. Schichler contrasts this connotation with that of the *hwales rode* ("whale's road"), which appears more frequently in deep-water, intimidating, and foreboding contexts (2002, 59–86). Although the whale, "proud roamer of the waters," to quote one translation (Bradley 1995, 356), is not in nature so aimless or unpredictable as the poetry may imply,[16] it is nevertheless unbound from land in a way humans (and gannets) cannot be and so is associated with the remoter parts of the sea (that whales emerge from its depths rather than fly or swim across its surface may also further estrange them). Gannets, which like ships return to land when coming to rest, signify much more clearly the parts of the ocean safe and familiar to sailors. Observation of the natural archipelagic environment, expressed through the navigation system, reveals greater depth in island literature.

PRACTICAL AND METHODOLOGICAL IMPLICATIONS

While it is important never to collapse the distinctions among them, bringing the literatures and cultures of archipelagoes from various places and times into conversation with one another is a productive exercise. Both encourage a wider flexibility in perspective and empathetic range, applied through both space and time. For many modern Western readers, the principles and preoccupations of the medieval period are often as unfamiliar as those from other cultures around the world. It is important, therefore, to keep in mind the particularities of the Middle Ages when trying to interpret its texts or recreate its culture; indeed, a large portion of medieval studies scholarship could best be characterized as attempts to better perceive the world from the point of view of our subjects. Yet despite the multiple and important differences among past and present archipelagic cultures, we see in their engagement with the sea the consequent development of some of the very same principles. Those principles, in turn, create worldviews that, in some cases, fall into alignment. For both the Pacific peoples and those of the medieval North Atlantic, their presence in an archipelago created cultures of exploration and expansion. And both, in turn, rebel against the efforts of outsiders to impose a contrary narrative upon them.

As a practical matter, such insights highlight the limits placed upon disciplines by their traditional boundaries. Medieval studies is a good example of the arbitrariness of a field. "Medieval" itself, as a descriptor, is usually only applied to Europe and the Middle East, with other parts of the world subject to other periodizations, both indigenous and those applied from without by Western scholars (O'Doherty 2017). The result is a Eurocentric discipline that discourages study outside a handful of cultures, one that has also, consequently, been seen (and, in many cases, proven) unwelcoming to students from non-European backgrounds (Medievalists of Color 2017). A white medievalist, like myself, who is put in a position to consider Pacific Island cultures finds in them potential applications for his own field; a Pacific Islander who is encouraged to study medieval literature would have very well made the same observations all the more quickly.[17] It is therefore imperative that medieval studies increase its diversity to provide a broader array of perspectives upon our topics, while those of us who represent more traditional backgrounds need to read more widely. Opening up the field to different worldviews and experiences can only provide more insight into material that is centuries old.

Expanding our horizons also recommends different types of categorization, beyond the temporal and cultural ones that usually dominate Western historical and literary research. In the case of archipelagic theory, we are prompted to think of peoples and practices in a geographical sense, to consider what societies subject to the same environmental factors may have in common. As a theoretical concept, the archipelago has only recently come to be employed by medievalists, with its full potential yet to be realized.[18] As I hope to have demonstrated, one of the most fruitful research avenues is more direct collaboration between medieval and modern scholarship.[19] The differences between such times and places are obvious, but this makes what they share all the more salient. That is, most prominently, their existence within an archipelago and the environment's influence on both them and their cultures. Peregrine Horden (2016, 570), an important scholar of the medieval Mediterranean, encourages medievalists (provisionally) to "Philippinize Britain." In the sense of highlighting what commonalities we can find between such disparate island literatures and to use these to gain greater insight into both, this is precisely what I believe we should do. In doing so—in combining geographical breadth with chronological depth—we may not only make new discoveries but also hone new arguments for the importance and applicability of our work in the academy today. For scholars, it is an opportunity to break new ground and settle new conceptual space. It is time, in the grand traditions of both of our subjects, to explore.

NOTES

1. An exception is Philip Hayward's (2012) concept of the "aquapelago," which "does not simply offer a surface model, it also encompasses the spatial depths of its waters" (5).

2. See, for example, Raley 1999, 68; Apter 2013, 326–29.

3. See also Bahun 2012, 376.

4. See, for example, the much-discussed comments of the vice chancellor of Queen's University Belfast, who stated, "Society doesn't need a 21-year-old who is a sixth century historian" (Johnston 2016). The lonely reality of many medievalists at their institutions is what has made online communities such as the Lone Medievalist popular as well (www.thelonemedievalist.com; "CARA News," 2017). On theory, see Cohen 2000, 1–18.

5. A well-known examination of this opinion is Freedman and Spiegel 1998, 677–704. There is also the medievalist criticism of Stephen Greenblatt's *The Swerve*, which identifies the primary fault of the book as precisely this attitude. See especially Hinch 2012; Miles 2016.

6. See Cunliffe 2001, 19–63, especially the maps on pages 20 and 35; Tompson 1986, 1–8. However, in adopting the term "archipelago," we are not reflecting medieval terminology and, perhaps, conceptualization. See Goldie and Sobecki 2016, 472, 475. For one example of a medieval characterization of the region and its unity—though not one without ulterior motive—see Sobecki 2011, 30.

7. See, for example, Greenfield and Calder 1986, 284; Klinck 1992, 31; Neville 2007, 203–4.

8. Ó Raithbheartaigh follows Meyer in translating *snámhach* as "floating," but *The Dictionary of the Irish Language* (1990, 552) records several metaphorical meanings of the word that derive from its association with water, both complimentary ("buoyant")

and derogatory ("creeping," "cunning"). Given this, it is unlikely that the poet intended *snámhach* to be neutral, especially since the *Saxon snámhach* are also credited with *dúire* ("obstinacy") in the same line. Whatever the connotations intended by the word choice, however, the literal meaning of the word stands.

9. For Rome, see Matthews 2007, 61–71. For France, see Levison 1946.

10. See also Hau'ofa 1998, 403–6. For a similar argument focused on the Caribbean, see Benítez-Rojo 1992, 1–4.

11. Matthew Boyd Goldie (2011, 7–11) has already brought Hau'ofa into conversation with premodern literature, with certain caveats.

12. In bringing these two wholly disparate traditions into conversation with each other, I am seeking to avoid the pitfalls of comparison as outlined by Juliet Hooker. Instead of implying a valuation, my intent is to uncover greater understanding through juxtaposing Anglo-Saxon and Pacific Islander societies (Hooker 2017, 11–14).

13. This tendency can also be seen in a comparison of the nominative plural forms of the Finnic peoples named in the *Orosius* to their Old Norse cognates: *Finnas*, *Beormas*, and *Cwenas* instead of *Finnar*, *Bjarmar*, and *Kvenir*. However, this was not the case for the Norse word *íss*, "ice," which gave Iceland its name, *Ísland*.

14. Some examples include Muirchú maccu Macthéni 1979, 62–63; Aldhelm 1961, 320; Aldhelm 1968, 455; Alcuin 1881, 198; Smaragdus 1881, 609. For a broader view, see Curtius 1948, 138–41; Dronke and Dronke 1997, 1–26.

15. Though see Goldsmith 1954, 225–35.

16. Whales follow migration routes and have favored feeding grounds, though neither of these are tied to land (Szabo 2008, 83).

17. Indeed, Adam Miyashiro, a Pacific Islander and medievalist, has demonstrated the failure of Old English studies to fully account for its past complicity in colonial projects and the structural racism that such a legacy perpetuates today. The result, as he points out, is a field that discriminates against minority perspectives and therefore rarely engages with them. When they are considered, it is often through the mediation of a white scholar such as myself rather than someone with lived experience (Miyashiro 2017).

18. See, for example, Cohen 2008, Goldie 2010, Sobecki 2011, Klein, Schipper, and Lewis-Simpson 2014, and the entire issue of *postmedieval* 7 (2016).

19. One of the few works of scholarship that pursues this angle is Hiatt 2016, 511–25.

REFERENCES

Alcuin. 1881. *Versus de sanctis Euboricensis ecclesiae*. In *Poetae latini aevi Carolini*, edited by Ernest Duemmler, 1:169–206. *MGH*, *PLAC* 1. Berlin: Weidmann.

Aldhelm. 1961. *De virginitate*. In *Aldhelmi opera*. Edited by Rudolf Ehwald, 211–323. *MGH*, *AA*, 15. Berlin: Weidmann.

———. 1968. *Aenigmata LIX*. In *Tatvini opera omnia; variae collectiones aenigmatvm merovingicae aetatis; anonymvs dedvbiis nominibvs*. Edited by Maria de Marco, 454–55. *CCSL* 133. Turnhout: Brepols.

Apter, Emily. 2013. *Against World Literature: On the Politics of Untranslatability*. New York: Verso.
Bahun, Sanja. 2012. "The Politics of World Literature." In *The Routledge Companion to World Literature*, edited by Theo D'haen, David Damrosch, and Djelal Kadir, 395–404. New York: Routledge.
Bately, Janet, ed. 1980. *The Old English Orosius*. New York: Oxford University Press.
Benítez-Rojo, Antonio. 1992. *The Repeating Island: The Caribbean and the Postmodern Perspective*. Translated by James Maraniss. Durham, NC: Duke University Press.
Bradley, S. A. J., ed. and trans. 1995. *The Whale*. In *Anglo-Saxon Poetry*, 355–57. London: Everyman.
"CARA News: The Lone Medievalist." 2017. *The Medieval Academy Blog*. May 24. http://www.themedievalacademyblog.org/cara-news-the-lone-medievalist.
Carver, M. O. H. 1990. "Pre-Viking Traffic in the North Sea." In *Maritime Celts, Frisians, and Saxons*, edited by Séan McGrail, 117–25. CBA Research Report 71. London: Council for British Archaeology.
Carver, Martin. 2010. "Four Windows on Early Britain." *Haskins Society Journal* 22: 1–24.
Clark, Gregory. 1998. "Writing as Travel, or Rhetoric on the Road." *College Composition and Communication* 49: 9–23.
Cohen, Jeffrey Jerome. 2000. "Introduction: Midcolonial." In *The Postcolonial Middle Ages*, ed. Jeffrey Jerome Cohen, 1–17. New York: Palgrave Macmillan.
———. 2008. *Cultural Diversity in the British Middle Ages: Archipelago, Island, England*. New York: Palgrave Macmillan.
Cubbin, G. P., ed. 1996. *The Anglo-Saxon Chronicle, MS D*. Cambridge: D. S. Brewer.
Cunliffe, Barry. 2001. *Facing the Ocean: The Atlantic and Its Peoples, 8000 BC–AD 1500*. Oxford: Oxford University Press.
Curtius, Ernst Robert. 1948. *Europäische Literatur und lateinisches Mittelalter*. Bern: Francke Verlag.
de Queirós, Pedro Fernandes. 1904. *The Voyages of Pedro Fernandez de Quiros, 1595–1606*. Edited and translated by Clements Markham. Hakluyt Society s.s. 14. 2 vols. London: Hakluyt Society.
Diaz, Vincent M. 2011. "Voyaging for Anti-colonial Recovery: Austronesian Seafaring, Archipelagic Rethinking, and the Re-mapping of Indigeneity." *Pacific Asia Inquiry* 2: 21–32.
Dobbie, Elliott Van Kirk, ed. 1942a. *The Death of Edgar*. In *The Anglo-Saxon Minor Poems*, 22–24. *ASPR* 6. New York: Columbia University Press.
———, ed. 1942b. *The Rune Poem*. In *The Anglo-Saxon Minor Poems*, 28–30. *ASPR* 6. New York: Columbia University Press.
Dronke, Peter, and Ursula Dronke. 1997. "Growth of Literature: The Sea and the God of the Sea." In *M. Chadwick Memorial Lectures* 8, 1–26. Cambridge: Department of Anglo-Saxon, Norse and Celtic.
Freedman, Paul, and Gabrielle M. Spiegel. 1998. "Medievalisms Old and New: The Rediscovery of Alterity in North American Medieval Studies." *American Historical Review* 103: 677–704.

Gladwin, Thomas. 1970. *East Is a Big Bird*. Cambridge, MA: Harvard University Press.

Goldie, Matthew Boyd. 2010. *The Idea of the Antipodes: Place, People, and Voices*. New York: Routledge.

———. 2011. "Island Theory: The Antipodes." In *Islanded Identities: Constructions of Postcolonial Cultural Insularity*, edited by Maeve McCusker and Anthony Soares, 1–40. New York: Rodopi.

Goldie, Matthew Boyd, and Sebastian Sobecki. 2016. "Editors' Introduction: Our Seas of Islands." *postmedieval* 7: 471–83.

Goldsmith, Margaret E. 1954. "The Seafarer and the Birds." *Review of English Studies* 5: 225–35.

Greene, Roland. 2000. "Island Logic." In *"The Tempest" and Its Travels*, edited by Peter Hulme and William H. Sherman, 138–45. Philadelphia: University of Pennsylvania Press.

Greenfield, Stanley B., and Daniel G. Calder. 1986. *A New Critical History of Old English Literature*. New York: New York University Press.

Hau'ofa, Epeli. 1993. "Our Sea of Islands." In *A New Oceania: Rediscovering Our Sea of Islands*, edited by Epeli Hau'ofa and Eric Waddell, 2–17. Suva, Fiji: School of Social and Economic Development, University of the South Pacific in association with Beake House.

———. 1998. "The Ocean in Us." *Contemporary Pacific* 10: 392–410.

Hayward, Philip. 2012. "Aquapelagos and Aquapelagic Assemblages: Toward an Integrated Study of Island Societies and Marine Environments." *Shima: The International Journal of Research into Island Cultures* 6: 1–11.

Hiatt, Alfred. 2016. "From Pliny to Brexit: Spatial Representation in the British Isles." *postmedieval* 7: 511–25.

Hinch, Jim. 2012. "Why Stephen Greenblatt Is Wrong—and Why It Matters." Review of *The Swerve*, by Stephen Greenblatt. *Los Angeles Review of Books*. December 1. https://lareviewofbooks.org/article/why-stephen-greenblatt-is-wrong-and-why-it-matters#!.

Hooker, Juliet. 2017. *Theorizing Race in the Americas*. Oxford: Oxford University Press.

Horden, Peregrine. 2016. "Afterword." *postmedieval* 7: 565–71.

Hornell, James. 1946. "The Role of Birds in Early Navigation." *Antiquity* 20: 142–49.

Howe, Nicholas. 1989. *Migration and Mythmaking in Anglo-Saxon England*. New Haven, CT: Yale University Press.

Johnston, Patrick. 2016. Interview by Rebecca Black. *Belfast Telegraph*. May 30. http://www.belfasttelegraph.co.uk/news/northern-ireland/more-than-a-third-of-students-leave-northern-ireland-at-18-we-cannot-afford-to-lose-such-talent-but-theyll-only-return-if-there-are-opportunities-34756003.html.

Klaeber, Fr., ed. 1950. *Beowulf and the Fight at Finnsburg*. 3rd ed. Lexington, MA: D. C. Heath and Co.

Klein, Stacy S., William Schipper, and Shannon Lewis-Simpson, eds. 2014. *The Maritime World of the Anglo-Saxons*. Tempe: Arizona Center for Medieval and Renaissance Studies.

Klinck, Anne L. 1992. *The Old English Elegies: A Critical Edition and Genre Study*. Buffalo, NY: McGill-Queen's University Press.

Lebecq, Stéphane. 2007. "Communication and Exchange in Northwest Europe." In *Ohthere's Voyages*, edited by Janet Bately and Anton Englert, 170–79. Roskilde: Viking Ship Museum.

Levison, Wilhelm. 1946. *England and the Continent in the Eighth Century*. Oxford: Clarendon Press.

Lewis, David. 1994. *We, the Navigators: The Ancient Art of Landfinding in the Pacific*. Honolulu: University of Hawai'i Press.

Malone, Kemp. 1930. "King Alfred's North: A Study in Mediaeval Geography." *Speculum* 5: 139–67.

Matthews, Stephen. 2007. *The Road to Rome: Travel and Travellers between England and Italy in the Anglo-Saxon Centuries*. Oxford, UK: Archaeopress.

Medievalists of Color. 2017. "On Race and Medieval Studies." *In the Middle*. August 1. http://www.inthemedievalmiddle.com/2017/08/on-race-and-medieval-studies.html.

Meyer, Kuno. 1897. "Two Middle-Irish Poems." *Zeitschrift für celtische Philologie* 1: 112–13.

Miles, Laura Saetveit. 2016. "Stephen Greenblatt's *The Swerve* Racked Up Prizes—and Completely Misled You about the Middle Ages." *Vox*. July 20. http://www.vox.com/2016/7/20/12216712/harvard-professor-the-swerve-greenblatt-middle-ages-false.

Miyashiro, Adam. 2017. "Decolonizing Anglo-Saxon Studies: A Response to ISAS in Honolulu." *In the Middle*. July 28. http://www.inthemedievalmiddle.com/2017/07/decolonizing-anglo-saxon-studies.html.

Morison, Samuel Eliot. 1942. *Admiral of the Ocean Sea: A Life of Christopher Columbus*. 2 vols. Boston: Little, Brown and Company.

Muirchú maccu Macthéni. 1979. *Vita patricii*. In *The Patrician Texts in the Book of Armaugh*. Edited by Ludwig Bieler, 60–123. *SLH* 10. Dublin: Dublin Institute for Advanced Studies.

Neville, Jennifer. 2007. "'None Shall Pass': Mental Barriers to Travel in Old English Poetry." In *Freedom of Movement in the Middle Ages: Proceedings of the 2003 Harlaxton Symposium*, edited by Peregrine Horden, 203–14. Donington, UK: Shaun Tyas.

Ó Raithbheartaigh, Toirdhealbhach, ed. and trans. 1932. *Genealogical Tracts I*. Dublin: Stationary Office.

O'Doherty, Marianne. 2017. "Where Were the Middle Ages?" *The Public Medievalist*. March 7. https://www.publicmedievalist.com/where-middle-ages.

Pinet, Simone. 2011. *Archipelagoes: Insular Fictions from Chivalric Romance to the Novel*. Minneapolis: University of Minnesota Press.

Pocock, J. G. A. 2005. "British History: A Plea for a New Subject." In *The Discovery of Islands: Essays in British History*, 24–43. Cambridge: Cambridge University Press.

Quin, E. G., ed. 1990. *Dictionary of the Irish Language: Compact Edition*. Dublin: Royal Irish Academy.

Rafael, Vicente L. 1994. "The Cultures of Area Studies in the United States." *Social Text* 41: 91–111.

Raley, Rita. 1999. "On Global English and the Transmutation of Postcolonial Studies into 'Literature in English.'" *Diaspora* 8: 51–80.

Robertson, George. 1948. *The Discovery of Tahiti, a Journal of the Second Voyage of H.M.S. Dolphin round the World, under the Command of Captain Wallis, R. N., in the Years 1766, 1767, and 1768*. Edited by Hugh Carrington. Hakluyt Society s.s. 97. London: Hakluyt Society.

Said, Edward W. 1994. "Identity, Authority, and Freedom: The Potentate and the Traveler." *Boundary 2* 21: 1–18.

Schichler, Robert L. 2002. "From 'Whale-Road' to 'Gannet's Bath': Images of Foreign Relations and Exchange in *Beowulf*." *Reading Medieval Studies* 28: 59–86.

Smaragdus. 1881. *Carmina*. In *Poetae latini aevi Carolini*. Edited by Ernest Duemmler, 1:605–19. *MGH*, *PLAC* 1. Berlin: Weidmann.

Sobecki, Sebastian. 2008. *The Sea and Medieval English Literature*. Cambridge, UK: D. S. Brewer.

———. 2011. "Introduction: Edgar's Archipelago." In *The Sea and Englishness in the Middle Ages: Maritime Narratives, Identity and Culture*, edited by Sebastian Sobecki, 1–30. Cambridge, UK: D. S. Brewer.

Stokoe, William C., Jr. 1957. "On Ohthere's Steorbord." *Speculum* 32: 299–306.

Szabo, Vicki Ellen. 2008. *Monstrous Fishes and the Mead-Dark Sea: Whaling in the Medieval North Atlantic*. Boston: Brill.

Tompson, Richard S. 1986. *The Atlantic Archipelago: A Political History of the British Isles*. Queenston, Ontario: Edwin Mellen Press.

Wooding, Jonathan M. 1996. *Communication and Commerce along the Western Sealanes,* AD *400–800*. BAR International Series 654. Oxford, UK: Tempvs Reparatvm.

Chapter Seven

Archipelago of the Maghreb

Mapping Mediterranean Movement from Transnational Migration to Transregional Mobility

Sarah DeMott

> The idea of the Archipelago is not that of a return to origins, but rather that of a counter reply to the history-destiny of Europe.
>
> —Massimo Cacciari, *L'arcipelago* (1997, 35)

> So, turning to my immediate location, the Mediterranean becomes the site for an experiment in a different form of history writing, an experiment in language and representation where it becomes possible to engage with "the outside of history of modernity" through points of resistance and refusal that continually relay us elsewhere, and leads to an inevitable "question of history as status quo."
>
> —Iain Chambers, *Mediterranean Crossings* (2008, 27)

In 1862 and 1863, 564 infants from the islands of the western Mediterranean were baptized in the Roman Catholic Church of Sainte-Croix in Tunis, Tunisia. The baptized children and their kin came to Tunisia from across the western Mediterranean and its islands of Pantelleria, Malta, Sardinia, Sicily, Favignana, Lampadusa, and Corsica. Sacramental records provide evidence about the movement of their families across the Mediterranean Sea and the construction of relationships between islands, inlets, and shorelines. The Mediterranean mobility of these families across the Strait of Sicily produces and represents a constellation or assemblage of individual connections made possible by close-distance island circulation, familial relationships, and shared ephemeral knowledge. Tracing familial relations, visualizing migration routes, and linking local life through island and shoreline connectivity, the cartographic reconstitution of regional relationships gives way to a Mediterranean archipelago.

Mapping the ways in which Mediterranean islanders moved in and out of Tunisia, as well as among other islands and across the Mediterranean Sea, makes visible an emerging archipelago as a unit of analysis. This chapter identifies the archipelago as a way of knowing how a people move, relate, connect—how they live in a complex environment. "An archipelago," as defined by social scientist Godfrey Baldacchino, "is a world of islands unto itself" (2015, 2). The archipelagic constitutes an interwoven way of life. Defining the islands of the Mediterranean as an archipelago thereby responds to Iain Chambers's call in the above epigraph "for an experiment in a different form of history writing."

Using baptismal records of the church of Sainte-Croix, I have mapped the genealogy of these Mediterranean children who were christened in Tunisia by plotting the longitudinal and latitudinal coordinates of both their place of birth and their paternal origin. Borrowing from literary critic and digital humanist Franco Moretti's (2013) computational approach to ways of reading, I have used a gradual scale of distant to close mapping to visualize and analyze the dataset. Keeping the data coordinates fixed and stable while reframing the cartography according to various scales and markers of legibility provides material to see correlations and variance. Using the same dataset, I varied the calibration by zooming in and out. Playing on Moretti's methodological distinction between distance reading and close reading, I have compared the series of maps through a process I refer to as "close mapping" of the cartography itself in order to critique patterns of difference in density and directionality. From this close mapping, we are able to use digital humanities techniques in elastic, undulating ways to highlight patterns of movement.

Southern Italian migration has largely been categorized as transnational migration, that is, as part of a global, diasporic phenomenon linked to familial networks in both sending and receiving communities. Archipelagic approaches provide a way in which to recontextualize the transnational view of migration. The series of maps and close mappings I offer here provides an alternative framework, using data visualization to challenge the history of migration to Tunisia as transnational migration. Through a comparative look at the empirical data and its corresponding cartographic visualizations, the process of close mapping explicates a system of everyday intimacies of family formation, relationships, and mobility.

The recorded movement of family members across the Mediterranean Sea is motivated by networks of local social ties, which exemplify island knowledge of the sea and distinct methods of moving through it. These epistemic ways of knowing the archipelago—as both an island and a sea space—illustrate a logic of the archipelago. Archipelago logics are an expression of the archipelago, not just the naming of a group of islands as a formal unit but a rooted expression of island connectivity, heterogeneity, communication, social networks, community connections, and relationships that circulates among islands. If bridges are the dominant paradigm for twentieth-century migration across national and continental borders, this chapter describes a more interstitial circulation within and among archipelagic spaces. Rather than a bridge, mobility and migration across the Mediterranean archipelago of islands, seas, and shorelines illustrates a constellation of connections and webs of close-distance exchange. Mediterranean inter-island mobility challenges migration narratives' reliance on the continental borders and accompanying geopolitical identities of Africa and Europe. When we look away from the continental framework and focus on archipelagic assemblages based on constellations of local relationships, a new vantage point, a new legible space, and an epistemology emerge.

AN EMERGENT ARCHIPELAGO

Between 1860 and 1930, approximately two hundred thousand Italians settled in Tunisia. From most accounts, the Italians' presence in Tunisia has been historicized as part of a larger history of transnational Italian migration (Choate 2008). The memory of migration to Tunisia circulates in the Sicilian cultural collective; yet its history is rarely acknowledged publicly. Tunisia as a destination, I argue elsewhere, served as a subversive space, an alternative space of escape, a refuge from precarious economic, political, and cultural conditions in Sicily and southern Italy (DeMott 2015). The proletariat classes of Sicily were dominated, as Antonio Gramsci described in *Prison Notebooks* and the essay "The Southern Question," by conditions of economic poverty, political oppression, and social subjugation (Hoare and Smith 1971). Migration from the southern half of the Italian peninsula and its islands was so intense at the turn of the twentieth century that whole villages were abandoned to be relocated across a global network of labor markets. Mediterranean migration between islands provided a source of relief from various continental pressures. Whether fleeing political unrest, economic oppression, or domestic violence, southern Italian migrants formed communities across the islands of the Mediterranean and Tunisia that opposed traditional social, political, and economic norms.

Sicilian settlement in Tunisia is largely regarded as influenced by the *transnational spirit of migration* that had taken place across the European continent. Yet a "close mapping" of these migrations suggests that a critical review of Mediterranean mobility is needed. Based on social practices of interaction and exchange between southern Italy, Mediterranean islands, and Tunisia,

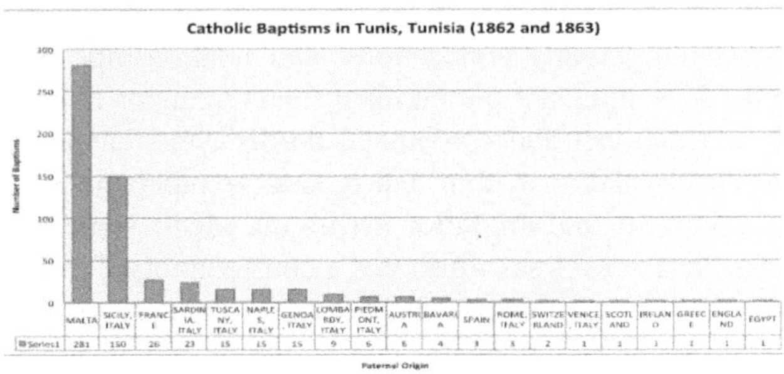

Figure 7.1. Origin of Children Baptized in Tunis, Tunisia, 1862 and 1863.

close mapping visualizes and reveals a "Maghreb archipelago" that recognizes an interwoven way of life between the Maghreb and the Mediterranean world.

Typically, a baptismal record outlines family ancestry through the record of parents' marital status and social networks in the form of testimonies from priest and godparents. The record also includes the status of the child's birth (i.e., stillbirth, illegitimacy, legitimacy) and occasionally the occupation of the father and origins of the godparents. The heritage of the baptized child is identified according to the father's origins in either national (Austria, France, Spain, Greece, Scotland, Ireland, England, Malta) or regional (Sicily, Genoa, Tuscany, Lombardy, Naples, Rome, Sardinia) territories.[1] Figure 7.1 illustrates the predominance of foreign-born parents through whom the overwhelming majority (99.8%) of children were tied to European heritage.

These vital records provide a dataset with which to examine how routes of mobility are imagined through four cartographic models. The following four maps were designed to move successively from a more traditional, continent-based geography toward one based on archipelagic thinking. The first two maps emphasize a continental perspective by describing movement as highly directional and spanning the continents. They provide a traditional picture of continental transit along shipping lanes between Europe and Africa. The latter two maps are nonlinear, spatially oriented, and geometric in their visual representations, getting us much closer to an "archipelagic logic." Through this nonlinear movement, an "archipelagic logic" provides the means to articulate spatial structures through which people order their knowledge of the world.

CONTINENTAL BRIDGE

On January 1, 1862, a Sicilian child, Antonino Errera, was christened in the medina of Tunis, Tunisia, at Sainte-Croix Roman Catholic Church. In attendance at Antonino's baptismal ceremony were his Sardinian mother, Maria

Granara; his Pantellerian father, Mariano Errera; and his Sicilian godparents, Antonio Prodo and Teresea Grecco. The record states,

> In the year of our Lord, one thousand eight hundred and sixty-two and at the first of January at 4 o'clock in the afternoon. I, signed, Camelliere Viarriale [priest's name] and the apostolic order of the Capuchins, I baptize a baby born in this city of Tunis on the 19th of last October 1861 at noon, the legitimate child of Mariano Errera of Pantelleria, construction worker, and Maria Granara of Cagliari [Sardinia], married as Sicilians this baby will take the name Antonino. The godparents, Antonio Prodo and Teresa Grecco, born in Palermo, Sicily.[2]

Figure 7.2 orients the reader to the space of the central Mediterranean corridor, as well as to the specific locations described in Antonino Errera's baptismal record. From the information on the baptismal record as outlined in Figure 7.3, Map A graphs the baptized child's place of birth (Tunis, Tunisia) as well as the place of birth of the mother (Cagliari, Sardinia), father (Pantelleria), and godfather and godmother (Palermo, Sicily.)

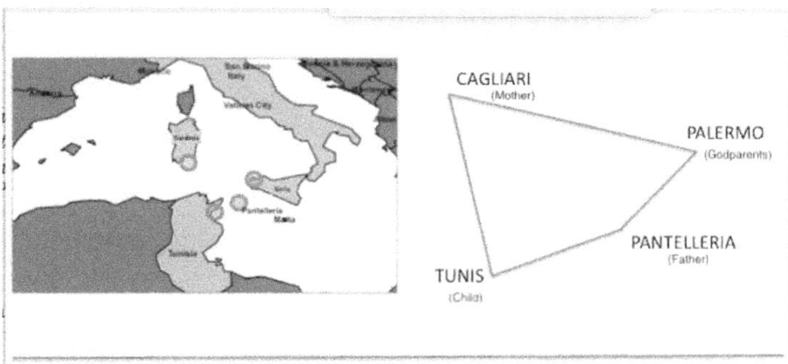

Figure 7.2. The Child Antonino, 1862.

Year	*Child Location*	*Parent (1) Location*	*Parent (2) Location*	*Godparent (1) Location*	*Godparent (2) Location*
1862	Antonio Errera Tunis, Tunisia	Mariano Errera Pantelleria	Maria Granara Calgliari, Sardinia	Antonio Prodo Palermo, Sicily	Teresa Grecco Palermo, Sicily

Figure 7.3. Dataset of the Child Antonino, 1862.

Figure 7.4 follows this pattern of plotting child and father for all 564 baptismal records from 1862 to 1863 in Tunis, Tunisia. This map emphasizes the aggregate of the children's ancestral connections between Tunisia and Europe. The lines of directionality expressed in this map appear as if erupting from the children's place of baptism (Tunisia) and radiating out to the sending nations of Europe. Yet this burst of lines coming from Tunisia and emanating east is not surprising. As demonstrated in Figure 7.1, all the children in this study are bap-

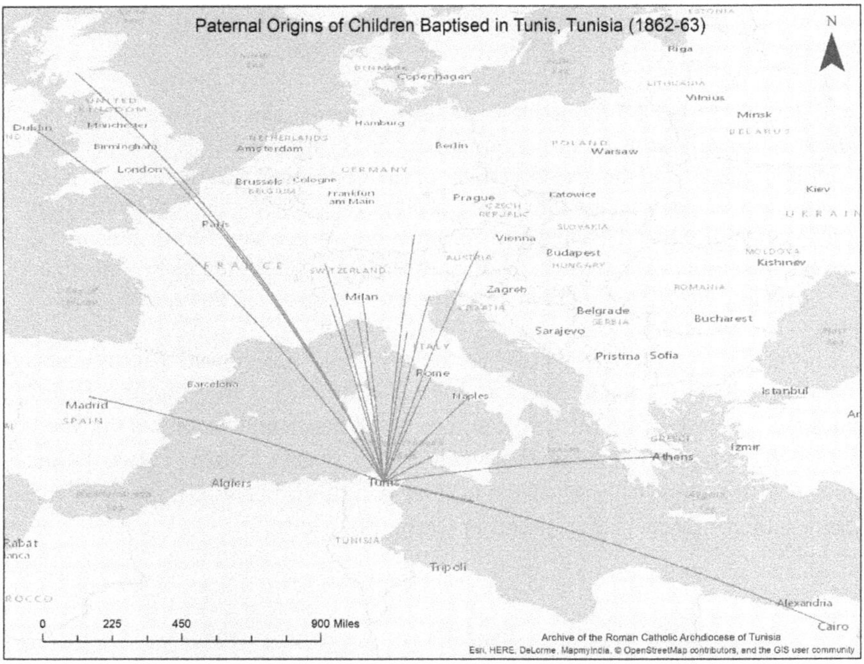

Figure 7.4. Continental Bridge: Paternal Origin of Children Baptized in Tunis, 1862 and 1863.

tized in Tunisia, and the majority of their parents are from Europe. Rather than from west to east, it is more curious if the map is read in reverse from east to west, or from Europe to Tunisia. From this vantage point, the directional lines depict a portrait of European immigration to North Africa. If one looks down on the map and reads it from its aerial perspective, Europeans appear to be arriving on the northern the tip of Africa. Here we also see "a colonizers view of the world," as J. M. Blaut puts it, "where progress is seen as flowing endlessly out of the center of Europe towards otherwise sterile periphery" (Lewis and Wigen 1984). It is the span of lines stretching between the two destinations that creates the image of a continental bridge between Africa and Europe. The child and parent are linked across this arching bridge connecting Europe and Africa. To frame migration to Tunisia, however, as transcontinental migration would be a misunderstanding of the centrality of the archipelago.

TRANSNATIONAL MIGRATION

Figure 7.5, the second map in this series, highlights a shift in gradation from Figure 7.4's explosive continental arch to Figure 7.5's more focused

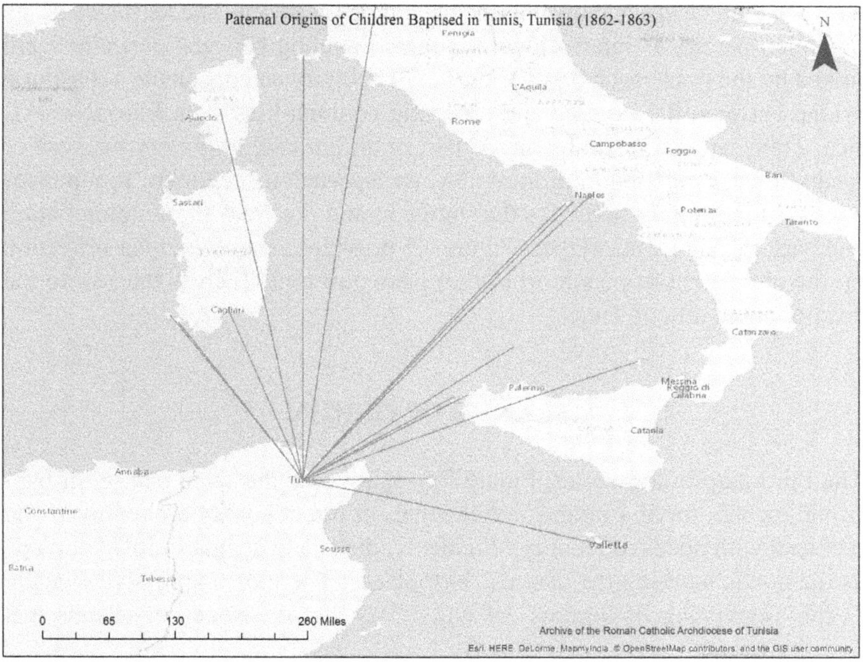

Figure 7.5. Transnational Migration: Paternal Origin of Children Baptized in Tunis, 1862 and 1863.

narrative between local points of origin and destinations. Figure 7.5 has the same data coordinates as Map 1, but with a greater focus on the most concentrated area of movement across the central Mediterranean corridor. As was evident in Figure 7.1, the largest percentage of families were from the Mediterranean islands of Malta, Sicily, and Sardinia (Baldacchino 2015, 2). This slight shift in cartography to a regional form changes the emphasis from maps of continental transit routes to transnational points of departure and arrival. We are able to identify regional port cities and zones of concentrated migratory movement. It is possible to imagine the schedule of transatlantic steamer ships making stops at the port cities of Naples, Cagliari, Palermo, and Tunis, before heading across the Atlantic Ocean. We can imagine that families' collective migration narratives are developed through systematic movement along a chain of family networks. Southern European migrants were assumed to travel in a direct route from their home of origin to their relatives' home abroad. A migrant's destination would be decided based on familial connections within his or her sending community. Both maps, however, maintain a portrait of a continental bridge stretching over the sea. Traversing the maritime space between continental landmasses, this portrait of migration emphasizes the port cities of departure and arrival and the space between them, a space that is divided by an ocean or sea. The heritages of the child and parent, as international migrants traveling between port cities, are linked by the lines reaching over the sea. Their movements create a diasporic bridge between the sending and receiving communities from Europe to Africa. Transnational directionality is part of an intricate migratory network of institutional regulation, migration law, transportation schedules, remittances deposits, and family networks that move in and out from the sending nation and back. Using identical data, Figure 7.5 provides a more granular reflection on the movement of people from European port cities, across the sea, to the cosmopolitan city of Tunis.

ISLANDS OF ORIGIN

The third map in this series, Figure 7.6, again uses the same dataset of baptismal records for 564 infants, but the lines of directionality are removed and replaced with nodes of identity.[3] Suddenly, the island chains visibly "pop up" as diamonds marking the parents' birthplaces. The most expressive element in this remapping of Figures 7.4 and 7.5 is the prevalence of islands that

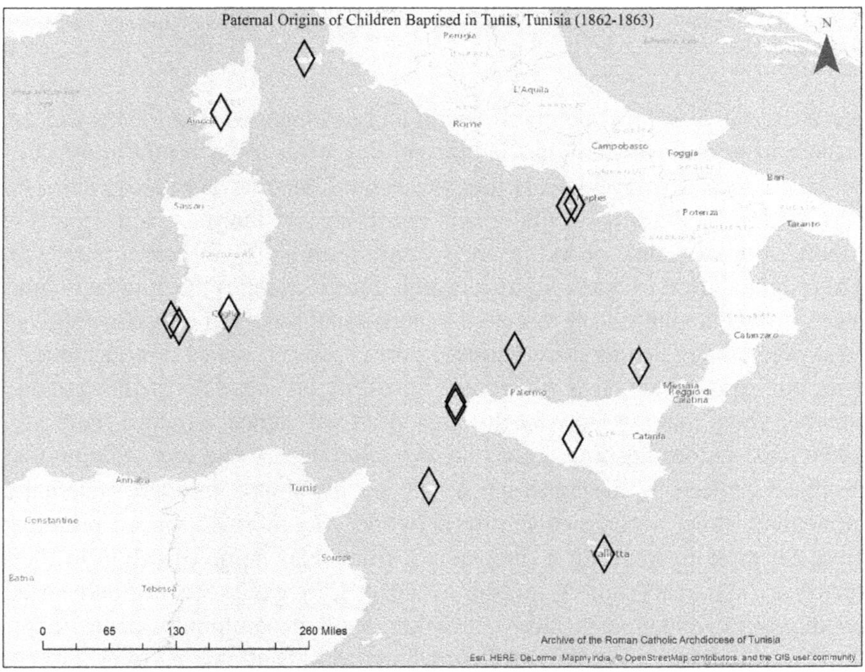

Figure 7.6. Islands of Origin: Paternal Origin of Children Baptized in Tunis, 1862 and 1863.

become visible. Islands, rather than bridges, shorelines, or port cities, are the dominant geographic unit. Through a nodal mapping, an island network appears within the central Mediterranean corridor. Island chain formations include the Sicilian islands of Pantelleria, Favignana, Levanzo, Ustica, and Lipari; the Neapolitan island chain of Procida and Ischia (missing only Capri); the Tuscan islands of Elba and Corsica; Sardinia and its smaller islands of San Pietro and Sant' Antioco; and the twin islands of Malta and Gozo. In these waters, the category of "place of origin" is transformed from a continental landmass to an island.

This remapping draws us further away from Europe's continental shoreline, further from the docks of the great maritime port cites, further from the metropolitan centers of empire. Figure 7.6 provides a decentralized portrait of mobility born of the island peripheries rather than urban centers—a portrait of island descent rather than migration from the Eurocentric model of nation-state or city center.[4] Nodes mark circumference, while lines tracing trajectories of directionality suggest a contrasting model of mobility. As Françoise Lionnet and Emmanuel Bruno Jean François explain, "The sea provides connection and links to multiple and simultaneous spaces of belonging rather than fixed roots" (Lionnet and François 2015, 15). Thus, in Figure 7.6 we move away from ideas of transnational migration to an understanding of island mobility as fluid.

ARCHIPELAGIC NETWORK

From the cartographer's position in the fourth (and final map) of the series (Figure 7.7a), the Mediterranean Sea is a focal point of the map. The viewer's gaze is no longer fixed on continental shorelines or islands but rests on the sea. Starting with the Tunisian baptismal record, this map links the birthplace of all five actors in the record. For the purposes of legibility, I did not use all the records but plotted only the 150 children of Sicilian descent baptized in Tunis between 1862 and 1863. The birthplace of the five principle actors in the baptismal record are linked between the child, mother, father, godfather, godmother, and back to the child in Tunis. In Figure 7.8, the five actors are represented as nodes and connect by rays or joining lines to form a geometric shape. When a similar mapping is done for the 150 children, a web of geometric shapes illustrates the multilayered portraits of the 150 records (see Figure 7.7b).

In the sea, the lines of directionality are not neatly traceable but appear overlapping and tangled in a familial web of connections. The map's various geometric shapes represent each family, layered one on top of the other. If

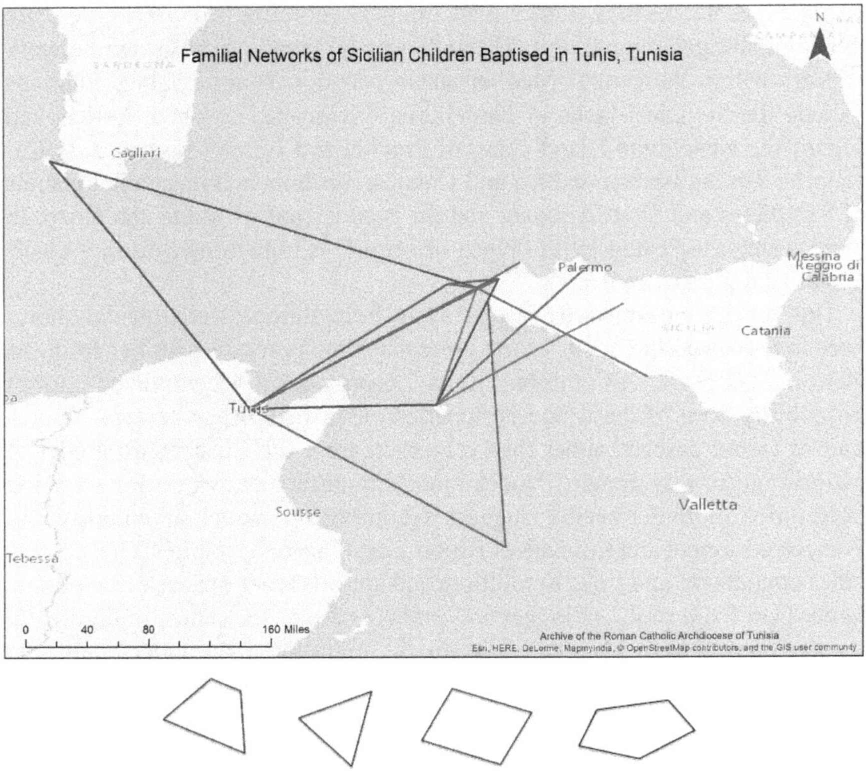

Figures 7.7a and 7.7b. Archipelagic Network: Familial Network of Sicilian Children Baptized in Tunis, Tunisia, 1862 and 1863.

Map design: Sarah DeMott

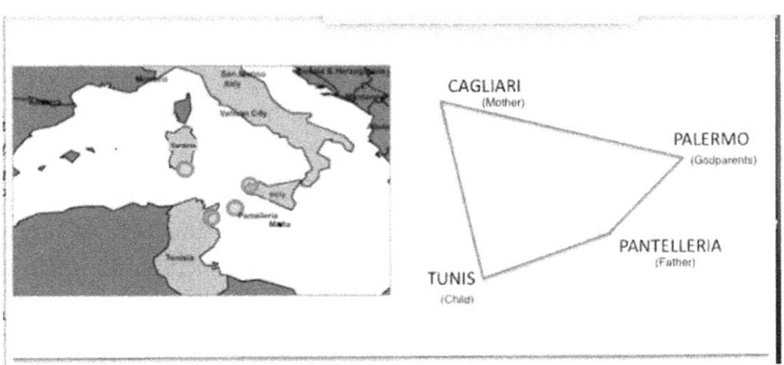

Figure 7.8. The Child Antonino, 1862.

we were to choose a single family and trace it, we would see that most of the nodes are not identical; therefore, the line is not straight but interrupted by various angles. The insets on the map show the geometric patterns created if you were to trace just one family's internal social network. Figure 7.7a is a composite of all the families' geometries, resulting from the compilation of all the family networks derived from the baptismal records, layered on top of each other. This map illustrates a geometric web—a spatial network, a kaleidoscope, a constellation, an archipelago.

Given the importance of the family in Mediterranean societies and the prevalence of familialism, one might easily assume that the lines of connectivity would be fairly direct, with all the family members descending from the same town (Schneider and Schneider 1996). Technically this would give us a solid straight line of directionality—much like in Figures 7.4 and 7.5. Logically, the husbands and wives would be from the same town, and the child's godparents most likely would be their parents' siblings or relatives from the same island. If the family was homogeneous with all five members as descendants of the same town, then all five nodes would appear on a single straight line:

(A__________B,C,D,E)

However, we see irregular shapes with multiple angles, including triangles, rectangles, and even pentagons, representing families with five different nodal points or places of origin. The geometric shapes lay on top of, rather than arch over, the sea. Compared to Figures 7.4 and 7.5, where the straight lines of directionality make a bridge spanning across the Mediterranean, Figure 7.7a's archipelagic portrait of irregular shapes expresses an inter-island heterogeneity that extends over the water. It expresses what I call an archipelago logic of relations at sea.

It is unclear from the baptismal records if the families' presence in the cosmopolitan port city of Tunis was permanent or temporary, but we do see that their mobility is regional, with deep knowledge of the shared waters between Sicily and Tunisia. Sicilian mobility is oriented to circulation among a cluster of Mediterranean islands and shorelines, which I refer to as the Maghrebin archipelago. I understand the Maghrebin archipelago to be the islands of the western Mediterranean, the Mediterranean waters of the Strait of Sicily, and the continental shoreline of North Africa and Sicily.

PRACTICAL AND METHODOLOGICAL IMPLICATIONS

The significance of close mapping across four cartographic models has been to walk us through the application of archipelago logic. Using the tools of digital humanities, combining archival research and mapping software, this chapter has employed a local dataset and, by isolating and remapping the same set of data over and over again, has demonstrated different map-making strategies, different map-reading literacies, and thus the differences between a continental and an archipelago logic of Mediterranean mobility. Mapping proletariat relationships across the central Mediterranean corridor defies traditional transnational accounts of southern Italian mobility.[5] Instead, the archipelagic account recognizes and names networks of islands, water, and shorelines as zones of circulation rather than land-centric movement across continents. Focusing on the social history (the baptismal record of a baby with parents and family from various isles) and this underhistoricized contact zone (a liminal border region of African and European contact), this microhistory provides a critical study for revisionist history (Lewis and Wigen 1984, ix). As historical experimentation, the child Antonino's baptismal record cuts against the grain of normative paradigms of southern Italian transnational migration history. The archipelago becomes a counter-reply as well as an intervention into writing the history of migration and reading the historical record. From this close-mapping methodology, the archipelagic emerges as a critical framework from which to read how people have historically circulated within the Mediterranean (Moretti 2013).

Another result of this study, with its paired archival and close-mapping methodologies, is the visualization of the interrelations between southern Italy, the islands of the Mediterranean, Tunisia, the Maghreb, and the Arab world through an archipelagic logic. The historiographical precedent for approaching the Mediterranean as a maritime space of networks and connectivity is well established (Horden and Purcell 2000). From the perspective "of the Sea, itself," Cambridge historian David Abulafia (2011) provides a framework for Mediterranean historiography by tracing cultural connection across the sea from shipping lanes based on trade, travel, and tourism. The Mediterranean is the sixth continent, according to historian Halikarnas Balikçisi (1991), who conceptualizes this body of water as a fixed geographic space from which to chart human mobility. For both Balikçisi and Abulafia, Mediterranean history is conceptualized as its own monolithic identity and exclusive spatial category. The Maghrebin archipelago is not distinct from Africa, Europe, or the Mediterranean; rather, it is a holistic and constituted element of the Mediterranean, one that conjoins spaces and histories. The Maghrebin archipelago reconceives both continental and maritime spaces in the Mediterranean in new ways, providing an alternative territory to historicize human interaction in the interstices of national territories, continental divisions, and even the ocean basin. In this process of exploring relationships visualized in archival records, this close-mapping methodology draws out the ways in which an archipelago logic can allow us to explore different spatial perceptions and different historical conceptions of human mobility and community formation.

NOTES

This chapter originated from the generous support and collective conversation of Rutgers University, Center for Cultural Analysis, Archipelagoes Seminar (2015–2017), led by Michelle A. Stephens and Yolanda Martínez-San Miguel, as well as Anjali Nerlekar.

1. Although Italian unification is officially in 1861, it is not until February 11, 1929, that the Vatican church officials recognize the Kingdom of Italy as a nation under the Lateran Pacts. For legal status regarding church/state battles of legitimacy, consider that between 1870 and 1929, marriages performed in churches without a municipal official witness were technically *illegitimate* in the eyes of the state.
2. Archive of the Roman Catholic Archdiocese, Tunis, Tunisia, December 2016. Translation by author.
3. For the strictest definition of archipelagos, see the *OED*: "Hence (as this is studded with many isles): Any sea, or sheet of water, in which there are numerous islands; and *transf.* a group of islands" (http://www.oed.com/view/Entry/10387 [accessed May 16, 2018]).
4. For further reading on continental theory, see Mignolo 2000, Hulme 1989, and Roberts and Stevens 2017.
5. For examples of literature representative of the history of Italian migration as transnational, see Gabaccia 2000; Choate 2008.

REFERENCES

Abulafia, David. 2011. *The Great Sea*. Cambridge: Cambridge University Press.

Baldacchino, Godfrey, ed. 2011. *Island Songs: A Global Repertoire*. Lanham, MD: Rowman & Littlefield.

———. 2015. "Lingering Colonial Outlier Yet Miniature Continent: Notes from the Sicilian Archipelago." *Shima: The International Journal of Research into Island Cultures* 9, no. 2.

Balikçisi, Halikarnas. 1991. *The Sixth Continent*. Ankara: Directorate General of Cultural Affairs of the Ministry of Foreign Affairs.

Bentley, Jerry, Renate Bridenthal, and Kären Wigen, eds. 2007. *Seascapes: Maritime Histories, Littoral Cultures, and Transoceanic Exchanges*. Honolulu: University of Hawai'i Press.

Ben-Yehoyada, Naor. 2017. *The Mediterranean Incarnate: Regional Formation between Sicily and Tunisia since World War II*. Chicago: University of Chicago Press.

Blaut, J. M. 1993. *The Colonizer's Model of the World; Geographic Difussionism and Eurocentric History*. New York: Guilford Press.
Braudel, Fernand. 1995. *The Mediterranean and the Mediterranean World in the Age of Philip II*. Vol. 1. Berkeley: University of California Press.
Cacciari, Massimo. 1997. *L'arcipelago*. Milan: Adelphi Press.
Chambers, Iain. 2008. *Mediterranean Crossings: The Politics of an Interrupted Modernity*. Durham, NC: Duke University Press.
Choate, Mark. 2008. *Emigrant Nation: The Making of Italy Abroad*. Cambridge, MA: Harvard University Press.
Clancy-Smith, Julia. 2011. *Mediterraneans: North Africa and Europe in an Age of Migration, c. 1800–1900*. Berkeley: University of California Press.
Costanza, Salvatore. 2005. *Tra Sicilian e Africa: Trapani. Storia di una città mediterranea*. Trapani, Italy: Corrao Editore.
Dainotto, Roberto. 2007. *Europe (in Theory)*. Durham, NC: Duke University Press.
Dawson, Helen. 2012. "Archeology, Aquapelagos, and Island Studies." *Shima: The International Journal of Research into Island Cultures* 6, no. 1.
De Certeau, Michel. 1984. *The Practice of Everyday Life*. Berkeley: University of California Press.
DeLoughrey, Elizabeth. 2007. *Routes and Roots: Navigating Caribbean and Pacific Island Literatures*. Honolulu: University of Hawai'i Press.
DeMott. Sarah. 2015. "Mediterranean Intersections: A History of Migration between Sicily and Tunisia, 1830–2015." PhD diss., New York University.
Dewhurst Lewis, Mary. 2014. *Divided Rule: Sovereignty and Empire in French Tunisia, 1881–1938*. Berkeley: University of California Press.
Gabaccia, Donna. 2000. *Italy's Many Diasporas*. Seattle: University of Washington Press.
Heim, Otto. 2015. "Locating Guam: The Cartography of the Pacific and Craig Santos Perez's Remapping of Unincorporated Territory." *New Directions in Travel Writing Studies*, edited by Julia Kuehn and Paul Smethurst, 180–98. Basingstoke, UK: Palgrave Macmillan.
Hoare, Quintin, and Geoffrey Norwell Smith, eds. 1971. *Selections from the Prison Notebooks of Antonio Gramsci*. New York: International Publishers.
Horden, Peregrine, and Nicholas Purcell. 2000. *The Corrupting Sea: A Study of Mediterranean History*. Malden, MA: Blackwell Press.
Hulme, Peter. 1989. "Subversive Archipelagos: Colonial Discourse and the Break-up of Continental Theory." *Dispositio* 14, nos. 36/38: 1–23.
LaTour, Bruno. 2005. *Reassembling the Social: An Introduction to Actor-Network Theory*. New York: Oxford University Press.
Lewis, Martin. 1999. "Dividing the Ocean Sea." *Geographical Review* 89, no. 2.
Lewis, Martin, and Kären Wigen. 1984. *The Myth of Continents: A Critique of Metageography*. Berkeley: University of California Press.
Lionnet, Françoise, and Emmanuel Bruno Jean François. 2015. "Literary Routes: Migration, Islands, and the Creative Economy." Unpublished manuscript.
Mignolo, Walter. 2000. *Local Histories/Global Designs: Coloniality, Subaltern Knowledge, and Border Thinking*. Princeton, NJ: Princeton University Press.

———. 2015. "Forward Yes, We Can." In *Can Non-Europeans Think?*, edited by Hamid Dabashi, viii–xiii. Chicago: University of Chicago Press.

Moretti, Franco. 2013. *Distant Reading*. New York: Verso Press.

Perez, Craig Santos. 2015. "Guam and Archipelagic American Studies." Unpublished manuscript.

Pugh, Jonathan. 2013. "Island Movement: Thinking with the Archipelagos." *Island Studies Journal* 8, no. 1: 9–24.

Purcell, Nicholas. 2003. "The Boundless Sea of Unlikeness? On Defining the Mediterranean." *Mediterranean Historical Review* 18, no. 2: 9–26.

Roberts, Brian Russell, and Michelle Stephens. 2013. "Archipelagic American Studies and the Caribbean." *Journal of Transnational American Studies* 5, no. 1: https://escholarship.org/uc/item/52f2966r.

Roberts, Brian Russell, and Michelle Ann Stephens. 2017. "Introduction." In *Archipelagic American Studies*, edited by Brian Russell Roberts and Michelle Ann Stephens. Durham, NC: Duke University Press.

Schneider, Jane, and Peter Schneider. 1996. *Festival of the Poor: Fertility Decline and the Ideology of Class in Sicily, 1860–1980*. Tucson: University of Arizona Press.

Stoller, Ann Laura. 2009. *Along the Archival Grain: Epistemic Anxieties and Colonial Common Sense*. Princeton, NJ: Princeton University Press.

Stratford, Elaine, Godfrey Baldacchino, Elizabeth McMahon, Carol Farbotko, and Andrew Harwood. 2011. "Envisioning the Archipelago." *Island Studies Journal* 6, no. 2: 113–30.

Tartamella, Enzo. 2011. *Emigranti anomaly: Italiani in Tunisia tra Otto e Novecento*. Palermo, Italy: Maroda.

Verdicchio, Pasquale, ed. 1995. *Antonio Gramsci: The Southern Question*. Boca Raton, FL: Bordighera Press.

Chapter Eight

Archipelagic Deformations and Decontinental Disability Studies

Mary Eyring

If we take the Earth's terraqueous spaces as our vantage point, bodies appear in altered form. No one populated a nineteenth-century maritime world more painstakingly than Herman Melville, who drew from his own experiences at sea to create narratives that did not so much disregard land-based spectrums of disability as precisely invert them. In *Billy Budd, Sailor* (published posthumously in 1924), perhaps the most complicated example of this inversion, the body of a preternaturally beautiful, young, and able sailor distinguishes him from the rest of his crewmates—only to make him a target of fatal political persecution. The novella offers an intriguing portrait of decontinentalized disability produced not by the ship's values (which in fact regularly honored and accommodated a sailor's injuries) but by the ship's archipelagic relationality to other ships within a fleet.

In this chapter I juxtapose Billy Budd's apparent disability, a speech impediment acknowledged by the narrator and recognized by other characters, with what I argue is his counterintuitive but more consequential disability—a body unmarked by the typical signs of sailors' hazardous work. Both disabilities are constructed within an archipelagic environment, where they acquire meaning and finally fatal significance. The two forms at the center of this interpretation—archipelagoes and disabled bodies—relate materially as well as metaphorically.[1] One of the grounding concerns of disability studies is how environments make demands on the human body that contribute to the construction of disability.[2] Because these demands, which can be physical, social, or emotional, differ from one environment to another, a person can be fully capable in one environment and incapacitated in another without undergoing any change. (Famously, for example, the resolute stillness of the scrivener Bartleby becomes a fatal disability only when it collides with the requirements of the urban, capital-driven American office in the story

Melville pointedly subtitled "A Story of Wall-Street.")[3] Sailors labored amid archipelagoes, among other geological forms, and if disability is constituted and understood against the backdrop of specific sites, their itinerant existence must have involved a series of jarring renovations of the very foundations of their selfhood. And crucially, sailors spent the majority of their careers on ships, and I posit that recognizing the archipelagic dimensions of seafaring fleets—the close and often constitutive relationships between individual vessels—gives us new and sorely needed terms in which to understand the (re) production of maritime disability.

While I engage the geographic logic of archipelagoes to explain the particularly fraught construction of sailors' disability amid transoceanic fleets, I find archipelagic thinking equally productive in illuminating the processes through which disability emerged as a metaphor in decontinental settings. Archipelagoes have always nested, as Brian Roberts and Michelle Stephens point out, at the "intersection of the Earth's materiality and humans' penchant for metaphoricity" (2017, 7), since their designation is culturally contingent even as it refers to material realities. The analogy between the designation of an archipelago and the designation of a disabled body or group is too striking to ignore. Both archipelagoes and disabled bodies are produced by a process of conceptual deformation. Here, I follow Roberts and Stephens in embracing the radical and constructive rather than pejorative potential of the term "deformation." Deformation is the first step in reimagining the earth and its elements in new relational configurations (11). When landmasses and waterways are de-formed as continents, islands, seas, and oceans and reconceived as archipelagic assemblages, new possibilities for theoretical understanding and practical cooperation open. And while dominant political, cultural, and religious paradigms have long framed physical and neurological diversity in terms of disability, human subjects can be conceptually deformed to more constructive ends. Imagine, for example, de-forming individuals isolated and disabled according to dominant paradigms in order to re-form them within contexts of social relationships and environments that render supposed disabilities typical and even valuable.[4] While it has radical potential, this process of conceptual deformation has a surprisingly long history. As a case in point, I contemplate the implications of a remarkable fact foregrounded in *Billy Budd, Sailor*: the common (one might even say *normal*) physical deformations of sailors injured during their careers provided the basis for their re-formation as embedded members of crews. As sailors moved from ship to ship, the visible marks of their commitment to a professional fraternity signified across linguistic, racial, cultural, or ethnic difference. In an archipelagic context, deformity was not a disability; it was a shibboleth.

As a basis for exploring these decontinental deformations, I will first consider the familiar metaphorical and material resonances of disability in the late eighteenth century, when *Billy Budd* is set, and the early nineteenth century, when Melville found inspiration for his writing during his own tenure at sea. Leaders in the burgeoning field of early American disability studies speculate that disability—as an idea and a term—is difficult to track in this period because it was so often invoked metaphorically.[5] The word "disabled" and its correlatives—"defective," "monstrous," "impotent," "mad," "lunatic," and so forth—signified a spiritual as much as a physical condition. As the introduction to a recent special issue of *Early American Literature* notes, "The term *disability* existed in a broad constellation of fields, including theology, law, and medicine. Many of these fields relied on the metaphorical weight of the word," which stressed "the moral dimensions of impairment" and "a distance from God or a threat to the body politic" (Altschuler and Silva 2017, 6). *Billy Budd* represents a perspective typical of the period of the novella's setting and Melville's early life: "The moral nature was seldom out of keeping with the physical make" (Melville [1924] 1986, 292). But of course, disability was not pure metaphor: early Americans, like every other group of humans, existed along a continuum of physical capacity, and many bodies were temporarily or permanently incapable of labor—or living.[6]

Over the past thirty years, theorists of disability have shown how statistical modeling, physical environments, and medical innovations combined with the force of literature to construct an image of the able body as a norm against which deviations—at both ends of the bell curve—could be measured.[7] This work tends to focus on the late nineteenth and twentieth centuries, when forces of industrialization were in full swing, and it understandably adopts a land-based perspective. But this model of disability organized around urban infrastructure did not obtain at sea during the Age of Sail. The ship has been compared to a factory, but the ship's orientation toward other ships rather than continental landmasses produced a wholly distinct set of social conditions and a strikingly different workforce. Like islands in an archipelago, individual ships were spatially and culturally distinct and yet intimately connected with other ships in and outside their fleet. In the paragraphs that follow, I engage the concept of the archipelago not only to describe a set of relationships among ships but also to highlight the ways these archipelagic formations allowed for identity formations and deformations unavailable in continental settings. Like archipelagoes, the merchant and naval fleets of eighteenth- and nineteenth-century imperial powers shared a geographic range and a cultural identity transmitted with the sailors who moved frequently among their ships. Also like archipelagoes, they shared values calibrated to local conditions and informed by collective anxieties and aspirations. If we accept a central premise of disability studies, that disability is constructed by cultural as well as physical environments, then how did imperial fleets in archipelagic formations produce alternatives to continental models of normalcy and deviance?

Billy Budd suggests intriguing interpretive possibilities, if not definitive answers. The novella tells the story of a guileless, exceptionally attractive young sailor on the English merchant ship *Rights-of-Man*, pressed into service on an English man-of-war, the HMS *Bellipotent*, in 1797. Like most sailors of the age, Billy moves between ships and fashions his identity in the context of imperial fleets rather than aboard any particular vessel. Billy's affability translates to new environments, but his speech impediment, inconsequential on the merchant ship, becomes a fatal disability on the warship when it prevents him from defending himself against a charge of conspiring to mutiny. Instead, in mute frustration, he strikes and kills his accuser, an officer, and is summarily tried and hanged for the offense. Although the action takes place entirely at sea, it is grounded upon the decks of ships that exist in relation to literal islands and archipelagoes (both England and the West Indies feature prominently) and function as archipelagoes, connected in imperial enterprise and yet so culturally specific that Billy thrives on one and dies on another. Ships are more than conveyance; like islands in an archipelago, they are complex, expansive worlds and the only homes Billy knows. The novel introduces him as a character "habitually living with the elements and knowing little more of the land than as a beach, or, rather, that portion of the terraqueous globe providentially set apart for dance-houses, doxies, and tapsters" (301). Shores are at the margins of Billy's consciousness; the sea and English merchant and naval fleets, like dynamic wooden archipelagoes, are at the center.

In a story that flips traditional notions of what is central and what is peripheral, it is not surprising that the expected poles of normalcy and deviance are exactly reversed. It is not Billy's obvious incapacity—his speech impediment—that first disables him on the *Bellipotent*. In *Billy Budd* the typical seaman is affectionately rendered as one "of few words, many wrinkles, and some honorable scars" (318–19). Scars, stutters (or "few words"), and "wizened faces" all abound aboard the man-of-war, suggesting that disability was expected rather than remarkable at sea (319). Records of seafaring life during the nineteenth century indicate that sailors were particularly prone to losing their eyesight, hearing, or an appendage in the course of their work on merchant ships. On naval ships, these threats were added to those of military engagements. All sailors labored under the threat of what Peter Linebaugh and Marcus Rediker call "slavish, hierarchical discipline" (2000, 150), so violent in nature that it inspired mutiny and spawned the Atlantic's golden age of piracy. Disability was, in other words, a norm. More importantly, like all norms, it bound seamen to a majority of the members of their archipelagic community—not just on a single vessel but on all the interconnected ships they would temporarily call home. Disability was something that sailors shared and came to regard as a mark of experience and commitment to seamen across the fleet. It was a manifestation of self-sacrifice in the line of duty, of a sublimation of the self for the good of the larger enterprise. The striking fact, then, is not that so many real and fictional sailors and especially *captains* (Melville's Ahab is the iconic example) were disabled but that a sailor could distinguish himself without this sign of his courage and allegiance to the seafaring fraternity.

What was deviant in the archipelagic context of imperial seafaring fleets was not physical disability but the lack of it. Billy Budd *is* disabled on both the *Rights-of-Man* and the *Bellipotent* in the familiar sense of being stigmatized and excluded, but in a way particular to his maritime existence. The so-called Handsome Sailor, unencumbered and undistinguished by a speech impediment, is repeatedly encumbered and distinguished on both ships by his aberrant physical beauty. Consider this passage:

> Cast in a mold peculiar to the finest physical examples of those Englishmen in whom the Saxon strain would seem not at all to partake of any Norman or other admixture, he showed in face that humane look of reposeful good nature which the Greek sculptor in some instances gave to his heroic strong man, Hercules. But this again was subtly modified by another and pervasive quality. The ear, small and shapely, the arch of the foot, the curve in mouth and nostril, . . . but above all, something in the mobile expression, and every chance attitude and movement, [was] suggestive of a mother eminently favored by Love and the Graces. (Melville [1924] 1986, 299–300)

The description goes on, but what becomes clear immediately is that Billy is repeatedly isolated and disadvantaged by precisely these physical characteristics. For example, he is the only crew member aboard the *Rights-of-Man* to be pressed into the king's service on the *Bellipotent*, the ship on which he eventually hangs.

> Plump upon Billy at first sight in the gangway the boarding officer, Lieutenant Ratcliffe, pounced, even before the merchantman's crew was formally mustered on the quarter-deck for his deliberate inspection. And him only he elected. For whether it was because the other men when ranged before him showed to ill advantage after Billy, or whether he had some scruples in view of the merchantman's being rather shorthanded, however it might be, the officer contented himself with his first spontaneous choice. (293)

Time and again, Billy's exceptional body makes him a target of this fatal favoritism. His appearance arouses the envy and ultimately the ire of John Claggart, the officer who accuses Billy and dies by his hand. Their relationship is defined by physical difference: "In view of the marked contrast between the persons of the twain, it is more than probable that when the master-at-arms applied to the sailor the proverb 'Handsome is as handsome does,' he there let escape an ironic inkling, not caught by the young sailors who heard it, as to what it was that had first moved him against Billy, namely, his significant personal beauty" (327). They are a study in contrasts: Billy possesses "a purity of natural complexion where, thanks to his seagoing, the lily was quite suppressed and the rose had some ado visibly to flush through the tan"; Claggart's pallid complexion "seemed to hint of something defective or abnormal in the constitution and blood," and his chin was disfigured by a "strange protuberant heaviness" (298, 314). But it is Claggart, not Billy, who holds the rank of officer, with all the privileges that implies. It would seem, then, that certain "defects" are associated with an ability to successfully integrate into and advance within the ships' community. Billy Budd, exceptionally abled, hangs aloft and alone at the end of the novel, an isolated object of pity. Claggart, typically disabled (which is to say, at sea, normal), operates well within the *Bellipotent*'s protective social and legal networks and obtains a position of such power that to strike him—whether the blow is fatal or not—is a capital offense.

Clearly, ships' networks of relationships reassigned the values of the standard bell curve to acknowledge the fact that most sailors were disabled over the course of an ordinary career at sea. Uncommonly serious injuries disqualified a sailor from the profession, so there was still a sense in which falling far below the norm excluded one from maritime communities. But to fall far above the norm, as Billy does, suggests not a hyperfitness for seafaring life but an incompatibility with it. For on ships, particularly on a military ship during a time of war, such as the *Bellipotent* in 1797, Billy's classically ideal physique signals a lack of experience with the hazards of maritime life that bound sailors together with a sense of shared sacrifice. The novel introduces him in ways that underscore this inexperience: he is "Billy Budd—or Baby Budd, as more familiarly, under circumstances hereafter to be given, he at last came to be called" (293). Billy is twenty-one when he is impressed—not unusually young—and yet he seems to grow younger as the novella wears on: he *at last* comes to be called Baby Budd. That is, his beauty increasingly seems to indicate professional immaturity. In contrast, the oldest crewman on the *Bellipotent* wears his experience on his face: "a long scar like a streak of dawn's light falling athwart the dark visage" (319). As his compatriots acquire these visible marks of their tenure, Billy's beauty becomes an alarming indication of his isolation from the crew, perhaps even his narrow self-interest. In a military context, his persistently able body is even more alienating. Lisa Herschbach points out that Confederate soldiers after the Civil War rejected Northern industrialists' attempts to re-member disabled veterans with mass-produced prosthetics. They sought instead to remember the war. "They held that showing, not hiding, their disfigurement was part of a broadly perceived social imperative against historical amnesia" (Herschbach 1997, 51). After armed combat, disability manifested commitment not just to a crew but to a *cause*. But when Billy fights, as when he lashes out at Claggart, it is to defend himself rather than to advance a communal objective. It is appropriate that we first meet the Handsome Sailor aboard the *Rights-of-Man*: as his unblemished body comes increasingly to signify, Billy is concerned with these rights in the singular rather than the plural.

In the literature that documents maritime experience, sailors frequently advance to positions of prominence not in spite of disability but precisely because their physical appearance manifests a familiarity with the perils of seafaring life and testifies to their inclination to subordinate self-interest to the concerns of vessels and crew. The accoutrements of the iconic sea captain—a wooden or metal prosthetic limb, a scar, or an eye patch—function as rank insignias. Surveying Billy's predicament from land, where shared language often forms a basis for national identification, the stutter that impedes his communication with Claggart looms large. But from an archipelagic perspective, we appreciate that sailors constantly found ways to transcend language barriers as ships picked up crewmembers around the globe. Surely, Claggart and Billy might have relied on these strategies to make themselves mutually comprehensible. Far more remarkable is that Billy finds himself pressed to defend himself against Claggart's baseless charge in the first place. Had he shared the prized deformations of the average sailor, he would never have been conscripted into service on the *Bellipotent* and would never have become an object of Claggart's toxic attention. In this context, to adapt a line from Lennard Davis (2010), to be "average"—that is, to be deformed before the mast—is, "paradoxically, a kind of ideal, a position devoutly to be wished" (5).

The specifically archipelagic context of Melville's work—which frames English ships like the *Bellipotent* always in relation to other English ships, like the *Rights-of-Man*—brings to the fore the ways deforming processes involved the body as well as the mind. Both archipelagic studies and disability studies advocate new methods of reading that reorient us toward more integrated geographies and epistemologies.[8] These reading practices encourage us to consider the sailors aboard naval and merchant fleets as products of archipelagic interchange as much as economic or continental political initiatives.[9] In a reading guided by these commitments, Billy Budd is disabled in his maritime context in ways that transcend the physical. Before the narrator mentions Billy's stutter, he describes the episode that establishes the environmental basis for Billy's exclusion and eventual execution. At the beginning of the novella, Billy is pressed into service from the *Rights-of-Man* onto the *Bellipotent*, but in an atypical and particularly crippling way: alone. Recall the description of the boarding officer "pounc[ing]" upon Billy and the fact that "[Billy] only he elected." The narrator stresses that the officer's actions here are unusual, even violent. "To the surprise of the ship's company, though much to the Lieutenant's satisfaction, Billy made no demur. But, indeed, any demur would have been as idle as the protest of a goldfinch popped into a cage" (Melville [1924] 1986, 293). Although Billy cannot resist impressment, both his mates and the boarding officer *expect* him to protest; their surprise alerts us to a break with maritime norms as the simile alerts us to an ecological disruption. The ship is by design an expansive and connected environment, operating archipelagically within a transglobal fleet, but for this "goldfinch popped into a cage," it becomes an unnaturally solitary and confining space.

Indeed, the boarding officer has nullified one of the chief virtues of seafaring life. The "motely retinue" is apparently meant to insulate the sailor as he circulates from ship to ship and ship to land (292). Even if this retinue comprises only a few sailors, these sailors provide key points of contact between a sailor's history and his new surroundings. John Claggart's mates, for example, transmit the "facts" of his multivessel professional career to counter salacious "gossip" and verify his habitual "sobriety," "deference," and "austere patriotism"—attributes that qualify him for a series of promotions (317). Because Billy is isolated as he is stripped from his environment aboard the *Rights*, he is unable to forge organically the social connections that should establish his reputation and protect him from a sham charge and a fatal sentencing. Billy has many character witnesses—but as it turns out, only aboard the *Rights*. The captain of that vessel rhapsodizes, "Before I shipped that young fellow, my forecastle was a rat-pit of quarrels. It was black times, I tell you, aboard the *Rights* here. . . . But Billy came; and it was like a Catholic priest striking peace in an Irish shindy. Not that he preached to them or said or did anything in particular; but a virtue went out of him, sugaring the sour ones. They took to him like hornets to treacle" (295). This testimony should have been conveyed through the human networks that connected ships across the globe, but Billy is wrenched from his. This development initiates all the disabling processes that follow. Billy's disability is first and ultimately his dislocation and disorientation in a maritime world organized around social and environmental relations.

This decontinentalized conception of disability, which relies on both a literal and a metaphorical reading of the word "deform," sits at the crosscurrent of archipelagic studies and disability studies. The forms of landmasses, like the forms of human bodies, acquire meaning in a clash of cultural and epistemological contests that determine the orientation of center and periphery—or normalcy and deviance. "Deformation," one of the terms through which archipelagoes are discursively constructed, creates a crucial conceptual link to the language and history of disability. In their introduction to *Archipelagic American Studies*, Brian Roberts and Michelle Stephens trace the geological origins and political fortunes of American islands across the globe, but their careful attention to geography anchors rather than eclipses more abstract approaches to the archipelagic:

> While topography involves the study of the surface shape and features of the earth's terrain, topology is concerned with more abstract relations between spatial entities. The level of abstraction available through topology means that spatial surfaces may take a variety of forms, or deformations. Taken to the extreme (e.g., in cases of extreme twisting or stretching), topology reveals the multiple shapes a single surface may take before undergoing, finally, a fundamental ontological shift. At these shift points (points of breaking or tearing), a shape or feature assumes a new topology. (2017, 11)

As Roberts and Stephens engage the term, a "deformation" is not the degradation of an existing form but the appearance, through a different epistemological lens, of a new form. Archipelagoes emerge when landmasses and waterways are seen to relate in some critical way—either materially (e.g., tectonically or spatially) or metaphorically (e.g., culturally, politically, economically, ethnically, or linguistically). An archipelago is an assemblage of land and water conceptually deformed so the constituent parts are revealed as operating systemically rather than in isolation, and the relationship reveals new possibilities for meaning and experience. In the history of disability, deformation has been an analogous process, involving the body as well as the mind: the physically and conceptually deformed sailor came into focus not as an isolated, disabled subject but as a weight-bearing joint in a complex network. A sailor injured in the course of his career was de-formed from a solitary individual into an imbricated member of a seafaring community instantiated in fleets of ships. As a paradigmatic shift well underway in American studies, island studies, and oceanic studies, the deformation of familiar geological forms (the continent chief among them) has brought the archipelagic toward the center of the frame. In disability studies, I posit, we might deform human subjects—even superlatively able-bodied ones like Billy Budd—to unsettle still-powerful myths of normalcy and more fully appreciate how disabilities are constructed and perceived relationally.

Even though recent publications in archipelagic studies do not take up disability as a central concern and the 2017 disability-focused issue of *Early American Studies* does not refer to archipelagoes, the fields can clearly be brought into productive conversation to illuminate the experiences of embodied individuals existing in relation to oceans, archipelagoes, and continents in America before 1840. Bringing archipelagic interchanges to the fore reveals a remarkable—but as yet unremarked—fact: disability did not operate (in a material sense) or signify (in a metaphorical sense) in archipelagic settings as it did in continental settings.

When we bring the concerns of archipelagic studies to bear on the study of disability, we can appreciate that Billy Budd is not a disabled sailor moving from vessel to vessel; he is a sailor aboard the *Rights-of-Man* who is disabled when he is wrenched from his position on one ship without the credibility of a deformed body or the insulation an archipelagic community should provide. Fleets only function archipelagically when the sailors who move among them act as conduits of social practice and cultural history that transcend local geographies. The boarding officer who takes Billy from his environment interrupts a circuit of archipelagic sympathy. As he is forced to reinvent himself, alone, on the *Bellipotent*, Budd's undeformed body suggests immaturity and even cowardice, and the demand that he speak forcefully and strategically disables and finally destroys him. Billy's ontological shift from able-bodied to disabled mirrors—in fact, is produced by—the violence of impressment that disregards archipelagic forms of community and assumes the disconnectedness and interchangeability of individual sailors and ships.

This analysis is a preliminary example of the insights available when archipelagic thinking is brought to bear on narratives of disability. Not only do these insights underscore the truly global sweep of early American and nineteenth-century studies, but they also help us understand theories of embodiment that run counter to those of other historical eras—including our own. The apparently topsy-turvy view of human ability that equates a sailor's deformities with his credibility and fraternal commitment assumes formal coherence and significance through the lens of archipelagic relationality.

PRACTICAL AND METHODOLOGICAL IMPLICATIONS

This chapter builds on a foundational premise of disability studies, that environments contribute to the construction of an individual's or group's ostensible disability, by tracing the construction of disability in archipelagic settings during the zenith of imperial shipping in the late eighteenth and early nineteenth centuries. This chapter proposes that the production of mariners' disability (a strikingly common phenomenon, given their occupational hazards) can only be fully understood if we foreground archipelagic relationality—or the close, often fraught relationship of ships to other ships in a fleet. This engagement not only brings archipelagic studies and disability studies into productive conversation but also illustrates the constructive potential of "de-forming" approaches to geological and human bodies. In archipelagic studies (and potentially, as I posit in this chapter, in disability studies), the move to deform a familiar construct is a first step toward reforming it more generously as part of a constitutive assemblage. Just as archipelagic studies has employed a process of deformation to decenter the continent and acknowledge the entanglements of continent, island, and water groupings, this chapter models an exercise in deforming bodies in order to unsettle the myth of the normal body and understand how specific sites and archipelagic assemblages reformed sailors' bodies according to decontinental values and thus produced distinctive versions of both able and disabled subjects.

NOTES

1. My goal here is to balance material and epistemic conceptions of disability by adapting approaches from archipelagic studies and, earlier, oceanic studies. Hester Blum (2010) begins an essay titled "The Prospect of Oceanic Studies" with a counterargument: "The sea is not a metaphor" (670). And yet her essay works with others in a theories-and-methodologies cluster in *PMLA* to demonstrate that the sea, like the disabled body, is not *merely* a metaphor, although it is undoubtedly a powerful and seductive one. For more examples of this approach in oceanic studies, see Cohen 2010.
2. In the words of Sarah Jaquette Ray and Sibara (2017), disability studies "takes as a starting point the contingency between environments and bodies" (1).
3. For an illuminating analysis of Bartleby's disability in the context of his nineteenth-century urban environment, see Murray 2008, 50–75.
4. My point here is that while every human being is physically incapacitated at some point in his or her life, only some are ever considered disabled. David Mitchell and Sharon Snyder (2000) have called disability "the master trope of human disqualification" (3). Rachel Adams, Benjamin Reiss, and David Serlin (2015) elaborate: "Disability encompasses a broad range of bodily, cognitive, and sensory differences and capacities. It is more fluid than most other forms of identity in that it can potentially happen to anyone at any time, giving rise to the insiders' acronym for the nondisabled, TAB (for temporarily able-bodied)" (5–6).
5. According to Helen Deutsch (2015), deformity "reigned supreme in the eighteenth century" (52). Keyword searches of early American texts on digital databases confirm this (Altschuler and Silva 2017, 1).

6. Recently Shaun Grech has made an analogous argument about the conception of "colonialism" in disability studies. In a special disability-focused issue of *Social Identities*, he argues, "The global colonial encounter is not simply a metaphor and cannot be bypassed in any global disability analysis" (Grech 2015, 6).

7. Here, I reference the work of Lennard Davis (2006) and Allan Sekula (1986), although Rosemarie Garland-Thomson, Harlan Lane, Sharon Snyder, David Mitchell, and many others have contributed in crucial ways to our understanding of the ways disability was constructed in European, British, and American narrative and history.

8. For more on "disordered epistemologies" in narratives and histories of colonial disability, see Siebers 2010; Altschuler and Silva 2017, 10.

9. Brian Russell Roberts and Michelle Stephens (2017) elaborate on the crucial difference between *island interchangeability* and *island interchange* in their introduction to *Archipelagic American Studies*, especially pages 29–35.

WORKS CITED

Adams, Rachel, Benjamin Reiss, and David Serlin. 2015. "Disability." In *Keywords for Disability Studies*, edited by Rachel Adams, Benjamin Reiss, and David Serlin, 5–11. New York: New York University Press.

Altschuler, Sari, and Cristobal Silva. 2017. "Early American Disability Studies." *Early American Literature* 52, no. 1: 1–27.

Blum, Hester. 2010. "The Prospect of Oceanic Studies." *PMLA* 125, no. 3: 670–77.

Cohen, Margaret. 2010. "Literary Studies on the Terraqueous Globe." *PMLA* 125, no. 3: 657–62.

Davis, Lennard J. 2010. "Constructing Normalcy: The Bell Curve, the Novel, and the Invention of the Disabled Body in the Nineteenth Century." In *The Disability Studies Reader*, edited by Lennard J. Davis, 3–16. 2nd ed. New York: Routledge.

Deutsch, Helen. 2015. "Deformity." In *Keywords for Disability Studies*, edited by Rachel Adams, Benjamin Reiss, and David Serlin, 52–54. New York: New York University Press.

Grech, Shaun. 2015. "Decolonizing Eurocentric Disability Studies: Why Colonialism Matters in the Disability and Global South Debate." *Social Identities* 21, no. 1: 6–21.

Herschbach, Lisa. 1997. "Prosthetic Reconstructions: Making the Industry, Re-making the Body, Modeling the Nation." *History Workshop Journal* 44, no. 1: 22–57.

Linebaugh, Peter, and Marcus Rediker. 2000. *The Many-Headed Hydra: Sailors, Slaves, Commoners, and the Hidden History of the Revolutionary Atlantic*. Boston: Beacon Press.

Melville, Herman. [1853] 2002. "Bartleby, the Scrivener: A Story of Wall-Street." In *Melville's Short Novels*. Edited by Dan McCall, 3–34. New York: Norton.

———. [1924] 1986. *Billy Budd, Sailor and Other Stories*. New York: Penguin.

Mitchell, David T., and Sharon L. Snyder, eds. 2000. *Narrative Prosthesis: Disability and the Dependence of Discourse*. Ann Arbor: University of Michigan Press.

Murray, Stuart. 2008. *Representing Autism: Culture, Narrative, Fascination*. Liverpool, UK: Liverpool University Press.

Ray, Sarah Jaquette, and Jay Sibara. 2017. "Introduction." In *Disability Studies and the Environmental Humanities: Toward an Eco-Crip Theory*, edited by Sarah Jaquette Ray and Jay Sibara, 1–28. Lincoln: University of Nebraska Press.

Roberts, Brian Russell, and Michelle Ann Stephens. 2017. "Archipelagic American Studies: Decontinentalizing the Study of American Culture." In *Archipelagic American Studies*, edited by Brian Russell Roberts and Michelle Stephens, 1–54. Durham, NC: Duke University Press.

Sekula, Allan. 1986. "The Body and the Archive." *MIT Press* 39 (winter): 3–64.

Siebers, Tobin. 2010. *Disability Aesthetics*. Ann Arbor: University of Michigan Press.

Chapter Nine

Digital Currents, Oceanic Drift, and the Evolving Ecology of the Temporary Autonomous Zone

Lisa Swanstrom

> Ours is the first century without *terra incognita*, without a frontier. Nationality is the highest principle of world governance—not one speck of rock in the South Seas can be left open, not one remote valley, not even the Moon and planets. This is the apotheosis of "territorial gangsterism." Not one square inch of Earth goes unpoliced or untaxed . . . in theory.
>
> —Hakim Bey, *The Temporary Autonomous Zone* (1991)

> Hail, Poseidon, Holder of the Earth, dark-haired lord! O blessed one, be kindly in heart and help those who voyage in ships!
>
> —Homeric hymn to Poseidon (seventh century)

In 2013, trident-wielding Poseidon—ancient Olympian god of the sea, brother of Zeus, father of Polyphemus, enemy of Odysseus, and (according to *The Onion*) rumored father of Michael Phelps—communicated a message to a group of artists who were spread across the Mediterranean Sea, waiting to receive his divine transmissions. As one might expect, and in keeping with the historic record of divine communiqués from other members of the Greek pantheon, the message was cryptic, difficult to decipher, and devoid of human syntax. For the artists who awaited it, however, it was a success. Created by Reza Safavi, Max Kazemzadeh, and an international team of collaborators, *Poseidon's Pull* was a multilayered and multimodal digital art installation meant to provide a communication interface between human beings and the ancient god. As initially comical as this project might seem, in this chapter I contend that works such as *Poseidon's Pull* highlight the potential of digital art to resist the totalizing authority of nations through their remediation of natural forces. In particular, by focusing on the power of underwater currents and oceanic drift, I argue that these works of art reorganize the logic of national borders, blur the boundaries of traditional aesthetic form, encourage transhistoric connections, and contribute to archipelagic discourse in a way that both sheds light on current methodological tensions within the digital humanities (DH) and updates and reinvigorates Hakim Bey's (1991) concept of the "Temporary Autonomous Zone."

ARCHIPELAGIC CURRENTS

As model, metaphor, and material reality, the archipelago provides a rich framework for rethinking assumptions about culturally contingent, materially specific, and naturally occurring organizing structures. In the field of geography, it emerges initially as a corrective to "continental" models of organization. In *The Myth of Continents*, one of the first important studies to identify this problem, Martin W. Lewis and Kären E. Wigen write, "Despite its ubiquity and commonsensical status . . . the . . . continental scheme . . . obscures more than it reveals. An obsolete formation, this framework is now wholly inadequate for the load it is routinely asked to carry" (1997, xiv). In their introduction to *Archipelagic American Studies*, "Decontinentalizing the Study of American Culture," Brian Russell Roberts and Michelle Stephens (2017) foreground the importance of the archipelago as a remedy to the model of continental organization. Not only has the continental model perpetuated an error in our understanding of geographic reality, Roberts and Stephens convincingly show, but it has also served to obscure historical relations and oversimplify the complex, porous, and shifting nature of national boundaries. As Roberts and Stephens demonstrate, this reorientation has far-reaching implications for how we conceive of everything from ecological systems to geological epochs to the formation of national boundaries to the foundation of individual identity. Their work encourages a reorientation of perspective and provides a malleable concept for thinking about the archipelago itself as a site of resistance, both as a metaphoric construct and material, geographic reality. As Roberts and Stephens mention in their introduction, the archipelagic perspective signifies "a push and pull between the metaphoric and the material" (6).

While our understandings of natural forces, which are also a result of such a "push and pull," are always provisional and culturally contingent, it seems to me that such forces are currently underexplored within both archipelagic thought and the digital humanities. If, as Roberts and Stephens suggest, a successful vision for a truly archipelagic perspective "involves attentiveness to what George B. Handley describes as 'the phenomenological encounter with natural forms'" (10), then it will be well worth the effort to provide more opportunities to investigate such encounters.

And yet a focus on the natural environment to the exclusion of human interest is equally problematic and has in fact been a self-prescribed limitation of my own research. In previous work I have attempted to resist the worn story that pits technology against nature and discussed at length the importance of digital aesthetics in reframing environmental discourse. Such topics continue to inspire in the context of archipelagic thought. For example, I am tempted to turn to recent works of electronic literature[1] that engage with the power of natural forces, particularly of seas and waterways, such as Jan Baeke and Alfred Marseille's (2012) "Channel of the North," a poem whose shape and size fluctuates "as a function of the tide in the Westerschelde river on the Dutch/Belgian border," or "Station 51000," a Twitter bot created by Mark Sample that "speaks" from the perspective of a National Oceanic and Atmospheric Administration's data buoy, "now lost at sea but still generating data" (Jørgensen 2015).[2] Both works are excellent examples of digital art's capacity to translate the powerful patterns of the natural world into human readable signs. In that they highlight primarily the agency of natural spaces, however, they are not consistent with the archipelagic tendency to foreground politics, to pay heed to historically fraught relations between cultural and natural ecologies. As Édouard Glissant writes in the *Poetics of Relation*, an environmental aesthetics must look honestly at the way the two are entangled: "An aesthetics of the earth? In the half-starved dust of Africas? In the mud of flooded Asias? In epidemics, masked forms of exploitation, flies buzz-bombing the skeleton skins of children? In the frozen silence of the Andes? In the rains uprooting favelas and shantytowns? In the scrub and scree of Bantu lands. . . . In city sewers? Yes. But an aesthetics of disruption and intrusion" (1997, 151). Instead, then, of isolating natural forces as if they could ever be wholly divorced from politics or free of cultural mediation, I shall keep Glissant's urgent admonition in mind as I proceed in my analysis of *Poseidon's Pull*, as well Reza Safavi's collaborative artistic practice in general.

Poseidon's Pull

Poseidon's Pull was an intricately staged, meticulously plotted, highly technological, and rigorously executed effort to create a communication interface between humankind and the ancient god of the Aegean and Ionian Seas. It was also nearly entirely ridiculous. Inspired to some extent by traditional modes of ancient divination, the artists mapped a constellation of Poseidon upon the sea—a celestial blueprint for a watery surface that would honor the timeless, sacred relationship between water and sky.[3] Since there was no

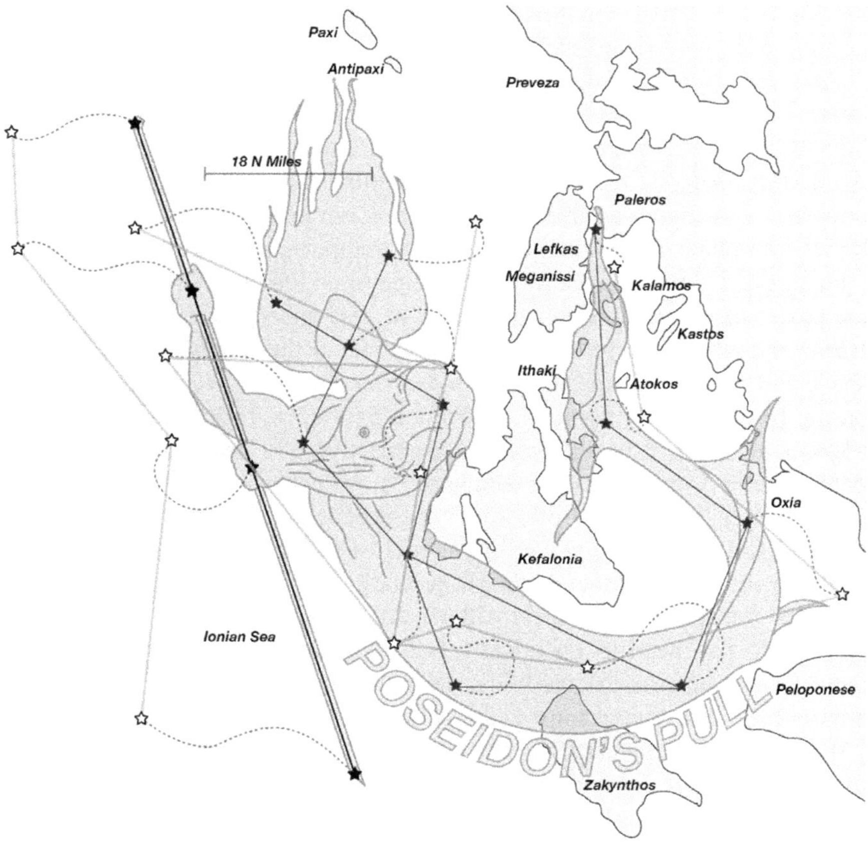

Figure 9.1. Constellation of Poseidon, from *Poseidon's Pull.* Initial Collaborators: Reza Safavi, Max Kazemzadeh, Joe Hicks, Hytham Nawar, Adnan Naseem. Additional Collaborators: Stavros Didakis, FOUNDLAND: Lauren Alexander and Ghalia Elsrakbi, Andrea Safavi, Mark Kazemzadeh, Manoutch Kazemzadeh, Jay Baek, Zengin, Princess Powers, Cavon Ahangarzadeh, Garric Simonsen

single constellation of Poseidon already in existence, however, they had to create one. Comprised of fifteen stars, the sea god's sidereal portrait, when fully illustrated, included his flowing tresses, a fishtail for a lower body, and an impressively sculpted masculine torso. It also featured a long spear, easily taken for a trident (see Figure 9.1).[4]

Once the constellation was created and established, as it were, in the firmament, the artists assigned each of its stars to latitude and longitude coordinates on the Mediterranean Sea. With these points in place, the team of artists, who refer to themselves in the project description as "Poseidon's titan mermen" (Safavi et al. 2013), traveled by boat to each of them, marked their respective locations, and then stopped, allowing themselves to drift on the sea for six hours from that initial point. Once the six hours were up, they took bearings again and measured them in relation to where they had begun. When they assembled this information, they remapped the constellation according to the shift. And it was through this differential—basically, a vector of ocean space created by the drift—that Poseidon's message was revealed. How so? The answer provided by the artists may well confound rather than dispel confusion:

> "Poseidon's Pull" is an international collaborative project with the purpose of creating a language that will reveal a message from the "God of the Sea," Poseidon, one of the twelve Olympian deities of the pantheon in Greek mythology, whose domain is the ocean. . . . Since Poseidon's domain is the ocean and waterways, he controls the currents and affects the GPS drifts of collaborators around the world, and therefore is controlling the morph of the constellation drawing in the Ionian Sea. (Safavi et al. 2013)[5]

A clear explication of Poseidon's divine will may never emerge, but what the project does manage to emphasize, with rather brilliant clarity, is the sea's power as a mediating agent. *Poseidon's Pull* involves several layers of mediation—the construction of a constellation, the assignation of stars to points on the sea, the decision to let the boat drift for six hours, and so forth. Each layer helps reframe the natural pull of the ocean current, reminds us that it is removed from human control, and places it squarely in the role of a naturally occurring force, even as it employs digital technology to do so. Digital technology, so often described in terms of its adherence to a rigid, rational taxonomy of abstraction (see McPherson 2012), in this case takes a subsidiary role to the power of the sea.

In later phases of the project, communication becomes even more elaborate, employing the newly formed constellations as a way to query Google images to indicate Poseidon's blessed intent (i.e., "messages from Poseidon that are communicated through his gesture, aka. the control of the sea currents"). In "*Poseidon's Pull* Revisited: Channeling the Gestures of a God", Safavi and Kazemzadeh (2018) have added another component: "a system that used EEG headsets to identify if they were in a meditative state. GPS tracking would only record while the artists were in that meditative state, and once they dropped out of that state, the GPS track would end." These meditative states, they argue, synchronize with Poseidon's own gestures and therefore both submit to and confirm the power of his drift (Safavi 2019).

These layers, admittedly arbitrary and shamelessly whimsical, create a Rube Goldberg machine for oceanic divination that nevertheless functions as an effective aesthetic trope.[6] By repurposing the Mediterranean and restoring it to an ancient authority, the artists mine its cultural and geographic history.

By calling attention to the pull of the current and charting the drift that results from their own *inactivity*, they highlight the power of the ocean, its measurable and forceful patterns, and their relative powerlessness against it. And by naming Poseidon as the author of this force the artists suggest, through a baroque communication system of their own invention, that it is this divine entity that controls the ocean. The drift they record is a watery rambling they observe rather than instigate. Their resultant map, revised according to the sea's own impulses, offers a compelling alternative to traditional aesthetic form.[7]

As Amitav Ghosh (2016) writes in *The Great Derangement*, his provocative study of modern literature's failure to confront climate change, traditional narrative is unable to address the power of natural agents precisely because it adheres to an overtly rational framework for understanding them—as measurable, knowable, conquerable. In modern literature, especially in works of realism, it is the "concealment of . . . exceptional moments that serve as the motor of narrative" (16). The realist novel is ill equipped to represent the chaotic, unruly, and unhuman forces of nature. "It is thus that the novel takes its modern form," Ghosh notes, "through 'the relocation of the unheard-of toward the background . . . while the everyday moves into the foreground'" (17). The reverse occurs in *Poseidon's Pull*, which is not beholden to the strictures of the modern novel or, for that matter, to any preestablished aesthetic. Rather than jamming natural forces into a canned narrative and trimming away the parts that do not fit, it instead functions overtly as space where the natural environment manifests its agency. The waters of the ocean are both mediated and mediating forces. They exceed conventional aesthetics. And yet the work that emerges nevertheless coheres into a form, as a network, a constellation of points that the ocean has itself connected, even as its original boundaries are repeatedly revised. It also creates a suggestive alignment between an arbitrarily constructed constellation comprised of stars and the social and political organization of island chains into archipelagoes.

This notion of shifting boundaries is a key concept in archipelagic thought. Philip Schwyzer and Simon Mealor put it succinctly in *Archipelagic Identities: Literature and Identity in the Atlantic Archipelago, 1550–1800*, when they assert that "the essence of what we might term the archipelagic perspective lies in a willingness to challenge traditional boundaries—boundaries, that is, between the histories of different nation-states, and also between academic disciplines" (2004, 4). *Poseidon's Pull* helps redraw the boundaries of the Mediterranean so that the sea speaks to the expansive, overflowing, global, and powerful forces that its waters connect and embody. The power of drift also recovers what many historians believe to have been an important factor in ancient trading routes in the Mediterranean, which were importantly not defined by national boundaries but "determined mostly by ancient Mediterranean geography, wind and current patterns, the locations of anchorages and ports, river mouths, and offshore islands" (Malkin 2013, 21). The power of the drift, of movement, is such that it dislocates known boundaries. By foregrounding the pull of currents and the force of tides, by calling attention to a shared space, controlled not by human law but rather by natural impulses and "divine" intent, *Poseidon's Pull* reframes the Ionian Sea and revives a history that defies a modern epistemological understanding of the region.[8]

As it disrupts known borders, *Poseidon's Pull* makes transhistoric connections, helping to recover something that linguistic evolution has obscured over time, namely, the significance of this region to the current conception of the archipelago. Roberts and Stephens call for precisely this task of making connections when they note the associative potential of the archipelago in terms of the ancient rhetorical technique of catachresis: "[For] Roman rhetorician Quintilian, catachresis is 'the practice of adapting the nearest available term to describe something for which no actual term exists,' as the tooth of a comb or the leg of a table. We want to frame archipelago-formation in terms of this trope of catachresis," they assert, "whereby archipelago itself becomes a term deployed in the attempt to name connections" (2017, 30).

Catachresis—then and now—is defined as a misuse in language (indeed, the term *catachresis* in Latin is synonymous with *abusio*, or the "improper use of a word," and the Greek *κατάχρησις* signifies "excess") ("Catachresis," n.d.). Yet perhaps this type of linguistic error is more fruitfully viewed as a process of errantry, etymologically closer to "traveling and wandering" than incorrect usage or linguistic abuse ("Errant," n.d.). Seen in this way, catachresis highlights linguistic evolution and provides a powerful way to think about the concept of the archipelago as a connecting agent. It also helps us think about the concept of drift, both as a naturally occurring consequence of ocean currents and as an aesthetic concept that, prompted by the movement of natural forces, connects points on an associative chain.

The use of catachresis in Saint Augustine is also suggestive. In "On the Christian Doctrine," Augustine writes, "Who does not use the word piscina [basin, pool, pond tank, or other large container of water] for something which neither contains fish nor was constructed for the use of fish, when the word itself is derived from *piscis* (fish)? This trope is called catachresis" (2001, 158). Augustine's use of the term in this passage is less about catachresis as error and more about the haunting nature of etymology, about linguistic skeumorphs that both supersede and preserve history. The etymology of the word "archipelago" itself is instructive along these lines; the term is a combination of two Greek roots, *archi-*, which means "primary" or "first," and *pelagos*, which means "sea." Although there is no record of the use of this portmanteau in the ancient world, by the turn of the sixteenth century it had come to signify the Aegean as the "chief sea"; by the turn of the seventeenth century it referred, by association—catachresis—to the many clusters of island chains that this ocean surrounded ("Archipelago," n.d.). A catachrestic excavation reveals that the concept of the archipelago was linguistically inspired first by the sea; only centuries later did this concept signify an organization of islands into chains, constellations, and nations.[9]

Poseidon's Pull employs catachresis through its associative, playful communication system, its comic excavation of ancient myth, and its literal, spatial drift from point to point, traversing land and sea. And in terms of archipelagic studies, the project helps highlight continuity—suggesting in every one of its iterations that land, sea, sky, and stars are connected, regardless of national boundaries.[10]

The project also points, however, to Greece's role today in the larger Aegean, a site of crisis in terms of the current influx of refugees that has affected the entire European continent. It reminds one of the special relation that Poseidon held to seafarers—as protector, as destroyer—in the ancient world. It is risky to situate such a playful work in a site fraught with such terror and vulnerability, lest the levity of the first undermine the gravitas of the second. While it is too severe to claim that "all discourses of nature that would offer visions of hope are deceptive or dangerous," it is important to acknowledge that "all apprehensions of nature—be they a pastoral landscape or the uninhabited sea—that erase the signs of labor, of political struggle and of historical accountability, must be assiduously critiqued" (Handley 2009, 655). *Poseidon's Pull* is deserving of assiduous critique for this very reason. The project on the surface is quite deliberately not about human beings, human politics, or human injustices. And yet by reframing the space of the Mediterranean so that it *precedes* contemporary politics, the work deterritorializes the Aegean, helps to recover its historically rich catachrestic chain, and removes any notion of contemporary nationality from its purview. From an archipelagic perspective, *Poseidon's Pull* succeeds in "making nonsense of all national and economic boundaries, borders that have been defined only recently, crisscrossing an ocean that had been boundless for ages before Captain Cook's apotheosis" (Hau'ofa 1994, 151).

Even so, if *Poseidon's Pull* were Reza Safavi's only work that engaged with this area, it would not be worth the risk to situate it within this context. Cultural identity, however, which includes but is not reducible to national identity, is one of Safavi's artistic preoccupations: "My research/artwork is informed by my own dual sense of identity. I was raised in Canada in the early years after passage of the Canadian Multiculturalism Act of 1988. . . . My father was from Iran and my mother's background was British" (Safavi

2017). Concerns about identity in the context of immigration, the search for political asylum, and the environmental landscape upon and within which such concepts unfold into hardship continue to drive Safavi's work. When read in the context of Safavi's larger artistic practice, the political implications of *Poseidon's Pull* become clearer, the stakes higher, and the reorientation of perspective more precise.

You/they are here/not

In 2013, the same year that "Poseidon's titan mermen" were communicating with their god, another of Safavi's works, *You/they are here/not*, debuted at the Can Serrat artist residency space in Spain. Can Serrat is an old farmhouse in the Spanish countryside in Catalonia, and in this installation, Safavi took advantage of the location to create a somber interactive experience. When viewers approached the exhibit, they were guided through the site's dark corridors, open-air peristyles with a semiconfining structure. The sole illumination stemmed from soft purple light, which evoked the feeling of the sea at night, with only the muted lights of distant vessels for guidance. Once visitors arrived at a central room, they were confronted with a congratulatory message on a large computer screen. This message was shaped like an award ribbon, written in both Catalan and English: "Vosaltres esteu dins! You are in! Nosaltres estem aquí! We are here!"

The sign then instructed each visitor to put his or her finger onto a button, which measured the visitor's pulse. The soft, marina-like lights of Can Serrat began to pulse in time with the individual heartbeat and, on a screen across the room, a map of Europe began to pulse in time as well. The map documented the locations and causes of death of anonymous refugees who had attempted and failed to find shelter in Europe. With every sixth heartbeat, the map would change, providing new information about different groups of unnamed people whose lives were lost in their attempt to emigrate to Europe. The installation sent a powerful message that connected the embodied touch of the viewer to the pulsing statistics on the screen. Since the viewer herself had just passed through dark, ocean-like spaces and been guided to move furtively through shadow-filled corridors, the installation put the viewer in the position of estrangement, one suggestive of vulnerable crossing. Although it presented the visitor as one of those who was fortunate enough to have made it across, it simultaneously documented those unknown, unnamed others—from Algeria, from Ceuta, from Syria—who were not. In one image, for example, the work calls attention to victims who drowned en route to Italy from Algeria by boat (see Figure 9.2).

Figure 9.2. Image from Reza Safavi's *You/they are here/not.*

You/they are here/not was inspired in part by Daniela Ortiz and Xose Quiroga's *NN15.518*, a project that attempts to keep track of unknown refugees who have died in their attempt to enter Europe. This project is a vital one, not only because it provides important, bleak statistics that are all too often omitted in normal news reportage but also because it situates these statistics within a history of indifference and complicity:

> NN 15.518 questions the current concept of citizenship where the state apparatus and the social body prevent the full recognition of migrants as citizens, in the face of the indifference of the European welfare society. The organization United Against Racism recently published a list of 16,264 documented refugee deaths through Fortress Europe. Of them, 15,518 people have been identified as NN (no name) on the list. . . . N 15.518 (NN 15,518) articulates a strategy to raise the visibility of the people that disappear at EU borders. (Ortiz and Quiroga 2013)

In the context of asylum seekers from the Middle East, where this same sea has become a site of international consternation, contention, and shameful abrogation of ethical responsibility, the importance of this intervention could not be higher. For the refugee, it is an inhospitable place that is nevertheless more hospitable than the spaces of terror and abuse that have preceded it. In the *Poetics of Relation*, Glissant focuses on the Caribbean "as one of the places in the world where Relation presents itself most visibly" (1997, 33). "Compared to the Mediterranean," he writes, the Caribbean "explodes the scattered islands into an arc. A sea that diffracts" (33). The current crisis, however, is forcing us to see a similar diffraction in this region, through tragic waves of forced mobility. These movements are not instances of empowering errancy, wandering nomadism, or liberating mobility. As Rita Raley writes in *Tactical Media*, "Mobility per se by no means endows the subject with an unconditional freedom. In fact, it is precisely the migrant's separation from the nation-state as the guarantor of human rights that places her at risk" (2009, 53). Safavi reminds us that the flow of water and the channels of transport exist independently of politics, even as national and international politics have propelled asylum seekers into such dangerous spaces in the first place.

Poseidon's Pull repurposes the Mediterranean so that it manages to exist, as it did in antiquity, before a clear sense of nationality had crystalized and, as a result, reframes the question of boundaries. In *You/they are here/not*, Safavi brings the problem of nationality to the center. As Philip Schwyzer and Simon Mealor note in *Archipelagic Identities*, "We are used to speaking of national 'identities' as if they were attributes belonging to individuals, no less personal, intimate, and essential than one's eye colour or blood type" (2004, 1). Yet the attribute of nationality is not an essential one. Instead, Schwyzer and Mealor attest, "it can hardly be a quality inhering in the individual subject. Rather, it would seem to be a field in which the subject is captured and made visible" (1). This statement becomes even more potent when read in light of the massive influx of refugees who have attempted to flee violent and chaotic political regimes. When one overtly flees one's nation, when one rejects that system of governance that has taken over one's homeland, the sense of nationality as a trap—that is, something that "captures" and "makes visible" a vulnerable identity—becomes literalized. *You/they are here/not*'s setting emphasizes the problem of nationality in this context.

"In fiction," Ghosh writes, "settings become the vessel for the exploration of that ultimate instance of discontinuity: the nation-state" (2016, 59).[11] The word "vessel" here is appropriate, for it suggests that the setting both contains the nation and entraps its citizens. *Poseidon's Pull* and *You/they are here/not* defy this convention. Safavi's exploration of national identity makes use of setting as a means of exploding this "vessel," refusing to allow nationality to be conflated with setting or naturalized *as* setting. By simulating a successful act of passage in a manner that both invites reflection and demands participation from its visitors, *You/they are here/not* reveals the fractured relations between setting and nation and calls attention to the burden that national identity imposes upon its citizens.

The refugees who have lost their lives in the Mediterranean speak to larger global patterns of migration that are tied to political upheaval and to environmental instability. Indeed, although the two are not often discussed together, they are not necessarily distinct categories. Of course, shipwrecks at sea are a consequence of both inclement weather and the lack of resources to confront it; in this context they are also interwoven with larger climate concerns. As Caitlin E. Werrell (2015) asserts, it would be an oversimplification to reduce the complex crisis in Syria to climate change, and yet it would be an equally egregious error not to consider the connections between the two: "evidence suggests that climate change has been a factor in creating the conditions underpinning Syria's fragility—and that this fragility is partly responsible for the situation the country finds itself in today."[12]

What is missing from our conversation about climate change is how it effects change at the national level. Conversely, what is absent from our discussion of the current refugee crisis is how climate change is a driving force of political unrest. Stated another way, the increasing friction between the two reminds us of the grave error in treating nature as a mere backdrop to history. As Dipesh Chakrabarty succinctly puts it in "The Climate of History: Four Theses," "The climate, and hence the overall environment, can sometimes reach a tipping point at which this slow and apparently timeless backdrop for human actions transforms itself with a speed that can only spell disaster for human beings" (2009, 205).

Safavi's artistic practice highlights the interconnections between vulnerable refugees, political instability, natural forces, and ecological precarity. As I have suggested, his work resonates with archipelagic thought for the way it destabilizes a commonsense understanding of national boundaries, repurposes the region of the Mediterranean so that it speaks to natural forces, and encourages connections across regional and historic registries. But it also contributes to and updates a related concept in media history and information technology that is worth reviving, one whose overt tactical impulses and playful energy have the potential to contribute usefully to both archipelagic thought and the digital humanities. This is the "Temporary Autonomous Zone."

A TAZ FOR THE TWENTY-FIRST CENTURY

In 1991 the countercultural critic Hakim Bey (2011) published a slim volume of writing titled *TAZ: The Temporary Autonomous Zone*. The book was as experimental as it was provocative, comprised of poetic meditations, lyric rants, and forays into critical theory that were at once brilliantly suggestive and maddeningly imprecise. Yet, with this energetic and kaleidoscopic screed, Bey gave name to a subversive information economy, a site of resistance that was both spatially and temporally bound: "All my research and speculation has crystallized around the concept of the TEMPORARY AUTONOMOUS

ZONE (hereafter abbreviated TAZ)" (68, caps in original). In Bey's explication, the TAZ arises spontaneously, flourishes briefly, and then disbands. While it lasts, it functions as an alternate sphere of experience that exists outside established hegemonic authorities of nation, state, or religion.

Bey's work is a product of its time. A specimen of the early 1990s, it responds to an atmosphere of increasing globalization, which solidified rather than challenged the economic claims of nations. As Bey put it then, "Nationality is the highest principle of world governance . . . the apotheosis of 'territorial gangsterism.' Not one square inch of Earth goes unpoliced or untaxed . . . in theory" (71). The phrase "in theory" is important, suggesting as it does that *in practice* the reach of globalism might not be totalizing. For even as this newly hatched system of global economic power was flexing its wings, an emerging technological innovation that enabled communication across a distributed digital network was providing opportunities for resistance, protest, and outright rebellion.

Although in the 1990s critical reception to the TAZ focused on the subversive potential of emerging digital technology, I would like here to call attention to one of its most important and under-acknowledged aspects: it pays equal heed to natural ecology. Indeed, the role of ecology, of natural forces, in facilitating these provisional spaces of resistance is as fundamental to the TAZ as are the burgeoning digital networks of the day. Bey calls for a relationship between the natural and the technological that complicates their traditional relationship as foils, suggesting that "the T.A.Z. may perhaps best find its own space by wrapping its head around two seemingly contradictory attitudes," embodied respectively by what he terms computer hackers and "Ultra-Green" dissidents (77). Bey's call to reconcile these two "seemingly" opposed viewpoints—of the "hackers" and the "greens"—remains as urgent now as it did then. In one of the most fully articulated examples of what form the TAZ might take, for example, Bey draws a suggestive relationship between the emerging information economy of the 1990s—then embodied by the internet and the World Wide Web—and a piratic, archipelagic past:

> THE SEA-ROVERS AND CORSAIRS of the 18th century created an "information network" that spanned the globe: primitive and devoted primarily to grim business, the net nevertheless functioned admirably. Scattered throughout the net were islands, remote hideouts where ships could be watered and provisioned, booty traded for luxuries and necessities. Some of these islands supported "intentional communities," whole mini-societies living consciously outside the law and determined to keep it up, even if only for a short but merry life. (67)[13]

Although Bey was writing well before the information economy of our current age reached its current frenzy, his analysis of an "information network that spanned the globe" and of "the sea-rovers and corsairs of the 18th century" that constructed and accessed it offers a prescient model for thinking about methods of trading, raiding, and resisting in our age of ubiquitous computing. Such clandestine spaces dotted the eighteenth-century seas, and pirates of this time made use of archipelagic trading—and raiding—routes as a means of both communicating among their fellows and resisting all manners of legal, colonial authority. The contemporary TAZ similarly resists globalization through clandestine actions that occur within cultural and ecological niches, enabled by emerging technological networks.[14]

And yet Bey's TAZ needs an update for the twenty-first century. His writing is extravagantly countercultural and anarchist, owing much of its rhetoric to Timothy Leary and the postpunk, cyberpunk aesthetics of the late 1980s and early 1990s, and his definition of the TAZ at times seems infused with the incompatible aromas of collectivism and libertarianism that were then brewing.[15] Safavi's artistic practice helps to update Bey's conception of the TAZ in an important way. Bey conceives of the Temporary Autonomous Zone as "in some sense a *tactic of disappearance*" that enables a short-lived reprieve from nationality's "territorial gangsterism" (91). Safavi's works are instead about *making visible* connections between national boundaries and what they obscure. Rather than hiding or disappearing to avoid censure, works such as *Poseidon's Pull* and *You/they are here/not* reveal what nationally organized boundaries often disguise—that is, a shared natural commons that the continental model elides and the archipelago embodies.[16] Roberts and Stephens note that "a general relation to the archipelagic Americas . . . might be described as a collective negative hallucination"—that is, "a hallucination that does not involve perceiving something that is not present, but rather a hallucination that involves the non-perception of something . . . that is present" (2017, 11). In contemporary versions of the TAZ, increased visibility can be a mode of resistance. By definition an unsettled time-space, the TAZ destabilizes the accretion of history. Unlike the Bakhtinian chronotope, in which "time, as it were, thickens, takes on flesh, becomes artistically visible" and "likewise, space becomes charged and responsive to the movements of time, plot and history" (Bakhtin 1982, 84), in Safavi's works the TAZ unpeels layers of calcified cultural assumptions and repurposes space, deterritorializes it, so that the whole enterprise, momentarily, disrupts the linear flow of history; it does so not by hiding from history or existing outside it but by playfully excavating and activating natural and cultural ecologies in a transhistoric fashion.

"If Bruno Latour is right," writes Ghosh, "then to be modern is to envision time as irreversible, to think of it as a progression that is forever propelled forward" (2016, 123). This modern sensibility, Ghosh argues, does us no favors in terms of national and environmental crises. In the *Poetics of Rela-*

tion, Glissant similarly criticizes the authority of linear and "progressive" momentum and stresses the importance of destabilizing it, not by returning the nation to the concept of the territory but by rethinking the entire colonial progressionist model of conquest: "under no circumstances could it ever be a question of transforming land into territory again. Territory is the basis for conquest" (1997, 151). The TAZ subverts the modern concept of linear progression and hence challenges both colonial expansion and globalization that such a model has historically encouraged and justified. It succeeds in deterritorializing long-held assumptions about the authority of the nation, as traditionally organized according to continental structure, and it takes advantage of both ecological and technological networks to do so. By exploiting digital technology's ease of visualization, Safavi's work defamiliarizes Can Serrat and deterritorializes the oceans that surround Europe, demonstrating through interactive experiences how "archipelagos laden with palpable death" (n.p.) can be made visible, how the unnamed dead can be upheld in living memory, even as memorializing each one of the unknown thousands is impossible.

In "Literary Routes: Migration, Islands, and the Creative Economy," Françoise Lionnet and Emmanuel Bruno Jean-François levy an important critique against digital technology: "We are in the age of big data, when a massive haul of statistics, captured every time we use the Internet, is mined to determine the contours of our social identities, to map our moves, to pry into our conscious and unconscious wishes"; moreover, they assert, "mastery of such computable, 'objective' evidence has also become the goal of much research and education" (2016, 1223). Lionnet and Jean-François make valid points. We have enough information, enough data. But data is not the same as art, as aesthetics. Works such as *Poseidon's Pull* and *You/they are here/not* are not about mining data or harvesting Big Data or presenting data as objective slices of reality. They are, on the contrary, about confronting that data in an experiential, embodied manner that cannot be dismissed as easily as abstract data dumps so readily seem to be. When the visitor at Can Serrat sees her pulse synchronize with lights and realizes that her heartbeat is causing a map to change before her eyes, and when the data on that map highlights the tragic loss of lives that is occurring in the same region of the world where she now stands, though safe on land, it creates a site and sense of engagement between the viewer and this complex, painful reality that blunt statistics seldom achieve. Indeed, compared to other data visualizations that attempt to show large-scale emigration patterns (see Metrocosm 2016) or complex weather systems (see Viégas and Wattenberg 2012), Safavi's work forces viewers into a physical relation with this data, rather than encouraging their distant contemplation. The resulting TAZ in such instances becomes experiential, as well as metaphorical and material.

The art installation is, of course, a safe space, far removed from the harsh realities that refugees confront. But because it depends upon the viewer's embodied interaction to operate, it does not allow the viewer to objectify this information in the same way. Does Safavi trivialize these hardships by creating such simulations? One might accuse him of doing so, I suppose, but I believe such a critique misses the mark. Rather, his work moves us toward the "critically reflective visualization tactics" that Raley hopes for in her analysis of works of digital media. With such tactics in place, she writes, "we might produce countersimulations and, as a result, new modes of understanding the past and imagining the futures of finance capitalism" (Raley 2009, 115). And, for what it is worth, Safavi certainly does not view his practice as trivial. After *You/they are here/not*, Safavi relates having had an artistic crisis that resulted in an installation called *Crossing*: "*Crossing* is a continuation of my ongoing research . . . that highlights migration issues and patterns in Europe. In *Crossing*, instead of focusing on creating an immersive situation for the viewer to experience and contemplate migration issues, I asked the question, how can I use my ideas to actually save lives?" (Safavi 2017). *Crossing* takes the simulation even further, demanding the user enact the risky procedure of traversing unfriendly waters, a challenge that Safavi himself undertook, literally, before settling on the installation's final form: "*Crossing* stemmed from an absurd sketch/diagram I created of a semi-functional transportation system that would allow migrants to safely cross the 15km strait," Safavi writes. When Safavi attempted it himself, it—fortunately—failed before he reached the open ocean. Yet the apparatus that he rigged up for the installation (see Figure 9.3) was consistent with the actual vessels that asylum seekers employ, "[consisting] of

Figure 9.3. Reza Safavi, *Crossing* (2014).

a dollar store raft that migrants commonly use to attempt the voyage, which is attached to a tow line connecting the two ends of the gallery space, allowing safe travel between sides/continents" (Safavi 2017).

The virtue and risk of the TAZ, in both Bey's time of writing and mine, is its performative nature, its playfulness, even if the subject of play holds the utmost gravitas. Yet the objective of play—to make systems of injustice visible in order to confound them—could not be more urgent. Similarly, the great strength of archipelagic discourse is its resistance not only to continental hegemony but to the colonial hegemony that continentalism justifies.

Only twenty years ago, around the time that Bey was writing, before the discipline known as the digital humanities cohered as such, similar sorts of things were being said about the potential of digital technology to disrupt the global information economy. This was partly due to its novelty. At that time, computers were still widely seen as expensive luxury items accessible only to a small number of specialists, and the "hacker" class that Bey identified, even though imbued with an almost mythical aura of technological wizardry, occupied a very small, very specialized labor niche. That has changed. Computers are no longer novel; they are ubiquitous, affordable, and easy to operate. The labor sector has grown in this area and is expected to continue to do so (see Csorny 2013). The digital humanities, formerly known as "humanities computing," once occupied a similarly rarefied niche and was similarly informed by a cyberpunk aesthetic. It now, however, appears to be following this larger trend of establishing itself as, well, an establishment. This is, in part, due to DH's efforts toward inclusion.

In 2002, humanities computing became subsumed under the "big tent" of the digital humanities. The tent metaphor was immediately problematic. Its welcoming and capacious connotations suggested an interdisciplinary cohesion that did not exist then and still does not exist. Although use of the big tent metaphor had the benefit of an immediate surge in numbers in DH—as people from humanities computing, media studies, history, philosophy, and so forth, scrambled to gather under it—it did not provide any means to overcome disciplinary isolation. Instead, the tension that ensued in some cases affirmed the big tent's relation to a circus spectacle—with all of the hilarious and at times nonsensical performances that comprise it—rather than to any unified field of inquiry. To be clear, "tension" is not the same as "hostility." Digital humanities is one of the most cooperative, friendly, intellectually curious, and wide-ranging fields of scholarly inquiry I am aware of. It prides itself on transparency, collaboration, and utility. But the disciplinary divide persists seventeen years later. Those interested in cultural studies and aesthetics, for example, are often dismayed by what they perceive as decontextualized "data dumps" coming from the humanities computing side of things, while those with more technical chops are often flummoxed by *qualitative* analyses about *quantitative* methods that offer no evidence of computational procedure. One alternative metaphor to emerge in the wake of this reorganization—and the subsequent methodological tension—was that of the archipelago.

PRACTICAL AND METHODOLOGICAL IMPLICATIONS

In 2004, Willard McCarty offered the archipelago—albeit in a sense that favored colonial exploration as a dominant trope—as a possibly fruitful concept to unite the field of digital humanities. In the republication of this essay six years later, however, his disenchantment with this possibility is palpable. He refers to it not as a connective chain but as a series of isolated landmasses, noting that when academics "profess their . . . 'interdisciplinarity,'" they instead confirm their disciplinary entrenchment: "The archipelago they reveal is their meta-island" (McCarty 2004, 118).

McCarty's critique is scathing but important, for it points to what is, from my perspective, a methodological impasse. Instead of resolving this problem—although there have been sincere attempts to do so—each field risks further entrenchment on its "meta-island." Each year the annual DH conference submission process serves to illuminate this divide. At the same time as this divide has become clearer, however, DH has benefitted from widespread infrastructural support, which has resulted in a different set of tensions regarding what is perceived by some as DH's complicity with (or resistance to) neoliberalism.[17] And yet these are—or could be—*productive* tensions, especially if we were to reconsider the material and metaphorical aspects of the archipelago as offering a connective matrix, if at times an unruly and shifting one. Those of us who identify as DHers have an opportunity to benefit from its perceived stability, even as we acknowledge and work to overcome these methodological divides—and even as we resist the depiction of our field as a naive instrument of capitalism.

Poseidon's Pull demonstrates a possible model of a DH practice and methodology that does not lopsidedly favor quantitative analysis on one hand or uninformed qualitative speculation on the other. Nor is the project complicit with the neoliberal, profit-driven ambitions of globalization. Far from it. Instead, with its playful and irreverent aesthetics, which depend upon computational technology, even as that same technology is subordinate to the sea's transhistoric power, *Poseidon's Pull* grounds the digital in aesthetic experience and ecological reality. It focuses on the archipelagic spaces of this region, through a formal approach that is similarly archipelagic, and the practical and methodological implications are profound. They remind us that digital technology, when coupled with cultural critique, has a role to play in challenging dominant forces and disrupting global systems of injustice.

NOTES

1. The Electronic Literature Organization defines electronic literature as "works with important literary aspects that take advantage of the capabilities and contexts provided by the stand-alone or networked computer" ("About the ELO," 2007).

2. To create the bot, "sample combined the environmental data collected by the buoy with Markov-chained content from Hermann Melville's *Moby-Dick*" (Jørgensen 2015).

3. Celestial navigation has linked sky and sea since antiquity (one of the first recorded instances occurs in book five of Homer's *The Odyssey*—in such cases the sky extends rather than departs from the sea).

4. The construction of this constellation calls attention to the culturally contingent nature of all constellations. For example, the now defunct constellation of the Argo Navis—named after the speaking ship from the story of Jason and the Argonauts—was once vast. In the Middle Ages, however, it was split into three: deck, keel, and sail (Eastlick, n.d.).

5. The surrender of human control here is consistent with ancient practices of divination (see Parke 1967). Additionally, it is worth noting that divination was a fundamental part of ancient governance (see Malkin 2013).

6. Rube Goldberg was an American inventor and cartoonist who excelled in imagining absurdly complicated systems for completing exceedingly simple tasks. By extension, a "Rube Goldberg machine" is any such mechanism, system, or device that is unnecessarily and comically complicated.

7. As Glissant notes, this type of movement—rambling, drifting—offers "an absolute challenge to narrative" (1997, 25).

8. For one example of the region's premodern past, consider a peculiar moment recounted in book seven of Herodotus's *Histories*: During the Second Persian War (480–479 BCE), Xerxes and his army stand before the Hellespont. Xerxes commands his army to build a bridge to cross this water, but as soon as his soldiers have completed the task, a violent storm dashes the bridge to bits. On learning of this, Herodotus (2013) recounts, Xerxes's "anger was so terrible that he ordered his men to give the Hellespont three hundred lashes of the whip, and to drop a pair of fetters into the sea." As his men administer this "punishment," they also lash the sea with verbal chastisements: "'O bitter water . . . King Xerxes will cross you, whether you wish it or not. How right people are not to offer sacrifice to you—turbid and briny river that you are!'" (462). Far from demonstrating the subversive potential to undermine hegemonic authority, Xerxes's relentless march provides an early model of colonization, emblematic of "arrowlike nomadism," the tendency to wander for the purpose of conquest (Glissant 1997, 12). Yet it is nevertheless instructive for the way the sea presents itself as a force outside human control. Xerxes's immoderate rage at the

Hellespont confirms the power of that narrow strait and its attendant natural features. It speaks to the limits of human power—even the power of kings. It returns us to a premodern relationship with natural spaces in which we are always subject to their forces. This episode reminds us that the contemporary conception of nature—as a force that can be known, measured, and subdued—is a modern fantasy.

9. Archipelagoes, like stars, cluster according to associative constellations informed by myth, geography, and narratives about national identity. The birth of Aphrodite, for example, according to Hesiod, occurs in ancient myth as a result of Ouranos's severed genitals foaming upon the coast of Cythera (or Cyprus); accordingly, this myth links oceans, islands, nations, the heavens, and the goddess of love in an associative chain.

10. These drifting routes also involved coming ashore. In an early plan, each team member was to land on a different island and collect clay deposits to be used in another ritual: "These five mermen will drive a boat from island to island in Greece . . . [where their] first task is to find clay or minerals from that island that can be used to sculpt a hydrai water jar" (Safavi et al. 2013). While this proved too complicated, in the end, to execute, maps of soil deposits remain on the project's home site, and the commitment to geographic continuity remains primary.

11. The victims of the current crisis, of course, do not exist "in fiction." Yet studies that focus on early accounts of colonization, both in fiction and in historic record, are helpful for providing continuity between then and now. Put another way, the people who comprise the roster of *NN15518* are not fictional characters, but neither were the members of the shipwrecked *Sea Venture* who lived, briefly, on Bermuda outside colonial control and whose story inspired *The Tempest* before their rebellious enterprise was halted, censured, and refolded into the Virginia Company (see Linebaugh and Rediker 2000).

12. This situation is not unique to Syria. As Ghosh (2016) notes, "The partial inundation of just one island in Bangladesh . . . has led to the displacement of more than half a million people" (88).

13. We might quibble about Bey's use of the word "merry." This blithe statement contributes to a reductive stereotype that is more consistent with the ride at Disneyland than historic reality: "In popular culture . . . piracy is apparently 1 per cent crime and 99 percent swashbuckling" (Campbell 2011, 11). The merriment of pirates' lives aside, Bey's larger point stands: spaces employed by such pirates functioned as sites of resistance precisely because they existed—temporarily, provisionally—outside the organizing structures of nation, capital, or other traditional hegemonic units of cultural organization (see Linebaugh and Rediker 2000).

14. In this the T. A. Z. has something in common with more overtly natural ecologies (see Gibson 1986).

15. For a thorough critique, see Morton 2011.

16. This is not to say that nationality cannot coincide with archipelagic land masses. It may be strategically prompted by geographic affinity. But it is one among many urgent relations.

17. To get a sense of this tension, see Posner 2016; Allington, Brouillette, and Golumbia 2016.

REFERENCES

"About the ELO." 2007. Electronic Literature Organization. https://eliterature.org/about.

Allington, Daniel, Sarah Brouillette, David Golumbia. 2016. "Neoliberal Tools (and Archives): A Political History of Digital Humanities." *Los Angeles Review of Books*. May 1. https://lareviewofbooks.org/article/neoliberal-tools-archives-political-history-digital-humanities.

"Archipelago." N.d. Etymology Online. https://www.etymonline.com/word/archipelago.

Baeke, Jan, and Alfred Marseille. 2012. "Channel of the North." Electronic Literature Collection Volume 3. http://collection.eliterature.org/3/work.html?work=channel-of-the-north.

Bakhtin, Mikhail. 1982. *The Dialogic Imagination*. Austin: University of Texas Press.

Bey, Hakim. 2011. *TAZ: The Temporary Autonomous Zone*. Seattle: Pacific Publishing Studio.

Campbell, Mel. 2011. "Pirate Chic: Tracing the Aesthetics of Literary Piracy." *Pirates and Mutineers of the Nineteenth Century: Swashbucklers and Swindlers*, edited by Grace Moore. London: Ashgate.

"Catachresis." N.d. *A Latin Dictionary*, by Charlton T. Lewis and Charles Short. Perseus Digital Library. http://www.perseus.tufts.edu/hopper/text?doc=Perseus%3Atext%3A1999.04.0059%3Aentry%3Dcatachresis.

Chakrabarty, Dipesh. 2009. "The Climate of History: Four Theses." *Critical Inquiry* 35, no. 2: 197–222.

Csorny, Lauren. 2013. "Careers in the Growing Field of Information Technology Services." *Beyond the Numbers* 2, no. 9 (April): https://www.bls.gov/opub/btn/volume-2/careers-in-growing-field-of-information-technology-services.htm.

Eastlick, Pam. n.d. "Argo Navis." University of Wisconsin-Madison Department of Astronomy. http://www.astro.wisc.edu/~dolan/constellations/extra/ArgoNavis.html.

"Errant." n.d. Etymology Online. https://www.etymonline.com/word/errant.

Ghosh, Amitav. 2016. *The Great Derangement: Climate Change and the Unthinkable*. Chicago: University of Chicago Press.

Gibson, James J. 1986. *The Ecological Approach to Visual Perception*. New York: Routledge.

Glissant, Éduoard. 1997. *The Poetics of Relation*. Ann Arbor: University of Michigan Press.

Handley, George B. 2009. "Towards an Environmental Phenomenology of Diaspora" (Review). *MFS Modern Fiction Studies* 55, no. 3: 649–57.

Hau'ofa, Epeli. 1994. "Our Sea of Islands." *Contemporary Pacific* 6, no. 1: 147–61.

Herodotus. 2013. *The Histories*. Translated by Tom Holland. London: Penguin Classics.

Jørgensen, Finn Arne. 2015. "Where Are All the Nature Bots?" *Ant, Spider, Bee*. April 14. http://www.antspiderbee.net/2015/04/14/where-are-all-the-nature-bots.

Lewis, Martin W., and Kären E. Wigen. 1997. *The Myth of Continents: A Critique of Metageography*. Berkeley: University of California Press.

Linebaugh, Peter, and Marcus Rediker. 2000. *The Many-Headed Hydra: Sailors, Slaves, Commoners, and the Hidden History of the Revolutionary Atlantic*. Boston: Beacon Press.

Lionnet, Françoise, and Emmanuel Bruno Jean-François. 2016. "Literary Routes: Migration, Islands, and the Creative Economy." *PMLA* 131, no. 5: 1222–37.

Malkin, Irad. 2013. *A Small Greek World: Networks in the Ancient Mediterranean*. Oxford: Oxford University Press.

McCarty, Willard. 2004. "Tree, Turf, Centre, Archipelago—or Wild Acre? Metaphors and Stories for Humanities Computing." In *Defining Digital Humanities*, edited by Melissa Terras, Julianne Nyhan, and Edward Vanhoutte. New York: Routledge.

McPherson, Tara. 2012. "Why Are the Digital Humanities So White? Or Thinking the Histories of Race and Computation." In *Debates in the Digital Humanities*, edited by Matthew K. Gold. Minneapolis: University of Minnesota Press.

Metrocosm. 2016. "All the World's Immigration Visualized in 1 Map." Metrocosm. June 29. http://metrocosm.com/global-immigration-map.

Morton, Timothy. 2011. "Objects as Temporary Autonomous Zones." *continent*. 1, no. 3: 149–55. http://www.continentcontinent.cc/index.php/continent/article/viewArticle/46From.

Ortiz, Daniela, and Xose Quiroga. 2013. *NN15.518*. Daniela Ortiz.http://daniela-ortiz.com/index.php?/projects/nn-15518.

Parke, H. W. 1967. *The Oracles of Zeus*. Cambridge, MA: Harvard University Press.

Posner, Miriam. 2016. "Money and Time." *Miriam Posner's Blog*. March 14. http://miriamposner.com/blog/money-and-time.

Raley, Rita. 2009. *Tactical Media*. Minneapolis: University of Minnesota Press.

Roberts, Brian Russell, and Michelle Stephens. 2017. *Archipelagic American Studies*. Durham, NC: Duke University Press.

Safavi, Reza. 2013. *You/they are here/not*. hi-reza.com. http://www.hi-reza.com/youthemherenot.html.

———. 2014. *Crossing*. hi-reza.com. http://www.hi-reza.com/crossing.html.

———. 2017. Email message to author. May 6.

———. 2019. Phone conversation with author. May 1.

Safavi, Reza, and Max Kazemzadeh. 2018. *Poseidon's Pull (Revisited, 2018): "Part 1: Channeling the Gestures of a God.*" Poseidonspull.com. http://www.poseidonspull.com/index.html.

Safavi, Reza, Max Kazemzadeh, Joe Hicks, Haytham Nawar, and Adnan Assim. 2013. *Poseidon's Pull*. Poseidonspull.com. http://www.poseidonspull.com/index.html.

Saint Augustine. 2001. "On the Christian Doctrine." In *The Norton Anthology of Theory and Criticism*. Edited by Vincent B. Leitch et al., 154–58. New York: Norton.

Schwyzer, Philip, and Simon Mealor. 2004. *Archipelagic Identities: Literature and Identity in the Atlantic Archipelago, 1550–1800*. Burlington, VT: Ashgate.

Swanstrom, Elizabeth. 2016. *Animal, Vegetable, Digital: Experiments in New Media Aesthetics and Environmental Poetics*. Tuscaloosa: University of Alabama Press.

Viégas, Fernanda, and Martin Wattenberg. 2012. "Wind." Hint.fm. http://hint.fm/projects/wind.

Werrell, Caitlin E. 2015. "Fragile States: The Nexus of Climate Change, State Fragility and Migration." *Angle*. November 24. https://anglejournal.com/article/2015-11-fragile-states-the-nexus-of-climate-change-state-fragility-and-migration.

Praise Song for Oceania

Craig Santos Perez

On June 8, 2016, World Oceans Day

"Ocean, we // had been your griot"

—Brenda Hillman[1]

~

praise
your capacity
for birth / your fluid
currents and trenchant
darkness / praise your contracting
waves & dilating
horizons / praise our briny
beginning, the source
of every breath / praise
your endless bio-
diversity / praise

your capacity
for renewal / your rise
into clouds and descent
into rain / praise your underground
aquifers / your rivers & lakes,
ice sheets & glaciers / praise
your watersheds &
hydrologic cycles / praise

your capacity
to endure / the violence

of those who claim dominion
over you / who map you
empty ocean to pillage / who divide you
into latitudes & longitudes /
who scar your middle
passages / who exploit
your economy[2] / praise

your capacity
to survive / our trawling
boats / breaching /
your open body /
& taking from your
collapsing depths / praise

your capacity
to dilute / our sewage
& radioactive waste /
our pollutants & plastics /
our heavy metals
& greenhouse gases / praise

your capacity
to bury / soldiers & terrorists,
slaves & refugees / to bury
our last breath
of despair / to bury
the ashes of our
loved ones / praise

your capacity
to remember / praise
your library of drowned
stories / praise your museum
of lost treasures / praise
our migrant routes
& submarine roots / praise

your capacity
to penetrate /
praise your rising tides
& relentless storms & towering
tsunamis & feverish
floods / praise

your capacity
to smother /
schools of fish & wash them
ashore to save them
from our cruelty /

to show us what we're
no longer allowed to take
/ to starve us like your corals
are being starved & bleached /
like your liquid lungs
choked of oxygen / praise

your capacity
to forgive / please
forgive our territorial hands
& acidic breath / please
forgive our nuclear arms
& naval bodies / please
forgive our concrete dams
& cabling veins / please
forgive our deafening sonar
& lustful tourisms / please
forgive our invasive drilling
& deep sea mining / please
forgive our extractions
& trespasses / praise

your capacity
for mercy / please
let our grandfathers and fathers
catch just one more fish / please
make it stop raining soon / please
make it rain soon / please
spare our fragile farms & fruit trees / please
spare our low-lying islands & atolls / please
spare our coastal villages & cities / please
let us cross safely to a land
without war / praise

your capacity
for hope /
praise your rainbow
warrior & peace
boat / your hokule'a
& sea shepherd / praise
your arctic sunrise & flotillas
of hope / praise your nuclear free
& independent pacific movement /
praise your marine stewardship
councils & sustainable
fisheries / praise your radical
seafarers & native navigators /
praise your sacred water walkers /
praise your activist kayaks

& canoes / praise your ocean
conservancies & surfrider foundations /
praise your aquanauts & hyrdolabs /
praise your coastal cleanups
& Google Oceans /
praise your whale hunting
& shark finning bans /
praise your sanctuaries
& no take zones / praise
your pharmacopeia of new
antibiotics / praise your wave
and tidal energy / praise your
#oceanoptimism & Ocean
Elders /praise

your capacity
for echo
location / our names for you /
that translate
into creation stories
& song maps
tasi & kai & tai & moana nui & vasa &
tahi & lik & wai tui & daob & wonsolwara /
praise

your capacity
for communion /
praise our common heritage /
praise our pathway
& promise to each other / praise
our endless saga / praise our most powerful
metaphor / praise this vision
of belonging / praise your horizon
of care / praise our blue planet,
one world ocean / praise our trans-oceanic
past, present & future flowing
through our blood /

NOTES

Phrases are quoted from or inspired by various scholars and poets, including Epeli Hau'ofa, Derek Walcott, Elizabeth DeLoughrey, Rob Wilson, Peter Neill, Sylvia Earle, Édouard Glissant, and Albert Wendt. The words chanted are the words for ocean in various Pacific languages.

1. Brenda Hillman, "The Pacific Ocean," *Practical Water* (Middletown, CT: Wesleyan University Press, 2009), 26.
2. The gross marine product of the ocean is $2.5 trillion.

Part III

ARCHIPELAGIC ENVIRONMENTS

Evolving Political Ecologies

Care

Craig Santos Perez

My 16-month old daughter wakes from her nap
and cries. I pick her up, press her against my chest

and rub her back until my palm warms
like an old family quilt. "Daddy's here, daddy's here,"

I whisper. Here is the island of O'ahu, 8,500 miles
from Syria. But what if Pacific trade winds suddenly

became helicopters? Flames, nails, and shrapnel
indiscriminately barreling towards us? What if shadows

cast against our windows aren't plumeria
tree branches, but soldiers and terrorists marching

in heat? Would we reach the desperate boats of
the Mediterranean in time? If we did, could I straighten

my legs into a mast, balanced against the pull and drift
of the current? "Daddy's here, daddy's here,"

I whisper. But am I strong enough to carry her across
the razor wires of sovereign borders and ethnic

hatred? Am I strong enough to plead: "please, help
us, please, just let us pass, please, we aren't

suicide bombs." Am I strong enough to keep walking
even after my feet crack like Halaby pepper fields after

five years of drought, after this drought of humanity.
Trains and buses rock back and forth to detention centers.

Yet what if we didn't make landfall? What if here
capsized? Could you inflate your body into a buoy

to hold your child above rising waters? "Daddy's
here, daddy's here," I whisper. Drowning is

the last lullaby of the sea. I lay my daughter
onto bed, her breath finally as calm as low tide.

To all the parents who brave the crossing:
you and your children matter. I hope

your love will teach the nations that emit
the most carbon and violence

that they should, instead, remit the most
compassion. I hope, soon, the only difference

between a legal refugee and an illegal migrant
will be how willing we are to open our homes,

offer refuge, and carry each other
towards the horizon of care.

Chapter Ten

Literary Archipelagraphies

Readings from the British-Irish Archipelago

Pippa Marland

Philip Schwyzer and Simon Mealor's coedited collection of essays, *Archipelagic Identities: Literature and Identity in the Atlantic Archipelago, 1550–1800*, was published in 2004 and spearheaded an archipelagic turn in British and Irish literary studies. In his introduction to the volume, Schwyzer poses a question that has continued to inform subsequent incarnations of archipelagic thought in these islands. He asks, "But when was Britain, and where?" (2). He goes on to describe the contingency of this name, along with its historical tendency to be interpreted in an Anglocentric fashion—a tendency in which both the discrete identities of, and the subtleties of the interrelationships between, the constituent nations have consistently been obscured. Thus, as Schwyzer also argues (citing Rebecca Bach), "These debates over geographical nomenclature were never mere antiquarian quibbling. Wherever cultures and languages come into contact, and above all where there is a 'colonial' dimension, acts of naming play a crucial role in establishing—and resisting—dominance and hegemony" (2). Given the continuance of such forms of dominance and hegemony in these isles, nomenclature has continued to be a vexed issue. J. G. A. Pocock, for example, whose 2005 collection *The Discovery of Islands: Essays in British History* gathers together essays written over a period of three decades, describes in his 1975 essay, "British History: A Plea for a New Subject," the attempt to introduce the term "Atlantic archipelago" as an alternative to "the British Isles." His declared motivation for this was his conviction that the terms "British" and "Britain" are not politically neutral: "their history in Scotland is unlike that they have in England, and both these are unlike that they have in Ireland, where there are good reasons in the republic for rejecting them altogether" (Pocock 2005, 293). "Atlantic archipelago" is a phrase, Schwyzer notes with approbation, that "by its very awkwardness

. . . does much to defamiliarize a geographical entity whose story we may imagine we already know all too well" (Schwyzer and Mealor 2004, 2).[1]

However, Pocock, writing at a distance of thirty years from his initial plea for a new subject, argues that the term "Atlantic archipelago" "has failed to catch on," both because "you cannot form a generic adjective from it" and as a result of "a general invective against naming or defining or having any identity at all, which is part of the politics of post-modernism" (2005, 293). Schwyzer also admits that "Pocock's plea went all but unanswered for some years" but argues that there has been a transformation: "Today it would seem nothing less than absurd to teach the history of England or any of its neighbours in the insular (or rather, profoundly uninsular) manner that was the norm just twenty years ago" (Schwyzer and Mealor 2004, 3). Schwyzer and Mealor's book itself bears witness to this shift in cultural attitudes, seeking, in its content, to extend the archipelagic project from the field of history into that of the study of literature. Schwyzer draws attention to the way in which the term "English literature" (as denoting both a literary canon and a subject for academic study) has tended to elide the complex interplay that has always existed between the component nations and regions of the archipelago, along with its connection to communities and nations that lie beyond its shores.

For Schwyzer, the potential of archipelagic thinking as a radical tool is clear, in terms of both disrupting dominant narratives and forging new, collaborative associations. He writes, "The essence of what we might term the archipelagic perspective lies in a willingness to challenge traditional boundaries—boundaries, that is, between the histories of different nation-states, and also between academic disciplines" (Schwyzer and Mealor 2004, 4). He ends his introduction with the suggestion that the scope of archipelagic thinking might be further augmented: "the real essence and value of this approach to literature lies in the relentless transgression of the very boundaries that seemed to define its scope" (7). Schwyzer's remarks have proved remarkably prescient. The publication of *Archipelagic Identities* marked the beginning of a sustained turn toward archipelagic thought in the British-Irish literary scene, and the "archipelagic perspective" he describes here has indeed breached its own boundaries, gaining in theoretical complexity and conceptual reach over the ensuing years.

These British-Irish archipelagic initiatives appear, at least initially, to have emerged independently of, and in certain cases to predate, similar developments in island studies, though they have much in common.[2] Both fields encourage approaches that explore the interconnections that exist between islands and islands, recognize the sea as a connective medium, challenge tropes of island isolation and remoteness, and decouple island and "peripheral" spaces from assumptions of their domination and definition by continental and/or colonial "centers" of power. Elizabeth DeLoughrey, writing in

2001, employed the neologism "archipelagraphy" to denote the field of study implied by such concerns, defining it as "a historiography that considers chains of islands in fluctuating relationship to their surrounding seas, islands and continents" (23). This term has been increasingly adopted in island studies (see, for example, Stratford et al. 2011; Pugh 2013) as recognition of the potential for archipelagic thinking to grow. Elaine Stratford et al. see the role of archipelagraphy as a form of "counter-mapping" and as "dislocating and de-territorializing the objects of study—the fixity of island difference and particularity—and constituting in their place a site or viewing platform by which they are perceived and analysed afresh and anew" (2011, 114).

Such acts of counter-mapping, deterritorialization, and recalibration are ones that are fulfilled by the examples of literary-critical writing discussed in this chapter. Hence my title, which deploys the term "literary archipelagraphies" in relation to these works. These archipelagraphies, through the new "viewing platform(s)" they create, also resonate beyond the immediate island spaces with which they engage and represent a broader challenge to received ontologies. In particular, they potentially offer a means of reassessing human-environmental relationships.

In recent articulations of British and Irish archipelagic thinking, environmentally oriented literary critics and creative writers have seen the archipelagic not only as a means of disrupting political and cultural hegemonies but also as a heuristic that might help to facilitate ideas of ecological entanglement at a range of different spatial scales. This is an insight shared by archipelagic theorists across the Atlantic. Brian Russell Roberts and Michelle Ann Stephens, in their introduction to *Archipelagic American Studies*, describe the emergence of "archipelagic imaginaries and reading practices that foreground the Americas' embeddedness within a planetary archipelago that holds in tension the supraregional and the microregional" (2017, 11). Thus, for Roberts and Stephens a "new world of archipelagic understanding" beckons (11).

This chapter begins with an account of the development of archipelagic perspectives in creative and literary-critical writing in the British-Irish archipelago in the years since the appearance of *Archipelagic Identities*, situating them in relation to comparable initiatives in island studies. Key stages include the founding of the creative writing journal *Archipelago* in 2007 and the corresponding burgeoning of literature from the "margins"; the publication of John Kerrigan's *Archipelagic English: Literature, History, and Politics, 1603–1707* and John Brannigan's monograph *Archipelagic Modernism: Literature in the Irish and British Isles, 1890–1970*, in 2008 and 2015, respectively; recent literary-critical articulations of the "archipelago" in Jos Smith's *The New Nature Writing: Rethinking the Literature of Place* and in the edited collection *Coastal Works: Cultures of the Atlantic Edge*, both published in

2017; and, finally, David Gange's hybrid work *The Frayed Atlantic Edge: A Historian's Journey from Shetland to the Channel*, published in 2019.

In tracing this trajectory of British-Irish archipelagraphy, the chapter both explores the ways in which these emanations of archipelagic thought begin to enact an ecological heuristic and articulates the case for incorporating the methodological and heuristic aspects of archipelagic thinking into a new discipline of "archipelagic ecocriticism." Here the archipelagic becomes not just a mode of engagement but an ecological paradigm—one that might suggest a means of addressing some of the theoretical dilemmas with which contemporary ecocriticism is wrangling. These include the perceived necessity to replace the traditional ecocritical valorization of local attachment with more global perspectives, the difficulty per se of engaging with the vast spatiotemporal scales of the Anthropocene, and the vexed question of how great the influence of cultural forms and cultural criticism can be in the face of powerful global economic and political systems.[3] In speaking to these concerns, archipelagic thinking offers the potential to foster valuable new perspectives on the human place within the planetary archipelago and thus to reframe and reinvigorate ecocriticism itself.

ARCHIPELAGO AND THE "NEW NATURE WRITING"

In June 2007 the first issue of the British-Irish literary journal *Archipelago* was published. In his introduction, editor Andrew McNeillie presented a manifesto that set out the overarching preoccupations of the journal. It would be concerned, he wrote,

> with landscape, with documentary and remembrance, with wilderness and wet, with natural and cultural histories, with language and languages, with the littoral and the vestigial, the geological and topographical, with climates, in terms of both meteorology, ecology and environment; and all these things as metaphor, liminal and subliminal, at the margins, in the unnameable constellation of islands on the Eastern Atlantic coast, known variously in other millennia as Britain, Great Britain, Britain and Ireland etc; even, too, too readily, the United Kingdom (including the North of partitioned Ireland, though no such thing ever existed, other than *in extremis* in wartime, but in the letter). (McNeillie 2007, vii)

As the latter part of the paragraph suggests, McNeillie, like Schwyzer and Pocock, is aware of the contingency of the naming of this "constellation of islands." His assertion that it is "unnameable"—his refusal to perpetuate historically and politically loaded terms—is a reflection of devolutionary concerns and is additionally informed by a desire to assert the importance of the "margins" against the traditional centers of political power and cultural force. Thus, in the pages of *Archipelago*, the "littoral and the vestigial" rise to prominence, and dominant Anglo- or urban-centric readings are disrupted. While the journal is in English, "it has reached out to a range of the different languages of the archipelago" and featured essays on, for example, Scottish and Irish Gaelic literature, as well as poetic translations from those traditions (Smith 2017b, 246).

Figure 10.1. Original cover artwork for Archipelago by Julian Bell. The image is the trademarked copyright of the Clutag Press and may not be reproduced, except for review purposes, without the permission of the Press.

The disruption to traditional geopolitical and cultural understandings implicit in the textual material is also figured in the artwork (see Figure 10.1). The cover of the journal features a deliberately disorienting illustration by Julian Bell, one that shows the "unnameable" archipelago tipped ninety degrees out of its north-south orientation, turned westwards on its axis such that Ireland is in the foreground (as the new South), and the Irish Sea and the western seaboard of Scotland, Wales, and England are central.[4] London and the South-East of England—the long-established sites of power and wealth in these isles—recede into the distance, partially obscured by cloud. Moreover, the Atlantic Ocean takes up more of the frame than the landmasses, in this way emphasizing the importance of the sea in relation to these islands. The editorial, along with establishing *Archipelago*'s interest in the particular, the local, and the liminal, also situates these concerns within a planetary context. McNeillie writes, "But while the unnameable archipelago is its subject, its vision is by implication global, and its concerns with the state of the planet could not be more of the hour" (2007, vii). As the mention of "the state of the planet" makes clear, the journal has an ecological sensibility. Again, this is a feature echoed in Bell's cover art, in which our attention is drawn to the presence of nonhuman animals. The sea is visibly populated by marine creatures, and the image as a whole seems to present a bird's-eye view of the archipelago. Three gannets are shown, one particularly highlighted, with the ocean and landmasses lying far below them. The view from this altitude seems to register the curvature of the earth, with the horizon describing an arc across the upper third of the picture, a feature that has the effect of situating this collection of islands and the writings they inspire within a planetary ecological imaginary.

It is evident that the concerns of *Archipelago* both anticipate and can be retrospectively mapped onto recent developments in island studies, particularly in terms of creating a new viewing platform from which the unnameable archipelago can be perceived afresh. The manifesto helpfully articulates the potential of the "archipelagic" as a methodology for such reframing—as a way of encountering the world based on a close, deep engagement with the materiality and histories of particular locales and an alertness to their interconnection with other "peripheral" places. This attitude is also indicative of the importance of the archipelagic as an ecological heuristic. Jos Smith draws attention to the way in which the journal's preoccupations with place "have extended into an ecology of the isles that is important in its own right," such that in the ten years since its inception, the journal has inflected existing archipelagic perspectives "in its own distinctive way" (2017b, 247). Smith aligns *Archipelago*'s orientation with writer (and frequent contributor to the journal) Robert Macfarlane's understanding of the archipelagic as "chthonic, marine, elemental and felt" (Macfarlane, cited in Smith 2017b, 247), a notion that corresponds with calls in island studies for perspectives rooted in phenomenological encounter (see Hay 2006; Roberts and Stephens 2017).

In fact, in a 2013 essay for the ecocritical journal *Green Letters*, Smith went as far as to suggest that the place-based creative writing currently (and arguably) termed the New Nature Writing might be more appropriately classified as "an archipelagic literature," because of its authors' attentiveness to "how this cluster of islands and ecological niches is related in complex ways to human communities at the local, regional and even national and global ways of life that are lived out across and within them" (6). Smith's suggestion has not been taken up, and he himself returned to the original term in his 2017 monograph *The New Nature Writing: Rethinking the Literature of Place*, though he continues to regard "the periphery" and "archipelago" as key themes. This lack of uptake perhaps reflects a dearth of awareness in the wider academic community of the manner in which archipelagic perspectives have augmented their early devolutionary concerns with an archipelagic imaginary that, as Schwyzer predicted in 2004, has breached its own boundaries.

Such conceptual expansion is a marker of the way in which archipelagic thought is itself archipelagic, representing a fluid mode of inquiry rather than a dogmatic framework for analysis. Considered in this light, the title of the journal—*Archipelago*—attains additional significance. Roberts and Stephens argue that "the wide-ranging human project of describing—and conjuring into existence—the coherence of groups of islands, has been a prime example of *catachresis* . . . whereby 'archipelago' itself becomes a term deployed in the attempt to name connections—the 'submarine' unities between land and sea, island and island, island and continent" (2017, 30). As catachresis, then, *Archipelago* is a title that signals its own unfathomability, its own refusal of fixed geopolitically entrenched understandings. In its displacement of normative meanings, it undermines "land-locked, above-ground, territorial epistemologies and ways of thinking" (30), opening a space for the proliferation of reconfigured narratives that have, as yet, no accepted terminology.

ARCHIPELAGIC CRITICISM

The appearance of *Archipelago* was complemented by the publication in 2008 of John Kerrigan's *Archipelagic English*. This literary-critical work builds on Schwyzer and Mealor's *Archipelagic Identities* (which included a chapter by Kerrigan himself) and continues with the project of challenging the Anglocentric bias that had become enshrined in the pedagogical field of "English literature" during the early twentieth century.[5] Kerrigan writes from the conviction that "the subject can neither be defined, nor Anglophone literature be historically understood, along purely national lines" (411), and he is motivated by a desire to "recover the long, braided histories played out across the British-Irish archipelago between three kingdoms [England and Wales, Scotland, and Ireland], four countries, divided regions, variable ethnicities and religiously determined allegiances" (2). Again, just as Julian Bell's image for *Archipelago* features a reoriented image of the archipelago, the cover of *Archipelagic English* shows the British Isles tilted to one side, this time ninety degrees eastward, such that, as Graham Perry notes, we see "Britannia on her back, with all her ragged western extremities pointing upwards" in a manner that encourages us to see the islands as an archipelago (2008, n.p.).

While his argument focuses more on devolutionary and relational themes than environmental concerns, a significant aspect of Kerrigan's archipelagic reframing is the recognition of the role of the sea in connecting the kingdoms and countries of the archipelago. He reminds us that in the seventeenth century, "the standard route from Edinburgh (Leith) to London was through coastal waters, not on horseback along difficult roads," and he goes on to assert that "the seas which we view on maps as surrounding and dividing the islands drew them together, and opened them to continental and Atlantic worlds" (2008, 48). "Stuart writing," Kerrigan notes, "is full of islands" (40). These interconnections, both local and international, provide Kerrigan with the rationale for his project, and he is careful to delineate the etymology of the title of his study: "My title is maritime, because *Archipelagic* derives from Greek *archi*, 'chief, primary,' and *pelagos*, a word for 'sea'" (48). He also identifies the point at which the word came into the English language, with its first recorded usage appearing in 1502 in reference to the Aegean. However, as he also notes, it was in Richard Hakluyt's *Voyages* ([1600] 1985) that the modern sense of "archipelago" was established, now connoting "any sea, or sheet of water, in which there are numerous islands; and *transf.* a group of islands" (48). He concludes, "It is as though the three kingdoms grew out of the watery medium of the seas" (48).

Like Schwyzer, Kerrigan also alludes to Pocock's use of the term "Atlantic archipelago," describing it as Pocock's attempt "to get away from inappropriate pan-national language" (2008, 77) and at the same time to explore the relationship between "national" and "extra-national" historiographies (xi). In relation to Pocock's observation, cited earlier, that the term "Atlantic Archipelago" failed to catch on, Kerrigan suggests that "once this book [i.e., Kerrigan's *Archipelagic English*] is published, the adjective *archipelagic* will carry all before it—context usually determining whether the Atlantic or the Pacific, or indeed the Aegean, is being discussed" (83). He also argues persuasively that while postmodernism might, on the one hand, resist naming, as Pocock suggested, its focus on interactivity might, on the other hand, make it particularly receptive to archipelagic perspectives. For Kerrigan, "the intellectual climate of the late twentieth century did more to advance than discourage archipelagic thinking" (83), and he sees the potential for the ongoing development of this approach.

The intervening years since the publication of Kerrigan's book have borne out this assertion, reinforcing the notion that the postmodernist imagination might be fertile ground for archipelagic thinking. John Brannigan's *Archipelagic Modernism: Literature in the Irish and British Isles, 1890–1970*, published in 2015, demonstrates that archipelagic literary-critical perspectives are still in the process of extending their reach in the postmodernist intellectual climate of the twenty-first century. Brannigan's work also shows how literature from the modernist period (which Brannigan sees as extending from the late nineteenth to the latter part of the twentieth century) manifested its own forms of archipelagic thinking. Written in Dublin and published in Edinburgh, *Archipelagic Modernism* is profoundly archipelagic in both its content and its production. The volume's subtitle has a similar effect as the cover images of *Archipelago* and *Archipelagic English*, which tilt the archipelago on its axis, thus disrupting the conventional view: the formulation "Irish and British Isles" draws attention to the significance of naming through its reversal of the usual ordering of the territories. Brannigan also wrestles explicitly with nomenclature within the text itself, adding his voice to Pocock, Schwyzer, McNeillie, and Kerrigan's critique of existing terms for the "unnameable archipelago." These terms, he writes, "have been bound to the legacies of imperialism, nationalism, and unionism" (Brannigan 2015, 7).

Given this difficulty, Brannigan is careful to establish the coordinates of his own archipelagic terminology: "To use the word 'archipelago' to talk about the relations between the constituent parts of the British and Irish Isles implies a plural and connective vision quite at odds with the cultural and political homogenisation which lay at the heart of the Unionist project" (6). However, while his approach is deeply rooted in what he calls the "devolutionary imaginary" (9), Brannigan is also wary of devolutionary discourses that privilege nationalist tropes of "exceptionalism and insularity" (6). At the same time, he is mindful that "the risk of an archipelagic analysis which stresses connection and interrelation is that it may appear to underwrite a tacit unionism" (8). His caution here—his concern that archipelagic perspectives can be harnessed equally by opposing arguments—alerts us to the notion that archipelagic thinking is not a simple, self-evident process but one that requires a self-reflexive, critical approach.

One of the ways in which Brannigan explores these entanglements without privileging either unionist "connection" or devolutionary "exceptionalism" is through his focus on the sea. He asserts the value of this approach, stating that one of the implications of the arguments in his book is that "the social and cultural connections of the people who live in the archipelago always exceed the limits of state or national formations, and that the spatial imagination of maritime zones may encompass some of those more fully than land zones" (10). This emphasis on maritime zones is in part an act of cultural recovery since, as Brannigan notes (citing Alan Sekula), the era that spawned the texts he studies also saw the advent of mass air travel—a development that rendered the sea "the forgotten space of modernity" (9). And with this focus on the importance of the sea comes a corollary interest in islands: like the Stuart writing in Kerrigan's book, modernist writing is full of islands. Indeed, several of the texts Brannigan explores are specifically island themed, and he devotes a whole chapter to the investigation of "folk revivals and island utopias," maintaining throughout the discussion a critical alertness to constructions of "islandness" in the literature he investigates.

It is in this context that Brannigan engages most explicitly with island studies, particularly in his dialogue with Pete Hay. Hay, while arguing for a phenomenology of islands in the launch edition of *Island Studies Journal*, expresses a distrust of literary evocations of islands, which, in his view, constitute an "appropriation of island realness" through their promulgation of unhelpful and misleading tropes of island boundedness and insularity (2006, 19)—tropes, moreover, often associated with dysfunction. While sympathetic to Hay's objection to "the absence of the physicality and the phenomenology of islands themselves" in literary accounts, Brannigan also challenges him, arguing that "Hay's wish for a way of seeing islands 'in themselves' outside of the means of cultural representation, is itself, of course, a mythical abstraction, an illusion upon which every island fantasy depends" (2015, 145). However, rather than asserting the value of literary metaphor per se, in *Archipelagic Modernism* Brannigan reads a range of modernist texts largely in terms of their relationship with seas and islands as material, felt spaces. He writes from a conviction that "despite the general cultural tendency towards seeing islands and seas in figurative terms, there has been a strong counter-tradition in twentieth-century anglophone literature . . . of reading them from resolutely material perspectives" (10). Nevertheless, he also demonstrates that a materialist approach need not preclude engagement with the texts' more imaginative elements. In his discussion of Virginia Woolf's *To the Lighthouse*, he investigates the way in which Woolf "recurrently fuses, transposes and mythologises one place into others" (109) (exemplified in the case of *To the Lighthouse* by the blending together of elements of the island of the Scottish island of Skye and the English Cornish coast). In Brannigan's view, this "inexactitude" undermines "gendered structures of power and representation," particularly in relation to landscape, "in which 'accuracy' is dependent upon aligning one's perspective with existing geopolitical and patriarchal knowledge" (111). Thus, for Brannigan, the literary imagination can assist in deterritorializing and reimagining landscapes in ways that subtly expose "underlying geopolitics of social, cultural, and economic division and conflict" (123).

Brannigan's own "resolutely material" reading of modernist texts and his corresponding investigation of how natural forms impact on and infiltrate literary production signal the relevance of this development in archipelagic literary criticism to the field of ecocriticism. He himself locates his book in a growing body of contemporary criticism that reassesses modernist literature in light of environmental considerations, adding his voice to the "greening of modernism" (108). In thinking about maritime zones, for example, Brannigan introduces specifically environmental and ecological concerns in his assertion that the relegation of the sea to a forgotten space has facilitated its conceptual designation as "a 'free space' without consequences, whether for waste dumping, overfishing, or most recently hydraulic fracture drilling" (13). He ends his introduction to *Archipelagic Modernism* with a quotation from Richard Jefferies's *The Story of My Heart*, in which the Victorian writer raises questions about why, on an earth that is so evidently bountiful, there are people dying of starvation. For Brannigan, Jefferies's questions can be seen as searching after a "post-theological and post-anthropocentric vision of humans-in-nature," a quest as relevant now, in Brannigan's view, as it was in 1883: "We share this struggle to acquire the necessary literacy to surpass the exhausted narratives of society and belonging which continue to bind us" (18). Brannigan's explorations of modernist fiction enter into this struggle and demonstrate the ways in which archipelagic literature might provide refreshed social-ecological narratives and offer more equitable and ecological modes of planetary belonging.

The most recent incarnations of British and Irish archipelagic thought—Smith's *The New Nature Writing* and the edited collection *Coastal Works*—though not archipelagic by name are certainly so in spirit, and both take the archipelagic imaginary forward. In addressing the theme of "archipelago," Smith draws on the theorizations of "place" in the work of Doreen Massey and of the "local" in the work of Arif Dirlik in order to outline his own concept of "archipelagic localism" (2017a, 159) as a means of denoting a progressive localism able to "celebrate complexity, divergence, difference and local distinctiveness" (163). For Smith, this is indicative of a "slippery indeterminacy" (62), reminiscent of Brannigan's sense of Woolf's creatively rich "inexactitude." He also insightfully identifies a characteristic in contemporary place-based literature, in which the environmental uncertainties of the Anthropocene have produced a simultaneous shift toward "the intensely local and the globally interconnected" (17).

Coastal Works brings together in one volume several of the writers and scholars already discussed in this chapter (including McNeillie, Brannigan, and Smith) and builds on the work of the Atlantic Archipelagos Research Consortium, an international network of scholars working in the humanities.[6] It also explicitly aligns itself with archipelagic perspectives in island studies. The introduction states, “Work in the field of island studies has helped to develop a theoretical and spatial framework of ‘archipelagic thinking’ that might be productively connected with archipelagic criticism’s devolved and interconnected account of nations, regions, and locales” (Allen, Groom, and Smith 2017, 12–13). The collection not only provides further recognition of the role of the archipelagic imaginary in creative writing but also features newly invigorated approaches to literary criticism itself, for example, Fiona Stafford’s extended mediation on the historical-cultural meanings of the Solway Firth and Nick Groom’s thought experiment in which the Irish Sea is drained and thus conceptually reframed as the “center” of the archipelago.

The volume also introduces new elements to the theorization of the archipelagic imagination. Damian Walford Davies explores Stratford et al.’s (2011, 117) discussion of archipelagic identities in the context of the British Isles as being “never quite at home,” applying this notion to the works of Welsh island author Ronald Lockley. For Walford Davies, this unhomeliness, in tandem with Lockley’s desire to “hold subjective and objective knowledges in productive balance (and tension)” (2017, 154), contributes to the writer’s “creatively unsettling” (154) modes of encounter and “hybrid discourses” (156). Walford Davies develops this observation into the concept of the “archipelagic uncanny” (154). He glosses this as “a way of preserving an agile, plural response to the world that recognizes ‘disjuncture’ and ‘disruption’ as well as ‘connection and entanglement’ as core features of island experience” (154). While these elements may be typical of island experience at a local level, they are also perhaps integral to broader human experience of the planetary archipelago, particularly in the context of current environmental uncertainties.

A new contribution to the archipelagic project, *The Frayed Atlantic Edge: A Historian’s Journey from Shetland to the Channel* by historian David Gange (2019), brings together the historical focus of early British and Irish archipelagic scholarship with contemporary environmental concerns. As a new form of archipelagraphy that crosses between narrative historiography and New Nature Writing, *The Frayed Atlantic Edge* constitutes a detailed study of the natural and cultural histories of the archipelago’s Atlantic seaboard. Gange’s research methodology involves navigating these oceanic and littoral spaces by kayak and on foot. Like the archipelagic thinkers already discussed in this chapter, Gange highlights the way in which the history of these isles has traditionally been constructed from the inside out, carrying with it the dominant narratives of the major cities and central power bases. In a process of gradual, inductive research, he offers a corrective to this dominance, developing an outside-in, periphery-core methodology that brings out very clearly the sense of islands and coasts offering counternarratives that disrupt and exceed received histories. For example, he argues, "The so-called Enlightenment . . . might best be interpreted as the triumph of a few cities—Dublin, Edinburgh, London, Birmingham—at the expense of other regions. For coastal regions it was the beginning, and the cause, of a lengthy dark age. In contrast, much of what were once referred to as the Dark Ages had been eras of great coastal strength and enlightenment, when the intellectual traditions of the Irish Atlantic were the most advanced in Europe" (4). Through challenging dominant historical accounts, Gange develops a form of "nissology"—of "studying island on their own terms" (McCall 1994, 2). His narrative archipelagraphy also reveals the importance of nonhuman actors: "seabirds, fish, and species of seaweed play roles as significant in this book as politicians and their institutions" (Gange 2019, xii), he writes, an observation that reveals the potential of this approach to contribute to ecological and posthumanist discourses, disrupting anthropocentric perspectives and gesturing toward an archipelagraphy of the nonhuman.

TOWARD AN ARCHIPELAGIC ECOCRITICISM

As much of the foregoing discussion has suggested, archipelagic perspectives in the British and Irish literary-critical scene have become more and more closely aligned with environmental and ecological concerns, and, as such, they speak to several of the theoretical issues that trouble contemporary ecocriticism. Ecocriticism is an umbrella term for a range of diverse approaches to exploring the ways in which culture conceives of, articulates, and dramatizes human relationships with the more-than-human world. Lawrence Buell's early framing of ecocriticism included the assertion that "if, as environmental philosophers contend, western ethics and metaphysics need revision before we can address today's environmental problems, then environmental crisis involves a crisis of the imagination the amelioration of which depends on finding better ways of imaging nature and humanity's relation to it" (1996, 2). While ecocriticism has done vital work in the intervening decades, critiquing existing ontologies of being-in-the-world and identifying alternative narratives, it seems that the crisis of the imagination has not receded but has instead accelerated along with the unfolding narratives of the Anthropocene. These effects and affects have resulted in three particular areas of conceptual difficulty for the ecocritical project: the need to replace ecocriticism's traditional focus on local affinities with a more global sensibility (Heise 2008); the challenge posed to existing cultural forms by the vast spatiotemporal scales and bewilderingly distributed agencies of the Anthropocene (Morton 2010, 2013; Clark 2015); and the question of whether ecocriticism can really influence other *autopoietic* functional units of human society and thus achieve any environmental praxis—in other words, whether it can effect any change in people's behavior in relation to the environment (Bergthaller 2011; Clark 2015).

Regarding the first dilemma, of ecocriticism's relationship to the local and the global, Ursula Heise argues that it is necessary for the environmental imagination to "envision how ecologically-based advocacy on behalf of the nonhuman world as well as on behalf of greater socioenvironmental justice might be formulated in terms that are premised no longer primarily on ties to local places but on ties to territories and systems that are understood to encompass the planet as a whole" (2008, 10). As we have seen, this is a question that has also featured in the framing of archipelagic perspectives within island studies and in the British-Irish literary studies explored in this chapter. In terms of the latter, both Brannigan and Smith engage explicitly with Heise's call for a planetary vision and respond with archipelagic interventions into ecocritical discourse that challenge the local-global binary: Brannigan's "inexactitude" (2015, 111) and Smith's "indeterminacy" (2017a, 62) and "archipelagic localism" (2017a, 159). These concepts, in tandem with Smith's suggestion that New Nature Writing turns to the intimate and the planetary simultaneously, offer a flexibility of perspective and, in Roberts's and Stephens's terms, enable us to hold the microregional and the supraregional in a kind of productive conceptual tension.

However, as Timothy Clark and Timothy Morton argue, even with an expanded sense of ecological interconnection between the local and the global, the Anthropocene still eludes full comprehension. For Morton, the Anthropocene is a "hyperobject" (2013), a term glossed by Clark as an entity "whose physical and temporal scale and complexity overwhelm both traditional notions of what a thing is and what 'understanding' it could mean" (2015, 8). Thus, in Clark's view, the Anthropocene's "unreadability" renders existing forms of culture and cultural critique inadequate and leaves in doubt whether new, more appropriate forms are likely to emerge (13): "The question arises, can its new demands be met by new forms of cultural and artistic innovation or, more darkly, are certain limits of the human imagination, artistic representation and the capacity of understanding now being reached?" (24). Walford Davies's suggestion that the complexities of archipelagic experience might entail the development of "hybrid discourses" (2017, 156) is pertinent here.

Gange's hybrid form combines an array of "ways of seeing"—historical, geological, social, ecological—that offer an expanded view of their subject matter and are able, in particular, to encompass discussion of deep time, a feature made visible in the geologies of the islands and coastlines he encounters.[7] His kayaking methodology also reveals the flexibility of spatial scales themselves: a coastline that represents only a few hundred miles of longitude on a map extends to thousands of miles when one traces all its intricate contours—estuaries, islets, and bays that together "would repay a lifetime's exploration" (Gange 2019, 3). This brings to mind the fractal geometries of Benoît Mandelbrot, which have come to play a significant part in the coastal and archipelagic imaginary (see Allen, Groom, and Smith 2017; Roberts and Stephens 2017). As Allen, Groom, and Smith note, fractal geometry "draws attention down into . . . minute details . . . while at the same time drawing it up towards an expanse that suggests a space almost planetary in scale" (2017, 1). It is an insight that reveals the availability of the concept of spatial vastness at the level of the intimate and local.

The hybrid, fractal nissology described above is not the only archipelagic innovation to emerge in the face of contemporary challenges to the imagination. As we have seen, Walford Davies, again, gestures toward an "archipelagic uncanny" (2017, 154), which might be usefully deployed as a means of interpreting and dramatizing the darker phenomenologies of the Anthropocene. At this point it might be worth dwelling in a little more detail on the way in which the Anthropocene has been interpreted by cultural critics. According to David Farrier, in the Anthropocene humans (with culpability unevenly distributed around the globe) have become a "sublime force . . . , the agents of a fearful something that is greater than ourselves" (2016, n.p.). He then corrects himself, rejecting the notion of the "sublime" and arguing that "the 'uncanny' might serve us better" (n.p.). He continues, "One of the most chilling traces of the Anthropocene is the global dispersal of radioactive isotopes since mass thermonuclear weapons testing began in the middle of the 20th century, which means that everyone born after 1963 has radioactive matter in their teeth" (n.p.). Given that much of this testing took place on coral atolls in the Pacific Ocean (see, for example, DeLoughrey 2013, on testing in the Marshall Islands), Farrier's observation highlights the literal sense in which the Anthropocene is archipelagic in its uncanny effects and draws our attention to the value to ecocriticism of a conceptualization of the "planetary archipelago" (Roberts and Stephens 2017, 11) alert to such strange and disturbing entanglements of colonialism, war, and global ecology. As Godfrey Baldacchino and Eric Clark note, while not underestimating the appalling effects of such testing on the actual inhabitants of the Pacific Islands, "ecologically, we are all islanders now" (2013, 131).

Finally, there is the "intractable question" (Clark 2015, 18) of whether cultural analysis of the ecocritical kind can actually effect any "real-world" changes. Clark asks, "How far does a change in knowledge and imagination entail a change in environmentally destructive modes of life?" (18). In other words, Buell's foundational statement that the amelioration of environmental crisis is dependent on the creation of new imaginaries is fundamentally called into question. Indeed, as Clark adds, the phrase "cultural *imaginary*" "is striking for almost conceding in advance its weakness as a sphere of agency . . . as compared with primacy of the power of material modes of production" (19). Ecocritic Hannes Bergthaller has also questioned the premise of the ecocritical project. Drawing on Niklas Luhmann's work on social systems theory and second-order cybernetics, he argues that modern human society is divided into *autopoietic* functional units (such as law, politics, science, religion, and the economy) that are open to flows of energy but closed at an operational level (Bergthaller 2011, 225). Thus, the implication is that ecocriticism might be best focused on exploring the discourse and blind spots of environmentalism itself rather than assuming that its influence might travel beyond this sphere. As we have seen, however, the "archipelagic imaginary," with its burgeoning interest in ecological perspectives, has already crossed borders between geographical areas and academic disciplines, embracing politics, history, social science, cultural studies, and so on, demonstrating its ability to influence at least the analysis of a range of different subjects.

Nevertheless, it remains a moot point as to whether there might be a direct relation between these recalibrations of the cultural imaginary and actual material effects and environmental praxis. At the same time, few would argue that the stories we have woven throughout human history, especially those emanating from the global North, about the earth and our relationship with our human and nonhuman earth others, have contributed to a range of socially and environmentally destructive behaviors. As the Anthropocene unfolds before our eyes, there is no doubt that we will need new narratives in order to negotiate a recalibrated understanding of the human place in this anthropogenically authored epoch. Stories, identified or fostered by an archipelagic perspective, that speak of complex social ecologies and larger planetary interconnections might serve us better than, to use Brannigan's phrase, the "exhausted narratives" that have brought us to this point. From its inception, ecocriticism has drawn attention to the inadequacy of existing ontologies and epistemologies surrounding the human relationship with nature while at the same time seeking new narratives, new ways of seeing. Similarly, archipelagic inquiry is a mode of resistance, a refusal of the discourses emanating from monolithic "centers of power" that begins by destabilizing existing narratives and looks, as Stratford et al. assert, for new viewing platforms. Both disciplines explore the material-semiotic tensions of their subject matter—"nature" in the case of ecocriticism, and islands and archipelagoes in the case of archipelagic perspectives. The gradual drawing together of the concerns of these two strands of contemporary thought around ecological understandings and environmental issues suggests the potential value of a new area of literary theory—archipelagic ecocriticism.

PRACTICAL AND METHODOLOGICAL IMPLICATIONS

The practical and methodological implications of an archipelagic ecocriticism are threefold. The first points toward a particular avenue of ecocritical inquiry that engages specifically with the literature of islands, coastlines, and seas and investigates the particular insights that arise from an encounter with such topographies. It is especially interested in the emergence of new narratives (or the restoration of occluded or suppressed local narratives) and the evidence of social and ecological connections that extend outward from these "peripheral" or littoral spaces, not least through the medium of their surrounding seas. The second implication draws on the more metaphorical dimensions of the archipelago and concerns a way of looking per se—one that can be applied to any cultural form. From a starting point of defamiliarization or catachresis, this approach marries ecocritique to a reading that works with texts (including with their more imaginative, "literary" aspects) to amplify the glimpses they provide of the complex interweavings of people, power relations, and economics with earth systems. Such a methodology might be seen as uniquely suited to the Anthropocene, which requires us to strive toward human species–wide thinking, while at the same time resisting wholly anthropocentric perspectives, instead forging more posthumanist understandings of our entanglement with the earth.

Finally, as a paradigm of interconnected inquiry and endeavor, archipelagic thinking might usefully shape not only an archipelagic ecocriticism but the discursive formation of the environmental humanities more broadly. The archipelagic model is able to carry simultaneously a sense of both individual distinctiveness and complex interconnectedness over a range of scales, ecosystems, social units, academic disciplines, and cultural manifestations. Roberts and Stephens use the term "network-assemblage" in relation to islands, figuring "the individual island . . . as a participant within a world genre of islands, which, in their insular interlinkings, emerge as a planet-spanning archipelagic assemblage" (2017, 29). Drawing out the metaphorical potential of such a formulation, archipelagic ecocriticism might be regarded as a new participant within a burgeoning, interlinked, and planet-spanning network-assemblage of archipelagic thought.

NOTES

1. In this, Schwyzer draws attention to the fact that the "geographical entity" in question is an extremely complex one. The collection of islands monolithically known as the "British Isles" encompasses five nations (England, Wales, Scotland, Northern Ireland, and the Republic of Ireland); two "crown dependencies" (the Isle of Man and the Channel Islands); and thousands of smaller islands, including the subarctic Shetland archipelago. Historically it has laid claim to far larger, even continental territories (for discussion of this in relation to the Americas, see Roberts and Stephens 2017, 12), and to the present day it is replete with its own intra-archipelagic devolutionary wrangles.

2. Indeed, Jos Smith has noted that John Kerrigan's book *Archipelagic English* (2008) has been "rather strangely overlooked in . . . recent articulations of archipelagic thought associated with island studies" (159).

3. The term "Anthropocene" involves the recognition that "humanity has come to play a decisive, if still largely incalculable, role in the planet's ecology and geology" (Clark 2015, 1).

4. According to Jos Smith (2017a), McNeillie asked the artist to carry out just such a reconfiguration: "I would like [a] somewhat tilting, distorting map, pushed to the lower right hand frame of the picture, with south-east England chopped off by the frame" (157–58).

5. This occurred particularly after 1921, a year in which, according to Kerrigan, the Newbolt Report instigated modern, institutionalized English teaching, and the partition of Ireland "made it easier to think of 'Britain' as England writ large" (2008, 5).

6. See the Atlantic Archipelagos Research Consortium: Identities, Cartographies and Cultural Ecologies (https://researchweb8.wpengine.com).

7. For fuller discussion of both Gange's work and a range of island-themed new nature writing in relation to the Anthropocene, see Marland, forthcoming.

REFERENCES

Allen, Nicholas, Nick Groom, and Jos Smith. 2017. "Introduction." In *Coastal Works: Cultures of the Atlantic Edge*, edited by Nicholas Allen, Nick Groom, and Jos Smith, 1–18. Oxford: Oxford University Press.

Baldacchino, Godfrey, and Eric Clark. 2013. "Guest Editorial Introduction: Islanding Cultural Geographies." *Cultural Geographies* 20, no. 2: 129–34.

Bergthaller, Hannes. 2011. "Cybernetics and Social Systems Theory." In *Ecocritical Theory: New European Approaches*, edited by Axel Goodbody and Kate Rigby, 217–29. Charlottesville: University of Virginia Press.

Brannigan, John. 2015. *Archipelagic Modernism: Literature in the Irish and British Isles, 1890–1970*. Edinburgh: Edinburgh University Press.

Buell, Lawrence. 1996. *The Environmental Imagination: Thoreau, Nature Writing and the Formation of American Culture*. Cambridge, MA: Harvard University Press.

Clark, Timothy. 2015. *Ecocriticism on the Edge: The Anthropocene as a Threshold Concept*. London: Bloomsbury Academic.

DeLoughrey, Elizabeth. 2001. "'The Litany of Islands, the Rosary of Archipelagoes': Caribbean and Pacific Archipelagraphy." *ARIEL: Review of International English Literature* 32, no. 1: 21–51.

———. 2013. "The Myth of Isolates: Ecosystem Ecologies in the Nuclear Pacific." *Cultural Geographies* 20, no. 2: 167–84.

Farrier, David. 2016. "How the Concept of Deep Time Is Changing." *The Atlantic*. October 31. https://www.theatlantic.com/science/archive/2016/10/aeon-deep-time/505922.

Gange, David. 2019. *The Frayed Atlantic Edge: A Historian's Journey from Shetland to the Channel*. London: HarperCollins.

Hakluyt, Richard. [1600] 1985. *Voyages and Discoveries*. London: Penguin.

Hay, Pete. 2006. "A Phenomenology of Islands." *Island Studies Journal* 1, no. 1: 19–42.

Heise, Ursula. 2008. *Sense of Place and Sense of Planet: The Environmental Imagination of the Global*. Oxford: Oxford University Press.

Jefferies, Richard. 2018. *The Story of My Heart (My Autobiography)*. n.p.: Tritech Digital Media.

Kerrigan, John. 2008. *Archipelagic English: Literature, History, and Politics, 1603–1707*. Oxford: Oxford University Press.

Luhmann, Niklas. [2002] 2013. *Introduction to Systems Theory*. Cambridge, UK: Polity Press.

Marland, Pippa. Forthcoming. "Deep Time Visible: New Nature Writing from the Scottish Islands." In *The Cambridge Companion to Literature and the Anthropocene*, edited by John Parham. Cambridge: Cambridge University Press.

McCall, Grant. 1994. "Nissology: A Proposal for Consideration." *Journal of the Pacific Society* 17, no. 2–3: 1–8.

McNeillie, Andrew. 2007. Editorial. *Archipelago* 1 (summer): vii–viii.

Morton, Timothy. 2010. *The Ecological Thought*. Cambridge, MA: Harvard University Press.

———. 2013. *Hyperobjects: Philosophy and Ecology after the End of the World*. Minneapolis: University of Minnesota Press.

Perry, Graham. 2008. Review of *Archipelagic English*. *Guardian*. March 15. https://www.theguardian.com/books/2008/mar/15/featuresreviews.guardianreview17.

Pocock, J. G. A. 2005. *The Discovery of Islands: Essays in British History*. Cambridge: Cambridge University Press.

Pugh, Jonathan. 2013. "Island Movements: Thinking with the Archipelago." *Island Studies Journal* 8, no. 1: 9–24.

Roberts, Brian Russell, and Michelle Ann Stephens. 2017. "Introduction." In *Archipelagic American Studies*, edited by Brian Russell Roberts and Michelle Ann Stephens, 1–54. Durham, NC: Duke University Press.

Schwyzer, Philip, and Simon Mealor, eds. 2004. *Archipelagic Identities: Literature and Identity in the Atlantic Archipelago, 1550–1800*. Aldershot, UK: Ashgate.

Smith, Jos. 2013. "An Archipelagic Literature: Reframing 'The New Nature Writing.'" *Green Letters: Studies in Ecocriticism* 17, no. 1: 5–15.

———. 2017a. *The New Nature Writing: Rethinking the Literature of Place*. London: Bloomsbury.

———. 2017b. "Fugitive Allegiances: The Good Ship *Archipelago* and the Atlantic Edge." In *Coastal Works: Cultures of the Atlantic Edge*, edited by Nicholas Allen, Nick Groom, and Jos Smith, 243–60. Oxford: Oxford University Press.

Stratford, Elaine, Godfrey Baldacchino, Elizabeth McMahon, Carol Farbotko, and Andrew Harwood. 2011. "Envisioning the Archipelago." *Island Studies Journal* 6, no. 2: 113–30.

Walford Davies, Damian. 2017. "Ronald Lockley and the Archipelagic Imagination." In *Coastal Works: Cultures of the Atlantic Edge*, edited by Nicholas Allen, Nick Groom, and Jos Smith, 131–60. Oxford: Oxford University Press.

Woolf, Virginia. [1927] 2000. *To the Lighthouse*. Oxford: Oxford University Press.

Chapter Eleven

Conservation Archipelago

Protecting Long-Distance Migratory Shorebirds along the Atlantic Flyway

Jenny R. Isaacs

Few people understand the coastal Atlantic Flyway as intimately as those conservation biologists who study and protect *rufa* Red Knots, an endangered shorebird with one of the longest migration routes on earth. The work of these biologists illustrates archipelagic thinking in practice. They know that in order to protect these incredible birds, they must understand and protect the entire shorebird migratory archipelago—from its smallest detail to its fullest, most expansive ideal, remembering each of its many dimensions, constitutive parts, and lively players. The biologists' managerial gaze extends far toward the poles in both directions. Their efforts are cross-cultural, multiscalar, and systems-based. *Thinking with shorebirds*, as these biologists do, offers new perspectives on connections between peoples and nonhumans across land, sea, and air.

Conservation success or failure is determined by biologists' ability to anticipate and adapt: to natural rhythms and flows, weather events, climate flux, a heterogenous mix of local circumstances across a host of sites, and the frequently surprising behaviors of the birds themselves. Phasing between widely distributed and culturally variable sites of human-shorebird contact, they negotiate conflicting social values, competing economic interests, and political differences. To coordinate their efforts between the poles, the biologists collaborate within the Western Hemisphere Shorebird Reserve Network (WHSRN), where they plan conservation strategy and advise governments on policy. The interventions proposed vary per site and are based upon a keen sense of place combined with the latest knowledge of where migratory birds go and what the birds need when they get there.

Shorebird migration is influenced by many variables. The birds' vertical and horizontal movements are launched at very specific moments and places, in response to shifting environmental conditions. For example, birds such as *rufa* Red Knots (*Calidris canutus rufa*) migrate twenty thousand miles annually to capitalize on temporary, optimal feeding, nesting, and daylight conditions available only at certain times of year in specific locations. Following the sun over the year, they prefer locations with the maximum hours of available daylight, enjoying both the austral summer in Tierra del Fuego and the boreal summer of the Canadian Arctic. On a range map, their migration appears as a two-dimensional constellation of point locations or "stopover sites"—an ecological network outlining the edges of continents. However, the birds' seasonally timed migration, combined with weather-sensitive bursts of flight upward, grants the archipelago a third and fourth dimension (of volume and time). Appreciating the physical dynamics of this complex assemblage is the responsibility of the *rufa* shorebird conservation biologists.

The biologists' work is daunting. They must find and follow, capture and tag elusive shorebirds across the hemisphere to understand where they go during the year. They must secure funding for future research. They plan and endure costly expeditions to remote and extreme habitats, such as the Arctic Circle. They tackle the logistical, political challenge of developing and executing a coordinated strategy for population recovery across all sites and multiple scales. They work with local political partners to battle development and commercial interests at multiple scales. They listen and respond to local residents' discrete concerns as they assemble a sprawling network of institutional partners. Organizing requires communication across language barriers and sensitivity to cultural differences regarding the value of nature. Partners coordinate their efforts across multiple, overlapping political jurisdictions and incongruent environmental regulations from the ground up. They know that for *rufa* to survive in any one location and for bird numbers to go up across the whole population, migratory shorebirds must be studied and protected transnationally—between and at each stop along their journeys—specifically, across forty US states and twenty-six countries. Though the *rufa* Red Knot subspecies recently won federal protection under the US Endangered Species Act in 2014, it has not made the project of shorebird conservation much easier. To manage this avian ecological stopover network, these scientists assemble what I call here the "shorebird conservation archipelago"—a network of reserves and shorebird conservation organizations that share knowledge and resources across different sites and scales, hoping to better understand and fortify the dynamics of interconnection. From decades of experiences in the field, these biologists have learned that one-size-fits-all approaches to shorebird conservation do not easily translate or deploy across the flyway. Like the shorebirds themselves, biologists and organizations like WHSRN are inevitably beholden to and must work *with* and *within* an archipelago (Hau'ofa

1999; Gillis 2004; Pugh 2013; Elaine Stratford et al. 2011), respecting its geography, history, cultural diversity, and local politics.

In this chapter, I argue that shorebird biologists are working within the contours and according to the rules of a massive, multidimensional archipelago. To protect shorebirds within this archipelago, they assemble what Richard Schroeder describes as "islands of protection in a sea of unprotected space" (personal correspondence). This hemispheric archipelago, actually comprised of many smaller archipelagoes, is really a "meta-archipelago" with "the virtue of having neither a boundary nor a center" (Rojo and Maraniss 1985, 432). Here, I *think with shorebirds* as a way of tracing the transnational conservation apparatus that operates across time, space, and scale, concretizing in particular moments of encounter, contact, and "conjuncture" (see Ghertner 2017). After establishing my multidisciplinary critical theoretical framework, I demonstrate how an archipelagic perspective shifts the conservation episteme toward alternative metrics of success and different research modes and models. The chapter is divided into two parts: Part One outlines how the archipelago and *archipelagraphy* (DeLoughrey 2001) might be utilized within critical environmental study. I review literatures that support connections between island and archipelagic studies, political ecology, and more-than-human geography. Part Two analyzes the challenges of long-distance migratory shorebird conservation through this frame. I specifically analyze WHSRN as my subject of archipelagraphy. I conclude with a reflection on the practical and methodological implications of this research for both archipelagic studies and conservation.

PART ONE: ARCHIPELAGRAPHY

In this section, I argue that a more-than-human, multispecies iteration of *archipelagraphy*, first described by DeLoughrey (2001) and adopted by Stratford et al. (2011), is a generative, interdisciplinary mode of environmental study that synthesizes research and methods across these and other fields. I review areas of convergence between (postcolonial) island and archipelagic studies, more-than-human geography, political ecology, and conservation.

Archipelagoes, Geography, and Political Ecology

Islands and archipelagoes are useful geographic tools for spatializing relations. Generally, the figure of the archipelago both suggests and enforces connection across space while preserving local difference. Sharrad describes the archipelago as a "loose system that does not homogenize its constituent islands," where "each is unique but all are interconnected and they owe their identity not just to what they individually contain but to the sea between them: sea here being not empty space, but road, history, cultural text" (1998, 103, in DeLoughrey 2001, 44). Clark similarly observes that although islands are "the embodiment of singularity and difference," prized as "particularly special or unique places" (2004, 293), island insularity also calls attention to what is outside the boundary; he explains, "Islands are by definition bounded off, but are nevertheless always connected. Being an island may be a matter of either/or, but insularity is a matter of degrees" (288, citing Biagini and Hoyle 1999). Stratford et al. (2011) offer a typology or "topology" of archipelagic research essentially organized by scale, including intra-island, inter-island, trans-island, island to continent, and one Earth island or "a watery planet that renders all landmasses into islands surrounded by the sea" (DeLoughrey 2007, 2). The archipelago in this topology exists at many scales, making it more than a "flat" network or territory. In other words, while the archipelago connects sites horizontally, it also holds "volume" in three dimensions (Steinberg and Peters 2015), suggesting more of a vertical topology (Allen 2011). It therefore retains constitution and gains definition across dimensions and scales when analyzed from different elevations and vantagepoints. For these reasons, the archipelago is a valuable tool within environmental geography and spatial theory.

Archipelagic texts often utilize spatial tools and geographic perspectives. Readers of island and archipelagic texts will note ubiquitous references to maps, mapping, counter-mapping, navigation, and cartography, especially tied to histories of colonization. Islands and archipelagoes are analyzed and imagined at different scales, shifting in representation and relationship to other objects (Godfrey Baldacchino 2006; Hall 2010; Depraetere 2008; Mountz 2015; Stephens 2015). Island studies scholars have noted how scale is manipulated in representation to minimize or disappear islands from continental, map, or historical views, to recast them as peripheral and marginal, or, conversely, to magnify them as laboratories or microcosms. For instance, David Quammen describes the island as "a simplified ecosystem, almost a caricature of nature's full complexity" (1996, 19). *Island Studies* journal founder Godfrey Baldacchino articulated how a focus on islands can stimulate new interdisciplinary modes of critical inquiry, explaining that "being on the edge, being out of sight and so out of mind, exposes the weakness of mainstream ideas, orthodoxies, and paradigms and foments alternatives to the status quo" (2007, 166). The current "relational turn" within island studies encourages scholars to "challenge the landlocked nature of geography and related disciplines" (Pugh 2016) and to "break out of stultifying and hackneyed binaries; privileging instead the power of cross-currents and connections, of complex assemblages of humans and other living things, technologies, artefacts and the physical scapes they inhabit" (Stratford et al. 2011, 125). Pugh, in this volume, recalling Hau'ofa (1999) and Gillis (2004), offers the archipelago as a contemporary metaphor for our state of ecological interdependence and mutual vulnerability in the time of the Anthropocene—positing the archipelago as a planetary frame or envelop *within* which we are all locked and cannot escape. In these ways, archipelagoes facilitate critical interventions by providing a fresh way of analyzing spatial relations of interconnection and interdependence across time, space, and scale, for instance between beings of the land, sea, and air.

Understanding the archipelago as multidimensional geographic space demands attention to those liminal places where land, sea, and air meet. Stratford et al. (2011) offer three identifying features in writings about islands: the island's "encircling" boundary of water, the island as defined by its dynamic interplay between elements of nature (land, water, etc.), and islands as often described in spatial relation to continents or other islands. Actual islands and archipelagoes have been described as hybrid, peripheral, coastal places, neither fully land nor sea—elsewhere called "aquapelagoes" (Dawson 2012; Hayward 2012). This makes them sites of research interest for scholars in ocean studies, who examine island shores as "wet" edge places extending from a sea center (Steinberg 2013), and island studies, which studies them as "dry" places with wet edges. From these contrasting perspectives, closer attention to the dynamic composition of the watery coastal environment is warranted. Because the environment is not a lifeless backdrop, it can be studied as a lively, multispecies ecological assemblage infused with history and politics.

Island and archipelagic studies scholars share with political ecology (PE) a critical focus on asymmetrical power relations in postcolonial environments (Mountz 2015; Sundberg 2015). Political ecology often focuses on postcolonial environments in the global South; for instance, foundational thinker Raymond Bryant explains that PE examines the political dynamics surrounding material and discursive struggles over the environment in the third world where "the role of unequal power relations in constituting a politicized environment is a central theme" and attention is given to the ways in which "conflict over access to environmental resources is linked to systems of political and economic control first elaborated during the colonial era" (1998, 79). This focus on postcolonial environments resonates with archipelagic and island studies' focus on "small places" and "minor literature" (Kincaid 1989; Lionnet and Shih 2005; Lionnet and Jean-François 2016). Examining the effects of colonialism is also a central element of DeLoughrey's project of *archipelagraphy*, which "seeks to undermine colonial discourses of island isolation and to fashion broader, anticolonial alliances" (2001, 46).

Geographers who write about archipelagoes and archipelagic relations might describe their work as "archipelagraphy." The term "archipelagraphy" was first used by Elizabeth DeLoughrey (2001). Remembering that "no island is an isolated isle," DeLoughrey offered "a system of archipelagraphy—that is, a historiography that considers chains of islands in fluctuating relationship to their surrounding seas, islands and continents" (23). Interested in "remapping" and "alternate mappings," DeLoughrey echoed the sentiments of Caribbean and Pacific Island writers seeking "a cartography of archipelagoes that maps the complex ebb and flow of immigration, arrival, and of island settlement" (23). She summarized that "archipelagraphy provides a local focus on 'particular sites within the pattern' while also serving as a 'structure existing only as a network of tracings of wind and tide, flight and quest, ancestors and arrivals, a dynamic of multiple anchorages and constant commuting amongst them'" (44-5, quoting Sharrad 1998, 205). This language of circulation, "tracing," and "networks" signals methods and figures shared between island/archipelagic studies, more-than-human geography, and political ecology, each mutually informed by actor-network theory from science and technology studies. For my case, this passage provides a framework for investigation of shorebird migration across a conservation archipelago comprised of particular reserve sites within a larger conservation "pattern" or structure, more often described as a "flyway" or "network" of "stopover sites." Stratford et al. recently used DeLoughrey's term, describing archipelagraphy as "counter-mapping" that facilitates "a double-destabilization: dislocating and de-territorializing the objects of study—the fixity of island difference and particularity—and constituting in their place a site or viewing platform by which they are perceived and analyzed afresh" (2011, 114). In these ways, DeLoughrey and Stratford et al. provide a critical framework in *archipelagraphy* through which I analyze WHSRN in part two.

Both island studies and political ecology focus on how foreign, continental, global actors and processes influence local island environments in uneven ways, in multiple directions. They highlight situated, subjugated, and subaltern voices, stories, and perspectives. Adding to Christian Depraetere's definition of *nissology* as the study of islands "on their own terms," Baldacchino explains, "The concluding phrase—'on their own terms'—suggests a process of empowerment, a reclaiming of island histories and cultures, particularly for those island people which have endured decades of colonialism" (2008, 37). We hear a parallel aim and sentiment expressed by Raymond Bryant and Lucy Jarosz, who assert that political ecology "privileges the rights and concerns (often livelihood-based) of the poor over those of powerful political and economic elites" (2004, 808), cognizant of its "obligations and responsibilities to 'distant strangers' near and far" (918, in Sundberg 2015, 118).

Stratford et al. offered a critique of early island studies scholarship, which they observed tended to problematically "produce dominant discourses *about* and *on* islands and islanders rather than *with, from* or *for* them" (2011, 114). We see then how island, archipelagic, and PE studies share a common intellectual commitment to ethical, politically engaged research and responsible methods that can amplify subaltern voices and serve their needs.

More-Than-Human, Multispecies Archipelagoes of Air, Land, and Sea

Working within a relational archipelagic framework, global shorebird conservation might be improved by refocusing on the local site, attending to how the whole is influenced and perceived by its many "small" players and parts. However, researchers are still exploring how to account more fully for nonhuman agents and their often "silent" subaltern voices, for example those of shorebirds. Using methods from more-than-human geography, ocean studies, and aerography, I demonstrate a more symmetrical research approach that takes seriously the role of nonhuman actants within the archipelago.

More-than-human geography and political ecology offer single analytical frames within which to study archipelagoes as multiscalar, multispecies, multidimensional assemblages. Scholars in these fields combine insights from posthumanism, new materialism, environmental philosophy, multispecies studies, animal geography, biopolitics, and actor-network theory to recenter focus on *nonhuman agency* (see Isaacs and Otruba 2019). For archipelagic studies this means taking the agency of nonhumans, elements, and forces more seriously, specifically by minding the dynamic rhythms and composition of, movements through, and changes on and in the land, the sea, and the air at different times. As Blum (2010) and Steinberg (2013) recognized for ocean studies and Jackson and Fannin (2011) and Adey (2008, 2015) recognized for aerogeography, it is not enough to study the ideas and actions of humans or other (shorebird) figures as they fly over or float at the surface of the water. In addition, the elements themselves have agency. As Bennett explains, "Humans are always in composition with nonhumanity, never outside of a sticky web of connections or an ecology" (2010, 32). Birds in flight offer a particular lens with which to study island archipelagoes and aquapelagoes in multiple dimensions, extending vertically into the air or flyway, the air being "that invisible sea within which we live" (Possony and Rosenzweig 1955, 1, in Adey 2008, 14) and "both a material domain, in that air changes in its composition spatially, and a social domain" (Wrigley 2018, 715). Scholars in these and other fields assert that the agency, composition, and affect of elements such as air, fire, and water must be better accounted for in analysis, especially in environmental contexts such as the Anthropocene and Sixth Mass Extinction (Lorimer 2015; Dooren, Kirksey, and Münster 2016; Peters, Steinberg, and Stratford 2018). Taking the more-than-human seriously in the case of shorebird conservation includes considering how the agency of tides, wind, rain, air and water molecules, birds, shrimp and crabs, buoys, oil tankers, climate change, sand, waves, and so forth, both determine and respond to conservation efforts articulated at multiple scales.

PART TWO: WITHIN THE CONSERVATION ARCHIPELAGO

In Part One, I presented overlapping intellectual concerns and shared foci between island and archipelagic studies, more-than-human geography, and political ecology. In Part Two, I demonstrate how one might think *with* shorebirds to imagine a multidimensional conservation archipelago.

Traveling Birds, Biologists, and Best Practices

Migrating birds shift the medium of archipelagic analysis and travel from water to air. Thinking with shorebirds provides a way to study the affect, volume, and interconnections of a hemisphere-wide meta-archipelago. Endangered shorebirds such as *rufa* traverse a multitude of coastal landscapes and locations, traveling along the "Atlantic Flyway," a three-dimensional, voluminous aerospace, ascending and descending. They offer the original bird's-eye view. Because they touch down and wade at the wet margins and liminal edges of continents, utilizing actual archipelagoes in North and South America as they forage on beaches and nest on frozen tundra, they challenge the land/water binary. With increasingly more sophisticated tracking devices, conservation professionals are constantly discovering new locations where birds such as *rufa* Red Knots go as they hop between far-flung sites, sometimes flying seven days nonstop and thousands of miles before touching down again (Newstead et al. 2013). The geographic contours of the archipelago are almost written into their biology as their bodies are built for long-distance flying (Colwell 2010). The need for such "long-distance" jaunts signals that the timing of migration between stopover sites is essential for shorebirds. As the birds land and biologists converge to study them, the conservation archipelago concretizes, materializing at specific times and places. Tracing shorebird movements across the Atlantic Flyway therefore does more than describe the environment (sea, air, land, or a hybrid mix of these) as a flat backdrop for migration or conservation. It brings the archipelago and its multiple dimensions into full relief through the embodied, mobile, shifting vantage points of birds—first aloft, then on the ground, running in and out of waves. Because they spend much of their time on the wind, their aeromobilities—their lives lived in the atmosphere—might also be studied to understand the affective qualities of the air, similarly to how Blum (2010) suggests focusing study on the experiences of sailors to get a better understanding of the sea. In these ways, thinking with shorebirds illustrates the benefits of better accounting for the agency of more-than-human agents, natural elements, and temporal rhythms when describing the conservation archipelago.

Coordinating long-distance migratory shorebird conservation is challenging because it is cross-cultural. Shorebirds are valued differently across the hemisphere. Biologists must respond to a host of site-specific challenges and respect the heterogeneous human-environment relations there, beholden to local authority and resident preferences (Burger and Niles 2017). Conservation professionals working with migratory species of concern in the field must sensitively attend to these local dynamics. For example, the Greater Caribbean region is a biodiversity hotspot and critical passage along the Atlantic Flyway. Because one-third of migratory birds in the region are under threat (American Bird Conservancy 2009), the archipelago is a high-priority site of conservation efforts and concern, featuring both new connections and clashes between local residents and the global conservation apparatus, organized around particular species. Many birds island-hop across the Leeward Islands as they stopover en route to distant nesting or breeding grounds. Recalling resource access conflicts studied in traditional political ecology, shorebird advocates must respond sensitively to local community concerns regarding conservation-affected livelihoods, subsistence, and small-scale shorebird hunting traditions (Andres 2011; Fowlie 2011; Wege, Burke, and Reed 2014; Niles, n.d.). Challenges arise when birds have uneven protections across the range: for instance, birds protected in Jamaica under the Wildlife Protection Act have been shot by the tens of thousands annually for sport in traditional recreational "shooting swamps" in Guadeloupe, a French territory, which lacks equal French environmental legal protections and enforcement. The case of shorebird conservation initiatives focused on the Caribbean displays the most extreme aspects and tensions between local and global perspectives, a subject of ongoing research beyond the scope of this chapter.

Further afield, birds and shorebird biologists confront different variables at other stopovers within the archipelago. For example, birds that stopover in northern Brazil, near Maranhão, are threatened by subsistence hunting as well as pollution and an encroaching shrimp-farming industry. At the extreme polar extents of *rufa* migration—the Mingan Archipelago in Nunavut territory, Canada, and Tierra del Fuego, Argentina—conservationists are grappling with how to face the emerging threats of climate change and new offshore oil production, challenging the wisdom of conventional conservation strategies such as purchasing land for reserves (Niles et al. 2007).

North American biologists confront instances of local resistance to state-based conservation laws and enforcement of beach closures to protect migrating birds along the heavily developed shorebird habitat of the Atlantic Seaboard. The densely populated flyway is characterized by highly variable yet interlocking local conditions, including the impacts of RV beach culture on Hatteras Island in North Carolina (Williams 2007), oil spills along the Gulf coast (Niles 2010), refuge opportunities on small islands such as Kiawah Island in South Carolina (Niles 2011), and the impact of tropical storms and vehicles on the beach along the Florida coast ("Resilient Birds," 2016). Even the scientific and stewardship "best practices" established on the Delaware Bay, WHSRN's first site, generate resistance. In New Jersey, over several seasons of participant observation fieldwork and talking with hundreds of beach-goers, I documented mixed opinions, approval as well as tension, around beach closures in that state. Navigating such a heterogeneity of conditions, marked by disagreement over the "appropriate" use of natural resources across the archipelago, requires local knowledge and sensitivity on the part of biologist/managers, who must appeal to, respect, and coordinate their efforts with and through local actors, cultures, conservation bureaus, and partner organizations. These contrasting examples from shorebird conservation are typical of the archipelagic challenges faced by transnational conservation operations.

At higher scales, shorebird conservation requires cooperation between advocates across borders and in different languages within durable, multinational institutions. Because *rufa* Red Knots perform one of the longest migrations of any animal, legal protections in the United States, for example, do not translate to protection across the full range as subspecies are not afforded conservation status from the International Union for the Conservation of Nature (a global conservation network that manages The Red List of Threatened Species) and international treaties like the Migratory Bird Conservation Act are difficult to enforce (Marzluff and Sallabanks 1998). One mechanism used to support conservation efforts is the preservation of habitat through the creation of protected area reserves; for example, Wilson (2010) describes the creation of a "Pacific Flyway" system of national wildlife refuges in the United States, under the authority of the federal government. An alternative system of reserves is the Western Hemisphere Shorebird Reserve Network (WHSRN).

WHSRN is a network of ninety-seven sites in fifteen countries, including more than 32.2 million acres of shorebird habitat within a voluntary system of recognized protected areas that meet certain criteria of biological importance (WHSRN, n.d.). It is comprised of over two hundred sublevel shorebird conservation partners and partner organizations across the western Atlantic Flyway. Headquartered at Manomet Conservation Science Center in Plymouth, Massachusetts, with staff and operations distributed across the Americas, WHSRN aims to restore shorebird population numbers to those in the 1970s (Castillo et al. 2011). At annual, floating WHSRN hemispheric meetings, conservation initiatives are discussed and strategies coordinated between global partners, voiced by representatives from widely distributed local communities of practice. As part of my graduate research that studied WHSRN as a conservation archipelago, I attended WHSRN's annual Hemispheric Council meeting, simulcast in English and Spanish, at a Washington, DC, federal building in 2015, where partners from across the range delivered updates on threats and productive local campaigns to protect shorebirds in their communities. In interviews and as a participant observer, I was impressed by the respect shown and solidarity evident among practitioners. In my ethnographic study, I focused on WHSRN's commonly circulated tools that allow for smooth collaboration: its discourses and narratives, site templates, best practices, range maps, and metaphors, such as the "Atlantic Flyway."

Problems with the "Flyway"

Metaphors of conservation organize the discourse and inform praxis, with real effects on policy, landscapes, and livelihoods. In the last century, bird biologists increasingly organized their management efforts along imagined flyways—perceived migration routes and networks of stopover sites essential to bird migration (Davidson et al. 1998; Boere, Galbraith, and Stroud 2006). However, Wilson (2010) and Whitney (2013) discuss the flyway concept as an oversimplification—depicting it more as a helpful conservation planning metaphor than an accurate representation of reality. But flyway maps are useful in that they reveal and illustrate the manager's gaze or "the analyst's worldview" mapped across an approximated biogeographic range (Rittel, Melvin, and Webber 1973). In other words, shorebird range maps illustrate the epistemology, ontology, and limits of current shorebird science. Like all maps, they prepare spaces for management.

Flat, two-dimensional global flyway maps used within avian conservation materials problematically show *rufa* migration routes as smooth lines running north and south. This misses the reality of shorebird migration as a precisely timed sequence of stopovers. Such a map is problematic for Red Knot conservation, firstly because site-specific migrants, such as *rufa*, do not migrate in "broad front" patterns (e.g., regular and unbroken waves) but rather island-hop between dozens of stopover and staging sites according to seasonal cues. Imagining or representing a *flyway* as a linear corridor lifted "above" a flat surface layer does double work: it animates the atmosphere as a place of transnational politics but also renders shorebird conservation groundless. When conservation is firstly or solely imagined in the air, it is neither local nor situated. Culturally, the flyway model obscures diversity and erases sovereign borders as well as the historical; it renders aerospace as equally flat, uninterrupted territory and, as such, serves to organize, depoliticize, mobilize, and legitimate expansive conservation interventions across a huge area. It facilitates the dismissal of particular places, concerns, experiences, contact zones, and often colonial histories. In other words, the Atlantic Flyway model is useful in extending ideas of governance upward (Feigenbaum and Kanngieser 2015), but as an idea in circulation within conservation, it is problematically indiscrete. Further, it aggregates and essentializes the disparate journeys taken by all migratory birds—comprised of hundreds of species, aggregating hundreds of thousands of divergent, idiosyncratic, individualized, historical "bio-geographies" (Jepson, Barua, and Buckingham 2011; Harrington et al. 2010)—into one abstracted, disembodied simplification to show how *all* birds migrate along the Atlantic. Thus, by representing species' range two-dimensionally, as either biogeographic territory below or apolitical flyway above, managers fix their sights too narrowly on "saving" and laterally pursuing particular species or subspecies across one or another horizontal layer of space. An archipelagic framing of shorebird conservation offers an alternative view.

Instead of *flyway* representations, the *archipelago* offers a simple alternative model for those manager/experts and concerned groups who seek to reimagine the project of migratory species conservation. Working with an *archipelagic* rather than a *flyway* spatial metaphor or ontology shifts thinking about migratory bird conservation in several critical ways. It grounds and recenters focus on liminal sites of contact and moments of encounter between species, air, sea, and land. It more accurately represents the multidimensional aeromobilties and time- and site-specific patterns of shorebird migration. It highlights the different ways that migrating birds utilize and are dependent upon an intact network of stopover sites across a constellation of locations. It captures both the heterogeneity of stopover sites as well as their interdependence. It recognizes the air, land, and sea as elements of a single, multiscalar political ecology, infused with nonhuman agency. Thusly, it does conservation work by highlighting the importance of quality habitat in each place, illustrating why degradation in one area generates ripple effects across the range and across time.

PRACTICAL AND METHODOLOGICAL IMPLICATIONS

The archipelago offers an interdisciplinary, conceptual framework for spatializing relations, working in and across many dimensions, places, and scales.

Using the case of *rufa* conservation, through a recentered focus on nonhuman agency, I have suggested the power of *thinking with shorebirds* for island and archipelagic studies, aerogeography, ocean studies, conservation, political ecology, and more-than-human geography. As *archipelagraphy,* I described how, where, when, and why *rufa* shorebird conservation is organized and staged globally. I illustrated how the archipelago might be used by environmental researchers to frame and represent connections across widely dispersed actors and places.

The methodological utility of the archipelago is its ability to present and hold together both materiality and theory, cases and frame, the whole and its parts, the land, sea, and air, at multiple scales and dimensions. For instance, the "conservation archipelago" offered in this chapter is at once an object of analysis, a critical framework and theoretical lens, a described geographical formation of literal islands and collection of real spaces of multispecies encounter, a social-professional network of scientists, a system of processes, a lived history, a metaphor for interconnection, a topology of relations and power (Allen 2011), a socioecological material assemblage, a spatial ontology, and more. These are held together through a tight analytical focus on tiny, mobile shorebirds, their attached tracking devices, and the conservation discourses, models, and managerial practices repeated across its many dimensions. That the archipelago can support these multiple meanings and functions simultaneously, without collapsing or losing substance, depth, or definition, is a testament to the versatility, capacity, and potential of archipelagic research.

This brief archipelagraphy offers many implications for conservation. It shows the conservation archipelago to be an unruly assemblage of places, lives, elements, and forces. This is consistent with writings from other archipelagic scholars who have argued recently that, especially in the Anthropocene, no one can be outside the archipelago. For their conservation mission, Northern conservation professionals and their counterparts across the global South must therefore work across, with, and always *within* the archipelago—a geography shaped by the agency of geologic forces and elements of nature, by residents, and by the shorebirds themselves. Rather than a fixed object to be ruled over, the conservation archipelago is shown here to be embodied, enacted, reproduced, and performed daily, in fleeting episodes and specific moments of contact and conjuncture between the human and the nonhuman, at particular sites and special times. While an archipelagic perspective might justify transnational conservation interventions, it also places ethical demands on professionals to fully respect these multiple dimensions, multispecies agents, natural rhythms, and dynamic local conditions.

WORKS CITED

Adey, Peter. 2008. "Aeromobilities: Geographies, Subjects and Vision." *Geography Compass* 2, no. 5: 1318–36. https://doi.org/10.1111/j.1749-8198.2008.00149.x.

———. 2015. "Accounting for the Elemental." *Dialogues in Human Geography* 5, no. 1: 98–101. https://doi.org/10.1177/2043820615571212.

Allen, John. 2011. "Topological Twists: Power's Shifting Geographies." *Dialogues in Human Geography* 1, no. 3: 283–98. https://doi.org/10.1177/2043820611421546.

American Bird Conservancy, comp. 2009. "Saving Migratory Birds for Future Generations: The Success of the Neotropical Migratory Bird Conservation Act." abcbirds.org. https://3pktan2l5dp043gw5f49lvhc-wpengine.netdna-ssl.com/wp-content/uploads/2015/05/act_songbirds.pdf.

Andres, Brad. 2011. "Shorebird Hunting Workshop Summary and Supplemental Information." US Shorebird Conservation Partnership. Denver. August 13. http://www.shorebirdplan.org/wp-content/uploads/2015/08/HuntingWorkshop.pdf.

Baldacchino, Godfrey. 2006. "Islands, Island Studies, Island Studies Journal." *Island Studies Journal* 1, no. 1: 3–18.

———. 2007. "Islands as Novelty Sites." *Geographical Review* 97, no. 2: 165–74.

———. 2008. "Studying Islands: On Whose Terms? Some Epistemological and Methodological Challenges to the Pursuit of Island Studies." *Island Studies Journal* 3, no. 1: 37–56.

Biagini, Emilio, and Brian Hoyle. "Insularity and Development on an Oceanic Planet." *Insularity and Development: International Perspectives on Islands*. London (1999): 2–14.

Bennett, Jane. 2010. *Vibrant Matter: A Political Ecology of Things*. Durham, NC: Duke University Press.

Blum, Hester. 2010. "The Prospect of Oceanic Studies." *PMLA* 125, no. 3: 670–77. https://doi.org/10.1632/pmla.2010.125.3.670.

Boere, G. C., Colin A. Galbraith, and D. A. Stroud. 2006. "Waterbirds around the World." Edinburgh, UK: Stationary Office.

Bryant, Raymond L. 1998. "Power, Knowledge and Political Ecology in the Third World: A Review." *Progress in Physical Geography* 22, no. 1: 79–94. https://doi.org/10.1177/030913339802200104.

Bryant, Raymond L., and Lucy Jarosz. 2004. "Ethics in Political Ecology: A Special Issue of Political Geography: Introduction: Thinking about Ethics in Political Ecology." *Political Ecology*, no. 23: 807–12.

Burger, Joanna, and Lawrence Niles. 2017. "Shorebirds, Stakeholders, and Competing Claims to the Beach and Intertidal Habitat in Delaware Bay, New Jersey, USA." *Natural Science* 9, no. 6: 181–205.

Castillo, Fernando, Ian Davidson, Charles Duncan, Deborah Hahn Hahn, Diego Luna Quevedo, Taej Mundkur, and Rosa María Vidal. 2011. "Five-Year Strategic Plan for the Western Hemispheric Shorebird Reserve Network, 2011–2015." Western Hemispheric Shorebird Reserve Network. http://www.whsrn.org/sites/default/files/WHSRN__2011-2015_Strategic_Plan_v1.pdf.

Clark, Eric. 2004. "The Ballad Dance of the Faeroese: Island Biocultural Geography in an Age of Globalisation." *Tijdschrift voorv Economische ene Sociale Geografie* 95, no. 3: 284–97. https://doi.org/10.1111/j.1467-9663.2004.00308.x.

Colwell, Mark A. 2010. *Shorebird Ecology, Conservation, and Management*. Berkeley: University of California Press.

Davidson, N. C., D. A. Stroud, P. I. Rothwell, and M. W. Pienkowski. 1998. "Towards a Flyway Conservation Strategy for Waders." *International Wader Studies* 10: 24–38.

Dawson, Helen. 2012. "Archaeology, Aquapelagos and Island Studies." *International Journal of Research into Island Cultures* 6, no. 1: 17–21.

DeLoughrey, Elizabeth. 2001. "'The Litany of Islands, the Rosary of Archipelagoes': Caribbean and Pacific Archipelagraphy." *ARIEL: A Review of International English Literature* 32, no. 1: 21–51.

DeLoughrey, Elizabeth M. 2007. *Routes and Roots: Navigating Caribbean and Pacific Island Literatures*. Honolulu: University of Hawai'i Press.

Depraetere, C. 2008. "The Challenge of Nissology: A Global Outlook on the World Archipelago. Part II: The Global and Scientific Vocation of Nissology." *Island Studies Journal* 3, no. 1: 17–36.

Dooren, Thom van, Eben Kirksey, and Ursula Münster. 2016. "Multispecies Studies." *Environmental Humanities* 8, no. 1: 1–23. https://doi.org/10.1215/22011919-3527695.

Feigenbaum, Anna, and Anja Kanngieser. 2015. "For a Politics of Atmospheric Governance." *Dialogues in Human Geography* 5, no. 1: 80–84. https://doi.org/10.1177/2043820614565873.

Fowlie, Martin. 2011. "Shooting of Whimbrels Sparks Calls for Regulation of Shorebird Hunting in the Caribbean." BirdLife International. http://www.birdlife.org/europe-and-central-asia/news/shooting-whimbrels-sparks-calls-regulation-shore bird-hunting-caribbean.

Ghertner, D. Asher. 2017. "When Is the State? Topology, Temporality, and the Navigation of Everyday State Space in Delhi." *Annals of the American Association of Geographers* 107, no. 3: 731–50. https://doi.org/10.1080/24694452.2016.1261680.

Gillis, J. R. 2004. *Islands of the Mind: How the Human Imagination Created the Atlantic World*. New York: Palgrave Macmillan.

Hall, C. Michael. 2010. "Island Destinations: A Natural Laboratory for Tourism: Introduction." *Asia Pacific Journal of Tourism Research* 15, no. 3: 245–49.

Harrington, Brian A., Stephanie Koch, Larry K. Niles, and Kevin Kalasz. 2010. "Red Knots with Different Winter Destinations: Differential Use of an Autumn Stopover Area." *Waterbirds* 33, no. 3: 357–63.

Hau'ofa, Epeli. 1999. "Our Sea of Islands." In *Inside Out: Literature, Cultural Politics, and Identity in the New Pacific*, edited by Vilsoni Hereniko and Rob Wilson, 27–38. Lanham, MD: Rowman & Littlefield.

Isaacs, Jenny R., and Ariel Otruba. "Guest Introduction: More-Than-Human Contact Zones." *Environment and Planning E: Nature and Space* 2, no. 4 (2019): 697–711.

Hayward, Phillip. 2012. "Aquapelagos and Aquapelagic Assemblages." *Shima: The International Journal of Research into Island Cultures* 6, no. 1: 1–10.

Jackson, Mark, and Maria Fannin. 2011. "Letting Geography Fall Where It May—Aerographies Address the Elemental." *Environment and Planning D: Society and Space* 29, no. 3: 435–44. https://doi.org/10.1068/d2903ed.

Jepson, Paul, Maan Barua, and Kathleen Buckingham. 2011. "What Is a Conservation Actor?" *Conservation and Society* 9, no. 3: 229.

Kincaid, J. 1989. *A Small Place.* New York: Plume-Penguin.

Lionnet, F., and E. B. Jean-François. 2016. "Literary Routes: Migration, Islands, and the Creative Economy." *PMLA* 131, no. 5: 1222–38.

Lionnet, Francoise, and Shu-mei Shih, eds. 2005. *Minor Transnationalism*. Durham, NC: Duke University Press.

Lorimer, Jamie. 2015. *Wildlife in the Anthropocene*. Minneapolis: University of Minnesota Press.

Marzluff, John M., and Rex Sallabanks, eds. 1998. *Avian Conservation: Research and Management*. Washington, DC: Island Press.

Mountz, A. 2015. "Political Geography II: Islands and Archipelagos." *Progress in Human Geography* 39, no. 5: 636–46. https://doi.org/10.1177/0309132514560958.

Newstead, David J., Lawrence J. Niles, Ronald R. Porter, Amanda D. Dey, Joanna Burger, and Owen N. Fitzsimmons. 2013. "Geolocation Reveals Mid-continent Migratory Routes and Texas Wintering Areas of Red Knots *Calidris Canutus Rufa*." *Wader Study Group Bulletin* 120: 53–59.

Niles, Lawrence J. n.d. "Tag: Guadeloupe." A Rube with a View. http://arubewithaview.com/tag/guadeloupe (accessed June 8, 2017).

———. 2010. "Potential Impact of Gulf Oil Spill on Migrant Shorebirds." A Rube with a View. June 4. http://arubewithaview.com/2010/06/04/potential-impact-of-gulf-oil-spill-on-migrant-shorebirds.

———. 2011. "Our Expedition to Capture Kiawah Island Knots." A Rube with a View. March 28. http://arubewithaview.com/2011/03/28/our-expedition-to-capture-kiawah-island-knots.

Niles, Lawrence J., Humphrey P. Sitters, Amanda D. Dey, Philip W. Atkinson, Allan J. Baker, Nigel A. Clark, Carmen Espoz, et al. 2007. *Status of the Red Knot (Calidris Canutus Rufa) in the Western Hemisphere*. Studies in Avian Biology 36. Riverside, CA: Cooper Ornithological Society.

Peters, Kimberley A., Philip E. Steinberg, and Elaine Stratford, eds. 2018. *Territory beyond Terra*. London: Rowman & Littlefield International.

Possony, Stefan T., and Leslie Rosenzweig. 1955. "The Geography of the Air." *Annals of the American Academy of Political and Social Science* 299, no. 1: 1–11. https://doi.org/10.1177/000271625529900102.

Pugh, Jonathan. 2013. "Island Movements: Thinking with the Archipelago." *Island Studies Journal* 8, no. 1: 9–24.

———. 2016. "The Relational Turn in Island Geographies: Bringing Together Island, Sea and Ship Relations and the Case of the Landship." *Social & Cultural Geography* 17, no. 8: 1040–59. https://doi.org/10.1080/14649365.2016.1147064.

Quammen, David. 1996. *The Song of the Dodo: Island Biogeography in an Age of Extinctions*. Maps by Kris Ellingsen. New York: Scribner.

"Resilient Birds, Devoted Advocates; 2016 Coastal Bird Conservation Results." fl.audubon.org. 2016. http://fl.audubon.org/sites/g/files/amh666/f/2016_coastal_program_update.pdf.

Rittel, H. W., M. M. Melvin, and M. Webber. 1973. "Planning Problems Are Wicked Problems." *Polity* 4: 155–69.

Rojo, Antonio Benítez, and James Maraniss. 1985. "The Repeating Island." *New England Review and Bread Loaf Quarterly* 7, no. 4: 430–52.

Sharrad, Paul. 1998. "Literary Archipelagoes." In *Literary Archipelagoes*, edited by Jean-Pierre Durix, 95–108. Dijon: Presses Universitaires de Dijon.

Steinberg, Philip E. 2013. "Of Other Seas: Metaphors and Materialities in Maritime Regions." *Atlantic Studies* 10, no. 2: 156–69. https://doi.org/10.1080/14788810.2013.785192.

Steinberg, Philip, and Kimberley Peters. 2015. "Wet Ontologies, Fluid Spaces: Giving Depth to Volume through Oceanic Thinking." *Environment and Planning D: Society and Space* 33, no. 2: 247–64. https://doi.org/10.1068/d14148p.

Stephens, Michelle. 2015. "Intimate Relations: Bodies, Texts, and Indigenizations in Archipelagic American Space." *American Quarterly* 67, no. 4: 1235–49. https://doi.org/10.1353/aq.2015.0064.

Stratford, Elaine, Godfrey Baldacchino, Elizabeth McMahon, Carol Farbotko, and Andrew Harwood. 2011. "Envisioning the Archipelago." *Island Studies Journal* 6, no. 2: 113–30.

Sundberg, Juanita. 2015. "Ethics, Entaglement, and Political Ecology." In *The Routledge Handbook of Political Ecology*, edited by Tom Perreault, Gavin Bridge, James McCarthy, and J McCarthy, 117–26. New York: Routledge.

Wege, David C., Wayne Burke, and Eric T. Reed. 2014. "Migratory Shorebirds in Barbados: Hunting, Management and Conservation." Cambridge, UK: Birdlife International.

Western Hemisphere Shorebird Reserve Network (WHSRN). N.d. "WHSRN Sites." WHSRN.org. http://www.whsrn.org/whsrn-sites (accessed June 10, 2017).

Whitney, Kristoffer. 2013. "A Century of Shorebirds—Public Participation and Conservation Science." *Wader Study Group Bulletin* 120, no. 2: 138.

Williams, Ted. 2007. "War Rages on Cape Hatteras." *Audubon Magazine.* January-February. http://www.audubon.org/magazine/january-february-2007/war-rages-cape-hatteras.

Wilson, Robert M. 2010. *Seeking Refuge: Birds and Landscapes of the Pacific Flyway.* Seattle: University of Washington Press.

Wrigley, Charlotte. 2018. "It's a Bird! It's a Plane! An Aerial Biopolitics for a Multispecies Sky." *Environment and Planning E: Nature and Space* 1, no. 4: 712–34. https://doi.org/10.1177/2514848618816991.

Chapter Twelve

The Debris of Caribbean History

Literature, Art, and Archipelagic Plastic

Lizabeth Paravisini-Gebert

Ever since 310 BC, when Greek philosopher Theophrastus, Aristotle's successor in the Peripatetic School, set sealed bottles afloat to prove his theory that the waters of the Atlantic Ocean flowed into the Mediterranean Sea, drift bottles have been used to chart ocean currents. As conveyors of messages to strangers on faraway coasts, they have long been part of the allure of the sea, whose currents—predictably for the likes of Theophrastus and the National Oceanic and Atmospheric Administration, yet unfathomably for so many—were central to the exchange of peoples, goods, and biota that so deeply marked the history of the Caribbean archipelago. The currents swirling around the Caribbean region "in Van Gogh–esque grandeur" (as we can see in Figure 12.1; see Nelson 2012)

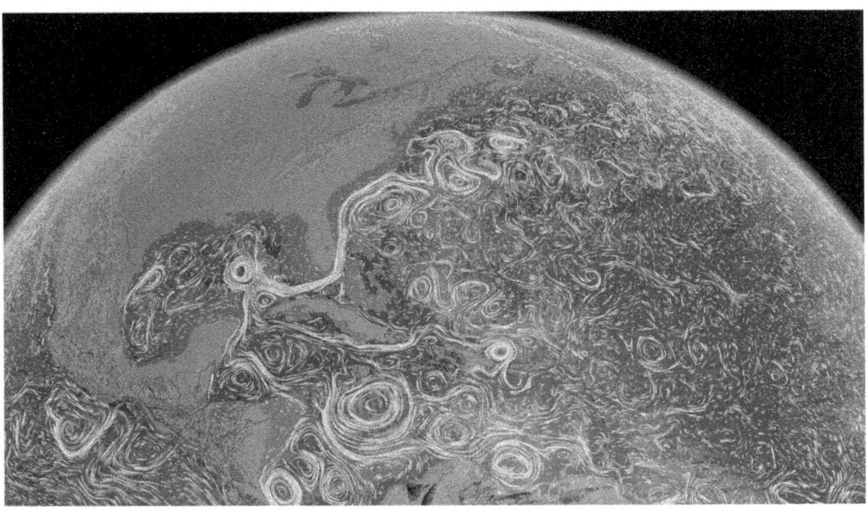

Figure 12.1. A still from the 2012 NASA/MIT video *Perpetual Ocean*.

give the archipelago a distinct current-driven identity, proposing archipelagic connections through winds and water flows that uniquely embrace the islands, "framing" them as a group.

These stunningly beautiful Van Gogh–esque swirls—visual renditions of a painterly sea—now sadly bear swarms of unwanted (and most decidedly nonalluring) drift bottles in the form of floating plastic debris that reaches the coasts of the Caribbean islands from far-flung places in the world, offering a plastic "frame" that once again embraces the islands as an archipelagic unit. The very currents that led Columbus and his vessels to the Caribbean Basin, making it the cradle of pan-American colonialism and planting the seeds for the significant environmental revolution that followed in their wake, now bring to the Caribbean shores a plethora of pollutants, from solid waste (mostly plastic waste) to sewage, hydrocarbons, and agricultural runoff. These pollutants, as Alice Te Punga Somerville has argued about the Great Pacific Patch, embody, as waste, a complex set of natural and cultural processes bound with colonialism and empire: "proximity, movement, disposability, invisibility, history, excess, destruction, reconfiguration, giant multimodal currents, and their life-changing effects on marine as well as human life" (2017, 324).[1] The result is that, according to the 2004 GIWA Regional Assessment of the Caribbean Islands, "pollution of aquatic ecosystems, including sensitive marine and coastal habitats, is the most severe and recurrent transboundary environmental concern in the region" (Villasol and Beltrán 2004, 36). I would read "transboundary," with its postnational implications, as "archipelagic," given the power of the pollution to join the territories of the Caribbean in yet another sea-driven cluster.

Ramón E. Soto-Crespo, writing about Jean Rhys and the Sargasso Sea—the portion of the North Atlantic region where ocean currents form a gyre that now contains a high concentration of nonbiodegradable plastic waste—describes it as "a living archipelago of unanchored trash forms" (2017, 309). Rhys herself, writing about waste in *Good Morning, Midnight* (1939)—before developments in chemical technology allowed for mass production of plastic in the 1940s and 1950s—proposed a "grotesque bilge theory of subjectivity" rooted in a preplastics understanding of the inevitability of ocean pollution (Zimring 2000, 227). Her protagonist, Sasha, speaks of the mind as a space where all memories of love, loss, and death wash about "like bilge in the hold of a ship . . . not in water-tight compartments . . . [but] all washing around in the same hold" (227). Hers is a "theory of the mind . . . as liquid, polluted and polluting . . . a deep container of watery garbage" that can be released into a psychological sea (227). Eighty years later, worrying about ocean contamination from bilge waters seems almost quaint, since "unanchored trash" now threatens to cover the Caribbean archipelago's shores, redefining coastal landscapes and turning the sea itself into a "living garbage heap" that bridges the islands (Soto-Crespo 2017, 310). Worsening the "perfect storm" of marine debris are the ever-growing challenges of garbage disposal in the small island territories (where most plastic ends up in the rivers and ultimately the sea) and waste and sewage discharges from hotels and countless cruise ships, all of which contribute to making the cumulative weight of waste on the fragile island ecologies significantly worse in the Caribbean than elsewhere in the world.[2]

Among the most eloquent voices inviting action against the seemingly insurmountable problem of accumulating plastic debris in the Caribbean have been those of regional and international artists whose waste-focused projects have brought attention to plastic contamination and proposed actions to mitigate or eliminate its impact. Through waste-based installations and wastescapes—which represent the antithesis of what Krista Thompson has called "the Caribbean picturesque" (2007, 27)—these projects intervene in the reimagining of "landscape" as it has been understood in the European/colonial/tourism imaginary while giving voice to deep concerns about the health of the region's coastal environments. Many of these projects offer excellent examples of postcolonial environmental-justice eco-art, artistic work that in addition to the goals of environmental art (the expectation that environmental art should help us "re-envision our relationship to nature, proposing new ways for us to co-exist with our environment" and restore ecosystems in deliberate and often aesthetic ways) also "takes on issues of race, class, gender and eco-colonialism in the unequal distribution of environmental problems and benefits within the US and around the globe" (Environmental Justice Cultural Studies, n.d.).

Eco-activist art, moreover, seeks to *change* the environment for the better, moving beyond "witnessing" or condemning environmental deterioration into specific, community-driven engagement with a clear path to environmental restoration or mitigation. Environmental work stemming from the Caribbean and addressing regional problems—whether produced by Caribbean or international artists—can aim, as T. V. Reed ([1997] 2010) explained in "Environmental Justice Ecocriticism: A Memo-Festo," to "bring to the fore the invasive, pervasive effects of corporate capitalism" on the creation of environmental pollution and "the racial-class dynamic" that has fostered it. Reed argues that the focus of ecocriticim on the "aesthetic appreciation of nature has precisely masked the effects of environmental degradation affecting the poor, protecting the privileged from seeing the toxic danger in front of them and allowing them to envision the 'airborne toxic event' as something that happens only to others, to lower class people in ghettos and such."[3] The projects discussed below explore the impact of environmental degradation and pollution through the prism of the class and racial categories imposed by capitalism and colonialism on the Caribbean. Their plastic wastescapes are deeply embedded in colonial and capitalistic global routes and currents, underlying the centrality of sea routes and winds in the history, the creativity, and (as I argue here) the very identity of the region as an archipelago. Elizabeth DeLoughrey has argued that "islands and their inhabitants are paradoxically positioned as contained' and 'isolated' yet this belies the consistent visitation by colonials, shipwreck, anthropology, and tourism" (2019, 301)—and, we must add now, plastic debris.

The projects I analyze here address plastic debris not as an isolated "tropical island" concern but as evidence of the archipelago's engagement in networks of global distribution and consumption that bring the "unnatural" to archetypal tropical landscapes.[4] In an installation deeply engaged with the suppressed history of slavery in Martinique, like Jean-François Boclé's *Tout doit disparaître/Everything Must Go* (2014), the use of cobalt plastic shopping bags as artistic material links environmental pollution to the pernicious and lasting impact of the institution of slavery, connecting global consumer exchanges and the disposal of trash to the discarding of African lives during the Middle Passage and the colonial history of racism, thus inserting human exploitation into the critical discourse on imperial debris. Another group of artists working with plastic debris, like Cuba's Tomás Sánchez, is particularly concerned with bringing attention to the visual and ecological pollution stemming from the failure to adequately dispose of solid waste in small island nations like those of the Caribbean. Sánchez, known primarily for the idealized forest landscapes through which he captures his flawlessly detailed renditions of the tropical sublime, has also produced hyperrealistic paintings of seascapes marred with accumulations of domestic garbage mixed with plastic flotsam. These photorealistic wastescapes juxtapose "pure" Edenic landscapes with the pollution represented by plastic bags overflowing with the debris of consumerist accumulation resulting from the region's engagement with global markets. Mexico's Alejandro Durán has focused instead on plastic debris as an international problem whose solution requires deep systemic changes in how we approach our understanding of the earth as a shared ecology. He has gained international attention for his *Washed Up* project, a multilevel artistic and eco-educational undertaking aimed at engaging the local and international communities in efforts to reduce the ever-increasing plastic-waste contamination coming to Mexico from countries as far away as Japan and Indonesia. His *Washed Up* series consists of photographs of ephemeral installations constructed from plastic trash collected by the artist from the shores of what should be an archetypal tropical Edenic landscape, the UNESCO World Heritage site at Sian Ka'an in Yucatán, Mexico. *Washed*

Up highlights the irony of fighting globalized plastic "colonization" of a landscape chosen as a UNESCO World Heritage site precisely because it is a remnant of the pristine landscapes inhabited by New World indigenous populations before the advent of colonialism.

Jean-François Boclé's *Everything Must Go*'s engagement with plastic debris is firmly anchored in the ideas of Martinican thinker Édouard Glissant, especially in their shared preoccupation with the impact of colonization and African slavery on Caribbean languages, identities, spaces, and histories (Figure 12.2). The installation, made entirely from a sea of blue supermarket plastic bags "animated" by pumped oxygen, is envisioned as "a quasi-memorial to lives lost at sea during the transatlantic slave trade" (Bonsu 2015), a plastic ode to absence and loss. In this striking work, installed in a pristine white room against which the tomb-like mound of cobalt blue bags appears as a massive force, the ubiquitous plastic of daily consumer interaction reminds us of the bags' original purpose as carriers, containers, objects with a capacity to hold "content" and meaning. Boclé's choice of plastic reminds us of the undisposability of seemingly disposable materials, of networks of exchanges and those consumed in the process of capitalistic enterprise, from the trade in human beings to the present reality of Martinique, its cementification (*bétonization*) and marked environmental degradation.

Boclé's allegorical installation "proposes a wasteland . . . rooted in the question of postcolonial consciousness and collective history" (Bonsu

Figure 12.2. Jean-François Boclé, Everything Must Go (2014). Copyright holder Justin Piperger.

2015). His "vertigo of blue"—the image of an ocean made/replaced by plastic bags—inserts itself into a Martinican environmentalist discourse put forth by the proponents of the *créolité* movement, Jean Bernabé, Raphaël Confiant, and Patrick Chamoiseau, who have repeatedly argued for an understanding of the island's marked environmental degradation as the most disturbing result of France's continued political control, as the disturbing byproduct of enduring colonialism (Gosson and Faden 2001). This shared postcolonial environmental anxiety bridges the gap between the local specificity of their movement's concerns and the increasing interconnectedness brought about by intensifying globalization, which Glissant (1997) wrote about in *Poétique de la relation*. Chamoiseau, Confiant, and Bernabé identify the environmental problems facing Martinique as those same issues confronting the rest of the Caribbean archipelago, as well as other archipelagoes around the world: acute food insecurity (Martinique produces only 2 percent of the food its population consumes); the building of hotels, supermarkets, shopping centers, and other tourism infrastructure at the cost of agricultural land; pollution of land and rivers with fertilizers and insecticides (some of them substances banned elsewhere); the problems of disposing of larger quantities of garbage than the island can absorb; and the destruction of mangroves and the wildlife they support.

Everything Must Go—sea, swimming pool, mass grave, abyss, vertigo, flood, *black (plasticized) Atlantic*—is above all "an accumulation of a serial sameness building out of an architecture of air" (Benítez Villanueva 2016), which in its repetition seems to approach infinity, while proposing a redefinition of the surrounding Caribbean Sea as "invaded" by the plastic of unleashed consumerism. This infiniteness resonates against the permanence of plastic, a malleable and seemingly vulnerable material that holds pockets of negative space ready to briefly bear objects of consumption and whose ubiquitous presence reminds us of its indestructibility, of the uninvited arrival of plastic debris to Caribbean shores bearing the sea air, the sounds of the sea, the voices of those submerged by history, the knowledge of the waves. It is the submersion of which Kendel Hippolyte writes in his poem "Night Vision," where he invokes

> The millions of Africans contorted into writhings of black coral on the sea floor,
> And the survivors living the other death, from the first lash of sunlight
> Till the cool, blessed dark dried out the whip and cutlass
> While they unburied, nightly, the still-warm, holy, undying dream. (2005, 36)

The permanence of plastic reminds us of the impossibility of keeping the violence and trauma of Caribbean history submerged forever, its indestructibility echoing the futile nature of attempting to clean up the mess of colonialism through submersion, the thankless task of trying to tidy up the debris of history, the toxicity of the slave trade, by amassing it in plastic carrier bags that are eminently floatable, whose nature compels them to rise to the surface.

This element of recovery and unveiling has been central to Boclé's collaboration with fellow Martinican composer Thierry Pécou in a project titled *Outre-mémoire* (Beyond/Besides Memory, with its play on *outré*/enraged), which has sought to create an audio and visual memorial to the black Atlantic. In this ongoing collaboration that began in 2004, the audience is invited to experience the two forms (music and art installation) separately while moving between the two, like mirroring events with the same conceptual current flowing between them. It is a collaboration in which the blue plastic of Boclé's installation is not forgotten. In one of his compositions, "Mulunga," the musicians produce sound by rubbing together a cobalt blue carrier bag like those that form Boclé's installation.

Boclé's work is best understood in the context of other Glissant-inspired installations in Martinique's recent artistic production that focus on debris as a metaphor for a history of slavery and human exploitation. His work inserts itself in this "conversation" about ways in which we can come to terms with a suppressed history and separates the project, despite the emphasis on plastics as materials and debris as a prominent focus, from the goals of what is normally recognized as eco-art.

The most iconic of these projects is the *Anse Cafard* monument, the work of sculptor Laurent Valère, installed in 1998, which commemorates the lives lost when a slave ship was wrecked against the rocks near Le Diamant in 1830 (Figures 12.3 and 12.4). Located on a slight promontory above the Atlantic a few kilometers south of Le Diamant, Martinique, fifteen massive statues forming a triangle rise from the earth, as revenants determined to step into the waves to march across the Atlantic in search of home and old roots. The large permanent installation was one of the earliest interventions in transforming the coastal landscape of the Caribbean into a site of historical memory and cultural significance. Referencing the triangular trade, the cast concrete figures stand at an angle of 110 degrees directly in line with the Gulf of Guinée, exposed to the winds and salt air, open to continued transformation by the elements. An early site-specific coastal monument, it sought to "re-landscape" the promontory through the incorporation of the massive figures growing out of the land itself. The monument addressed multiple levels of the history and culture of Martinique, echoing the efforts of historians and artists seeking to bring attention to the untold stories of slavery and the triangular trade as they had impacted the island's development. One of these,

Figures 12.3 and 12.4. Anse Cafard monument by Laurent Valère (1998). Photo: Lizabeth Paravisini-Gebert.

Patrick Chamoiseau's "Man vs Mastiff," an excerpt from his 2018 novel *Slave Old Man*, speaks to the narrator's rejection of the sea as a place of escape and freedom, given the sea's power to turn the bodies of young black men and women into jetsam washed onto the shore. In this complex short story, a young slave pursued by a relentless mastiff toward death or the killing sea ponders his limited options:

> Loads of times I had, on buoyant bois-flot rafts, faced high waves to deliver barrels or casks of sugar to merchants' ships. Heading into the waves, negotiating them exact, using their opposing unleashed energies to head up and across. An ancient intoxication found again there, intact in the depths of those Great Woods. My boutou-bludgeon in hand, I'd wound up a hunter. Back to me came attack cries on bright savannas. Many bled-out elephants and wild beasts roaring. Tracking crocodiles in exhausted mires. Dances for the courage of the brave. A blogodo-hullabaloo of peoples and very angry gods. A dementia of four million years illuminated by towering flames. I was going back toward the monster. I no longer saw any of the earlier impediments. I felt myself a warrior. (n.p.)

Valère and Boclé invoke the language of ruination and the specter of solid-waste and plastic pollution in ways that link exploitation in the name of global markets and capitalistic currents across the centuries. The transport of merchandise across the seas, whether human cargo or consumer products whose built-in obsolescence and resulting debris represent a new form of market enslavement to former metropolitan centers, as their work argues, connects their preoccupations with the suppression of the history of slavery in Martinican culture and literature to an emerging conversation on the role of waste and/in postcolonial ecologies in an increasingly archipelagic reformulation of what constitutes Caribbean art.

We normally do not associate the name of Cuban artist Tomás Sánchez (b. 1948) with garbage, as his landscape work is sought primarily for its spiritual focus on "the forest as a site of holiness, a place of energy and power" (Sullivan 2014). Frequently compared to the German Caspar David Friedrich (1774–1840) and the painters of the Hudson River School (particularly Frederic Church, 1826–1900) for his idealized landscapes that conjure images of a lost paradise, Sánchez fluctuates between painstakingly (re)created pristine and luminescent landscapes he imbues with deep spiritual significance and his equally realistic but significantly more disturbing images of steaming garbage dumps and seashores covered in metal and plastic flotsam. Once considered a "virtuoso of hyperrealism," he has turned its techniques to laying bare "his profound longing for vanished landscapes, the virgin forests, the natural state before the arrival of the Spaniards, the destructive action of mankind on nature and the preservation of the environment as the motif of choice for his creations" (Montero 2010).

Sánchez "thinks" garbage from a different conceptual perspective than that of Boclé—as a locally produced consumerist accumulation that is slowly drowning coastal landscapes. His focus is primarily the archipelagic Caribbean's seashore landfill, where the debris of global consumerism and built-in manufacturing obsolescence meets the archetypal tropical landscape that is already the site of accumulation of floating international refuse, highlighting environmental degradation as both a local and international problem.

In Sánchez's landfills and coastal dumpsites, the debris is different from what we see in Boclé's installation and what we will see in our discussion of Durán. Here plastic debris (floatable and transportable by sea currents) is mixed with weightier postconsumer garbage, heavier rusting metal household appliances, construction and packaging materials, metal cans and discarded glass bottles. This accumulation of garbage precludes thinking of islands as isolated, remote locations, forcing a reconsideration of notions of "islandness" as pristine other-worlds to which we can escape as tourists in search of a Robinson Crusoe experience, since such fantasies of escape involves committing the islands to globalized flows of tourists demanding increasingly more complex "authentic" island experiences that perpetuate the local population's exploitation and poverty at the cost of the sacrifice of their ecologies. Sánchez's *basurero* (garbage dump) paintings are focused precisely on the twenty-first-century island as a place flooded with debris, no longer "entire of itself" but as one with the debris-bound first-world continents, as "parts of the main." Sánchez's work fits perfectly into Elizabeth DeLoughrey's "tidalectic" approach, one that, following Kamau Brathwaite, "considers the relationships between lands and sea, settlement and migration, indigeneity and displacement, nation and diaspora, in short, between 'roots and routes'" (L. Thompson 2017, 62). Brathwaite is, in fact, the poet to turn to when acknowledging debris as a new expression of colonial trauma. In "Trench Town Rock" (1992), which follows the ordeal of an armed assault on his house in 1990, Brathwaite returns to the devastation caused by Hurricane Gilbert in 1988 to describe the trauma of facing work turned into debris: "my room like the sea—the debris and litter all over the beaches—bibliography files and my poetry manuscript folders—all trampled and curled by the breakers—books hit by like a hurricane—what more can I tell you?" (188).

Sánchez's concern with the overwhelming quality of garbage besieging the archetypical island paradise, a frequent subject of his more conventional landscape paintings, is eloquently illustrated in *Mirage* (1991). Here the "sea" of garbage threatens to drown a deeply forested island, the lost primeval paradise of the pre-Columbian Caribbean. The viewers in each case are forced into complicity in this postcolonial invasion, as the wave of debris dominates the foreground of the painting, providing an uneasy bridge of hyperrealistically detailed postindustrialist, postconsumerist garbagescape between them and the *mirage* of a lost environmental wholeness that collapsed upon contact with colonialism and incipient capitalism and consumerism. The apocalyptic image of the destruction of the Caribbean's Eden also conjures the vision of an archipelagic invasion in which waste replaces the sea as the connecting element that gives the region archipelagic continuity.

Figure 12.5. Tony Capellán, Mar Invadido (detail), 2015. Photo: Lizabeth Paravisini-Gebert

Installation artist Tony Capellán's concern with notions of "invasion" of the sea through the proliferation of local plastic debris in his native Dominican Republic focuses, however, on the "narratives" embodied in plastic objects as "evidence" of the loss and sorrow of poverty and economic marginalization (see Figure 12.5). The focus of his installations, from *Mar Caribe* (1997) to *Flotando* (2013) to *Mar Invadido* (2015), is debris he recovered from the beaches of the capital city of Santo Domingo, especially from those near the mouth of the Ozama River, the location of the city's bustling port. Santo Domingo's poorest, most marginalized populations, pushed by rapid urbanization to the most vulnerable riverside land, live in substandard housing in overcrowded *barrios* along the Ozama. "These severely polluted waterways," as Sletto and Díaz argue, are consistently represented as "decaying," "filthy," and "dangerous" (2015, 1680). Persistent flooding from heavy rains threatens lives and property and brings residents into dangerous contact with the river's highly polluted waters, bearing harmful bacteria and toxic concentrations of metals like thallium. The bed of the Ozama, moreover, is below sea level, so as tidal flooding and coastal erosion from storm surges grow ever stronger due to climate change, the sea penetrates deep into the Ozama's watershed, its saltwater infiltration adding to the population's vulnerability to flooding and further contaminating the already deeply compromised freshwater supply. The Dominican poor living along the Ozama are, as a result, among the world's "most endangered people." Capellán's work engages these lives lived at the mercy of the tides amid clear indications of rising sea levels as—in Stefanie Hessler's own reading of Brathwaite's *tidalectics*—"dissolving purportedly terrestrial modes of thinking and living, attempting to coalesce steady land with the rhythmic fluidity of water and the incessant swelling and receding of the tides" (2018, 31).

The objects Capellán collected and arranged into what he called "stories, metaphors, visions" are those washed away by floodwaters from these endangered people's homes into the Caribbean Sea, where the currents return them to its shores at high tide, its "gift" to the artist. They first claimed his attention because they were not (like the plastic debris featured in Sánchez's paintings) the kinds of objects you would put in a plastic trash bag but "useful" objects dragged away by the floods—perhaps beloved objects whose loss could be felt, objects perhaps missed, mourned, but now indistinguishable from other plastic objects thrown away after a single use. Through their incorporation into Capellán's installations, these "eloquent materials" become "visions about the reality of these people, but also visions that encompass many countries, many situations similar to those of the Dominican Republic, in the same geographical area" (Capellán 2015). Capellán's work, as a chronicle of the plight of the people of the Ozama, acknowledges a dialogue with Domingo Liz—a painter known as "the magician of the Ozama"—whose paintings capture the accumulative complexity of the history of the slums that grew along the river's banks in the second half of the twentieth century, chronicling the transformation of a once-green riverside belt into a precarious pile of tumbling houses.

For Capellán, the proliferation of plastic debris on the beaches of Santo Domingo speaks as eloquently to the viewer about environmental pollution (of which his audience is aware) as about the environmental vulnerability of a population that has been doubly displaced; he described the flip-flops that make up *Mar Caribe* as encompassing the stories of those who used to wear them—"farmers without land who migrated to the city where everything has been taken from them and sealed off by barbed wire" and who live now at the mercy of a river and a sea who invade their makeshift dwellings at will, taking away meager possessions and needed everyday implements: combs, buckets, cups, plates, toys. As art, they "reaffirm and deconstruct our idea of the Caribbean" (Capellán 2015). As Elizabeth DeLoughrey has written about Capellán's *Mar invadido*, "Turning to Caribbean allegories of waste, we can interpret Capellan's installations in terms of creating not a colonial archive but, rather, a site of witnessing, rendering the 'secret' of wasted lives visible to the more privileged classes who benefit from the labor and the sacrifices made by the undifferentiated poor" (2019, 121). Capellán, like Sánchez in his rising sea in *Espejismo*, addressed the growing impacts of climate change in the Caribbean region in his awareness of the tidal nature of the increasing vulnerability of the Dominican poor and its connection to rising sea levels.

As the discussion above suggests, artists working with plastic debris throughout the Caribbean region are responding to a broad variety of motivations and conceptual imperatives. I want to turn here to a project conceived primarily as an environmental intervention "to raise awareness and change our relationship to consumption and waste" (Durán 2016), as it is one of the salient examples of eco-art in the Caribbean engaged in a consistent relationship with a specific site set aside as a biosphere because of its natural beauty, a vulnerable community with few resources for cleanup and management, and a belief in the power of art to propel international environmental change.

Mexican "retired poet" Alejandro Durán's photographic series *Washed Up* addresses plastic pollution reaching the UNESCO World Heritage site of Sian Ka'an on the Caribbean coast of Yucatán, Mexico's largest federally protected reserve, from at least fifty nations around the world. Sian Ka'an (Maya for "Origin of the Sky") covers roughly two thousand square miles of an area of great importance during the pre-Columbian and early colonial period. Its early abandonment after the consolidation of Spanish power over Mexico due to frequent flooding and poor soil quality allowed for centuries of regeneration of its ecosystems and habitats. The reserve, inscribed into UNESCO's program in 1987, is home to more than twenty pre-Columbian archeological sites and to a remarkable vast array of flora and fauna (more than 100 documented mammal and 330 bird species) and the world's second-largest coastal barrier reef. The biosphere's rich habitats include tropical forests, mangroves, a complex hydrological system featuring *cenotes* (deep sinkholes that serve as habitats for endemic fauna), and (most importantly in our context) a complex coastline harboring richly diverse ecosystems with a wealth of marine life.[5]

Sharing the coastline with this flora and fauna are thousands of cubic feet of plastic deposited on its shores by ocean currents. Sian Ka'an is unfortunately located in an ideal geography for the deposit of marine debris, a reality not lost on Durán and which played a role in his selection of the site as the focus of his work. From his perspective, the "continental" site of Yucatán is archipelagic in ways that connect his project with Yolanda Martínez-San Miguel's definition of the archipelago as "integrating the study of seemingly disparate island chains (and their corresponding networks of ports, fortifications, plantations, and cities, as well as their social, cultural and productive systems) to complicate our conceptualization of the Caribbean in conversation with other regions that share a similar set of conditions" (2017, 155). Sian Ka'an illustrates perfectly the interconnectedness of the Caribbean Sea and the rest of the planet.

The route of the Caribbean Current (see NASA/MIT illustration in Figure 12.1) shows the vulnerability of the Sian Ka'an coastline to floating plastic debris riding the region's sea currents. This oceanic circulation in the Caribbean Sea is largely caused by trade winds that flow from east to west. The main surface circulation, the Caribbean Current, runs from the southeast near Grenada, St. Vincent, and St. Lucia until it reaches the Gulf of Mexico, where some of the current forms the Gulf's loop current and the rest creates the Florida Gulf Stream (Rivera-Monroy et al. 2004). The northwestward running Caribbean Current deposits water into the Gulf of Mexico through the Yucatán current. Wind currents also aid in the spread of plastic debris in the Yucatán Peninsula, playing a significant role in the spread of light plastic debris and other floatables. Studies have shown that beaches exposed to the wind, like the beaches of Sian Ka'an, receive larger amounts of foreign plastic debris and higher levels of contamination.

I linger on the flow of the currents here because Durán's project is inconceivable without an understanding of how the Caribbean archipelago cannot be disconnected from the water flows that first linked the Spanish global empire and now continue to hold the region in its vise. His project focuses on the depiction, primarily through the photography of the ephemeral sculptures and installations he had constructed with marine debris on the beaches of Sian Ka'an, of a landscape tainted by plastic being brought by sea and wind currents from far afield. The work originated during a visit to Cancún in February 2010, where the artist was struck by the volume of plastic garbage on the beaches of "one of the most beautiful places in the world," believing at first that the community was using the site as a garbage dump before learning from residents of the largest coastal community in the area, Punta Allen (population five hundred), the extent and source of the problem. His response, his "instinct," was to create art, to work on "landscapes" that would draw attention both to the global problem of ocean plastic pollution and to the specific vulnerabilities of small, resourceless coastal communities (many of them communities of fishermen dependent on the health of the reefs for their livelihood) to a problem created in an "elsewhere" that implicates all of us. As he has explained, "One of the main pillars of the project [is that] the material that I am using is all of our material. We are all implicated in this. If you ever drank water out of a plastic bottle then you're implicated, so it's basically every one of us. It's not about pointing fingers at one particular person or one particular corporation, it's humanity that's causing this" (Durán 2016). His aesthetic response has been to work with light and composition to create ephemeral sculptures made through blurring the separation between natural and unnatural elements in a process he calls "alchemizing the ugliness." The main concept behind the project is that "these synthetic objects are infiltrating and mimicking nature," and the work should reflect the process through which the plastic is washed up on the shore by the currents. His work has also led him into the role of archivist, since in taking inventory of his "harvest of refuse," he notes the country of origin and nature of the debris. His conceptual work is mirrored by a growing archive of straightforward photographs of individual cans and bottles from distant countries, which Durán records in situ.

Durán claims British artist Andy Goldsworthy, whose ephemeral installations have made him the "master of anthropocentric beauty" (Weintraub 2012, 185), as a "huge influence." Goldsworthy's work is praised for its ability to convey the transient quality of natural beauty through the imposition onto the landscape of an ephemeral, conceptualized form achieved through the skillful (some would say virtuoso) manipulation of naturally found materials that will revert to their natural form after the work of art is achieved and photographed. This "arranging of unstable elements in precarious patterns within ever-shifting environs" has become the characteristic mark of his art (Weintraub 2012, 186).

Their work, indeed, has many points of correspondence, from its commitment to restoration ecology to its engagement with landscape as one element of a larger ecosystem. Like Durán, Goldsworthy is a photographer whose central subjects are his own site-specific sculptures and installations. Goldsworthy, however, works solely with natural materials, while Durán has focused on the invasion of the natural landscape by "unnatural" plastic. They are also separated by the extent to which Durán wants his artistic process to be apparent in the photographs. Speaking of Goldsworthy, he has said that although the British artist has had a great influence on his work, his work is anthropocentric: "In a lot of his pieces you see the hand of man. You can tell there is an artist working with the materials, whereas my work, although I organize materials by color and form, really mimics nature," prompting people to ask if the objects he has photographed "were found like that." Durán would like us to look at a composition like *Mar/Sea* as the work of the sea currents themselves (if the currents were able to select cobalt blue plastic and compose it accordingly) (Figure 12.6).

Blue exceeds all the other colors of plastic washing up to Sian Ka'an, a fact that gave birth to *Mar/Sea* (Figure 12.6), the first piece Durán conceived. The piece, so central to the developing concept of the project, responded to two key influences for Durán, both of them poets, or more precisely, poems: Elizabeth Bishop's "The Monument" (1955) and John Ashbery's "The Painter" (1956). Durán, who holds a master of fine arts in poetry writing, links the conceptual genesis of the project to Bishop's description of an artwork made of wood that projects its image on the rest of the physical world, turning everything, sea and sky, into wood. Bishop's focus on constructing an artwork out of a humble material like wood resonated with Durán, whose available

Figure 12.6. *Mar/Sea* (2013). Copyright holder: Alejandro Durán.

material was the humblest—plastic in the process of degrading upon prolonged contact with the sea—and whose interest was centered on a desire for art to mirror natural forces in their natural configuration. Bishop's poem is concerned with the moment humble materials transform into art through the viewer's perception, the alchemy that turns materials into art. For Durán, the first test shot for the composition of *Mar/Sea* became a galvanizing moment in which he saw in his image echoes of the totalizing power of wood in "The Monument," a moment in which all the elements of his composition, shore covered in plastic debris, sand, sea, and sky, "became plastic":

> It is an artifact
> of wood. Wood holds together better
> than sea or cloud or sand could by itself,
> much better than real sea or sand or cloud.
> It chose that way to grow and not to move.
> The monument's an object, yet those decorations,
> carelessly nailed, looking like nothing at all,
> give it away as having life, and wishing;
> wanting to be a monument, to cherish something.
>
> —Elizabeth Bishop, "The Monument," from *North and South* (1955)

For Durán, the process of gathering the blue plastic, organizing the composition, and facing in his first photographic image the extension, through color, of plastic onto the natural elements (blue sea and sky) turned into the moment he conceived his series: "Everything became plastic, either the plastic becomes the sea and the sky, or vice versa, but it all became this one material, all is one. All the materials reflected each other, symbolized each other" (Durán 2016).

Ashbery's "The Painter," included in *Some Trees* (1956), ponders in turn an artist's path toward creative independence and echoes Durán's embrace of visual art when first confronted with the need for a form from which to speak of the environmental degradation of Sian Ka'an.

> Sitting between the sea and the buildings
> He enjoyed painting the sea's portrait.
> But just as children imagine a prayer
> Is merely silence, he expected his subject
> To rush up the sand, and, seizing a brush,
> Plaster its own portrait on the canvas.
> . . .
> How could he explain to them his prayer
> That nature, not art, might usurp the canvas?
> . . .
> Slightly encouraged, he dipped his brush
> In the sea, murmuring a heartfelt prayer:
> "My soul, when I paint this next portrait
> Let it be you who wrecks the canvas."

Durán responded to Ashbery's painter's desire for the sea to "usurp his brush, to paint the canvas for him" by experimenting with gathering plastics (blues, primarily, at the onset of the project) and leaving them on the beach for the sea to rearrange during the night, hoping for compositions that reflected minimum impact on his part as artist, letting the ocean—the currents responsible for bringing the plastic to the shore—create the sculpture, which he would then photograph. Like Ashbery's painter, he sought a collaboration with nature, a conversation, leading him to want "to create works that looked like they have been created by nature" (Durán 2016). Since that initial "collaborative" moment, Durán has returned to his own manipulation of the materials instead of simply capturing the debris in its unadulterated state, seeking instead to "mirror what is happening in nature." The abandonment of the brief "collaboration" with the sea prompted deep reflection about the nature of the materials he was working with, as "one of the main pillars of the project is that the materials I am using are mimicking nature," and his mirroring of nature's processes through in-site installations and photographs has as a goal the spread of the message that "the world is being covered with this junk."

If Durán's conceptual conversation with Ashbery and Bishop provides a foundation for conveying the fraught links between currents, coastlines, and invading plastic, his art is also committed to exploring the creative alchemy through which the images created from ephemeral sculptures could bring attention to the environmental quandary facing the land and people of Sian Ka'an. For Durán, "the photos demonstrate something a documentary photo can't"—the aesthetic power to draw the attention of viewers who could become interested, and potentially also engaged, in helping address the plastic problem in Sian Ka'an. As a self-described "interventionist and advocate," Durán sees his art as a first step in fund-raising to bring awareness to the plight of Sian Ka'an and to create environmental awareness projects for the children of local communities to enhance efforts to clean the ocean debris, in recognition that the local authorities lack the funding necessary to emphasize the cleaning of the beaches in a sustainable way.[6] A secondary goal is that of underscoring through his work the shared responsibility of those using and discarding plastic—regardless of their location in the world—for the situation in Sian Ka'an. His photographs, striking and engaging as one may find them, never lose sight of their central aim of bringing attention to the larger issues of "our stewardship of the planet and the consequences of convenience" (Durán 2016). His work is that of a serendipitous artist—he discovered art through his search for expressing the realities of Sian Ka'an—whose art is committed to the eco-art goals of denunciation and education. Durán has captured, through his attention to surface and wind currents as a factor in the plastic pollution in Sian Ka'an, the complex connections between science, environmentalism, and community goals so central to environmental artist/educators. Durán's concern with the provenance of the plastic materials that ride the world's currents to gather on the shores of Yucatán recognizes the interconnectedness between islands, archipelagoes, and continental shores far and wide as established through currents, colonial histories, and, most recently, the debris of continental consumerism.

PRACTICAL AND METHODOLOGICAL IMPLICATIONS

Disruptive of old theoretical parameters for examining the environmental crises facing vulnerable island nations as strictly local, archipelagic approaches invite us to inscribe the threats to islands and archipelagoes brought by climate change—from rising sea levels and the intensification of storms to floating debris—materially and metaphorically, as belonging to newly acknowledged multidirectional flows of oceanic forces. In "Archipelagic Accretions," Paul Giles argues that "if postcolonialism focused on the power structures, and transnationalism on power exchanges of one kind or another, then the archipelagic imaginary might be said to involve a more fluid system, one that disrupts solid foundations and reinscribes them as ontologically evanescent" (2017, 432). Giles's concern with gradual accretion and "relational flows" is central to the analysis of the work of the Caribbean artists discussed above in their response to the local impacts of planetary forces that have become erratic and unknowable in the Anthropocene. The archipelagic imaginary, as seen in the work of these artists, from Boclé to Durán, allows us entry into the seemingly endless possibilities of reading plastic debris as a multivalent material through whose relational flows we can access the ways in which histories, geographies, ocean currents, and environmental quandaries overlap and echo against earlier representations and obsolete values.

Durán's work—with its focus on plastic bottles blown to the Caribbean from far-flung geographies by wind and sea currents that have moved large masses of water around the world's oceans since the last ice age—has brought us quite far from Theophrastus's original drift bottles. Not very far, however, from the focus on wind and ocean currents that cemented drift bottles and their messages in the popular imagination. For Durán, as for the other artists discussed above, the Caribbean archipelago is defined through the currents of wind and ocean that flow through it, currents that brought us, among other problematic flotsam, European conquerors and the horrors of the Middle Passage and now countless pieces of plastic debris that remind us of our connection to a global capitalism that began with the global enterprise that was our discovery by Spanish sailors. In the scourge of plastic—as tide-wrack on our beaches, pollution in our ecosystems, and a threat to our marine fauna—we have a daily reminder that our archipelago and its history have been bound by ocean currents, framed by the swirling Van Gogh–esque flows that define and embrace the islands of the Caribbean in a twirling frame. Now that arabesque is marked by plastic, offering plastic debris as a new form of archipelagic unity resisted by artists across the region.

NOTES

1. Most notable among the artists bringing attention to the Great Pacific Patch is British photographer Mandy Barker, whose eye-catching images capture plastic debris as suspended in ocean waters; New Caledonia's Ito Waïa, a multimedia artist and poet whose project *Eau, Centre de l'Ocean* speaks to the need to protect the Coral Sea Natural Park; and Māori sculptor George Nuku, whose *Whale Skull Cube* is constructed from plastic bottles, Plexiglas, and polystyrene. For a literary response to the Pacific plastics crisis, see Guam's Craig Santos-Pérez's poem "The Age of Plastic" (2017, 164).

2. Dominica, Grenada, and the Bahamas have announced the banning of single-use plastics by 2020 as part of a green initiative focused on addressing the problems of management and disposal of solid waste. Throughout the region, only 40 to 50 percent of all solid waste (including plastics and other floatables) reaches official landfills or dumpsites, with the rest being openly dumped onto river banks, into ravines, or onto the seashore, from which it washes into the sea, leaving the islands "swimming in excessive garbage" (see Coe and Rogers 1997).

3. Recent Caribbean literature, like Rita Indiana's prescient 2019 novel *Tentacles* (*La mucama de Omicunlé*), posits a world where the poor are threatened by sea-level rise, devastating viruses, rapid desertification, and crippling food shortages.

4. The commitment to art as the means to engage audiences with the plight of threatened archipelagos around the world is not unique to the Caribbean, as we've seen through Tuvalu's entry in the 2013 Venice Biennale, *Destiny. Intertwined*, or Kirivati's 2019 pavilion, *Pacific Time-Time Flies*, both of which draw attention to climate change and the fragility of small island nations. Drawing on Plato and seeking to engage the potential disappearance of the archipelago and the nation founded on it, the 2019 Kiribati Pavilion underscores how "in contrast to the impermanent nature of our world and objects within our world, forms and ideas are permanent" (see http://institute-ergosum.org/kiribati-pavilion-2019).

5. In early 2018 it was revealed by archeologists using a revolutionary technology known as LiDar (Light Detection and Ranging) that what had been previously thought to be "virginal territory" hides the ruins of more than sixty thousand houses, palaces, and other complex infrastructure hidden under the jungle for centuries, revealing a sprawling pre-Columbian civilization just beginning to be studied.

6. Durán's project engages the community, particularly the children, through workshops and an annual project called the Museum of Garbage, which he describes thus: "I want to empower the youth of the region to have their voices be heard and enable them to express themselves through co-created plastic installations. . . . It is my hope that the children's messages will bring about greater respect for individuals who are unjustly bearing the weight of other nations' pollution, encouraging change and social responsibility" (see http://www.alejandroduran.com/museo).

WORKS CITED

Ashbery, John. 1956. "The Painter." From *Some Trees*. Poetry Foundation. https://www.poetryfoundation.org/poetrymagazine/poems/26810/the-painter.

Benítez Villanueva, Sofía. 2016. "Notes on Jean-François Boclé's *Everything Must Go!*." Personal communication.

Bishop, Elizabeth. 1955. "The Monument." From *North and South*. PoemHunter.com. https://www.poemhunter.com/poem/the-monument.

Bonsu, Osei. 2015. "Artist Profile: Jean-François Boclé." *Osei Bonsu*. https://www.oseibonsu.com/pangaea-ii/jean-francois-bocle.

Brathwaite, Kamau. 1992. "Trench Town Rock." *Hambone* 10: 123–201.

Capellán, Tony. 2015. "Watch Tony Capellán Discuss His Works in *Poetics of Relation*." Video posted to YouTube by Pérez Art Museum on May 29. https://www.youtube.com/watch?v=zT5WnhJqx9A.

Chamoiseau, Patrick. 2018. "Man vs Mastiff." Translated by Linda Cloverdale. *Aesop* 22 (March): https://www.aesop.com/au/r/the-fabulist/man-vs-mastiff.

Coe, James M., and Donald Rogers, eds. 1997. *Marine Debris: Sources, Impacts, and Solutions*. Berlin: Springer.

Deloughrey, Elizabeth. 2019. *Allegories of the Anthropocene*. Durham, NC: Duke University Press.

Durán, Alejandro. 2016. Interview with Sofía Benítez and Lizabeth Paravisini-Gebert. June.

Environmental Justice Cultural Studies (EJCS). N.d. "Environmental Justice Eco-Art." Cultural Politics. http://culturalpolitics.net/environmental_justice/art.

Giles, Paul. 2017. "The Archipelagic Accretion." In *Archipelagic American Studies*, edited by Brian Russell Roberts and Michelle Ann Stephens, 427–36. Durham, NC: Duke University Press.

Glissant, Édouard. 1997. *Poetics of Relation*. Translated by Betsy Wing. Ann Arbor: University of Michigan Press.

Gosson, René, and Eric Faden, dirs. 2001. *Landscape and Memory: Martinican Land-People-History*. Toronto: CaribbeanTales World Distribution.

Hessler, Stefanie. 2018. "*Tidalectics*: Imagining and Oceanic Worldview through Art and Science." In *Tidalectics*: *Imagining and Oceanic Worldview through Art and Science*, 31–81. Cambridge, MA: MIT Press.

Hippolyte, Kendel. 2005. *Night Vision*. Evanston, IL: Northwestern University Press.

Indiana, Rita. 2019. *Tentacle*. Translated by Achy Obejas. Sheffield, UK: And Other Stories.

Martínez-San Miguel, Yolanda. 2017. "Colonial and Mexican Archipelagos: Reimagining Colonial Caribbean Studies." In *Archipelagic American Studies*, edited by Brian Russell Roberts and Michelle Ann Stephens, 155–73. Durham, NC: Duke University Press.

Montero, Hortensia. 2010. "Tomás Sánchez: Going beyond Time." *Arte por Excelencias* 4. http://www.revistasexcelencias.com/en/arte-por-excelencias/editorial-4/reportaje/tomas-sanchez-going-beyond-time.

Nelson, Bryan. 2012. "NASA Presents Mesmerizing Visualization of Earth's Ocean Currents." *Mother Nature Network*. April 1. https://www.mnn.com/earth-matters/climate-weather/stories/nasa-presents-mesmerizing-visualization-of-earths-ocean-curre-0.

Poupeye, Veerle. 1998. *Caribbean Art*. London: Thames & Hudson.

Reed, T. V. [1997] 2010. "Environmental Justice Ecocriticism: A Memo-Festo." *Cultural Politics*. http://culturalpolitics.net/index/environmental_justice/memo.

Rhys, Jean. *Good Morning, Midnight*. London: Constable, 1939.

Rivera-Monroy, V., R. Twilley, D. Bone, D. Childers, C. Coronado-Molina, I. Feller, J. Herrera-Silveira, et al. 2004. "A Conceptual Framework to Develop Long-Term Ecological Research and Management Objectives in the Wider Caribbean Region." *Bioscience* 54: 843–56.

Santos-Perez, Craig. 2017. "The Age of Plastic." In *Big Energy Poets: Ecopoetry Thinks Climate Change*, edited by Heidi Lynn and Amy King, 164. Kenmore, NY: BlazeVOX.

Sletto, Bjørn, and Oscar Omar Díaz. 2015. "Inventing Space in La Cañada: Tracing Children's Agency in Los Platanitos, Santo Domingo, Dominican Republic." *Environment and Planning A* 47: 1680–96.

Soto-Crespo, Ramón. 2017. "Archipelagic Trash: Despised Forms in the Cultural History of the Americas." In *Archipelagic American Studies*, edited by Brian Russell Roberts and Michelle Ann Stephens, 302–19. Durham, NC: Duke University Press.

Stoler, Ann Laura. 2008. "Imperial Debris: Reflections on Ruins and Ruination." *Cultural Anthropology* 23, no. 2: 191–219.

Sullivan, Edward J. 2003. *Tomás Sánchez*. Milan: Skira.

———. 2014. *A Vision of Grandeur: Masterworks from the Collection of Lorenzo H. Zambrano*. Sotheby's. November 24. http://www.sothebys.com/en/auctions/2014/collection-of-lorenzo-h-zambrano-n09230.html.

Te Punga Somerville, Alice. 2017. "The Great Pacific Garbage Patch as Metaphor: The (American) Pacific You Can't See." In *Archipelagic American Studies*, edited by Brian Russell Roberts and Michelle Ann Stephens, 320–39. Durham, NC: Duke University Press.

Thompson, Krista. 2007. *An Eye for the Tropics*. Durham, NC: Duke University Press.

Thompson, Lanny. 2017. "Heuristic Geographies: Territories and Areas, Islands and Archipelagoes." In *Archipelagic American Studies*, edited by Brian Russell Roberts and Michelle Ann Stephens, 57–73. Durham, NC: Duke University Press.

Villasol, A., and J. Beltrán. 2004. *Caribbean Islands: Bahamas, Cuba, Dominican Republic, Haiti, Jamaica, Puerto Rico*. GIWA Regional Assessment 4. Kalmar, Sweden: University of Kalmar/United Nations Environment Programme.

Walmsley, Anne, and Stanley Greaves. 2010. *Art of the Caribbean: An Introduction*. London: New Beacon Books.

Weintraub, Linda. 2012. *To Life! Eco Art in Pursuit of a Sustainable Planet*. Berkeley: University of California Press.

Zimring, Rishona. 2000. "The Make-up of Jean Rhys's Fiction." *Novel: A Forum on Fiction* 33, no. 2 (spring): 212–34.

Part IV

RELATIONAL ARCHIPELAGICS

Redefining Imperial and Postcolonial Studies

Family Trees

Craig Santos Perez

Written for the 2016 Guam Educators Symposium
on Soil and Water Conservation

1

Before we enter the jungle, my dad
asks permission of the spirits who dwell
within. He walks slowly, with care,
to teach me, like his father taught him,
how to show respect. Then he stops
and closes his eyes to teach me
how to *listen. Ekungok*, as the winds
exhale and billow the canopy, tremble
the understory, and conduct the wild
orchestra of all breathing things.

2

"Niyok, Lemmai, Ifit, Yoga', Nunu," he chants
in a tone of reverence, calling forth the names
of each tree, each elder, who has provided us
with food and medicine, clothes and tools,
canoes and shelter. Like us, they grew in dark
wombs, sprouted from seeds, were nourished
by the light. Like us, they survived the storms
of conquest. Like us, roots anchor them to this
island, giving breath, giving strength to reach
towards the Pacific sky and blossom.

3

“When you take,” my dad says, “Take with
gratitude, and never more than what you need.”
He teaches me the phrase, “eminent domain,”
which means “theft,” means “to turn a place
of abundance into a base of destruction.”
The military uprooted trees with bulldozers,
paved the fertile earth with concrete, and planted
toxic chemicals and ordnances in the ground.
Barbed wire fences spread like invasive vines,
whose only fruit are the cancerous tumors
that bloom on every branch of our family tree.

4

Today, the military invites us to collect
plants and trees within areas of Litekyan
slated to be cleared for impending
construction. Fill out the appropriate forms
and wait 14 business days for a background
and security check. If we receive their
permission, they’ll escort us to the site
so we can mark and claim what we want
delivered to us after removal. They say
this is a benevolent gesture, but why
does it feel like a cruel reaping?

5

One tree my dad never showed me is
the endangered hayun lågu, the last
of which is struggling to survive in Litekyan
its only home. Today, the military plans to clear
the surrounding area for a live firing range,
making the tree even more vulnerable
to violent winds, invasive pests, and stray
bullets. Don’t worry, they say. We’ll build
a fence around the tree. They say this is an act
of mitigation, but why does it feel like
the disturbed edge of extinction?

6

Ekungok, ancient whispers rouse the jungle!
Listen, oceanic waves stir against the rocks!
Ekungok, i taotaomo'na call us to rise!
Listen, i tronkon Yoga' calls us to stand tall!
Ekungok, i tronkon Lemmai calls us to spread our arms wide!
Listen, i tronkon Nunu calls us to link our hands!
Ekungok, i tronkon Ifit calls us to be firm!
Listen, i tronkon Niyok calls us to never break!
Ekungok, i halom tano' calls us to surround
i hayun lågu and chant: "We are the seeds
of the last fire tree! We are the seeds of the last
fire tree! We are the seeds of the last fire tree!
Ahe'! No! We do not give you permission!"

Chapter Thirteen

Archipelagoes as the Fractal Fringe of Coloniality

Demilitarizing Caribbean and Pacific Islands

Mimi Sheller

Archipelagoes are assemblages of seascapes and islands that are not only physical but also imagined, symbolic, social, and communicative. Archipelagoes, suggests Philip Vannini et al., "are inventions whose validity and usefulness is contingent on the dynamics of their formations and the particulars of their contexts. Archipelagoes, in other words, are not essential properties of space but instead are fluid cultural processes dependent on changing conditions of articulation or connection."[1] They are also a kind of flexible infrastructure, a multiplier, which has served the purpose of managing various kinds of uneven and unequal mobilities. Roberts and Stephens have drawn attention to the ways in which anti-imperialist and postcolonial transnational American studies has "increasingly tended to highlight a view of the United States as imbricated with insular and archipelagic spaces."[2] Within these changing articulations and connections, what new insights can we learn from archipelagic politics?

Islands have long served as a pivotal space in the overlapping geographies of military bases and tourism developments, becoming places where both the romance of landscape and the violence of war can be waged "offshore."[3] The same Caribbean and Pacific archipelagoes often "host" military bases, weapons-testing ranges, or migrant detention facilities, as well as plans for, or realities of, gated hotels and residential developments, tourist resorts and privatized beaches, and nature preserves or marine protected areas with limited access. They operate as what Keller Easterling calls "realms of exemption," with transient populations, temporary status, and impermanent installations, often predicated on landgrabs, sea grabs, and eviction of inhabitants. In serving these overlapping purposes, island territories generate friction and resistance. Here I want to highlight how island archipelagoes not only enable extra-state "infrastructure space" to persist, to expand, to mutate, and to metastasize but can also become pivotal points of politicization and countermobilization.[4]

This chapter interrogates the conception of offshore archipelagoes, in which they are imagined and articulated as liquid spaces of mutating imperialism, in which sovereignty, citizenship, and rights are repeatedly suspended. This is not the only imaginary of archipelagic formations, for they might equally be imagined as spaces for counterimperial alternatives, subaltern trajectories of belonging, and subversive itineraries—spaces for indigeneity, locality, communality, piracy, marronage, freebooting, or hacking. My aim is not to adjudicate between these two views, for I believe they both exist in tension with each other. Instead I want to focus on the ways in which struggles over archipelagic spaces may lead to changing articulations of their connections and meanings, offering opportunities to challenge the very codes of imperialism, coloniality, violence, and occupation by reimagining scale, connection, and alliances.

In that sense, resistance to the militarization of archipelagic spaces—and sometimes just making the violence of such secretive military occupations visible—can rearticulate their meanings. Resistance can begin small but have large implications. Calling into question the right of the world's largest military to occupy a remote, small, off-offshore island can call into question it's right to occupy anywhere. Thus the archipelago claimant who counteroccupies islands can become a kind of fractal fringe of political willfulness. Roberts and Stephens have beautifully delineated the fractal geographies of archipelagic formations, showing how the "corrugated edges" of island coastlines suggest a geographically infinite perimeter when modeled using Mandelbrot's fractal mathematics. If extensive military power can occupy planetary archipelagoes and assert "global" power, then a fractal counterpolitics suggests that small points of friction that generate attention to minute scales can powerfully magnify across the whole formation, calling into question its legitimacy, its narratives, and its smooth workings: "the very small becomes highly significant."[5]

My analysis focuses on a small scale "fractal" comparison of two places: Vieques Island, Puerto Rico, and Tinian and Pågan islands in the Commonwealth of the Northern Mariana Islands. These far-flung archipelagic islands have been intensely locally contested within competing imaginaries of territory, belonging, sovereignty, and dependency that have infinite implications for the end(s) of empire. From these relatively remote locations people have raised their voices against colonial/military occupation and myths of the preservation of nature, drawing the link between temporary states of military usage and fleeting economies of tourism, offshoring, and extraction, in contrast to more equitable, holistic, communal, and locally grounded living.

In this chapter I ask, How does tourism rearticulate or reiterate the historical relation of the United States and its military projects to particular kinds of "offshore" islands and "unincorporated territories"? How do "infrastructuring" and "islanding" processes position especially nonindependent (or semidependent) Caribbean and Pacific archipelagic territories in terms of geopolitical relations to wartime mobilities, to temporary security regimes, but also to tourism developments and zones of temporary status? Lastly, what forms of terraqueous resistance have taken root against such shapeshifting dynamics of the offshore archipelago, and is it possible to turn these small, marginal, fractal spaces into cogs that stop the global machine of military occupation, colonialism, and imperialism and bring to a halt all their managed (im)mobilities?

In this repeating archipelagic perspective of a positively magnified insularity, the protection of a single beach, access to a coastal strand, or assertion of rights to indigenous fishing grounds and food sovereignty can be a large politics writ small, with powerful reverberations. As Craig Santos Perez argues, US marine national monuments that are declared around archipelagic territories often serve to "bluewash" the colonization and militarization of the ocean itself. Unilaterally designated without public input, these "ocean and submerged land grab[s] by the federal government" simultaneously displace indigenous inhabitants, exempt military users from environmental regulations, and open vast ocean regions to US fishing rights and federal concessions to for-profit industries such as tourism, deep-sea mining, and geothermal energy.[6] Revelations of the slippage between military occupation, nature preserves, tourism development, and other regimes of uneven (im)mobilities can thus become significant sites of a local politics with global implications—what theorist Ricardo Dominguez of b.a.n.g. lab calls a "lobal" politics.[7]

ARCHIPELAGOES OF MANAGED (IM)MOBILITIES

In the making of US empire, there are distinctive "unincorporated territories" that share certain characteristics as "geopolitical gray zones": Puerto Rico, the Virgin Islands, and Guantánamo in the Caribbean, and in the Pacific, the Mariana Islands, including Guam or, on a smaller scale, Tinian and Pågan islands. With sliding scales of citizenship and inside/outside status, they share both belonging to and externality to the United States, a duality that colonial historian Ann Laura Stoler argues is a "shared feature of imperial forms."[8]

Building on Stoler's understanding of the "carceral archipelago," we can think through the genealogy of contemporary tourism in relation to particular kinds of "offshore" islands and "unincorporated territories" linked to the United States and its military projects. Stoler elaborates upon Foucault's concept of the "carceral archipelago" to examine the ongoing colonial present. Colonial penal colonies, such as those in New Caledonia, French Indochina, and French Guiana, were not marginal spaces on the edge of empire, she argues, but were central technologies for producing it.[9] From the internment camps of the Spanish in Cuba in 1898 to the US prison camp at Guantánamo Bay, these spaces "are not outside imperial networks of security, surveillance, and intelligence or the visionary bounds of governing bodies. In many ways, camps and colonies are precisely where regimes of security have taken their modern forms."[10] As a place "marked by unsettledness, and forced migration," a colony is, in Stoler's description, "a political concept" that is based on "*a principle of managed mobilities*, mobilizing and immobilizing populations according to a set of changing rules and hierarchies that orders social kinds: those eligible for recruitment, for resettlement, for disposal, for aid, or for coerced labor and those who are forcibly confined" (emphasis in original).[11]

Colonial archipelagoes have thus served as experiments in empire building with their mobile regimes of military security and indigenous unsettling. Today, the citizens of US unincorporated territories continue to live in an ambiguous state of uncertain mobilities, between belonging and exclusion, yet with their land made available to US military use, or tourism development, or infrastructural disposition. This more mobile perspective (employing the plasticity of a *dispositif* as "mobile and mutating connectivities") offers us important insights into archipelagic space as a colonial project both physical and symbolic for managing mobilities and immobilities. Crucially, Stoler argues, "Ambiguous zones, partial sovereignty, temporary suspensions of what Arendt called 'the right to have rights,' provisional impositions of states of emergency, promissory notes for elections, deferred or contingent independence, and 'temporary' occupations—these are conditions at the heart of imperial projects and present in nearly all of them."[12] The archipelago here serves as a specific kind of plastic space for the temporary suspension of territorial sovereignty and rights, including rights to circulation across the sea and enforced incarceration on islands, as well as eviction and deportation from island homes.

Alison Mountz, Jenna Loyd, and their collaborators furthermore argue that the militarization and carceralization of immigration "enforcement"—whether through militarized border enforcement, noncitizen detention and deportation, or migrant incarceration—increasingly erodes human rights and specifically does so using "offshore" liminal spaces on former military bases.[13] There is a "long-standing connection between US military operations abroad and US immigration at home," with military bases such as

Guantánamo, in Cuba, long being used "to police the mobility of migrants and asylum-seekers."[14] So we can begin to envision an extended carceral archipelago of former zones of military occupation, which, on the one hand, can become free trade zones overlaid with tourism, shopping, and entertainment and, on the other hand, may just as easily become detention camps for migrant interception.

Compare the use of Nauru and Manus Island as detention centers by the Australian government, where the "Regional Processing Centre" in Manus is located at the Papua New Guinea Navy Base Lombrum (previously a Royal Australian Navy Base). The transfer of "unlawful maritime arrivals" to these facilities was based on the "excision" of Australian "external territories" (Christmas Island, Ashmore and Cartier Islands, and Cocos Island) from the "Australian migration zone," a decision that has faced legal challenges and extensive protests by detainees and charges of inhumane conditions for detainees. Once we center these carceral archipelagoes we can begin to see how this liminal positioning makes use of the staging of proximity and distance, accessibility and inaccessibility, to control the movement of both residents and visitors, prisoners/detainees and security forces, colonial occupiers and those dispossessed of their land.

In all these cases, archipelagoes are a growing complex of what Easterling calls extra-state spaces, special zones crafted to serve the escape from state regulation and rule of law. The zone "has transformed itself from a penal colony and strategic military position," argues Easterling, eventually turning "into a 'free economic city,'" often on an island retreat where it merges the offshore resort with the offshore financial center.[15] Such unincorporated extra-state territories serve multiple, fluid purposes. They are open to reconfiguration and temporary deployments. What is the relationship between the carceral archipelago, the isolated base or camp, the island imaginary, and the contemporary (im)mobilities of tourism and militarism? How are the very forms of camp and colony that were conscripted for these imperial projects precisely those that lend themselves to tourism mobilities today? How has the offshoring of the carceral archipelago mutated into the offshoring of tourism, property development, and island tax havens that sell off residency rights? And what forms of archipelagic resistance can interrupt such extra-state multipliers?

As Vernadette Vicuña Gonzalez demonstrates, particular forms of aeromobility (such as helicopter tours) and road building (such as the H-3 Interstate in Hawaii) connect histories of military occupation and the tropical island tourist gaze through embodied mobilities and spatial practices. Indigenous lands are seized, local mobilities are severely restricted, and access is controlled for the benefit of a few; and those who resist such processes today "remind the tourist that his or her ability to get around, to have access to a sublime vista, to penetrate a hidden interior, or save on precious time—each benefit is contingent on another's historical or immediate unfreedom."[16] This reminds us of the imperial formations that remain central to tourism practices and of the ways in which tourist mobilities are enabled by the restrictions of other forms of circulation and other claims to the land and ocean.

Tourist mobilities are secured by imperial formations and indigenous immobilities, often built on historical plantation systems and their violent forms of land occupation and labor control. Military bases, weapons-testing grounds, marine national monuments, and protected areas remain at the heart of such processes, in the Caribbean as much as in the Pacific. Sometimes it is those living on these fringes of empire who, faced with the loss of these seemingly small places, are best able to call attention to vastly distributed global processes. In the following sections I will follow the fractal pathways of one Caribbean and one Pacific site, understanding that they echo many other corrugated island histories.

VIEQUES ISLAND: "IT'S A PARADISE, AND IT'S A WAR ZONE"

The imperial uses of Vieques Island, Puerto Rico, exemplify how an "offshore" island used as a US Navy bombing range for over sixty years was reinvented as an "untouched" natural paradise ripe for tourism development. In Vieques we can see a particularly acute example of the spatial restructuring of land, infrastructure, and media representations that produce a new performance of place—one that draws on uneven mobilities to support tourism rather than militarization, yet implicitly still facilitates US military power by normalizing destructive processes of island occupation, first for sugar plantation slavery under Spanish colonial rule, then for US military training and weapons testing, and subsequently via tourism and displacement. This case is not unique and is representative of others throughout the world.

As Santos Perez makes clear, "U.S. Marine National Monuments protect environmentally harmful U.S. military bases throughout the Pacific and the world" because they use a unilateral federal landgrab to effectively "further colonize, militarize and privatize the Pacific."[17] The same can be said for Vieques Island, which offers a remarkable example of extreme place remaking by the US military and federal government. When the US Navy finally closed its long-existing bombing range and weapons facility there in 2003 in response to a well-publicized local resistance movement against it, the island was suddenly promoted for its potential tourism development predicated on its now federally protected nature reserve, bioluminescent bay, and lack of population. Through processes of eviction and expropriation, this Caribbean space was unpeopled, "developed," abandoned, and then represented as "natural," "pristine," and "untouched." The more recent decimation of Vieques by Hurricane Maria threatens further depopulation and expropriation of land for future development, part of a "landgrab" that is feared across the hurricane-affected Caribbean.[18]

After Spain's defeat in the Spanish-American War, Puerto Rico was occupied by the United States in 1898, along with other Spanish "possessions" in the Pacific such as Guam. Congress established Puerto Rico's civil government in 1917, but only with the 1952 constitution did it became a commonwealth, or "free associated State," and no longer a colony. Residents of Puerto Rico (like those in Guam) are US citizens who serve in the US Armed Forces, but they have only a resident commissioner, no voting representative in Congress, and no electoral college vote in presidential elections.[19] More recently, the debt crisis fueled by Puerto Rico's "triple-tax-exempt bonds" led to a $74 billion debt and the imposition through the PROMESA Act of an unelected Financial Oversight and Management Board to run the island's finances. This brought massive cuts to pensions, education, and other public services and a sense of crisis around Puerto Rico's structure of governance, especially since the failure of federal response and rebuilding after Hurricane Maria.[20]

After World War II Puerto Rico "became the hub of the US military presence in the Caribbean . . . [and] US security became identified with military hegemony in the Caribbean."[21] In 1941 the US Navy arrived in Vieques, a twenty-one-mile-long island just off the east coast of Puerto Rico, and purchased "by demand" twenty-six thousand acres, or 72 percent of the territory, to use for maneuvers, bombing practice, and storage of military explosives. In 1947 the US Interior Department was defeated in its plan to forcibly relocate the entire population of Vieques to St. Croix, so a population of about nine thousand remained squeezed between a live firing range and a weapons storage area of the Atlantic Fleet Weapons Training Facility. The navy used the island "for training in Marine amphibious landings, naval surface fire support from offshore, and air-to-ground bombing from Navy and Marine Corps aircrafts launched from carriers."[22] The island was part of the Roosevelt Roads Naval Base, which was the heart of LANTCOM, the navy's unified command for the entire Western Hemisphere.[23] Vieques was a staging ground for US interventions, including Guatemala in 1954, the Dominican Republic in 1965, and preparations for the Balkans, Haiti, Iraq, Afghanistan, Vietnam, Korea, the Persian Gulf, and Somalia, among others.[24] The land was also marketed to the North Atlantic Treaty Organization (NATO) and US allies, who were charged up to $80 million for its use in training aircraft carrier battle groups.

As recently as 1998, up to twenty-three thousand bombs were dropped in the nine-hundred-acre live impact zone.[25] The navy effectively "controlled the fate of the entire island," and its plight "dramatically exposed Puerto Rico's lack of sovereignty and subordinate status within the US orbit of power."[26]

Over the years a protest movement grew, especially when concern over environmental destruction and restrictions on use of fishing areas led local fishermen and their supporters to physically block NATO naval maneuvers in 1978 and 1979. Thousands of islanders and their supporters in the United States were mobilized, but according to Katherine McCaffrey, the local framing of the movement, alongside external events that increased military entrenchment and tensions over Puerto Rican independence, undermined its success. Post–Cold War restructuring of US military strategies and moves to close bases kept the issue alive, as did secretive plans to build a major radar station for the "war on drugs." Then, in April 1999, two bombs dropped by F-18 airplanes outside their target area killed David Sanes Rodríguez, a civilian security guard, and injured four others. That event provoked the resistance movement to establish civil disobedience camps inside the bombing range, and in February 2000 the largest mass demonstration in Puerto Rican history (up to then) was organized in San Juan, the Marcha para la Paz de Vieques.[27] Drawing on a more widely accessible "peace" discourse and nonviolent tactics, according to McCaffrey, that protest became "a public relations disaster for the service, rippling back into cities with large Puerto Rican populations like New York."[28] Through "new and old media technologies, from Internet sites to film festivals, newspapers to television, the civil disobedience campaigns of the late 1990s and early 2000s made the 'Paz para Vieques' struggle for peace internationally known."[29]

In January 2000, President Bill Clinton finally issued an order for the return of the entire naval ammunition facility to civilian use and implied devolution of the land to the municipality of Vieques or to the Commonwealth of Puerto Rico. The process of withdrawal began in May 2001 and was finally completed in May 2003. After sixty years of US Navy presence in Vieques—which allegedly included testing live ammunition, depleted uranium shells, napalm, and germ and chemical warfare—the resistance movement had driven out the navy.[30] In any case, prevailing military practice had turned toward discourses of "precision bombing" supported by Global Positioning System (GPS) technologies.[31] Bombing targets could be simulated as imaginary islands, and weapons training could use Virtual At-Sea Training (VAST), a program that was accelerated in 2002 after the decision to pull out of Vieques. As Caren Kaplan argues, the rise of GPS and geographic information systems (GIS) has supported changing cultural practices of targeting within the military-industrial-media-entertainment complex.[32] Now seascapes, landscapes, and aerial-scapes were ready for reappropriation into another infrastructure space: the managed mobilities of tourism.

Most of the US Navy land was returned not to the government of Puerto Rico or to the municipality of Vieques but to the US federal government. Large portions were designated as a nature reserve and wildlife refuge under the direction of the US Fish and Wildlife Service.[33] Presumably that gesture of natural "conservation" was supposed to create a beneficent image of the military; yet it also served to conceal the damage to the island under the cover of "nature" and benefits to "wildlife." Much as Santos Perez argues for the "bluewashing" of marine preserves in the Pacific, this kind of greenwashing "keeps the land under federal control as opposed to public (and indigenous) trust. So if the military ever wants to use the land in the future, it can simply be converted from the Department of the Interior or Commerce to the Department of Defense. This is the 'logic of military conservation.'"[34] Now blighted with thousands of remnants of unexploded ordnance, high levels of heavy metal concentrations in the soil, and other forms of contamination, Vieques is represented by the US Department of Interior and the Fish and Wildlife Service as the largest wildlife refuge in the Caribbean.[35]

Ironically, the designation of the land as a "wildlife refuge" shields it from the environmental cleanup that would be necessary were it to revert to local control and to human residential use. After extensive community mobilization, the land and waters surrounding Vieques were federally designated as a "Superfund site" in February 2005. Without cleanup, however, the land cannot be devolved from federal government control, and that ensures a loophole by which it can later be reclaimed for military purposes.[36] It is in another state of suspension, of waiting, of temporary inconclusive occupation. Today, Viequenses continue to protest the status of the island. In April 2019, a group of five women's organizations (Vide Viequenses Valen, Bieke Microbusiness Incubator, Vieques Women of El Timón, Vieques en Rescate, and El Panal) denounced "the detonations of bombs, the burning of toxic materials, the intentional delay of cleaning and the delivery of lands to the Viequenses by the Navy" and the "negligent abandonment" of the island in favor of foreign investors.[37]

As an island-off-an-island, Vieques has a somewhat separate identity from Puerto Rico itself and is marketed as a distinct destination. Moreover, the continuing exclusion of the local population from the "conservation zone" has set the scene for a new round of development in the name of tourism expansion and new employment opportunities. The new threat faced by Viequenses is the rapid emergence of the apparatus of tourist development on the island. As Vicuña Gonzalez argues for Hawaii, the tourist is "cast as the innocent subject of leisure whose right to move freely and safely exemplifie[s] the ideologies of neoliberal governance"; yet "because of their radically unequal natures," these encounters "are always contestable in violent, creative, or other means. That is, they are themselves generative of critique and alternative ways of seeing."[38] A critical archipelagic perspective can help us to generate other ways of seeing and imagining Vieques's future.

As Sherrie Baver noted in 2006 and is equally pertinent today, Viequenses are considering contending futures for the island in terms of three related questions: Who will control Vieques's development, outsiders or the local community? Will tourism be the large-scale enterprise common throughout the region or a smaller-scale, ecologically and culturally more sensitive version that will provide more benefits to longtime residents? How "green" will Vieques be in the future?[39] The artist duo Allora and Calzadilla called attention to the barrage of major media articles that promoted the island as "the Navy's 'priceless gift' of 'untouched land' [which,] protected from crude development by an accident of history[,] shouldn't be squandered."[40] The American tourism press began describing Vieques as "a prime vacation destination," a "low-profile" and "undiscovered frontier," and thanked the navy for keeping it "off-limits" to tourists and developers.[41] Others described it as "unspoiled" and "a pristine world," despite the toxic contamination.[42] "Bombs have paradoxically preserved much of the island's natural beauty and delicate tropical ecosystems by preventing the unchecked land speculation and slowing the pace of modernization."[43] Bombing and the trampling of local rights ironically become a mode of wilderness conservation, supporting "natural beauty" and "delicate tropical ecosystems." Some foreign journalists even claimed that "the island's preservation can actually be attributed to its gringo oppressors. . . . 'Vieques would be a huge slum if it weren't for the Navy,' says Carlos Latimer, a lawyer from San Juan who has had a second home there since the 1980s. 'The Navy kept it pristine.'"[44]

In contrast, the Committee for the Rescue and Development of Vieques, which claimed one thousand local members in 2003, was calling for four things: "devolution of lands to Viequenses, demilitarization, decontamination, and ecologically just and sustainable development."[45] The costs of tourism have long been known in the Caribbean.[46] Baver argued in 2009 that "Puerto Ricans and other Caribbean citizens are not rejecting tourism development through their struggles but are saying yes to tourism that respects their environment and culture and understands that scale matters on small islands. Caribbean citizens must have meaningful participation in tourism planning or future projects will not go forward."[47] The destructive impact of Hurricane Maria on Puerto Rico has further exacerbated these already existing tensions. Questions of what kind of recovery is needed throw into question the entire shape of all forms of future development.[48] Elizabeth Yeampierre and Naomi Klein call for strengthening local projects such as Resilient Power Puerto

Rico, which distributes solar-power generators to remote areas, or Boricua Organization for Ecological Agriculture, which has "agroecology brigades" distributing seeds and soil to support community-led development alternatives.[49] In Vieques itself crucial grassroots organizations are emerging like La Colmena Cimarrona (Maroon Hive), which "practices community based agroecology and beekeeping" while "working towards food sovereignty, racial and economic justice, strengthening networks of solidarity between poor and landless Viequense, and other folks throughout the Antilles."[50]

The question of alternative futures is fought out at a local level, yet connected across the corrugated fractal politics of islanding around the world, hence "lobal." These actions connect the Vieques Libre movement to wider Caribbean resistance movements, cultural interventions, and cultures of resistance, including protests against inappropriate tourism development by a growing Caribbean environmental movement. Against hundreds of years of colonialism and militarism, as Angelique Nixon's work demonstrates, Caribbean writers and artists have been doing "the work of challenging dominant narratives of history, critiquing exploitation, asserting subjectivity, and creating alternative narratives [that] are central for resistance, liberation, and decolonization."[51] In her book *Resisting Paradise*, Nixon engages with artists and writers who critique the overdependence of the Caribbean on tourism and shows how those critiques are "grounded in a resistance to paradise: defined as exposing the lie and burden of creating and sustaining notions of paradise for tourism and the extent to which this drastically affects people."[52]

Her analysis shows the ways in which tourism economies are an extension of colonialism and in many ways therefore also an extension of military occupation. Resisting such tourism occupation may call for envisioning alternative modes of development and alternative forms of economy, circulation, agriculture, and renewable energy. Such anticolonial, antimilitary archipelagoes of resistance that begin small, in remote places, can have a much larger significance when they connect topologically across dispersed island geographies. To resist locally is to call into question the logics of militarization and offshore landgrabs of all kinds. By showing the connections across space of colonial/military/tourist landgrabs, these archipelagoes of island resistance represent a fractal politics that might extend around the world, from the micropolitics of one tiny island to the macropolitics of the global military-industrial-logistical economy.

RESISTANCE CAN BEGIN SMALL: TINIAN AND PÅGAN

Meanwhile, halfway around the world, another battle over a US weapons-testing range has been unfolding in Pågan in the US Commonwealth of the Northern Mariana Islands, which is made up of fourteen islands. After fleeing a volcanic eruption in 1981, the three hundred people of this small island were relocated to Saipan, but they have long wanted to return to their home island (and a few in fact have). However, during the Barack Obama administration, there were negotiations with Japan to close the major US military base at Okinawa and relocate eight thousand marines to Guam and Hawaii. As part of this transfer, the US Department of Defense outlined plans to use Pågan and the nearby inhabited island of Tinian to conduct war games. Alarmed by this announcement, "dozens of former residents have joined forces with environmental campaigners to launch a lawsuit they hope will expose the folly of the Pentagon's plans to transform Pågan and Tinian, an inhabited island 200 miles to the south, into simulated theatres of war."[53] Local opposition and protests have been building since the plan became public in 2013, and a lawsuit to stop the plan was filed in July 2016 by Earthjustice, a nonprofit environmental law organization representing the Tinian Women's Association, Guardians of Gani, PåganWatch, and the Center for Biological Diversity.

The suit argues that the navy failed to consider all the impacts associated with the plan and did not carry out a sufficient environmental impact statement, violating the National Environmental Policy Act (NEPA). As an attorney for Earthjustice put it, "People feel deep connections to Pågan, and to deprive them of the opportunity to return to this beautiful place is wrong. The thought that this tropical paradise will be destroyed by 2,000-pound bombs and ship-to-shore bombardments is unconscionable." In October 2017 US District Judge Ramona Manglona denied a motion to dismiss a lawsuit seeking to challenge the navy plans, allowing the suit to go ahead. The plaintiffs charge that

> the training proposed for Tinian and Pågan would be intense and destructive. War games would include artillery, mortars, rockets, amphibious assaults, attack helicopters and warplanes and, on Pågan, ship-to-shore naval bombardment. The training would destroy native forests and coral reefs, kill native wildlife—including endangered species—and destroy prime farmland. . . . Families evacuated from Pågan in 1981 would never be able to return, their former home turned into a militarized wasteland.[54]

The Department of Defense and the Navy, as defendants, asked the court to dismiss the lawsuit, arguing that "the court lacks jurisdiction, and that the lawsuit presents a political question because the U.S. executive branch decided to relocate the Marines as part of a treaty negotiated with Japan."[55]

Nevertheless, in view of such opposition, the Department of Defense had already slowed the process, revising its study and announcing that no final decision would occur before 2020.[56] The outcome, still pending as of this writing, will also depend on the Donald Trump administration's decisions, which have thus far undercut Obama's "Asian Pivot" policies and also will likely seek to gut NEPA. Yet the logistics of extending US military presence across the Pacific dictate that the Northern Marianas will remain an important military outpost whatever the outcome. In the wake of the suit, the *Marine Times* continued to report that "two-thirds of Tinian, the uninhabited island of Farallon de Medinilla and parts of Tanapag are now used for live-fire training and bombing. The military wants to expand live-fire training and take over the entire island of Pågan for a variety of Navy and Marine combat training. About 2,500 Marines are expected to move to a proposed Marine Corps base in Guam in 2021, with the rest arriving by 2026."[57] The case echoes the situation in Vieques, although here the opposition is occurring before the weapons-testing range and war games are fully operational rather than after the fact. But it raises similar issues of citizenship, environmental protection, military power, and the rights of the populations of small islands, alongside competing discourses of tourism and "paradise."

In exchange for US citizenship and federal funding, the Northern Marianas Islanders had voted in 1975 to lease land to the US military for training exercises and bombing practice (including fifty-year leases on the entire island of Farallon de Medinilla and the northern two-thirds of Tinian island).[58] Yet, like Puerto Rican citizens, they do not vote in federal elections or pay federal taxes. They too exist in a kind of suspended temporary status with certain promises attached to it, which may or may not be fulfilled. They share, in other words, what Stoler called the ambiguous state of partial sovereignty, "provisional impositions of states of emergency, promissory notes for elections, deferred or contingent independence, and 'temporary' occupations," which she argued is central to all imperial projects. But rather than seeing these conditions as leftovers or outliers of old colonial histories, the work of Stoler, Easterling, and Vicuña Gonzalez underlines that these managed (im)mobilities are the workings of imperial extra-statecraft, models for contemporary regimes of security, and premises for the multiplication of landgrabs across expanding offshore zones.

Tourism is one of the key appropriations of offshore territories, which folds islands into a generalized imagery of sun, sea, and sand, where beaches replace bases. The expansion of air travel especially in Asia brings the Pacific Islands ever more closely into tourist itineraries and tourism's imagined geographies. The official travel site of the United States, VisitTheUSA.com, promotes the Northern Mariana Islands for their "endless ocean views and endless island fun": "The Northern Mariana Islands are an easy warm-weather escape northeast of Guam. Once you're on the islands, the only difficult decision you'll have to make is whether to sunbathe on a beautiful beach, try your luck at a casino, play golf with dazzling views of the Philippine Sea or go scuba diving in a World War II shipwreck. Then again, there is no reason why you can't do them all."[59] The tourism site notes not only the islands' natural beauty and World War II historical heritage but also "the indigenous Chamorro culture," which it says is "alive everywhere you look," noting the "Ancient Chamorro latte stones" that dot the landscape and once supported buildings and palaces. The site also hosts a ten-minute video called "Saipan Aerial Tour," which features various kinds of mobile video shots, many taken from the air, with a drone rotor and shadow showing in some shots. The drone view has become a new way of experiencing the kinds of aerial gaze described by Vicuña Gonzalez and brings another kind of military technology to remote island locations. The craggy cliffs, rocky shorelines, and rugged edges of the Northern Marianas, once only seen by navy pilots on practice bombing runs, now are accessible to the tourist gaze.

Behind these territorial possessions are processes of occupation, invisibilization, and erasure, with the Mariana islands often literally missing from maps. As the poet Craig Santos Perez, a native Chamoru from the Pacific Island of Guahan (Guam), writes, "When the body of Guam does become visible, it is mapped as a tourist destination, a migratory stopover, and a military possession. In these ways, global imperialism has mapped and remapped Guam in violently visible ways."[60] The US military occupies about one-third of the island of Guam, and it is variously described as "Where America's Western Frontier Begins!" or "Where America's Day Begins!" as Santos Perez writes in his poem "(The Birth of Guam)." Seemingly on the edge of American time and space, Guam and outer islands like Tinian and Pågan must be recentered as meaningful places through conscious acts of local appropriation and meaning-making.

The Visit the USA website describes Tinian as "a tiny haven of diving and World War II history" and "probably not somewhere you'd want to spend more than a night or two." Most remarkably, it describes an unusual feature of "this lovely little island" as "the Manhattan Connection":

> In World War II, when much of the island was used as an air base by the US Navy, the local streets were laid out and named after those in New York City. You can still drive on Broadway, 42nd Street, Lenox Avenue, Riverside Drive and Eighth Avenue, even though you're 13,000 kilometers away. The abandoned runways can be seen in North Field, where plaques mark the pits where the Enola Gay and Bockscar planes were loaded with atomic bombs before heading off to Hiroshima and Nagasaki.[61]

And so, from this "tiny" place we come full circle, from the "Big Apple" of Manhattan to its miniature streets in Tinian, from the Manhattan Project to the launch of the *Enola Gay* carrying its nuclear payload, from a remote US military base in the Pacific to the devastation of Hiroshima and Nagasaki. Little places, big consequences. As the prospective tourist dwells on this thought, a link at the bottom of the page will take you to another experience: "Puerto Rico: Discover a Caribbean Paradise in the U.S."

CONCLUSION: FRACTAL POLITICS FROM SMALL PLACES

Resistance movements against military occupations, against colonialism, and even against inappropriate tourism development can serve as reminders of the importance of small outlying places in an archipelagic fractal politics. Amid the shape-shifting liquid regimes of military power, offshore "bluewashing," and flighty tourism investment, naming local claims to place and making marginalized peoples visible can be a powerful political act. Crucial to such fractal political movements are not only the archipelagoes of land that they name and claim as their own but the archipelagoes of solidarities that they mobilize. Indigenous people's movements have been crucial to building such solidarities around the world, linking from place to place to assert local sovereignty and communal rights to land, water, and life. And equally important are the political movements of the poor, of the marginal, of the former plantation workers and their descendants scattered across the global plantation archipelago and its carceral echoes.

Viequenses have continued to advocate for local ways of life such as agroecology and small farming to take precedence over military occupation and tourism occupation, despite efforts at greenwashing military "nature" preservation. People of Tinian/Pågan have continued to call for rights of return to their land and protection from weapons testing, despite military "bluewashing" of oceanic national monuments. Building bridges of connectivity not only across islands but also between these small movements in the Caribbean and the Pacific is a crucial way for imagining and enacting new archipelagic geographies beyond coloniality, militarism, and tourist occupation. While Vieques and Tinian/Pågan are unique examples, they are indicative of how communities throughout the entire Caribbean and Pacific are engaging in reclaiming their own lands and protecting their environments through local cultural heritage and revaluation of "small places" in their own terms, to borrow Jamaica Kincaid's title *A Small Place*.

Island survival in the future will rest on the remaking not only of ecological and technological systems but, as Sylvia Wynter teaches us, of the human within them, the "Human after Man," as she puts it. Caribbean theory has provided some important insights here, as Aaron Kamugisha notes, offering "us a compelling critique of coloniality, a hope of a different future to come, and a charting of the space that future would have to occupy to create a new Caribbean, and world."[62] The critique of coloniality and its ongoing violence is just as urgent in the Pacific and is being advanced by indigenous theorist/activists like Santos Perez. It is these kinds of transits through past and present, between islands, and across generations of radical archipelagic social thought that give hope for livable Caribbean and Pacific futures. These fractal archipelagoes of radical thought, connected across time, carry across terraqueous borders to imagine a common world beyond the occupied zones of militarism, tourism, and offshoring. In conclusion, I hope to have shown how these small places form crucial and widely distributed fringes, moving beyond militarism and coloniality in ways that corrugate and conjugate Caribbean and Pacific archipelagoes as spatiotemporal memory-spaces of shared political action and fractal future alliance.

PRACTICAL AND METHODOLOGICAL IMPLICATIONS

Throughout this chapter I have argued for one view of islands as "multipliers" that distribute functions of offshoring across imperial and postimperial spaces. They have parallel roles in military logistical flows, in the traveling circulation of tourists, and in being used as sites of "no-man's-land," whether for weapons testing, incarceration, or migrant detention. In particular, I highlight the mutation of military occupations into tourism occupations, in which the management of (im)mobilities for military purposes leads directly into the managed (im)mobilities of colonization for tourism but never stops there. Under the "logic of military conservation," supposedly "protected" areas house strategic military bases, for example the US bases on Guam, Tinian, Saipan, Rota, Farallon de Medinilla, Wake Island, and Johnston Island. Providing a mechanism to displace indigenous peoples and severely limit public access—such as the Afro-Indigenous Garifuna people who were displaced from the Cayos Cochinos island group in the Caribbean off Honduras—"bluewashing" seizes small islands and occupies vast oceans for both military use and tourism development.[63]

Through a comparative methodology I explore how local challengers can create more corrugated and rough edges that intervene in the smooth mobilities of both tourism and military occupation, undermining these offshore infrastructures of exploitation, extraction, and natural destruction. Through these resistant "terraqueous" formations, as Stephens calls them, islanders have seized upon "the possibility for new spatial configurations of Caribbean [or Pacific] contemporaneity," asserting their own trajectories of "criss-crossing, decentralized motion."[64] Ultimately, in rejecting the regionalism of area studies and the linear temporalities of Eurocentric colonial histories, this "lobal" archipelagic approach enacts local resistance to global forces and seeks to show how a fractal countergeography can arise at the furthest fringe of empire.

NOTES

1. P. Vannini et al., "Recontinentalizing Canada: Arctic Ice's Liquid Modernity and the Imagining of a Canadian Archipelago," *Island Studies Journal* 4, no. 2 (2009): 124.
2. Brian R. Roberts and Michelle A. Stephens, "Introduction: Archipelagic American Studies: Decontinentalizing the Study of American Culture," in *Archipelagic American Studies*, ed. Brian R. Roberts and Michelle A. Stephens (Durham, NC: Duke University Press, 2017), 3.
3. Cynthia Enloe, *Bananas, Beaches and Bases*, 2nd ed. (Berkeley: University of California Press, 2014); John Urry, *Offshoring* (London: Polity Press, 2014).
4. Keller Easterling, *Extrastatecraft: The Power of Infrastructure Space* (New York: Verso, 2015).
5. Roberts and Stephens, "Introduction," 23, 26.
6. Craig Santos Perez, "Bluewashing the Colonization and Militarization of Our Ocean: How U.S. Marine National Monuments Protect Environmentally Harmful U.S. Military Bases throughout the Pacific and the World," *Hawaii Independent*, June 26, 2014.
7. Rachel Fabian and Hannah Goodwin, "An Interview with Ricardo Dominguez," in "Access/Trespass," special conference issue, *Media Fields Journal* (2014).
8. Ann Stoler, *Duress: Imperial Durabilities in Our Times* (Durham, NC: Duke University Press), 88.
9. Ibid., 88. On French Guiana, see Peter Redfield, *Space in the Tropics: From Convicts to Rockets in French Guiana* (Berkeley: University of California Press, 2000); Miranda Frances Spieler, *Empire and the Underworld: Captivity in French Guiana* (Cambridge, MA: Harvard University Press, 2011).
10. Stoler, *Duress*, 116. On Guantánamo, see Claudia Aradau, "Law Transformed: Guantánamo and the 'Other' Exception," *Third World Quarterly* 28, no. 3 (January 1, 2007): 489–501.
11. Stoler, *Duress*, 117.
12. Ibid., 195.
13. A. Mountz and J. Loyd, "Transnational Productions of Remoteness: Building Onshore and Offshore Carceral Regimes across Borders," *Geographica Helvetica* 69 (2014): 389–98; J. Loyd and A. Mountz, "Managing Migration, Scaling Sovereignty," *Island Studies* 9, no. 1 (2014): 23–42; J. Loyd, E. Mitchel-Eaton, and A. Mountz, "The Militarization of Islands and Migration: Tracing Human Mobility through US Bases in the Caribbean and the Pacific," *Political Geography* 53 (2016); A. Mountz et al., "Conceptualizing Detention: Mobility, Containment, Bordering and Exclusion," *Progress in Human Geography* 37, no. 4 (2012): 522–41.

14. Loyd, Mitchel-Eaton, and Mountz, "The Militarization of Islands and Migration," 1.

15. Easterling, *Extrastatecraft*, 59–60.

16. Vernadette Vicuña Gonzalez, *Securing Paradise: Tourism and Militarism in Hawai'i and the Philippines* (Durham, NC: Duke University Press, 2013), 80.

17. Santos Perez, "Bluewashing the Colonization and Militarization of Our Ocean."

18. Bill Moyers, interview with Yarimar Bonilla, "Vulture Capitalists Circle above Puerto Rico," BillMoyers.com, October 3, 2017, http://billmoyers.com/story/vulture-capitalists-circlepuerto-rico-prey.

19. For discussion of Puerto Rican citizenship and national identity specifically in relation to Vieques, see Kathryn McCaffrey, *Military Power and Popular Power: The US Navy in Vieques, Puerto Rico* (New Brunswick, NJ: Rutgers University Press, 2002); Kathryn McCaffrey, "Social Struggle against the US Navy in Vieques, Puerto Rico: Two Movements in History," *Latin American Perspectives* 33, no. 1 (146) (2006): 83–101.

20. Yarimar Bonilla, talk presented at "States of Crisis Symposium: Disaster, Recovery, and Possibility in the Caribbean," Heyman Center for the Humanities at Columbia University, New York, May 3–4, 2019.

21. Humberto Garcia Muñoz and Jorge R. Beruff, "US Military Policy toward the Caribbean in the 1990's," *Annals of the American Academy of Political and Social Science* 533 (1994): 116.

22. Gabriel Marcella, "Review of *Vieques, the Navy and Puerto Rican Politics* by Amilcar Antonio Barreto," *Journal of Military History* 67, no. 1 (2003): 315.

23. Garcia Muñoz and Beruff, "US Military Policy toward the Caribbean."

24. McCaffrey, *Military Power and Popular Power*.

25. Jennifer Allora and Guillermo Calzadilla, *Land Mark*, exhibition viewed by the author at the Tate Modern, London, 2003, including a large graphic-and-text pamphlet.

26. McCaffrey, "Social Struggle against the US Navy," 88. See Amilcar A. Barreto, *Vieques, the Navy and Puerto Rican Politics* (Gainesville: University Press of Florida, 2002).

27. Sherrie Baver, "Peace Is More Than the End of Bombing: The Second Stage of the Vieques Struggle," *Latin American Perspectives* 33, no. 1 (146) (2006): 102–15. See full account in McCaffrey, *Military Power and Popular Power*.

28. Eric Adams, "The Navy's Phantom Bombing Range," *Popular Science* 262, no. 2 (2003): 36.

29. Allora and Calzadilla, *Land Mark*.

30. Baver, "Peace Is More Than the End of Bombing."

31. Caren Kaplan, "Precision Targets: GPS and the Militarization of US Consumer Identity," *American Quarterly* 58, no. 3 (2006): 693–713.

32. Ibid.

33. Baver, "Peace Is More Than the End of Bombing."

34. Santos Perez, "Bluewashing the Colonization and Militarization of Our Ocean."

35. A study by epidemiologist Carmen Ortiz-Roque of the College of Physicians and Surgeons of Puerto Rico found that inhabitants had elevated levels of explosive residues, including mercury, aluminum, cadmium, and lead, in their bodies. The Vieques cancer mortality rate is also 30 percent higher than in the rest of Puerto Rico. The navy refused to acknowledge any causation of these health problems. Shane DuBow, "Vieques on the Verge," *Smithsonian* 34, no. 10 (2004): 82.

36. Baver, "Peace Is More Than the End of Bombing," 108, 112.

37. "Women from Vieques Denounce That the Navy Is Still Attacking Nena Island," *El Calce*, April 29, 2019, https://elcalce.com/contexto/mujeres-viequenses-denuncian-la-marina-sigue-atacando-la-isla-nena.

38. Vicuña Gonzalez, *Securing Paradise*, 218, 222.

39. Baver, "Peace Is More Than the End of Bombing," 109.

40. Allora and Calzadilla, *Land Mark*, 2003.

41. Amy Laughinghouse, "Unspoiled? In the Caribbean?," *Philadelphia Inquirer*, December 4, 2005.

42. Steve Stephens, "Vieques Unspoiled: Tourists Discover Pristine World of Island the Navy Once Bombed," *Columbus Dispatch*, 2004, http://www.columbusdispatch.com/default.php?sec=features&subsec=travel (accessed November 25, 2006).

43. Pableaux Johnson, "Vieques, Far from the Lounge Chair Crowd," *New York Times*, March 11, 2005, http://www.nytimes.com/2005/03/11/travel/escapes/11vieques.htm (accessed October 16, 2006).

44. Mark Healy, "The Coast Is Clear: The Navy Pulled Up Anchor, and Now the Developers Are Circling the Quiet Caribbean Island of Vieques," *New York Times Style Magazine* (winter 2006): 136.

45. Mimi Sheller. "Retouching the 'Untouched Island': Post-military Tourism in Vieques, Puerto Rico," in "Tourisme dans la Caraïbe," ed. Olivier Dehoorne, special issue, *Téoros* 26 no. 1 (2007): https://journals.openedition.org/teoros/2823.

46. Polly Pattullo, *Last Resorts: The Cost of Tourism in the Caribbean* (London: Cassell/Latin American Bureau, 1996).

47. Sherrie Baver, "Environmental Politics in Paradise: Resistance to the Selling of Vieques," NACLA, August 21, 2009, https://nacla.org/news/environmental-politics-paradise-resistance-selling-vieques (accessed June 5, 2017).

48. Kate Aronoff, "Republicans Plan to Turn Puerto Rico into a Theme Park for Fossil-Fuel Corporations," *The Intercept*, November 9, 2017, https://theintercept.com/2017/11/09/puerto-rico-hurricane-fossil-fuels-congress.

49. Elizabeth Yeampierre and Naomi Klein, "Imagine a Puerto Rico Recovery Designed by Puerto Ricans," *The Intercept*, October 20, 2017, https://theintercept.com/2017/10/20/puerto-rico-hurricane-debt-relief.

50. "Colmena Cimarrona," HASER, http://www.hasercambio.org/en/allies/colmenacimarrona (accessed May 6, 2019). Thanks to Adriana Garriga-Lopez, who is researching the Maroon Hive and presented on related issues at "States of Crisis Symposium: Disaster, Recovery, and Possibility in the Caribbean," Heyman Center for the Humanities at Columbia University, New York, May 3–4, 2019.

51. Angelique Nixon, *Resisting Paradise: Tourism, Diaspora, and Sexuality in Caribbean Culture* (Jackson: University Press of Mississippi Jackson, 2015), 32.

52. Ibid., 4.

53. Justin McCurry and Daniel Lin, "Pågan: The Tropical Paradise the US Wants to Turn into a War Zone," *Guardian*, September 13, 2016, https://www.theguardian.com/world/2016/sep/13/pagan-marianas-tropical-paradise-us-war-zone-bombing-practice (accessed June 5, 2017).

54. "Federal Judge Denies Motion to Dismiss Lawsuit over Pågan, Tinian Training," Earthjustice, 2017, https://earthjustice.org/news/press/2017/federal-judge-denies-motion-to-dismiss-lawsuit-over-p-gan-tinian-training.

55. Bryan Manabat, "US Court Denies Motion to Dismiss Lawsuit over Live-Fire Training in CNMI," *Guam Daily Post*, October 17, 2017, https://www.postguam.com/news/cnmi/us-court-denies-motion-to-dismiss-lawsuit-over-live-fire/article_fac1fb24-b232-11e7-a49d-5f8b763c1391.html.

56. Anita Hofschneider, "The US Military Won't Bomb Pagan or Tinian Just Yet," *Honolulu Civil Beat*, March 9, 2017, http://www.civilbeat.org/2017/03/the-u-s-military-wont-bomb-pagan-or-tinian-just-yet (accessed June 5, 2017).

57. Kent Miller, "Plans to Move Marines to Guam, Train Them in Northern Marianas, Hits Snag," *Marine Times*, October 14, 2017, https://www.marinecorpstimes.com/news/your-marine-corps/2017/10/14/plans-to-move-marines-to-guam-train-them-in-northern-marianas-hits-snag.

58. Anita Hofschneider, "More Political Power for the Marianas?," *Honolulu Civil Beat*, December 15, 2016, http://www.civilbeat.org/2016/12/more-political-power-for-the-marianas (accessed June 5, 2017).

59. "Northern Mariana Islands," VisitTheUSA.com, https://www.visittheusa.com/state/northern-mariana-islands.

60. Craig Santos Perez, "The Poetics of Mapping Diaspora, Navigating Culture, and Being From," Doveglion Press, April 20, 2011, http://www.doveglion.com/2011/04/craig-santos-perez-the-poetics-of-mapping-diaspora-navigating-culture-and-being-from-part-4.

61. "Tinian and San Jose," VisitTheUSA.com, https://www.visittheusa.com/destination/tinian-and-san-jose.

62. Aaron Kamugisha, "On the Idea of a Caribbean Cultural Studies," *Small Axe* 41 (July 2013): 56–57.

63. Santos Perez, "Bluewashing the Colonization and Militarization of Our Ocean."

64. Michelle Stephens, "What Is an Island? Caribbean Studies and the Contemporary Visual Arts," *Small Axe* 41 (July 2013): 25.

Chapter Fourteen

Sardinia "Lost between Europe and Africa"

Archaeology and Archipelagic Theory

Thomas P. Leppard, Elizabeth A. Murphy, and Andrea Roppa

> But [Sardinia] still reminds me of Malta: lost between Europe and Africa and belonging to nowhere. Belonging to nowhere, never having belonged to anywhere. To Spain and the Arabs and the Phoenicians most. But as if it had never really had a fate. No fate. Left outside of time and history.
>
> —D. H. Lawrence, 1921

What does it mean, for D. H. Lawrence and for Sardinia, to be lost between Europe and Africa? At best suggesting only a partial view of what it means to be Sardinian, Lawrence's optic and emphasis nonetheless challenge—demanding a response that is contextual while simultaneously asserting something intrinsically vital about Sardinia itself, circumventing Spanish, Phoenician, or "European" perspectives. An archaeological approach provides the former; as regards decentering external framings of Sardinia and the Sardinian past, archipelagic theory now offers a powerful tool for those of us also working comparatively in the interstices of power, geography, time, and inequity (e.g., DeLoughrey 2001; Baldacchino 2006; Pugh 2013; Roberts and Stephens 2013, 2017). In this chapter we aim to adopt a simultaneously archipelagic and archaeological perspective, exploring how archipelagic approaches allow us to view and approach archaeology's characteristic interest in landscapes, materials, and political action. Glimpsing human action through materials and archaeological landscapes rather than texts, we explore attempts to subvert and disperse mainland strategies of domination and cultural hegemony through an archipelagic lens.

We expressly do not intend to understand Sardinia (or, indeed, Sardinians) as somehow intrinsically archipelagic, doing violence as this would to the myriad of ways that these islanders have framed and constituted themselves over the millennia. Rather, we aim to use archipelagic approaches as an interrogative tool: for various reasons, not least the peculiarly focused continental imperial interest in Sardinia, we suggest that a body of theory that invites us to consider how subaltern relationships are established, curated, contested, resisted, and sometimes strategically ignored, through the lens of island-island rather than island-mainland relationships, has a clear utility in the fractured (and fractal) Mediterranean. While archipelagic approaches resist, in part, attempts to frame and delimit (Baldacchino 2006), we see utility in the possibility that the mainland—as imperial, hegemonic, and centralized—may be challenged by the archipelago as postcolonial, heterarchical, and dispersed (cf. van Dommelen 2011). As archaeologists working in an island context, we see such approaches as useful correctives to center-dominated narratives. This is, then, an exploration: as archaeologists who invest time in thinking about how social and state power is expressed and maintained in landscapes as a field of action, we investigate how a novel, archipelagic viewpoint invigorates or challenges received understandings, which too often render islanders as passive nonparticipants in the sweep of history (as per Lawrence). We choose our avenues into this new territory partially but, we think, relevantly and perhaps illuminatingly.

In this contribution we explore the human landscapes and temporalities of southwest Sardinia as refracted through archipelagic thinking, focusing on space and time as distributed, dispersed, and confusing entities. We show how an interpretation informed by archipelagic thinking allows us to approach the material records and human pasts of Sardinia in highly local terms and to interpret how the distribution of social power in a disintegrating landscape has been a recurrent strategy to avoid, circumvent, and disappear, in the face of imperial and colonial incorporation. We structure our intervention in terms of three discrete themes that we nonetheless attempt to interweave into one another. Under the sign of *archipelagoes* and *fractality*, we emphasize how Sardinia complicates and blurs mainland categories, showing its insularity to be protean and contingent. We then turn to the ongoing and iterative creation of Sardinian landscapes as *palimpsests*, exploring how imperial-continental landscapes of extraction and profit can be understood as overwritten by subversive and transgressive local commentary and action. Throughout, our emphasis is on mainland attempts to integrate through normative structures (nonintegrated societies are "isolated," for example) and local attempts to define the local outside the island-mainland binary.

Our tangential contention is that archaeology has something unique to offer. Archaeology, broadly conceived, is the study of human and material worlds and how they change over the *longue durée*, and, as practitioners, we see this potential in terms both of demonstrating how archipelagic worlds are constituted materially and how the remit of archipelagic thinking (*sensu* Roberts and Stephens 2013, 2017) can be traced through long rhythms of time. We may never have been modern (Latour 1991), but archaeologists have striven toward meaningful comparison, a goal that matches the broader implicit aims of archipelagic approaches. An archaeological perspective reaches back into broader stretches of time, and we are comfortable when dealing with behaviors and convergent lived experiences that recur over centuries and millennia. Finally, and despite an apparent rediscovery of material ontologies under the many banners of the New Materialisms (Barrett 2016; Hamilakis and Jones 2017; Thomas 2015; Witmore 2014; see now Knapp 2017), archaeologists have always been deeply interested in *stuff*—how it constrains and enables humans, how it impinges upon us, and how we arrange it around ourselves as we pursue social strategies. In this chapter, we aim to interknit these materialist and temporal threads of archaeology with the incisive social critique offered by archipelagic thinking, suggesting that archipelagic approaches have wide applicability beyond the solely textual.

Throughout this chapter, we refer broadly to the archaeological and historical record of Sardinia, selecting vignettes that capture, in microcosm, many of the broader themes we allege to identify. While some may rightly see these issues explored through a collection of examples that verges on the motley, the temporal span and varied nature of the archaeological record presented, if anything, underscore the exceptionally rich inscription of the historical encounters between metropole-archipelago onto the landscapes of Sardinia.

FRACTAL ARCHIPELAGICS, PALIMPSESTS, AND ISSUES OF DEFINITION

The categorization of fragments of land into "islands" and "mainlands" is less, we suggest, an enduring truth about the physical organization of the universe and more a structuring principle, at best metastable but fraying toward its edges. A fundamental assumption of our ensuing discussion is that Sardinia is a place that frustrates this simple dichotomy (cf. DeLoughrey 2001; Pugh 2013; and emphasis on the subversion of mainland/nonmainland polarities), and in this frustration various interests and stakeholders have found cultural ammunition. There is, of course, a substantial literature on the representation of the islander in a subaltern role, a literature that we do not discuss here (although we note that archaeology has incorporated, to some extent, the notion that distance and isolation can be profoundly relative in the sense that they are socially constructed and culturally specific; e.g., Boomert and Bright 2008; Rainbird 2007; Terrell, Hunt, and Gosden 1997). We wish to start, instead and simply, with physical geography: What work is achieved—what is gained and lost—when we define Sardinia as an island? How defensible is this position, and does an archipelagic reconsideration open up new avenues of understanding?

The Archipelago

The intuitive but resilient definition of insularity—*an island is a fragment of land surrounded by water*—is evidently unhelpful here. All fragments of land are surrounded by water, at least on this planet. Afro-Eurasia is, by these lights, as insular as Tristan da Cunha, so the observation does not provide a useful analytic lens unless we focus on relative types of organization and on scale. Here, too, islands become uncompromisingly difficult. We can agree that Sardinia is, in some sense, insular to and in respect of mainland Italy, but does it then itself become a mainland to Corsica or to Sant'Antioco? Is it meaningful to talk of Sardinia and the countless small rocks that break the sea's surface just off its coasts as the same class of phenomenon? Insularity emerges here not as a means of capturing meaningful variability but as a structuring device that creates explicit power relations. Insularity is best considered as part of a wider epistemology of dominance that constructs deliberate and value-laden distinctions between the metropole and the (island) hinterland. To subscribe to Sardinia-as-island vis-à-vis Italy is in part to endorse this power dynamic, and here we would rather subvert or bypass this discourse in favor of viewing Sardinia as both archipelagic and fractal.

Sardinia is clearly archipelagic in some more literal senses—as one among many islands in the Tyrrhenian, most clearly. But we can expand this optic outward—Sardinia as one component of the greater Mediterranean archipelago, more readily (if capriciously) linked to Valetta, Mahon, and Patras than to the high Italian interior. Focusing inward, however, Sardinia also fades into the archipelagic, comprising as it does a series of small, enclosed basins and plains interspersed with a rugged and vertiginous topography that demarcates and segments. The area that we discuss in detail, the Santadi Plain in Sulcis, is exemplary of this phenomenon: separated from the coast and divided from its neighbors, studded with modern settlements that form their own archipelago in the midst of the sea of wheat and rolling vineyards, it might be considered archetypically Mediterranean in its heterogeneity. Even with the advent of modern, distance-compressing metropolitan technologies of communication and connectivity (e.g., Castells 1996), journeys across and around this island, between its lowlands and its uplands, are necessarily circuitous, skirting mountainsides, badlands, rocky headlands, and saline marsh. The intrinsic sinusoidality that the Sardinian countryside imposes on movement is a theme to which we return, but even without considering the temporal depth of this pattern, the observation suggests that it may be more

Figure 14.1. Graffito in the Sulcis mountains on the side of a disused railroad building.

fruitful to consider Sardinia as essentially distributed, at times cohering under the metropolitan gaze, but always under centripetal pressure to revert to a dynamic, archipelagic organization. Accordingly, it might be productive to understand this landscape as relentlessly and necessarily archipelagic, intrinsically resistant, regardless of topographic specifics. This tendency is underscored both when considering raw distribution of settlement from an archaeological perspective (initial fieldwork tentatively suggests a cyclical reversion to settlement dispersal as power structures, represented by "central places" as agglomerations of nonlocal or elite hegemony, decline) and in modern conceptions: a prominent and repeated *graffito* in the mountain pass that leads to Santadi reads, "Attention, tourists, Sardinia is not Italy" (Figure 14.1). The language of this message, being neither Italian nor Sard, hints at a broader intended audience; Sardinia gravitates away and apart.

The Fractal

Before moving to specifics, we note finally that the intrinsically nodal, networked archipelagic qualities of Sardinia suggest to us that the physical-cultural geography of the island may be usefully understood in fractal terms. A fractal structure can, for our purposes here, be defined as a morphology that is in essence isomorphic across scales; as the observer moves closer or further away, the structure remains the same (Mandelbrot 1975). We are certainly not the first to suggest that parts of the Mediterranean exemplify fractality (Broodbank 2000), but we would suggest that the fractal model is especially apposite for Sardinia—as the scale of analysis shrinks, the landscape breaks apart into different types of archipelagic nodes, from the organization of the mountains and the plains down to the dispersed farmsteads of a remote hillside village. We do not consider here the extent to which fractality is characteristic of all archipelagic formations, but we do propose that the local capacity to resist the metropole is in part a function of this fractal morphology, not least in the capacity of shattered, scalar landscapes to resist and frustrate the projection of state/hegemonic institutional power (Scott 2009, 2017) (we will return to distributed social power as an alternative to and dilution of imperial structures). As the center aims to demarcate and circumscribe (from the nation-state down to the unit of taxation), the scale-free organization of the landscape allows these goals to be frustrated and avoided.

The Palimpsest

Landscape is conceptually neither exclusively natural nor exclusively cultural; rather, it emerges from the interplay between natural cycles and cultural action. Some such interactions leave their scars on the surface of the earth: people quarry into the mountain face; families dig field boundaries; cities level the terrain to make way for new development. Other processes come to conceal these interactions: hill slopes erode and cover over abandoned buildings; storms and earthquakes degrade and destroy structures. Moreover, humans engage not only with the natural environment but with the cultural features: old buildings and monuments are torn down, renovated, or carefully curated in "authentic" form; property lines are redefined according to new social orders; cultural features acquire or lose significance according to contemporary practices and values.

In this dynamic process, the landscape becomes a record of repeated cycles of people who encountered and engaged with the environment. The landscape, then, has been likened to a *palimpsest*—a living document upon and into which text is written, crossed out, erased, written over, and altered, with marginalia commenting on or subverting the text; these dynamics occur cyclically and iteratively in an ongoing ontological process of "becoming" (Bailey 2007). Following the interweaving rhythms of time and the lived experience of place-making, meaning of landscape is *gathered* from material traces across space (Ingold 1993, 155). Consequently, it is more than a simple document of the layered traces left by history; it is likewise not a static record of a cultural past. Instead, these earlier material remnants of past peoples come to be part of a new present, engaged and interwoven into new narratives about the past, present, and future. Landscape thereby becomes more than a mere backdrop for the actions of history; it is the manifestation of history, and it is the space within which history and meaning are gathered. Landscape-as-palimpsest represents, we suggest, a *temporal* analog for the spatiality of the archipelago. The palimpsest allows us to conceive of time less as the linear progression of the metropole and more as a distributed series of points in which competing narratives have been inscribed (then altered, then erased) on the landscape. It is this dynamic—the attempts to impose structure from without and subaltern strategies of resisting from within the archipelago—that we now explore, accessing these deeper questions via a material vignette and the reflections it prompts on the imposition of systems of governance.

LANDSCAPES OF ISOLATION, EXTRACTION, AND INSCRIPTION

What does utilizing archipelagoes, fractals, and palimpsests as toolkits offer us if we are deeply interested in structures of integration, resistance, and localization? Sardinia, with its diverse resources and agricultural fertility, has repeatedly been an object of attention for various Mediterranean metropolitan, "mainland" powers. Even as early as the Neolithic, the island was already exploited for nonlocal consumption. However, it was during the Bronze and Iron Ages (3000–200 BCE) that more formalized exploitation of the island became institutionalized by pan-Mediterranean powers. This exploitation is typically attributed first to Phoenicians, intensified by Carthage and Rome (and during the Byzantine reconquest), and continued in its union with Spain (CE 1323–1708) and later Italy during the nineteenth-century *risorgimento*. Through these phases of integration and disintegration (or perhaps Mediterraneanization and localization; although see Herzfeld 2005), Sardinia's history and archaeology have been defined by an unequal ebb and flow between local insular communities and external continental powers.

We take, as our point of departure, a colonial narrative abstracted in an individual object: a third century BCE bronze coin probably minted on Sardinia during the Second Punic War (Figure 14.2). This issue (from approximately 216 BCE) presents a diademed male head on obverse and an ear of grain on the reverse (Dyson and Rowland 2007, 129–130), cast in bronze and possibly intended by the former imperial power (Carthage) to assist in fomenting resistance to the growing power (Rome). Grain, imperial fungibility, contestation, appeals to local identity, tacit subordination, and

Figure 14.2. Bronze issue from the second half of the third century BC.

copper: this complex object captures the main themes of nonlocal, metropolitan interest and how these themes have often been interknit into stories of Sardinian liminality and dependence.

Sardinia is endowed with two resources that have made it especially, enduringly, and perhaps unfortunately desirable to metropolitan and imperial centers—agriculturally fertile soils and rich mineral sources. Both are considered what Scott has referred to as "state-accessible products"—that is, materials of high value per weight or volume (to justify the accruement of transportation costs) that were easily tracked and measured (2009, 73). As types of desirable material that lend themselves to quantification, "state-accessible products" offered ancient, as well as modern (Scott 2017), metropolitan centers desirable opportunities in terms of fungibility and metrology (systems of weight and measurement), facilitating the construction of taxable economic systems. Moreover, "state-accessible products" such as grain and precious metals present a means not only to store wealth but also to deploy and disburse wealth, affording themselves to market types of organization. Agricultural products and metals nonetheless represent very different forms of extraction and alteration, and the scars of these actions leave very different traces on the landscape. Here, we explore how metropolitan ideologies of extraction interdigitate with mainland-island narratives; in the next section, we consider how locals have resisted or subverted via the archipelago and the palimpsest's margin.

Cereal agriculture was essential to ancient Mediterranean political economy and society. Control over labor; the investment of disbursable wealth in production, processing, and storage; the means to coercively extract food staples as taxes and rents—all were central concerns for ancient Mediterranean states. Early prehistoric farming was likely small-scale and mixed; it is only with the burgeoning interest in Sardinia of mainland urban polities that large-scale systems of agrarian production seem to have been implemented. We focus here on the site of Pani Loriga, in the Santadi Plain, as exemplary of metropolitan interest in quantification, centralization, and storage. Indigenous types of settlement organization, until very late in the second millennium BCE, betray little interest in the accumulation of state-accessible products. Yet, with the incorporation of coastal Sardinia into, first, larger Mediterranean economic systems and, then, formally into the inchoate, quasi-dynastic polity that we call the Carthaginian Empire, the architecture of extraction becomes increasingly evident. This is apparent not only at newly founded coastal sites but also at inland Pani Loriga, where a Bronze Age settlement is succeeded by a large Phoenician-Punic site (Botto, Dessena, and Finocchi 2014). Pani Loriga, with its ordered galleries (indexing, among other activities, storage of surplus) (Botto 2016, 2017), surely signified an altered dynamic of accumulation in an imperial context in which taxes are believed to have been acquired in kind (Crawford 1985, 103). From a dispersed and scattered pattern of settlement, on a central hill with commanding views over the nearby fields as well as the road to the coast, a large facility with the capacity for storage of many metric tons appears. We might assume that this process was not innocent, and the central interest in *storing* and *counting* suggests that claims on land and stored produce were not recognized to be equal. The pattern is intensified as the island's North African metropole's power ebbed and as Sardinia was absorbed into the ambit of the closer sphere of Roman *imperium*: Roman-period agricultural organization was, by the first century CE, based on large villa estates and connected to the imperial ports at Caralis, Nora, Turris Libissonis, and Sulcis via stone-built roads (for a broader discussion of Punic and Roman exploitation of rural Sardinia, see van Dommelen and Gómez Bellard 2008; Roppa and van Dommelen 2012). Such systems of property, infrastructure, and human capital reflected less local topographical nuance than attempts, if sometimes faltering, to build efficient extractive systems. If Byzantine and Aragonese systems of extraction were less rigorous, they nonetheless relied on a now-familiar apparatus of dominance: the tax collector, the road, the storehouse, and the judge (the *judike* of the medieval *judicadu*).

These extractive works and the construction of imperial agrodeserts (cf. Scott 2009) can be usefully understood as an extension of imperial writing about and onto the island—imposing a certain way of understanding landscape. This is the enactment of the mainland-island relationship, building the island into productive and measured subordinance. We consider below how local reaction, in the face of overwhelming continental economic and military dominance, was enacted in the "margins" of the landscape; before doing so, however, we address the materials that have often been an engine for nonlocal/local interactions on Sardinia: its stones and ores.

The extraction of Sardinia's rich mineral resources has left enduring gouges across the landscape—from Neolithic chipped-stone quarrying of obsidian at Monte Arci since the Neolithic to the expansion of shaft mining for coal and the establishment, by Mussolini, of the town of Carbonia. Metal resources, including ores of copper, silver, lead, iron, and zinc, were particularly sought after in different periods. Attestation of quarrying and mining is often known today archaeologically through indirect means—via isotopic attribution of ancient artifacts to known ore sources (early imperial Roman bronze coins; Klein et al. 2004), textual references to imperial overseers of mines (the third century CE Roman altar dedicated by a *procurator* of the mines; Hirt 2010, 125), or legal codes regulating mines (Book IV of the fourteenth century CE *Brevi di Villa di Chiesa*; D'Arienzo 1978). These practices indexed by this evidence, we contend, should be understood as variously destructive acts of imperial place-making. Repeated phases of mineral extraction obliterate one another in succession as larger-scale operations and more destructive mining technologies supersede one another at the sources, gnawing at the landscape. These activities, whether conducted in the Roman period or the nineteenth century CE, were noxiously conspicuous: often loud, frequently dirty, and dangerous for the locals involved, they uprooted settled endemic ecologies and produced talus slopes of rubble and toxins. The modern mining operations are most obvious in this regard, with the rusting sidings of Carbonia and the dismantled railroads of the Santadi Plain offering grim testament to the unsustainability of a mode of economic exploitation painfully exposed to the vagaries of the global market; most notably, after major industrial expansions that peaked in the 1950s, global trade outcompeted Sardinian mining, with "clean" energy sources offering alternative sources. The lasting detrimental effects of such mining are well known, lingering in local soils and percolating into water sources (Boni et al. 1999). Yet geochemical studies suggest that this is not a modern phenomenon, and Roman mining was (by contemporary standards) a likewise substantial pollutant, with minute traces of Sardinian pollution finding their way into the atmosphere, becoming encased in Greenland ice cores, and offering lasting geochemical testament to these imperially driven actions (Rosman et al. 1997). As much as large-scale agrarian exploitation, but more viscerally so, mineral acquisition by imperial managers and for the benefit of the metropole imprints itself onto the landscape in boldface. While agricultural works stretched across the rolling hills and valleys of Sulcis, mining becomes an exercise in autophagy: local miners consume their own hillsides, for this ore mash to be transferred in bulk to the center and rendered into bullion.

We return, then, to our coin, in which the insular wealth and the mainland imperial project (with its gaze upon the passive island landscape) are enmeshed. This object is more than a representation of imperial interests in a subjected landscape; this coinage played an active role in the ongoing imperial dialogue that reinforced the relationship between island producer and continental consumer. It is both the product and perpetuator of the metropolitan narrative, whereby the riches of the Sardinian landscape are materially abstracted; insular, subordinate, written-over, and without its own capacity to inscribe, it becomes emblemized as grain on metal through the imperial, metropolitan mint. It is these very categories, however—the written and the insular—that have been used by locals to navigate ways to subvert and critique external definitions.

RESISTANCE AND SUBVERSION: IN THE ARCHIPELAGO, IN THE MARGINS

The metropolitan and imperial gaze values demarcation and decision-making hierarchies expressed in various modes of landscape organization. In contrast, local communities engaged with the palimpsestic and archipelagic aspects of their material worlds to emphasize difference and to subvert mainland-island power dynamics. In the enduring symbolism of the Nuragic Bronze Age in Sardinia, one can interpret this endurance as an outcome of attempts to decenter local notions of self away from types of landscape organization associated with imperial regimes and toward a distributed indigeneity. Punic-Phoenician-Sard identity is a complex proposition, and the material realties of this cultural experience frustrate a dichotomous approach to local/nonlocal, insider/outsider (van Dommelen 1998, 2011), renegotiated as they were over the long term and in the face of cyclical metropolitan integration (see Roppa 2014 for a detailed discussion). At least for the Late Bronze Age, however, it is difficult to trace obvious, durable concentrations of power, via either foodways, accumulation of material wealth, or—especially—distribution of settlement.

Settlement in Late Bronze Age Sardinia was dominated by small, stone-built towers called *nuraghi* (Figure 14.3), which ranged in size from single-tower structures to large, complex fortified buildings with associated settlements and evidence for/of craft production (Webster 2016). Emma Blake (2014) has characterized this political organization as "acephalous cohesion"—that is, a mode of occupying and using the landscape that is remarkably consistent (and potentially cohesive) in the absence of evidence for formal, or even de facto, capacity to project governance or coercive power

Figure 14.3. A well-preserved *nuraghe*.

into the landscape. Here, we are interested less in whether this characterization is correct and more in how modern interest in nuragic heritage engages this acephalousness within a dispersed framework that frustrates imperial and nonlocal discourse about identity and the nature of governance. When dealing with cultural heritage, Sardinian interest is largely focused on these *nuraghi* and the type of indigenous lifeways they index. Indeed, reuse, celebration, and a general intense focus on these monumental constructions seem to have been a feature of Sardinian history more generally, as Smith and van Dommelen (2017) have shown, and this has included metropolitan attempts to co-opt and incorporate; the *nuraghe* tower of Ploaghe was incorporated into the construction of a Roman building (Rowland 1977, 465), for example. We would like to move beyond the specifics of celebration of nuragic culture and settlement and address instead the interest in types of social power typified by *nuraghi* and that would come to be referenced in subsequent political dialogues with mainland powers.

While recognizing that the distribution and maintenance of social power in Bronze Age Sardinia is hard to rationalize in modern terms (Broodbank 2013, 423–25), the archaeology points to something of a paradox. On the one hand, the capacity to exercise and demonstrate wealth and political power close to *nuraghi* (themselves demonstrable statements of power and intent, often seemingly associated with feasting and *materiël*, which celebrates an ethos focused on display and weaponry) seems to have been a goal of elites. Yet, on the other, the archaeology appears to hint that this power decayed extremely rapidly as one moved out into the hinterland. That is, the ability of elite families to project power laterally and at a distance across a landscape seems substantially circumscribed (when compared to, for example, other types of social organization in the Late Bronze Age Mediterranean). We suggest that local interest in accentuating and activating a nuragic heritage may be significant not only in its indigenous bent but in how it endorses a local approach to power and the abrogation of power (also Blake 1998). Late Bronze Age social power was dispersed and constrained, although not within egalitarian structures, and this was in part facilitated by the fragmented topography of the island. Settlement and power are compartmentalized and heterarchical, executive capacity stretching only as far as the next ridge or dry riverbed. When viewed from a material perspective, the lived human landscape was archipelagically distributed, with power ebbing along lines of community connection. Witness, for example, the scattered distribution (in a range of contexts) of high-value bronze oxhide ingots, as well as finished products (Russell and Knapp 2017). "Foreign" objects generally valued in other Mediterranean contexts as exemplars and articulators of status do not find themselves concentrating in incipient centers of power but rather experience flow and movement and, taken as a whole, are quite limited in number (Russell and Knapp 2017; also Blake 2008), with this pattern repeated when it comes to other media (such as imported, or made-to-*appear*-imported, pottery; Webster 2016). This trend suggests not that exotica were not valued (in the context of strategies of social advancement and consolidation, they surely were) but rather that extant systems of power militated against exaggerated wealth inequality. In short, Sardinians were clearly active participants in trans-Mediterranean "trading" (possibly anachronistic) activity but also actively disseminated potentially powerful and thereby socially dangerous objects. The tacit endorsement of this vision of Sardinia, rather than the hierarchical pyramid of metropolitan taxation and hegemonic quantification, challenges the imperial gaze: authentic social power is not attached to an office but is spread throughout and embedded in the landscape and its local topographies (both social and geographic).

What, then, can be said of the margins? Marginality has been an enduring concept in Mediterranean archaeology (Walsh 2008), but here we would like to interpret the margin more literally. Metropolitan inscriptive practices—the *autostrada*, the railroad, the wheat field, the open-cast mine—leave the marks of empire legible on the landscape, part of a wider narrative about mainland-island dynamics and the dominance of the center. We can find Sardinian commentary and critique in the literal margins: it is in the uplands and the hillsides where subversive practices occur. Intrinsically nonstate or antistate spaces (Scott 2009), the broken upland topography is a zone that permits a broad latitude of practices that exist on a spectrum from (in the view of the metropole) curiously archaic to outright illegal. We focus here on just one example, but one which is illustrative: goat pastoralism and attendant activities.

We should be skeptical of models that purport to show that goat pastoralism has an ancient and intrinsically indigenous quality about it; there are good reasons to suppose that specialized pastoral lifeways can only exist in symbiosis with state-level agrarian organization (Cherry 1988). Yet pastoral herding on the *comunale*, the commons, is ingrained with the Sardinian psyche as a native activity—not only because it takes place in zones and areas that continental governments have struggled to incorporate fully into the body politic, with goatherds disappearing into scrubby trails through the *macchia* along which carts and infantry cannot follow, referencing an intrinsic mobility that frustrates the cartographic state impulse; but also because of the difficulty of translating into state terms the metric utility of storage on the hoof. Heads of livestock resist measurement and quantification (and thereby taxation) in their mobility, and their products (overwhelmingly dairy) are highly variable in quality and hard to store, in a manner that militates against bureaucratic control. Consequently, for the metropole (from ancient to modern Rome), the uplands are beyond the bounds of law, the haunt of the barbarian and the *bandito sardo* as well as the *pastore*, categories that can elide into one another depending on perspective (Moss 1979 highlights how Piedmontese and pre-Piedmontese legal codices rendered the transition into banditry comparatively easy, as viewed from the state). Not that this inhibits new attempts to integrate and reframe the uplands as part of the state project; as Heatherington (2010) explores in detail, the environmental movement—and the Italian state's interest in the preservation of biodiversity—can be and frequently is read as an attempt to disrupt and intervene against what are conceived to be traditional lifeways.

Antistate practices are politically active marginalia. Away from the metropolitan infrastructure, and in the face of overweening capacity for Foucauldian social violence, the state is critiqued and strategically ignored in the reiteration of what it means to be Sardinian. This, then, is a very specific type of palimpsest, in which text and erasure are the domain of the state while elaboration—in the margins or between lines—encodes resistance and alterity. This activity is archipelagic in its distributedness: as the metropole tries to consolidate and organize, the local fades away into dispersed networks of social power, sometimes literally disappearing into the hills. During moments of integration into wider Mediterranean power structures, we speculate that the Sardinian archipelago (both real and less real) has always offered a refuge for retreat and consolidation, whether democratizing and ubiquitous appearance of monumental settlement across the uplands as well as the lowlands during the Late Bronze Age into wider Mediterranean exchange structures, or ideological and literal refuge in *banditismo* during the incorporation into the Italian state (Moss 1979).

PRACTICAL AND METHODOLOGICAL IMPLICATIONS

What are the broader implications, deriving from these observations, in practical and methodological terms, both for the archaeology of Sardinia and more broadly? The study of Sardinia over the very long term has been bedeviled by a primary fixation: how to parse the recurrent, overwhelmingly one-sided power dynamic between mainlanders and Sardinians. Distant mainlands become a source of culture and change in tacit contrast to a timeless insular backwardness, with this dynamic even inscribed genetically, with Sardinia as *recipient* (Fernandes et al. 2019). Several scholars have attempted to move the debate beyond or around this (van Dommelen 1998), and approaches that orbit postcolonial theory are clearly useful correctives, but focusing on the archipelago as a tool allows us to break up a monolithic "Sardinia" into a fragmented world with different actors pursuing disparate strategies in dispersed landscapes. Reorienting the discussion away from mainland-island dynamics complicates a simple subaltern relationship, underscoring diverse local responses to imperial hegemons, and allows us as archaeologists to navigate around geopolitics and focus on the local and the contextual. This also, happily, suggests a means of complicating unidirectional and teleological models of Mediterranean "connectivity," models that often make implicit assumptions about intrinsic qualities of power relations (cf. Morris 2003).

More generally, it would seem unwise to reflect on the other side of the disciplinary conversation—the utility of archaeology for archipelagic theory. Suffice it to say here that, to the extent that humanities disciplines are interested in exploring and accentuating difference (and how archipelagic theory allows them to gain traction on that difference), then archaeology can emphasize the extent of that difference in the temporal dimension.

CONCLUDING REMARKS

Sardinia is not an archipelago (dependent on perspective, perhaps), and identifying archipelagic strategies over the very long term would be a fruitless exercise. Archipelagic thinking allows a corrective and comprehensive panorama, however: by decoupling and moving beyond dichotomies, we can view the Sardinian landscape as lending itself toward dispersed and distributed social power, with this dispersal a counterweight to imperial narratives as well as providing conceptual and physical retreat in the face of state power. We see this as an initial foray; interested as we are in the material expression and curation of social power, bodies of theory that deal explicitly with how this power is articulated are likely to be very fruitful. This, then, is more of an exploration than anything definitive, an attempt to link how archaeologists think about landscapes of political and social action with a body of theory that, in part, subverts and avoids continental perceptions of "the island."

In exploring what has been gleaned, we wonder whether approaching Sardinia in an island-mainland context is both less effective as a means of analysis and also a tacit endorsement of cyclical attempts to build the island into unequal and chauvinist structures. We suggest above that weaving together the material focus and deep-time perspective offered by archaeology with the incisive social critique offered by archipelagic thinking allows us to circumvent these unequal power dynamics. If island-mainland dynamics can be interpreted as part of a broader attempt to rationalize away difference in the face of the homogeneous (nation-)state, then transferring focus to the archipelagic allows a more sympathetic and less jarring reading of Sardinian cultural behavior over deep time. Emphasizing dispersal and circumscription of power, intrinsic heterogeneity, and complex networks of social and landscape interdependence, through drawing on the postcolonial and heterarchical heritage of archipelagic thinking, frees us to approach Sardinia and its meanings in immediately local terms. These archipelagic identity structures are, we stress, highly localized and hugely resilient in their localization. Longitudinal, if specific, collective memory is a source of strength in the face of powerful but fleeting metropolitan interlopers. By way of concluding, we wonder whether a focus on iterative and networked strategies of self-preservation in the face of cycling integration and disintegration may be a generally prevalent aspect of Mediterranean societies "between Europe and Africa" of the sort whose existence has been critiqued by Herzfeld (2005); strategies that promote long-term cultural resilience, and their archaeological indices, are, we suggest, worthy of more substantive investigation.

ACKNOWLEDGMENTS

We would like to thank the editors of this volume for their initial invitation to participate and for their forbearance. Peter van Dommelen's work on culture, power, and postcolonial perspectives on Sardinia has been immensely influential. Toby Wilkinson drew our attention to vital and illuminating literature, the existence of which we had hitherto been unaware. Under the aegis of the Landscape Archaeology of Southwest Sardinia (LASS) project, we would like to thank the Comune di Santadi, the Soprintendenza per i Beni Archeologici per le province di Cagliari e Oristano, Prof. Massimo Botto and the Missione Archeologica di Pani Loriga, and Dott. Remo Forresu. LASS has been funded by Florida State University, the McDonald Institute for Archaeological Research (via the D. M. McDonald Fieldwork Fund), the Prehistoric Society, and Homerton College, University of Cambridge.

REFERENCES

Bailey, Geoff. 2007. "Time Perspectives, Palimpsests and the Archaeology of Time." *Journal of Anthropological Archaeology* 26: 198–223.

Baldacchino, Godfrey. 2006. "Islands, Island Studies, Island Studies Journal." *Island Studies Journal* 1: 3–18.

Barrett, John. 2016. "The New Antiquarianism?" *Antiquity* 90: 1681–86.

Blake, Emma. 1998. "Sardinia's Nuraghi: Four Millennia of Becoming." *World Archaeology* 30: 59–71.

———. 2008. "The Mycenaeans in Italy: A Minimalist Position." *Papers of the British School at Rome* 76: 1–34.

———. 2014. "Late Bronze Age Sardinia: Acephalous Cohesion." In *The Cambridge Prehistory of the Bronze and Iron Age Mediterranean*, edited by A. Bernard Knapp and Peter van Dommelen, 96–108. Cambridge: Cambridge University Press.

Boni, M., S. Costabile, B. De Vivo, and M. Gasparrini. 1999. "Potential Environmental Hazard in the Mining District of Southern Iglesiente (SW Sardinia, Italy)." *Journal of Geochemical Exploration* 67: 417–30.

Boomert, Arie, and Alistair J. Bright. 2008. "Island Archaeology: In Search of a New Horizon." *Island Studies Journal* 2, no. 1: 3–26.

Botto, Massimo, ed. 2016. *Il complesso archeologico di Pani Loriga*. Sassari: Sardinia Archaeological Guides and Itineraries.

———. 2017. "The Punic Settlement of Pani Loriga in the Light of Recent Discoveries." *Fasti Online*. http://www.fastionline.org/docs/FOLDER-it-2017-393.pdf.

Botto, Massimo, Fabio Dessena, and Stefano Finocchi. 2014. "Indigeni e Fenici nel Sulcis: le forme dell incontro, i processi di integrazione." In *Materiali e Contesti nell'eta del Ferro Sarda: Atti della giornata di studi, Museo civico di San Vero Milis (Oristano), 25 maggio 2012*. Edited by P. van Dommelen and A. Roppa, 97–110. Pisa: F. Serra.

Broodbank, Cyprian. 2000. *An Island Archaeology of the Early Cyclades*. Cambridge: Cambridge University Press.

———. 2013. *The Making of the Middle Sea: A History of the Mediterranean from the Beginning to the Emergence of the Classical World*. London: Thames & Hudson.

Castells, Manuel. 1996. *The Rise of the Network Society*. Vol. 1 of *The Information Age: Economy, Society and Culture*. Oxford, UK: Blackwell.

Cherry, John F. 1988. "Pastoralism and the Role of Animals in the Pre- and Protohistoric Economies of the Aegean." In *Pastoral Economies in Classical Antiquity*, edited by C. R. Whittaker, 6–34. Cambridge: Cambridge Philological Society.

Crawford, Michael H. 1985. *Coinage and Money under the Roman Republic: Italy and the Mediterranean Economy*. Berkeley: University of California Press.

D'Arienzo, L. 1978. "Il codice del 'Breve' pisano-aragonese di Iglesias. *Medioevo: Saggi e Rassegne* 4: 67–89.

DeLoughrey, Elizabeth. 2001. "'The Litany of Islands, the Rosary of Archipelagoes': Caribbean and Pacific Archipelagraphy." *ARIEL: A Review of International English Literature* 32, no. 1: 21–51.

Dyson, Stephen L., and Robert J. Rowland Jr. 2007. *Archaeology and History in Sardinia from the Stone Age to the Middle Ages: Shepherds, Sailors, and Conquerors*. Philadelphia: University of Pennsylvania Museum of Archaeology and Anthropology.

Fernandes, D. M., Mittnik, A., Olalde, I. et al. 2020. "The Spread of Steppe and Iranian-Related Ancestry in the Islands of the Western Mediterranean." *Nat. Ecol. Evol.* 4: 334–45.

Hamilakis, Yannis, and Andrew M. Jones. 2017. "Archaeology and Assemblage." *Cambridge Archaeological Journal* 27: 77–84.

Heatherington, Tracey. 2010. *Wild Sardinia: Indigeneity and the Global Dreamtimes of Environmentalism*. Seattle: University of Washington Press.

Herzfeld, Michael. 2005. "Taking Stereotypes Seriously: 'Mediterraneanism' Reconsidered." In *The Mediterranean Reconsidered: Representations, Emergences, Recompositions*, edited by M. Peressini and R. Hadj-Moussa, 25–38. Gatineau: Canadian Museum of Civilization.

Hirt, A. M. 2010. *Imperial Mines and Quarries in the Roman World: Organizational Aspects, 27 BC–AD 235*. Oxford: Oxford University Press.

Ingold, Tim. 1993. "The Temporality of Landscape." *World Archaeology* 25, no. 2: 152–74.

Klein, S., Y. Lahaye, G. P. Brey, and H.-M. Von Kaenel. 2014. "The Early Roman Imperial AES Coinage II: Tracing the Copper Sources by Analysis of Lead and Copper Isotopes—Copper Coins of Augustus and Tiberius." *Archaeometry* 46, no. 3: 469–80.

Knapp, A. Bernard. 2017. "The Way Things Are . . ." In *Regional Approaches to Society and Complexity: Studies in Honor of John F. Cherry*, edited by Alex Knodell and Thomas P. Leppard, 288–308. Sheffield, UK: Equinox.

Latour, B. 1991. *We Have Never Been Modern*. Cambridge, MA: Harvard University Press.

Lawrence, D. H. 1921. *Sea and Sardinia*. New York: Seltzer.

Mandelbrot, Benoît. 1975. *Les objets fractals: Forme, hasard et dimension*. Paris: Flammarion.

Morris, Ian. 2003. "Mediterraneanization." *Mediterranean Historical Review* 18, no. 2: 30–55.

Moss, David. 1979. "Bandits and Boundaries in Sardinia." *Man* 14: 477–96.

Pugh, Jonathan. 2013. "Island Movements: Thinking with the Archipelago." *Island Studies Journal* 8: 9–24.

Rainbird, Paul. 2007. *The Archaeology of Islands*. Cambridge: Cambridge University Press.

Roberts, Brian R., and Michelle Stephens. 2013. "Archipelagic American Studies and the Caribbean." *Journal of Transnational American Studies* 5, no. 1: 1–20.

———, eds. 2017. *Archipelagic American Studies: Decontinentalizing the Study of American Culture*. Durham, NC: Duke University Press.

Roppa, Andrea. 2014. "Identifying Punic Sardinia: Local Communities and Cultural Identities." In *The Punic Mediterranean: Identities and Identification from Phoenician Settlement to Roman Rule*, edited by Josephine C. Quinn and Nicholas Vella, 257–81. Cambridge: Cambridge University Press.

Roppa, Andrea, and Peter van Dommelen. 2012. "Rural Settlement and Land-Use in Punic and Roman Republican Sardinia." *Journal of Roman Archaeology* 25: 49–65.

Rosman, Kevin J. R., Warrick Chisholm, Sungmin Hong, Jean-Pierre Candelone, and Claude F. Boutron. 1997. "Lead from Carthaginian and Roman Spanish Mines Isotopically Identified in Greenland Ice Dated from 600 BC to 300 AD." *Environmental Science and Technology* 31: 3413–16.

Rowland, Robert J., Jr. 1977. "Aspetti di continuità culturale nella Sardegna romana." *Latomus* 36: 460–70.

Russell, Anthony, and A. Bernard Knapp. 2017. "Sardinia and Cyprus: An Alternative View of Cypriotes in the Western Mediterranean." *Papers of the British School at Rome* 85: 1–35.

Scott, James C. 2009. *The Art of Not Being Governed: An Anarchist History of Upland Southeast Asia*. New Haven, CT: Yale University Press.

———. 2017. *Against the Grain: A Deep History of the Earliest States*. New Haven, CT: Yale University Press.

Smith, Alexander, and Peter van Dommelen. 2017. "Monumental Engagements: Cultural Interaction and Island Traditions in the West Mediterranean." In *Regional Approaches to Society and Complexity: Studies in Honor of John F. Cherry*, edited by Alex Knodell and Thomas P. Leppard, 288–308. Sheffield, UK: Equinox.

Terrell, John E., Terry L. Hunt, and Chris Gosden. 1997. "Human Diversity and the Myth of the Primitive Isolate." *Current Anthropology* 38: 155–95.

Thomas, Julian. 2015. "The Future of Archaeological Theory." *Antiquity* 89: 1287–96.

Van Dommelen, Peter. 1998. *On Colonial Grounds: A Comparative Study of Colonialism and Rural Settlement in 1st Millennium B.C. West Central Sardinia*. Archaeological Studies Leiden University 2. Leiden: Faculty of Archaeology, Leiden University.

———. 2011. "Postcolonial Archaeologies between Discourse and Practice." *World Archaeology* 43: 1–6.

van Dommelen, Peter, and Carlos Gómez Bellard. 2008. *Rural Landscapes of the Punic World*. Sheffield, UK: Equinox.

Walsh, Kevin. 2008. "Mediterranean Landscape Archaeology: Marginality and the Culture-Nature 'Divide.'" *Landscape Research* 33, no. 5: 547–64.

Webster, Gary. 2016. *The Archaeology of Nuragic Sardinia*. Sheffield, UK: Equinox.

Witmore, Chris. 2014. "Archaeology and the New Materialisms." *Journal of Contemporary Archaeology* 1, no. 2: 1–44.

Chapter Fifteen

Sovereignty between Empire and Nation-State

The Archipelago as Postcolonial Format

Christopher J. Lee

This short critical reflection is concerned with the archipelago as an anti-imperial political form. In contrast to Benedict Anderson's positioning of nationalism and the nation-state as preeminent modular forms of the twentieth century, this chapter asserts that the archipelago is *the* principal state paradigm of the postcolonial Cold War period.[1] It does so by pondering the structural resemblance between twentieth-century postcolonial state movements and international institutional formations—from the Non-Aligned Movement (NAM) to the United Nations (UN)—with what can be called the archipelago format. These political communities comprised of noncontiguous nation-states flourished during the postcolonial Cold War era. They marked transnational attempts at generating political, economic, and cultural solidarities after an era of European empires. Indeed, these groupings often embraced a geographic scale and political ambition that echoed past imperial formations. Though these political archipelagoes varied in purpose and power—and at times were provisional, rather than permanent, in scope—they nonetheless point to the ways in which new states sought to create new assemblages of group empowerment against the remaining vestiges of Western imperialism, as well as against emergent threats of global dominance through the foreign policies of the United States and the Soviet Union. The archipelago as a postcolonial form consequently underscores how the nation-state was not a culminating position for liberation struggles and anticolonial sentiment during the era of decolonization. Rather, new nation-states continued anticolonial momentum through collective arrangements as a means of protecting sovereignty. The political archipelago served as an intermediate political format between delegitimized empires and newly independent countries with their respective vulnerabilities.

One starting point for thinking through the role of the archipelago model in shaping the institutional and spatial possibilities of postcolonial sovereignty is the case of an actual archipelago—the country of Indonesia and its territory of 17,508 islands. In April 1955, twenty-nine countries from Africa and Asia convened in the city of Bandung on Java to address pressing issues their continents faced during the early Cold War period. Formally called the Asian-African Conference, the Bandung Conference, as it is more commonly known, was the largest diplomatic meeting of its kind at that point, ostensibly representing 1.4 billion people, or almost two-thirds of the world's population by popular estimates. Only the UN, which had seventy-six members in 1955, was larger in numeric representation and in terms of geographic and political magnitude. The conference, sponsored by Indonesia, Burma (present-day Myanmar), Ceylon (present-day Sri Lanka), India, and Pakistan, offered a global stage for such statesmen as Sukarno (1901–1970) of Indonesia, Jawaharlal Nehru (1889–1964) of India, Gamal Abdel Nasser (1918–1970) of Egypt, and Zhou Enlai (1898–1976) of the People's Republic of China (PRC), all of whom promoted personal, national, and international interests. Official delegations in attendance came from the PRC, Egypt, Turkey, Japan, Libya, Lebanon, Jordan, Syria, Iran, Iraq, Saudi Arabia, Yemen, Afghanistan, Nepal, Laos, Cambodia, Thailand, North and South Vietnam, the Philippines, Ethiopia, the Gold Coast (present-day Ghana), Sudan, and Liberia. Though regional conflict in Southeast Asia between North and South Vietnam provided a catalyst for holding the conference—the Bandung meeting can partly be seen as a regional response to the 1954 Geneva Conference organized by the United States, the USSR, and their European partners—the program ultimately included broader issues regarding American and Soviet influence in Asia and Africa, the consequent importance of postcolonial sovereignty, and remaining questions about the surge of decolonization then occurring, particularly in Africa. The origins and purposes of the meeting were therefore multifaceted, geographically and politically, reflective of the expansive continental representation at hand and the political changes then occurring across the world.[2]

The Asian-African Conference also symbolizes one occasion for the emergence of the archipelago as a postcolonial form, both literally and figuratively. The choice of Bandung as the site for the meeting is unusual in retrospect, given that Indonesia was an outlier among the other four sponsors, which were former members of the British Empire. This planning decision can be attributed to the diplomatic ambitions of Indonesian prime minister Ali Sastroamidjojo, who proposed the meeting, as well as the importance of Indonesia and its relative neutrality within the region, in contrast to its cohosts India and Pakistan, who rivaled one another. The archipelago context and format nonetheless proved fateful. As Michelle Ann Stephens has written, the Bandung meeting challenged conceptions of liberation and sovereignty, particularly among black intellectuals, that had remained, up to that point, largely fixed to the nation-state.[3] The African American writer Richard Wright (1908–1960) captured these new opportunities and tensions that the conference presented in his firsthand account of the event, *The Color Curtain* (1956).[4] The vision he laid out of the emergent possibilities of interracial solidarity against past and present Western aggression contributed to the so-called Bandung Spirit, which itself soon circulated to have an immediate impact on the First Congress of Negro Writers and Artists held in Paris in 1956, as described by James Baldwin.[5] But beyond the role and importance of attitudes and political will, the latent political structure present at Bandung shared affinities with ideas of federation that circulated in the Caribbean and West Africa among such figures as C. L. R. James, Eric Williams, Aimé Césaire, and Leopold Senghor, who debated the possibilities of this option for political futures after decolonization.[6] At Bandung, Sukarno declared "Unity in Diversity" ("Bhinneka Tunggal Ika") to be the slogan of postcolonial Indonesia and a model for the nations at hand.[7] This rhetorical gesture vis-à-vis Indonesia's archipelago status offered a broadly applicable strategic template for conceptualizing postcolonial sovereignty elsewhere.

It must be stressed that archipelagoes could equally serve imperial intentions. Indeed, from a historical standpoint, they have arguably served this purpose more often. Archipelagic territories account for much of the United States's imperial reach in the Pacific, the Caribbean, and the Indian Ocean—be it through the American-controlled Philippines of the early twentieth century, the US trusteeship of Micronesia after World War II, or the military installation at Diego Garcia.[8] Brian Russell Roberts and Michelle Ann Stephens have recently situated the archipelago as a neglected, yet vital, territorial means for rethinking the spatial expansiveness of the US empire, American studies, and, by extension, global political geographies.[9] Building upon these recent critical interventions, the archipelago as a political framework might also stimulate a rethinking of anti-imperial endeavors as well. Indonesia is, of course, a geological archipelago in the conventional sense of being a collection of closely grouped islands. But through the prism of Bandung, it also offers a symbolic framework for approaching and interpreting postcolonial futures, as experimented with on different occasions during the global Cold War period. The Afro-Asian meeting reflected the transformation of global anticolonialism from a technique of insurgent grassroots politics to a method of diplomatic statecraft, which in turn frequently manifested institutional assemblages that sought to coalesce political and economic solidarities among postcolonial countries that would otherwise remain vulnerable individually.

Indeed, the continuation of anticolonial politics after political independence is important to recognize, given that many members of the first generation of postcolonial leaders, such as Nehru and Sukarno, actively led liberation struggles. Anticolonialism as a political sentiment and worldview did not subside and disappear but instead took new shape to inform the international politics of the postcolonial period. Continuities must be stressed. Diplomatic occasions like Bandung occurred during a critical period of transition between colonial and postcolonial eras, amid the passing of modern European imperialism and the emergence of new Cold War rivalries and neo-imperial interventions. Furthermore, global decolonization was an uneven process, with many liberation movements, like those in southern Africa, continuing for decades up to the 1980s and, in the case of apartheid South Africa, the early 1990s. Postcolonial diplomacy and statecraft as political practices consequently materialized during a time of uncertainty and opportunity both. These conditions perpetuated the continuation of an anticolonial discourse by Asian and African countries in order to secure sovereignty, resist foreign intervention, and participate as equal members in the global politics then taking shape. In his opening address at Bandung, Sukarno pointed to these tensions of continued anticolonial anxieties in a postcolonial world, at once celebrating the start of a new epochal period in world history guided by Asian and African countries, while also warning of the new dangers presented by the United States and the Soviet Union.

These global conditions that bore the threat of neocolonialism through trade agreements, security arrangements, and other political alignments with Cold War powers not only encouraged the persistence of anticolonial rhetoric to protect the meaning of political independence but also demanded collaborative efforts and institutions to substantiate such discourse. A number of forums before and after Bandung were committed to such a task. Preceding meetings like the 1947 Asian Relations Conference in Delhi and the 1953 Asian Socialist Conference in Rangoon (present-day Yangon) raised matters of sovereignty and collective diplomacy against neocolonial trends, as did later events like the 1956 Asian Socialist Conference in Bombay (present-day Mumbai) and the 1958, 1960, and 1961 All-African People's Conferences in Accra, Tunis, and Cairo, respectively.[10] But more than identifying problems being faced, these events enabled new patterns of political networking among states and ongoing liberation movements during the early postcolonial period—archipelagoes of solidarity and resistance to remaining forms of empire. Indeed, the locations of these conferences in cities across Asia and Africa highlighted a new political geography that had emerged from the shadows of Western colonial rule, yet also a diplomatic geography that was not always contiguous, thus necessitating an inventive format.

In this sense, these Cold War–era meetings continued a sequence of conferences and networks that had started during the early twentieth century, including such precursors as the Pan-African Congresses and the 1927 League against Imperialism meeting in Brussels, Belgium. These earlier efforts did not represent cohesive territorial unity as such but instead utilized conditions of diaspora and political dispersal for subversion and dissensus, to disrupt political expectations and geographical limits regarding definitions of political community. While these preceding efforts failed to achieve their objectives in the short term, the postcolonial period renewed their energies in a fundamentally different environment, with African and Asian diplomats representing sovereign nation-states rather than liberation movements and like-minded intellectuals. The earlier meetings enabled geographically dispersed people to meet and generate political solidarities transnational and often intercontinental in orientation. Though attended by persons without official capacity, they engaged with self-determination and acquired important symbolic value through political ambitions and imaginations anchored in shared experiences of racial discrimination, cultural prejudice, and political repression. The growing involvement of autonomous postcolonial states allowed a new set of possibilities to come into view that reflected political independence and the resources that went with it, in contrast to previous conferences where those gathered had little actual political power.

As proposed at the beginning of this chapter, the Bandung Conference played a foundational role in renewing these efforts while initiating the archipelago as a model of political solidarity for noncontiguous states. Indeed, conferences among postcolonial states proved to be generative in scope. Though the geographic balance of delegates at the 1955 meeting tilted toward Asia, the future of Asia-Africa relations soon shifted to the African continent. Nasser positioned himself as a leader of the third world, a status enhanced by the support Egypt garnered during the 1956 Suez Crisis. In December 1957, the Afro-Asian Peoples' Solidarity Organization (AAPSO) was established in Cairo, marking a new intercontinental endeavor that embraced the Bandung Spirit and the multinational solidarity it commenced. The AAPSO had wider involvement, including a range of organizations, rather than solely official state delegations, from Asian and African countries. The conferences it assembled continued an archipelagic sensibility of "unity in diversity" through professional exchanges, cultural promotion, women's coalitions, and youth participation. Its meetings were held within an expanding range of geographic locales, including Ghana, Guinea, and Tanzania, which further articulated a set of ambitions that noted the collective strength that could be attained through the involvement of many nation-states. The archipelago format once again presented an intermediate political model between the hegemonic empires of the past and politically vulnerable nation-states of the present.

Arguably the most significant example of this new configuration was the Non-Aligned Movement—its name indicating an anticolonial stance and collective grouping redefined for the Cold War era. The Belgrade Conference of Nonaligned States that convened in Yugoslavia in September 1961 formalized this political coalition. Key figures included former Bandung participants such as Nehru, Nasser, and Sukarno, along with new figures such as Kwame Nkrumah (1909–1972) of Ghana and host Josip Broz Tito (1892–1980) of Yugoslavia.[11] Zhou and the PRC were notably absent—an indication that political solidarities did not necessarily persist. In this case, tensions and eventually war—specifically, the 1962 Sino-Indian War—put an end to good relations between India and the PRC. A second NAM conference was held in Cairo in October 1964 with delegations from forty-seven states in attendance, a numeric growth assigned to the wave of decolonization in sub-Saharan Africa. The NAM ultimately superseded the 1955 Asian-African Conference—drawing upon it symbolically but inaugurating a different configuration of nation-states that would meet routinely in the decades ahead up to the present.[12]

The upshot is that the NAM and other postcolonial transnational collectives—such as the Organization of African Unity (1963–2002), the Southern African Development Community (established in 1980), and the Organization of Petroleum Exporting Countries (founded in 1960)—signaled a convergence of concerns for postcolonial sovereignty wedded to pragmatic practices of statecraft. These efforts were at times underpinned by more specific ideological positions. In addition to the Afro-Asianism articulated at Bandung, ideologies of Pan-Africanism, promoted by Nkrumah, and Pan-Arabism, promoted by Nasser, represented parallel efforts to unite different states under the banner of broadly construed identities—geographic, cultural, and unavoidably archipelagic in scope—positioned against the Euro-American West. The 1963 founding of the Organization of African Unity and the establishment of the United Arab Republic (1958–1961), which briefly unified Egypt and Syria, manifested these ideas, institutionally and territorially. Leaders such as Nkrumah, Julius Nyerere (1922–1999) of Tanzania, and Léopold Senghor (1906–2001) of Senegal also pursued variations of African socialism that blended local ideas with introduced ideologies such as Marxism and Maoism, as in the case of *ujamaa* ("familyhood") in Tanzania, which aimed to make their respective countries economically independent from former colonial powers.[13] Nonalignment itself had circulated earlier as an idea before it became a state practice through the diplomatic efforts of Nehru and India's UN ambassador V. K. Krishna

Menon. It partly originated from Nehru's experience with the anticolonial strategy of noncooperation before India's independence.[14]

These ideological developments gained strength through collective institutions. However, third-world solidarities enabled through the continuation of an anticolonial ethos during the Cold War must also be understood as more complex and limited than public rhetoric might suggest. Despite claims of neutrality, a number of NAM members had formal relations with the United States and the Soviet Union. In fact, such alignments had developed prior to the Bandung meeting with the Manila Pact (1954) and the Baghdad Pact (1955), underscoring the need to avoid oversimplified political views that would have third worldism and its proponents share uniform policies or political consistency. Though they would continue to be applied to politically fraught situations in South Africa and Israel during the 1970s and 1980s, the terms "anti-imperialism" and "anticolonialism" became increasingly rhetorical, rather than radical, in scope during the postcolonial period—a tool for declaring a critical position, citing a patriotic history, and creating political solidarity, in contrast to identifying an immediate threat, galvanizing a revolutionary movement, or achieving nation-state sovereignty as such. The archipelagic model as a postcolonial form has therefore had limits as a force for political transformation. Nonetheless, this flexibility in function explains its endurance. Though scholars such as Gary Wilder and Frederick Cooper have recently examined the idea of federation during the era of decolonization as an alternative to nation-state sovereignty, the archipelago provides a potential proxy for examination and scrutiny beyond this failed option.[15]

The archipelago as a political form remains in the present, if often concealed by a continued fixation on the nation-state. It eludes conventional narration in the same way that it has eluded forms of state power. Indeed, the two are connected. Like the third world and its successor, the global South, which themselves are noncontiguous groups of states and political communities, the emancipatory capacity of the archipelago format is through its resistance to any normative geography and hence its powerful ability to reimagine alternative histories and different aspirational futures. Rather than conveying disunity or political impossibility, the fragmented, allusive shape of the postcolonial archipelago is the very source of its radical potential.

PRACTICAL AND METHODOLOGICAL IMPLICATIONS

This chapter positions the archipelago as a principal state paradigm of the postcolonial Cold War period. It should be contrasted with and used to critique the dominance of nationalism and the nation-state as addressed by

Benedict Anderson and others. The archipelago format can be observed in the multinational structures of postcolonial state movements and institutions, such as the AAPSO and the NAM. These political communities comprised of noncontiguous nation-states flourished after mid-century decolonization. They exemplify the kinds of political, economic, and cultural solidarities that emerged after an era of European empires. Scholars should embrace the archipelago format in order to render visible these transnational formations that resist the conformities of the nation-state model and the linear historical narratives that often result.

NOTES

1. Anderson [1983] 2006, 4, 137.
2. On the Bandung meeting, see, for example, Abraham 2008, Ampiah 2007, Kahin 1956, Lee 2010, Mackie 2005, Pham and Shilliam 2016, Tan and Acharya 2008.
3. Stephens 2015, 222.
4. Wright 1956. For a critical study and additional documentation of Wright's trip, see Roberts and Foulcher 2016.
5. Baldwin [1961] 1989, 14, 23.
6. For recent studies of federation, see Cooper 2014; Wilder 2015.
7. Abdulgani 1981, 90.
8. For a recent study of this phenomenon, see Immerwahr 2019.
9. Roberts and Stephens 2017.
10. On the these conferences, see McCallum 1947a, 13–17; McCallum 1947b, 39–44; Jack 1959, 11–17; Houser 1961, 11–13; "All-African People's Conferences," 1962; Jansen 1966, 265–66.
11. On the NAM, see Mišković, Fischer-Tiné, and Boškovska 2014; Rakove 2012.
12. Lüthi 2016.
13. On the challenges of *ujamaa*, see Lal 2015; Schneider 2014.
14. On the challenges of dating nonalignment, see Abraham 2008; Brecher 1968, 3–15; Chaudhuri 2014, 13; Lüthi 2016; McMahon 1994, 37.
15. Cooper 2014; Wilder 2015.

REFERENCES

"All-African People's Conferences." 1962. *International Organization* 16, no. 2: 429–34.

Abdulgani, Roeslan. 1981. *The Bandung Connection: The Asia-Africa Conference in Bandung in 1955*. Translated by Molly Bondan. Singapore: Gunung Agung.

Abraham, Itty. 2008. "From Bandung to NAM: Non-alignment and Indian Foreign Policy, 1947–65," *Commonwealth & Comparative Politics* 46, no. 2: 195–219.

Ampiah, Kweku. 2007. *The Political and Moral Imperatives of the Bandung Conference of 1955: The Reactions of the US, UK and Japan*. Leiden: Brill/Global Oriental.

Anderson, Benedict. [1983] 2006. *Imagined Communities: Reflections on the Origin and Spread of Nationalism*. Rev. ed. London: Verso.

Baldwin, James. [1961] 1989. *Nobody Knows My Name*. New York: Vintage.

Brecher, Michael. 1968. *India and World Politics: Krishna Menon's View of the World*. Oxford: Oxford University Press.

Chaudhuri, Rudra. 2014. *Forged in Crisis: India and the United States since 1947*. New York: Oxford University Press.

Cooper, Frederick. 2014. *Citizenship between Empire and Nation: Remaking France and French Africa, 1945–1960*. Princeton, NJ: Princeton University Press.

Houser, George M. 1961. "At Cairo: The Third All-African People's Conference." *Africa Today* 8, no. 4: 11–13.

Immerwahr, Daniel. 2019. *How to Hide an Empire: A History of the Greater United States*. New York: Farrar, Straus and Giroux.

Jack, Homer A. 1959. "Ideological Conflicts." *Africa Today* 6, no. 1: 11–17.

Jansen, G. H. 1966. *Nonalignment and the Afro-Asian States*. New York: Praeger.

Kahin, George McTurnan. 1956. *The Asian-African Conference, Bandung, Indonesia, April 1955*. Ithaca, NY: Cornell University Press.

Lal, Priya. 2015. *African Socialism in Postcolonial Tanzania: Between the Village and the World*. New York: Cambridge University Press.

Lee, Christopher J., ed. 2010. *Making a World after Empire: The Bandung Moment and Its Political Afterlives*. Athens: Ohio University Press.

Lüthi, Lorenz M. 2016. "Non-alignment, 1946–1965: Its Establishment and Struggle against Afro-Asianism," *Humanity* 7, no. 2: 201–23.

Mackie, Jamie. 2005. *Bandung 1955: Non-alignment and Afro-Asian Solidarity*. Singapore: Editions Didier Millet.

McCallum, J. A. 1947a. "The Asian Relations Conference." *Australian Quarterly* 19, no. 2: 13–17.

———. 1947b. "Personalities at the Asian Relations Conference." *Australian Quarterly* 19, no. 3: 39–44.

McMahon, Robert J. 1994. *The Cold War on the Periphery: The United States, India, and Pakistan, 1947–1965*. New York: Columbia University Press.

Mišković, Nataša, Harald Fischer-Tiné, and Nada Boškovska, eds. 2014. *The Non-aligned Movement and the Cold War: Delhi-Bandung-Belgrade*. New York: Routledge.

Pham, Quynh N., and Robbie Shilliam, eds. 2016. *Meanings of Bandung: Postcolonial Orders and Decolonial Visions*. London: Rowman & Littlefield.

Rakove, Robert B. 2012. *Kennedy, Johnson, and the Nonaligned World*. New York: Cambridge University Press.

Roberts, Brian Russell, and Keith Foulcher, eds. 2016. *Indonesian Notebook: A Sourcebook on Richard Wright and the Bandung Conference*. Durham, NC: Duke University Press.

Roberts, Brian Russell, and Michelle Ann Stephens. 2017. "Introduction. Archipelagic American Studies: Decontinentalizing the Study of American Culture." In *Archipelagic American Studies*, edited by Brian Russell Roberts and Michelle Ann Stephens, 1–54. Durham, NC: Duke University Press.

Schneider, Leander. 2014. *Government of Development: Peasants and Politicians in Postcolonial Tanzania*. Bloomington: Indiana University Press.

Stephens, Michelle A. 2015. "Federated Ocean States: Archipelagic Visions of the Third World at Midcentury." In *Beyond Windrush: Rethinking Postwar Anglophone Caribbean Literature*, edited by J. Dillon Brown and Leah Reade Rosenberg, 222–37. Jackson: University Press of Mississippi.

Tan, See Seng, and Amitav Acharya, eds. 2008. *Bandung Revisited: The Legacy of the 1955 Asian-African Conference for International Order*. Singapore: National University of Singapore Press.

Wilder, Gary. 2015. *Freedom Time: Negritude, Decolonization, and the Future of the World*. Durham, NC: Duke University Press.

Wright, Richard. 1956. *The Color Curtain: A Report on the Bandung Conference*. New York: World Publishing Company.

Chapter Sixteen

Archipelagic Feeling

Counter-Mapping Indigeneity and Diaspora in the Trans-Pacific

Haruki Eda

In this chapter, I explore the analytical potential of archipelagic studies for inspiring decolonial struggles in the Trans-Pacific region. Specifically, I examine the archipelagic formation of Japanese imperialism, along with the indigenous and diasporic subjectivities that are subsumed under the myth of ethnic homogeneity promoted by the Japanese nation-state. As I aim to show, subjugation of ethnic groups in Japan, including the Ainu, Ryūkyūans, and Zainichi Koreans, hinges on not only the homogeneity myth but also the dominant geographic imaginary of Japanese territories. Neither "archipelago" nor "Trans-Pacific" is an inherently subversive geographical term, but I aim to demonstrate how both can animate alternative imaginations of space and place, as well as life and death, that move beyond the territoriality and subjectivity undergirding hegemonic Japanese nationalism. While the Japanese nation-state is already an archipelago, archipelagic thinking draws attention to the relationalities between land and sea spaces as multiscale sites of geophysical and geosocial processes. In short, the geoformal concept of archipelagoes allows me to examine the relational formations of ethnicity, nation, and empire in Northeast Asia and the Trans-Pacific, with a dual analytical emphasis on material flows and discursive forces.

The "Trans-Pacific," as a discourse, emerged partly in response to earlier geographical frameworks such as "Asia Pacific" and "Pacific Rim" to emphasize the historical and cultural proximities and geopolitical entanglements spanning the Pacific Ocean (Hoskins and Nguyen 2014; Yoneyama 2016). As a key driving force behind transpacific studies, Asian American studies has developed enormous insights into the dynamic relationships between Asia, Pacific Islands, and the Americas through the lived experiences of Asian Americans. These transdisciplinary fields challenge the nation-centered structure of mainstream East Asian studies, which was originally established to fulfill Cold War–era intelligence needs for legitimizing imperialist knowledge of China, Japan, and Korea (Mirsepassi, Basu, and Weaver 2003; Miyoshi and Harootunian 2002). From its birth in the Third World Liberation movement, Asian American studies has always prioritized critiques of US imperialism, global capitalism, and white supremacy that shape the lives of Asians and Pacific Islanders in the Americas. In addition, researchers have emphasized the tenacity of Asian immigrants by studying vibrant cultural productions and community mobilizations, often blurring the boundaries between the academy and the community.

Although Asian American studies offers rigorous analyses of diasporic community formations, some geographical assumptions underlying Asian ethnic identity categories largely remain unaddressed in these fields. In particular, the ethnic homogeneity of the Japanese nation-state tends to be taken for granted, whereas scholarship on ethnic formations in Northeast Asia often falls short of investigating the confluences of Japanese and Western imperialisms. As a result, although much work on Asian diasporas has focused on the deterritorialization of ethnic subjectivities, the geosocial formation of the very territoriality of the "homelands" is only partially scrutinized. Here, the convergence of transpacific studies and archipelagic studies promises a more nuanced analysis of spatialized power that flows between materiality and discourse. In other words, archipelagic thinking inspires a geopoetics that renders subjectivity and territoriality mobile, relational, and affective. Connecting the psychic to the planetary, I interrogate how dominant geographic knowledge production underscores the ongoing hegemony of imperialist operations in the Trans-Pacific. Central to such operations are the erasure and displacement of ethnic minorities, indigenous peoples, and diasporic communities. To disrupt such knowledge, I propose *counter-mapping* and *archipelagic feeling* as key methodological tools for cultivating decolonial solidarity. In the following section, I draw inspiration from previous research that attends to geographic discourse by denaturalizing space. Later, I illuminate an archipelagic historiography of imperial powers in the Trans-Pacific, before discussing how the methodological tools employed here can challenge such dominance.

GEOPOETICS OF WATER: INDIGENOUS AND DIASPORIC CONFLUENCES

As critical geographers have established, space and place are socially produced as humans engage with, embody, remember, and imagine them interactively (Cresswell 2004; Massey 1994; Mills 2012). What may appear to be "natural" ways of recognizing and representing the material contours of space and place are innately informed by social norms. Katherine McKittrick explains that dominant geographic knowledge "materially and philosophically arrange[s] the planet according to a seemingly stable white, heterosexual, classed vantage point . . . naturaliz[ing] both identity and place, repetitively spatializing where nondominant groups 'naturally' belong" (2006, xv). Thus, particular spatial imaginations that uphold the economic and cultural hegemony of the dominant group become privileged, institutionalized, and naturalized, trampling over marginalized groups' material presence and embodied experience.

Such geographic domination, as McKittrick terms it, deploys maps and place-names as technologies for epistemic manipulation. At the level of representation, cartographic renditions of material space can often aid imperial expansion by stabilizing the geophysical dynamics and naturalizing the geosocial constructions (Akerman 2008). Central to this operation is data production that objectifies materiality while privileging the dominant worldview of the researcher as the most legitimate mode of knowing. In the context of such imperial cartography, the emphasis on scientific accuracy in mapmaking is not so much representing the material world most truthfully as universalizing the mapmaker's situated gaze. In addition to visualizing, the naming and remembering of space also contribute to the construction of place as a socially imbued process. As Amy Mills (2012) points out, place and memory rely on each other within both individual and collective consciousness. Be it crossroads or continents, place-names thus serve the function of institutionalizing a particular historical narrative as public memory.

Furthermore, these epistemological issues also shape methodological debates. Methodological nationalism occurs when nation-centered thinking pervades in social research and nation-states are uncritically used as comparable units of analysis (Wimmer and Schiller 2002). This obscures the historical and geographical contingencies of social relations that traverse political boundaries. As these scholars suggest, imperial and colonial domination thus relies on the disciplinary regime of cartographic, toponymic, and geospatial knowledge production. Through this regime of knowledge, geophysical and geosocial processes become conflated, obscured, and essentialized. Thus, territory formation and subject formation complement one another in consolidating national identity. Shared spatial imaginary and collective memory of the nation become assembled, disseminated, and inhabited as a state-sanctioned sense of self, place, and belonging (Anderson 1983; Zerubavel 2003, 2012).

To challenge essentialized national identities, drawing attention to indigenous and diasporic experiences can reveal the fluid relations of territoriality and subjectivity. As academic discourses, diaspora studies and indigenous studies have often produced theoretical tensions. This is mainly because diasporic cultural production is often emphasized for deterritorializing ethnicity while overlooking the material relations of mobilities and immobilities in which indigenous peoples' sovereignty is implicated (see Anthias 1998; Fujikane and Okamura 2008; Smith 2011). As Shona Jackson's (2012) study of creole indigeneity in Guyana reveals, even material production in the context of slave and indentured labor can be incorporated into the claims of indigeneity by devising new creolized settler identities. Examining native Pacific societies through Stuart Hall's theory of articulations (indicating connectivity), James Clifford holds that conceptualizing indigeneity as articulations instead of authenticity "offers a nonreductive way to think about cultural transformation and the apparent coming and going of 'traditional' forms" (2001, 478). These attempts to reconcile indigeneity and diaspora achieve partial success by providing nuanced accounts of cultural processes and subjectivities that relativize diasporic and indigenous articulations. Nevertheless, they do not fully address the issue of territoriality underpinning the geopolitics of ethnic formations. Building on these debates on subject formation, I aim to show how territory formation is embedded in the entanglements of geophysical and geosocial processes.

By infusing the subject with an embodied sense of territorial belonging, geographic domination naturalizes the material and spatial configurations of power. By contrast, marginalized people's gendered, racialized, and socially stratified sense of place can be mobilized for cultivating alternative knowledge of the planet. I draw on the concept of *geopoetics* as a methodological framework for delineating how archipelagic studies contributes to a critical geopolitics of the Trans-Pacific region. A number of writers and researchers have deployed geopoetics to highlight the relationality between the geophysical and the geosocial (e.g., Balasopoulos 2008; Bouvet and Posthumus 2016; Italiano 2008; White 1992). In particular, Angela Last provides the following meditation by reading the Guadeloupean writer Daniel Maximin's (2006) articulation of humans as the "fruit of the cyclone": "Geopoetics appear as a poetics that takes geographical features and geophysical forces seriously as an element of geopolitics, while seeking to constructively reinscribe them as a means to counter imperialist aspirations and hegemonic worldviews. In short, they represent a materialist, decolonial process of rewriting geopolitics" (2015, 57). My conceptualization of geopoetics resonates with Maximin's and Last's approaches because I foreground decolonization as the axiology of geopoetic offerings. I firmly situate humans in the materiality of the Earth so as not to exaggerate human agency in contrast to objects, matters, and racialized bodies that dominant epistemology renders inanimate (see Chen 2012). Geopoetics enables an alternative imagination of the planet in which power resides in the very relationship between geophysical structures and geosocial engagements.

Envisioned as a "sea of islands" by Epeli Hau'ofa (1994), archipelagoes offer a geopoetics of water that elucidates the material confluences between indigeneity and diaspora. By shifting the lens from landmasses to bodies of water that embrace the lands, archipelagoes evoke relationality and connectivity (Baldacchino 2015; Stratford et al. 2011). According to Astrida Neimanis (2017), refiguring humans as literally "bodies of water" enables a transcorporeal critique of discrete individualism, anthropocentrism, and phallogocentrism undergirding hegemonic, and I would say mediocre, imaginations of life and living. Attending to both human and nonhuman bodies of water, archipelagic thinking thus unleashes nonbinary methodological sensibilities for water-mediated social dynamics, including differences and intimacies, mobilities and dwellings, and fluidity and solidarity. Practicing such sensibilities, the following three sections examine the emergence of Japanese and Western imperialisms in the northwestern Pacific to illuminate an archipelagic historiography of indigeneity and diaspora.

MAPMAKING AND EMPIRE BUILDING IN AINU MOSIR

The modern Japanese nation-state consists of four main islands, Honshū, Kyushū, Shikoku, and Hokkaidō, as well as 6,848 other islands. Neither such a grouping of those islands nor a cultural identity of "Japanese" ethnicity is essential or even primordial. Japan's current territorial and maritime claims emerged from the context of imperial expansions and competitions in the nineteenth century, through which non-Japanese ethnic societies of the indigenous Ainu and the Ryūkyū Kingdom were conquered, colonized, and folded into the territoriality and subjectivity of the Japanese nation. This national discourse was nascent during the feudal Tokugawa shogunate period (1603–1868) among elite scholars. When the emperor Meiji became the sovereign of the Empire of Japan in 1868, the modern centralized government established the universal education system and began to institutionalize the Japanese national identity. Here, I refer to the dominant ethnic group that became "Japanese" as Yamato, a group whose genealogy is entangled with the imperial clan's rule. Briefly, the Japanese nation-state includes at least three autochthonous ethnic groups: the dominant Yamato of Honshū, Kyushū, and Shikoku, the indigenous Ainu of Ainu Mosir (Hokkaidō/Ezo), and the Ryūkyū people of today's Okinawa prefecture.

As archipelagic thinking would illustrate, the boundaries of culture, language, and ethnic identity among these groups were not demarcated neatly

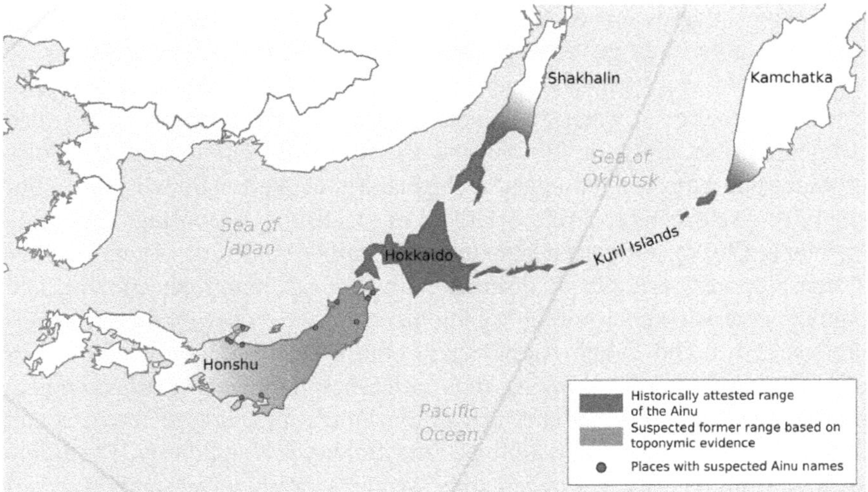

Figure 16.1. Historical expanse of the Ainu. ArnoldPlaton, 2013. *Source*: Wikimedia Commons (https://commons.wikimedia.org/wiki/File:Historical_expanse_of_the_Ainu.svg)

into these island spaces. Based on linguistic evidence in place-names, researchers argue that the northern part of Honshū was at least partially under the Ainu cultural sphere (Vovin 2009). In fact, the Ainu is indigenous not only in Hokkaidō, which the Yamato previously called Ezo, but also in Sakhalin and the Kuril Islands (see Figure 16.1).

In the Ainu cosmology, *ainu* means "people"/"humans" as distinguished from spirits (*kamuy*), which reside not only in nonhuman animals but also in objects, elements, and matter, such as house, fire, water, and trees/mountains. This cosmology centers on the relationship between the human world (*ainu mosir*) and the spirit world (*kamuy mosir*). Thus, the land on which the Ainu live, including Hokkaidō/Ezo, is called Ainu Mosir. Here, I refer to Hokkaidō as Ainu Mosir, although it does not preclude other Ainu lands such as Sakhalin and the Kurils.

The Yamato gradually settled into southwestern Ainu Mosir, starting in the 1200s. While they initially traded with the Ainu rather peacefully, tensions percolated in the fifteenth century. In 1457, an Ainu chief Koshamain (Kosamaynu) of today's Oshima Peninsula led an unsuccessful revolt against the Yamato settlers, who called themselves Wajin in relation to the Ainu. In 1669, another Ainu chief, Shakushain (Samkusaynu), organized an island-wide independence war against the Yamato, but it was ultimately suppressed. After this failed revolution, the Tokugawa shogunate gained control of most of the island and began to exploit the Ainu further by

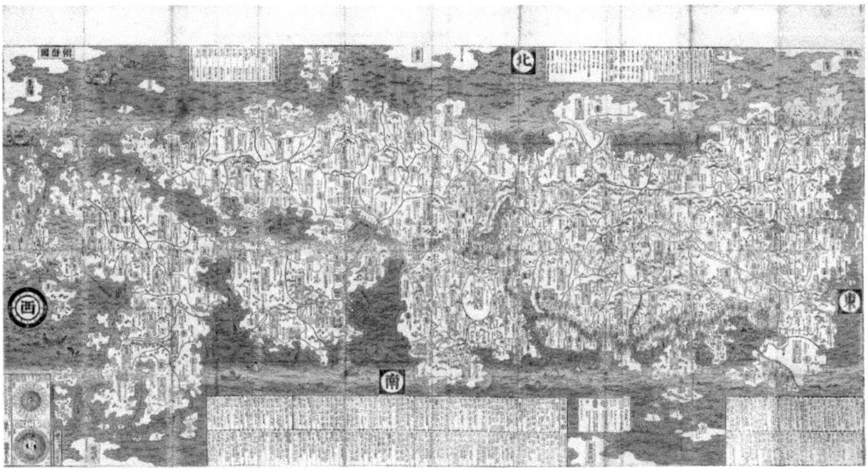

Figure 16.2. Nihon Kaisanchōriku-zu. Ishikawa Tomonobu, 1691. *Source*: Geospatial Information Authority of Japan (https://kochizu.gsi.go.jp/items/266).

facilitating the settlement of more Wajin traders. The Yamato capitalists treated the Ainu brutally, forcing them to relocate and perform harsh labor; physical and sexual violence was rampant, and starvation was common. Paralleling the decimation of indigenous peoples of the Americas, diseases brought by the settlers also devastated the Ainu population (Shinya 2015; Walker 2001).

In the seventeenth and eighteenth centuries, however, Ainu Mosir was not part of the Yamato nation in terms of cartographic representations or international treaties. For instance, *Nihon kaisanchōriku-zu* created by Ishikawa Tomonobu in 1691 centers on Honshū, Kyushū, and Shikoku, while land spaces representing Ezo, the Ryūkyū Kingdom, and Joseon (Korea) appear at the periphery (see Figure 16.2). While this map-like drawing depicts major land and sea travel routes throughout the Japanese archipelago, it also maps out local feudal lords, suggesting the cognitive horizon of the networked political and economic sphere in the isolationist Tokugawa period. Between 1639 and 1854, foreign relations were restricted to diplomacy with Joseon Korea, Ryūkyū, the Ming and Qing dynasties, and the Dutch East India Company. This map was widely disseminated, and its popularity lasted for a century until more precise maps were created. In fact, Dutch scholar Adriaan Reland obtained this map and reproduced it into a map of sixty-six regions of Japan around 1720 (see Figure 16.3).

As European imperial powers loomed closer and closer in the nineteenth century, Ainu Mosir became deeply entrenched in larger geopolitical dynamics, and its cartographic representations directly linked to the production

Figure 16.3. Dōkoku nihon rokujū-rokushū-zu. Adriaan Reland, circa 1720. *Source*: Geospatial Information Authority of Japan (https://kochizu.gsi.go.jp/items/211).

of Japan's territoriality. In 1800, Inō Tadataka created the first empirically measured and extremely precise map of Ezo, funded partially by the shogunate. In the 1850s, major Western powers challenged the Dutch monopoly of the Japanese market and demanded Japan open its ports. Consequently, the shogunate signed treaties with the United States and the United Kingdom in 1854, the Russian Empire in 1855, and the Netherlands in 1856. In this 1855 Treaty of Shimoda, Russia and Japan agreed on the territorial demarcation in the Kurils, between Iturup (Etorofu) and Urup islands, while leaving the status of Sakhalin undetermined. Neither Russia nor Japan recognized the territorial sovereignty of the Ainu in Hokkaidō, Sakhalin, and the Kurils; this was the moment when the Ainu world became entrapped and erased in the discourse of modern nation-states and international treaties. Russia's activities in the Far East urged cartographical needs and curiosity among the Japanese, particularly after the opening of the port of Hakodate in the Oshima Peninsula. Various cartographic renditions of Ainu Mosir were produced in the latter half of the nineteenth century, including the 1854 *Ezo kōkyōyochi zenzu* (Figure 16.4).

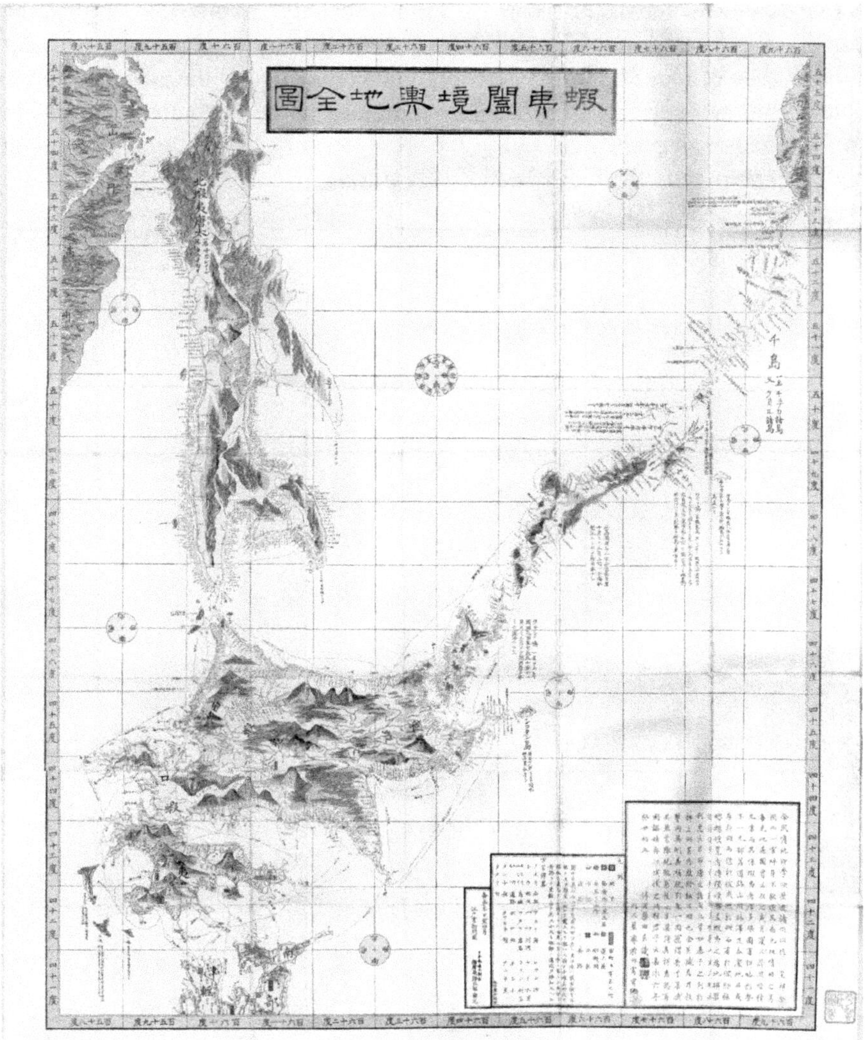

Figure 16.4. Ezo kōkyōyochi zenzu. Fujita Junsai and Hashimoto Gyokuransai (Utagawa Sadahide), 1854. *Source*: Geospatial Information Authority of Japan (https://kochizu.gsi.go.jp/items/142).

Created by a well-known calligrapher and artist pair, this map depicts all the islands on which the Ainu lived. While this and other maps of Ainu Mosir feature detailed place-names in the Ainu language, which had no writing system, the imperial Meiji government began to impose Japanese place-names with Chinese characters on Ainu places. The modern government immediately renamed Ezo as Hokkaidō and produced highly accurate maps of Ainu

Mosir based on the 1800 Inō map. Throughout the nineteenth century, the Japanese expanded and fortified their territorial control of this northern region as they produced dominant geographic knowledge of the islands, shores, rivers, and sea currents of what they imagined as new frontiers. The 1875 *Shinsen nihon zenzu* depicts all the Japanese territories at that time, including Ezo, Sakhalin, and the Kuril Islands (see Figure 16.5).

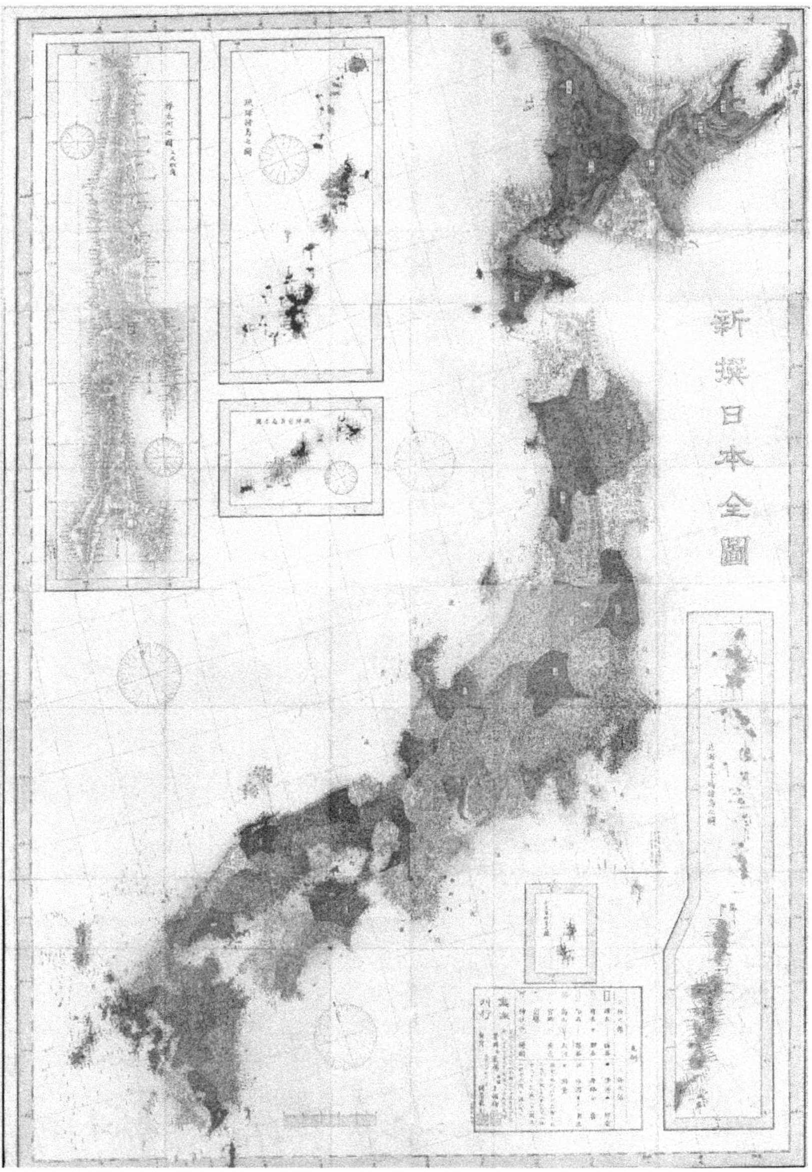

Figure 16.5. Shinsen nihon zenzu. Urabe Seiichi, 1875. *Source*: Geospatial Information Authority of Japan (https://kochizu.gsi.go.jp/items/186).

A month after this map was published, Japan gave up its control of Sakhalin to Russia while obtaining all the Kuril Islands in the 1875 Treaty of Saint Petersburg. In light of the rising geopolitical tension with Russia, the Yamato nation-state sought to assimilate the Ainu as a buffer. The Meiji government then forcibly relocated the Sakhalin Ainu to inland Hokkaidō and the Kuril Ainu to an island closer to Hokkaidō. To complete the assimilation, the Japanese destroyed Ainu livelihoods and undermined the traditional Ainu culture while giving them imperial citizenship (Shinya 2015; Siddle 1996). The Ainu had subsisted on salmon fishing and deer hunting for food and material items, but the government banned these activities. Although the Ainu did not have the concept of land ownership, the government stole and redistributed their lands among the settlers of the former samurai class. As a method of taxation and state control, this land policy forced the Ainu to cultivate land as sharecroppers even though they had no experiential knowledge of farming. Meanwhile, traditional Ainu spiritual practices such as tattooing and piercing were banned, and the Ainu language was forbidden in schools, where Ainu children were taught to be ashamed of their culture. The Japanese government did not recognize the Ainu as an indigenous population until 2008. As I have shown here, the dual historical processes of mapmaking and empire building have consolidated dominant geographical knowledge to subsume Ainu Mosir and the Ainu under Japanese territory and national subjectivity.

COLLIDING EMPIRES IN THE RYŪKYŪ ARC

In contrast to the Ainu, who had no centralized system of governance, Ryūkyūans (Lewchewans) had formed a kingdom by the time the Yamato power began to engulf its territories spanning across the Ryūkyū Arc. This island chain lies in the southwest of the Japanese archipelago, and it consists of 198 islands forming multiple archipelagoes between Kyūshū and Taiwan. Historical evidence indicates that the people of Ryūkyū Islands had maritime trade with people in the Chinese continent, Korean Peninsula, Japanese archipelago, Luzon, Siam, and Malacca Strait from the thirteenth century (Sakamaki 1964). The Ryūkyū Kingdom was established in 1429 under Shō Hashi, who unified the main island of the Ryūkyū Arc, Okinawa Island. The second dynasty emerged in 1469 and expanded its control to the majority of the islands in the arc in 1571. The maritime kingdom thrived by controlling foreign trade as an official enterprise, achieving economic viability in a kingdom otherwise scarce in natural resources. While the first dynasty focused on trading with Southeast Asia, the second prioritized its relationship with the Chinese continent. It maintained a tributary and suzerain relationship with the Ming dynasty, which sought to maintain its power in the western Pacific at a low military cost. A similar diplomatic relationship continued with the Qing dynasty after the ethnic Jurchen (later renamed Manchu) defeated the Ming dynasty in 1644.

In the Tokugawa era, the shogunate permitted the feudal lord of Satsuma (today's Kagoshima prefecture) to invade the Ryūkyū Kingdom and turn it into a vassal state in 1609, although the Ryūkyūans' de facto sovereignty in the arc continued. A 1785 cartographic illustration of the Ryūkyū Kingdom and its islands, *Ryūkyū sanshō narabini sanjūrokutō no zu*, shows sea routes that connect Naha, the maritime hub of the kingdom, to Fujian on the Chinese continent as well as other islands across the Ryūkyū Arc, up to Yamato Japan (see Figure 16.6). This map was drawn by Hayashi Shihei, a Yamato military scholar, as part of a book on three countries: Joseon, Ezo, and Ryūkyū. Such a grouping of adjacent non-Yamato territories implies the shogunate's increasing imperial interest. In fact, when the Western powers approached Japan to negotiate unequal treaties in the 1850s, the Ryūkyū courts signed treaties as an independent kingdom with the Americans (1854), French (1855), and the Dutch (1859), despite being dually subordinate to Qing and Tokugawa.

Within a couple of decades, however, these treaties lost any effectivity when the Ryūkyū Kingdom lost its sovereignty in 1887; the Meiji government annexed the Ryūkyū Islands as a colony and established Okinawa

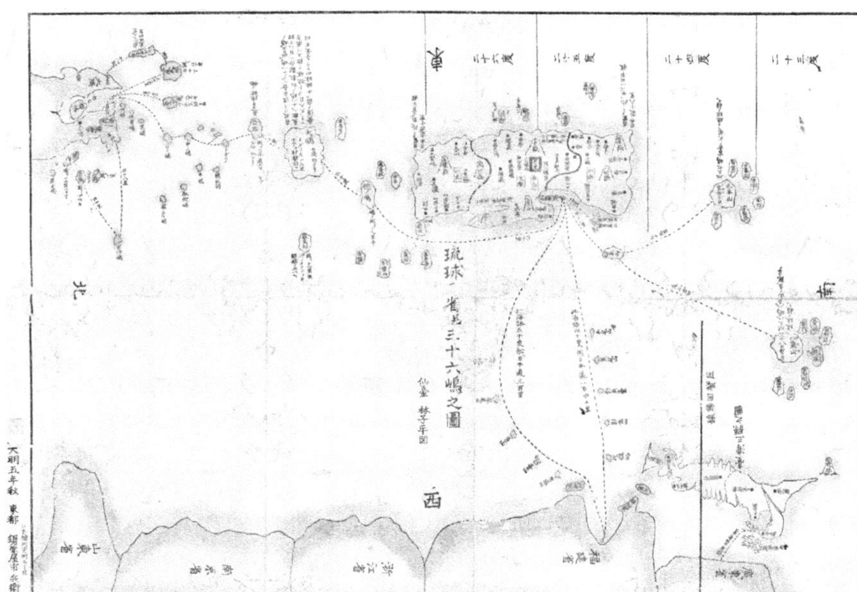

Figure 16.6. Ryūkyū sanshō narabini sanjūrokutō no zu. Hayahi Shihei, 1785. *Source*: University of British Columbia Library–Rare Books and Special Collections. Japanese Maps of the Tokugawa Era. (https://www.flickr.com/photos/ubclibrary_digicentre/14612998062/).

prefecture. Juridical and military control was paired with forced cultural assimilation, particularly through the education system. Punishing Ryūkyūan language speakers proved very effective in drawing these islands culturally closer to the Yamato-controlled Japanese archipelagoes, away from the Chinese sphere of influence (Matsushima 2014). The colonization of Ryūkyū by Japan can be contextualized within the expansion of Western imperial control in Asia, much like the competition between Russia and Japan that placed Ainu sovereignty in a vulnerable geopolitical location. The major blow came from the Qing dynasty's defeat to the British in the First and Second Opium Wars (1842 and 1860). This catalyzed a series of unequal treaties between the Qing dynasty and the United Kingdom, France, and the United States, which forcibly opened major ports and transferred the control of Hong Kong Island to the United Kingdom. Consequently, the weakened dynasty was in no shape to defend the Ryūkyū Kingdom from the Japanese invasion. In response to Western imperialisms, Japan sought to expand its territory further and waged the First Sino-Japanese War. As a result of Japan's victory in 1895, Korea's Joseon dynasty became independent from Qing's suzerain control and established itself as the Empire of Korea. In addition, Qing lost important strategic territories such as Taiwan (inhabited by Han Chinese and indigenous people) to the Japanese Empire. The southwestward expansion of imperial Japan into the Ryūkyū and Taiwan archipelagoes emerged from such global competitions for the Chinese market.

Overpowering the regional hegemony of China in East Asia for the first time, the Japanese Empire then quickly consolidated its territorial control over the Korean Peninsula by waging another war in 1904. The enemy target was now the Russian Empire. By winning this war, Japan became the first non-Western nation powerful enough to pose a real threat to the European global ascendancy. With the Treaty of Portsmouth, Japan gained firm imperial access to the Korean Peninsula, parts of Manchuria, and Sakhalin—in addition to the Ryūkyū and Taiwan islands. Thus, territorial and cultural collisions between imperial powers, both "Western" and "Eastern," led to the demise of Chinese continental hegemony and the rise of the archipelagic Yamato Japanese Empire in the age of maritime trade and militarization. From the perspectives of the Ainu and Ryūkyūans, the emergence of the Japanese nation-state has since interpellated them as indigenous subjects struggling for cultural survival and self-determination.

COLLUDING EMPIRES IN THE TRANS-PACIFIC

Empires did not only collide into each other in the Pacific; they also colluded with one another to establish hegemony across the ocean. Japanese and US imperialisms faced one another at Pearl Harbor in 1941, exchanging fire across the western Pacific until the Hiroshima and Nagasaki atomic bombings in 1945. Aside from these four years of belligerence, the two hegemonic powers colluded to control the Pacific region through territorial occupation and targeted migration restrictions. Perceiving the surge of Asian immigrants during the gold rush as a threat, the United States passed the infamous Chinese Exclusion Act in 1882. Meanwhile, US capital flowed into Hawai'ian sugar plantations, which imported cheap labor from southern China (Qing), Japan (Tokugawa/Meiji), Korea (Joseon), and the Philippines (Spanish colony). Capitalizing on the overthrow of the Kingdom of Hawai'i in 1893, the United States annexed the archipelago in 1898. Winning the Spanish-American War in 1898 and the subsequent Philippine-American War in 1902, the United States also acquired Cuba, Puerto Rico, Guam, and the Philippines, establishing itself as an archipelagic empire (Thompson 2010). During the Russo-Japanese War, Japan and the United States sought to avoid conflict through the 1905 Taft-Katsura Memorandum by mutually agreeing that Japan should control Korea and the United States should occupy the Philippines. Furthermore, the two archipelagic empires also sought to diffuse tensions by restricting migration from Japan, and by extension Korea, to the mainland United States in a 1907 Gentlemen's Agreement. However, Japan continued to issue passports for immigrants to Hawai'i until all Asian immigration to the United States was banned in the 1924 Immigration Act.

Endorsed by the United States, the Japanese Empire turned the newborn Empire of Korea into its protectorate in 1905 and formally annexed the Korean Peninsula in 1910. A peninsula-wide independence movement erupted on March 1, 1919, after a handful of Korean activists based in Tokyo drafted the first declaration of independence on February 8 and smuggled it to Seoul. The March First Independence Movement was violently suppressed, and some Korean communists and anarchists came to regard violence and assassination as a primary tactic. The Japanese media actively constructed the figure of "unruly Koreans" (*futei senjin*), which was used to justify the massacre of thousands of Koreans in the aftermath of the Great Kanto Earthquake in 1923 (Eda 2015). The Yamato Japanese justified their colonization of Korea by claiming that they were protecting Koreans, with whom they allegedly shared the same ethnic roots, from looming Russian dominance. This rhetoric was coupled with harsh and widespread discrimination against Koreans, some of whom internalized the oppression and actively sought to assimilate with the Japanese. Like the Ainu and Ryūkyūans, Koreans were forced to speak Japanese at school, and it became common for Koreans in Japan to use Japanese aliases to avoid discrimination. Ultimately, however, the colonial assimilation policy failed to integrate Koreans into the Japanese empire (Caprio 2009).

During the colonial period, millions of Koreans came to Japan to escape poverty. After the Second Sino-Japanese War began in 1937 without an official declaration, Koreans left the peninsula, voluntarily as well as through deception and coercion, as laborers in coal mines and construction, soldiers in the Imperial Japanese Army, and wartime sex slaves euphemistically called "Comfort Women" (Soh 2008). This colonial mass migration included some of my ancestors. By the end of colonization, some 2 million Koreans had migrated to the Japanese archipelago. The defeat of the Japanese Empire, however, did not result in Korean sovereignty because within weeks, the United States and Soviet Union agreed on a temporary demarcation line along the thirty-eighth parallel north. The Western Allied powers deemed Koreans incapable of self-governance, and Koreans' efforts to establish an independent republic were denied because the United States, United Kingdom, Soviet Union, and Republic of China had already planned for trusteeship of the Korean Peninsula before defeating Japan. In 1948 communist activists on Jeju Island began an uprising for Korean reunification on April 3; under US control, the newly emerging South Korean police and military massacred as many as eighty thousand Jeju Islanders (Ryang 2013). In the same year, the Republic of Korea (ROK) in the south and the Democratic People's Republic of Korea (DPRK) in the north were successively established, characterized by their mutual nonrecognition. The temporary division has since been tragically prolonged because of the Korean War, which remains officially ongoing today with only a cease-fire agreement between the DPRK and the United States.

This postcolonial national division created a stateless diasporic community of Koreans in Japan (Lie 2008). The majority of Korean colonial migrants returned to the peninsula after Japan's defeat, but travel restrictions, lack of resources, and homeland chaos prevented some six hundred thousand Koreans from crossing the Korea Strait, one of the most traveled water passages despite its dangerous currents. Many Jeju Islanders fled the massacre to Japan, constituting at least 15 percent of the quasi-refugee Korean population in Japan. This was the birth of the postcolonial exiles, Zainichi Koreans (*zainichi* literally means "residing in Japan"). After they formally lost Japanese citizenship in the 1951 Treaty of San Francisco, Zainichi Koreans were registered as nationals of the defunct "Chōsen" (Joseon), which commonly indicates predivision Korea. Including those who had been born in Japan, ethnic Koreans were now special permanent residents. Zainichi Koreans immediately organized themselves and established schools for their children to learn their own history in their own language, with their Korean names. This effort faced violent repression by the Japanese state under US occupation, which saw these schools as communist breeding grounds. The Japanese police also sought to regain the symbolic authority they had lost in the defeat and military occupation by suppressing Korean livelihoods (Kim 1978).

Territorial conflict in the homeland extended into the diaspora and fractured the Zainichi Korean community. Some aligned themselves with the DPRK and established the socialist Zainichi organization Chongryun (General Association of Korean Residents in Japan), while those aligned with the ROK founded the capitalist Mindan (Korean Residents Union in Japan). Chongryun has since maintained an ethnic education system from primary to higher education despite political repression from the Japanese (Ryang 1997). In comparison, Mindan has tended to be more assimilationist, especially after Japan and the ROK normalized their relations in 1965, enabling the stateless Zainichi Koreans to obtain ROK nationality (but not citizenship). Because Japan takes the *jus sanguinis* approach, Zainichi Korean children cannot have Japanese citizenship by birth unless one of their parents is a Japanese citizen. Thus, four or five generations into this diasporic exile, many Zainichi Koreans remain technically stateless and disenfranchised from mainstream Japanese society.

Ryūkyūans were forced to take a similar path of compromised sovereignty under military occupation in the context of imperialist collusion. The Treaty of San Francisco indicated that the Ryūkyū Islands would be governed under a potential US strategic trusteeship, which ultimately did not materialize. When this treaty came into effect in 1952, the security treaty between Japan and the United States also became effective, restoring Japan's sovereignty and placing the Ryūkyū Islands under US military occupation—as proposed by the Japanese emperor himself (McCormack and Norimatsu 2012). The Soviet Union and China had suggested Ryūkyū be governed by China, but the internal division of China made way for the United States to establish an island chain anchored in Okinawa to contain the socialist expansion (Matsushima 2014). Thus, Ryūkyūans remained under US military occupation until 1972, with no legal or practical power to address the land theft, sexual and physical violence, vehicular and aircraft accidents, environmental destruction, and discrimination conducted by US military personnel. In 1972, the Ryūkyū Islands reverted to Japanese control, but the US military bases continued to occupy nearly 20 percent of the total area of the main Okinawa Island. Even though Okinawa prefecture comprises only 0.6 percent of the total area of Japan, more than 70 percent of the US bases in Japan are located in Okinawa (Okinawa Prefectural Government 2016). The US military takes advantage of the colonial status of Ryūkyū under Yamato Japanese rule today, controlling the land, sea, and skies of the Ryūkyū Islands with ample funds donated by the Japanese government.

COUNTER-MAPPING THE NORTHWESTERN PACIFIC

Thus far, I have illustrated how Japanese, European, and US imperial interests have ensnared the lands and lives of the Ainu Mosir, Ryūkyū Arc, and Trans-Pacific by deploying geographic domination to naturalize Japaneseness as a subjectivity and territoriality. My archipelagic reading of these processes reveals the historically and geographically contingent formations of indigenous and diasporic subjectivities in relation to empire building. To articulate an alternative sense of space and place from the perspectives of indigenous and diasporic peoples, I propose counter-mapping as the first methodological tool for cultivating decolonial solidarity. The original uses of the term "counter-mapping" emerged from indigenous and community efforts to manage natural resources, address environmental racism, and revitalize traditional knowledge of sacred places, and these efforts usually involved participatory processes (Hodgson and Schroeder 2002). My use of this term here is broader, encompassing alternative cartographic renditions and imaginations generally. In the Trans-Pacific region, such counter-mapping entails an archipelagic attention to the bodies of water, not as maritime extensions of territorial control but as geophysical mediators of culture. This re-visioning, I argue, has the potential to disrupt the neoliberal capitalist rhetoric of land ownership and offer instead a geopoetics of sharing the sea.

Non-Yamato counter-mapping can begin with the largest island of the Japanese archipelago, Honshū. Constituting approximately 60 percent of Japanese territory and hosting about 80 percent of Japan's population, Honshū has been the central stage of Yamato history. However, the Yamato are not the sole autochthonous ethnic group of this island. Considering mythological and linguistic variations for delineating collective ethnic identity, Okamoto Masataka (2014) argues that the notion of Yamato ethnicity, or that of ethnicity and nation altogether, did not exist in Northeast Asia until 1888. According to Okamoto, this concept was deployed for the Meiji restoration of imperial sovereignty to invent the continuity of imperial reign since the mythical ancient past written in the seventh century. Briefly, the narrative legitimacy of the Yamato emperor comes from his alleged genealogical connection to the mythical solar-celestial deity (Amaterasu). Meanwhile, Okamoto draws attention to the Izumo region in western Honshū (today's Shimane prefecture), where Izumo Ōyashiro (Izumo-taisha) is located. Along with the Grand Shrine of Ise (Ise Jingū)—the highest-ranking Shinto shrine closely associated with the emperor, Izumo Ōyashiro is one of the most historically significant shrines in Japan. While the Yamato mythology enshrines the solar deity as the sovereign of Japan, the Izumo people have historically worshipped a different deity more closely associated with the sea than the sun. In Izumo mythology, the deity who created the world between the heavens and the netherworld (Ōkuninushi) ceded his control of the world to the solar deity in exchange for the establishment of Izumo Ōyashiro. This myth of territory ceding is said to reflect the actual political takeover of the Izumo region by the Yamato power. To accomplish the assimilation of Izumo at the spiritual level, the Izumo deity had to surrender to the Yamato deity in the nation-building mythology. Rereading the history of those who did not acquiesce to Yamato control, Okamoto thus suggests that the Izumo people, including himself, maintained their sense of difference from the Yamato at least until the Meiji era.

Examining the Japanese archipelago through the Izumo cultural sphere further reveals the intense and complex relationalities across the northern coasts of Honshū and Kyushū as well as the Korean Peninsula. In addition to the mythological analysis, Okamoto (2014) further points out material and linguistic evidence of expansive trade and migration routes along the oceanic currents across the East Sea/Sea of Japan. In the times of maritime transportation, the northern coasts facing the continent had flourished from economic and cultural exchanges, before the Meiji government designated the southern coasts facing the Pacific as the locus of development. Therefore, as Okamoto asserts, the popular rhetoric of insularity that compels the national illusion of ethnic homogeneity does not actually reflect the material connectivities of communities across land and sea spaces surrounding the Japanese archipelago and the Korean Peninsula. Okamoto's in-depth critical analysis demonstrates the archipelagic relationalities of peoplehood in the region, including the Izumo, Yamato, Ainu, Ryūkyūans, and Koreans.

To illustrate this material and spatial intimacy of Northeast Asia, Okamoto (2014) introduces a regional map that looks upside down, placing the Pacific Ocean at the top and the Eurasian continent at the bottom (Figure 16.7). In the middle are the archipelagic chains of the Kuril Islands, Sakhalin, Ainu Mosir, Japanese archipelago, Korean Peninsula, Ryūkyū Arc, and Taiwanese and Philippine archipelagoes. At the crossroads of hegemonic powers such as China, Russia, Japan, and the United States, these northwestern Pacific archipelagoes have long been the site of imperial domination as well as grassroots resistance. In contrast to the conventional mapping, this alternative cartographic schema renders the archipelagoes more embedded and interconnected, even closer to the continent, by centering them in between the continent and the ocean. Moreover, this mapping highlights the role of the inner seas as maritime passages rather than geophysical boundaries. Although the content of the map may not immediately challenge the dominant geographic discourse, flipping the directionality can animate a different sense of space that attends to the material and cultural intimacy of these land spaces con-

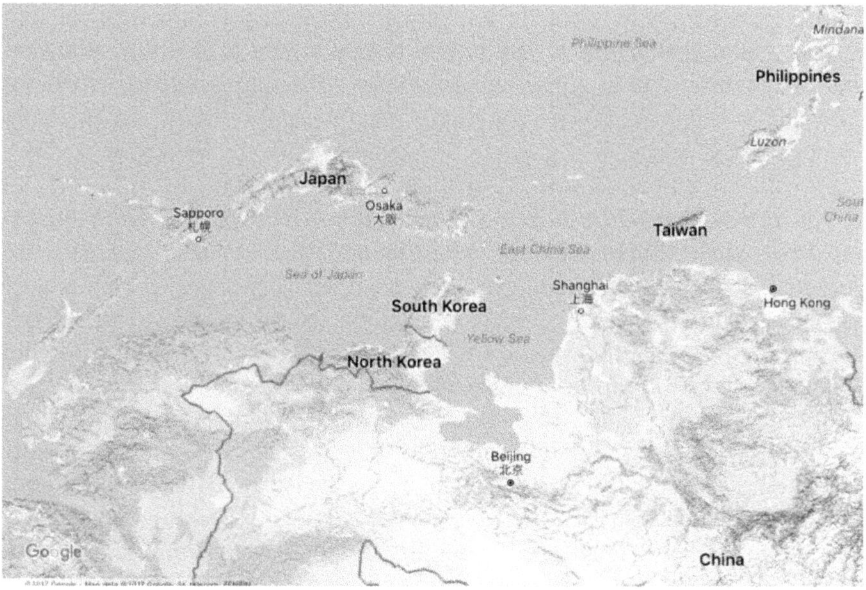

Figure 16.7. Archipelagoes of the Northwest Pacific. Created by the author using Google Maps based on Okamoto (2014)

nected by the sea. As Okamoto suggests, such "upside-down" maps were common in Japan before the modern cartographic conventions converged with the Meiji-era societal reorientation toward Western modernity, away from Asian backwardness. While more precisely indigenous and community-based mapping remains to be conducted, this cognitive counter-mapping can challenge the Japanese insular-imperial cartography and activate a spatial imagination of continental Asia and the archipelagic northwestern Pacific as full of pathways, engagements, and intimacies.

ARCHIPELAGIC FEELING

A cognitive shift alone is not sufficient to challenge the hegemonic geographic knowledge rooted in the false binary between the mind and the body (and the spirit). To augment counter-mapping as a decolonial methodology, I focus on emotions, affects, and sensations that animate the geopoetics of water. To riff on this anthology's titular focus on thinking: What would it mean to practice archipelagic feeling? How might we feel archipelagic? Although emotions may commonly appear as reactions to external factors, I think of the gerund "feeling" as a mindful and embodied practice of engaging with the body's situatedness in space. To do feeling and sensing well, in other words, requires simultaneous input and output, reception and expression, precision and elaboration, and humility and generosity. It requires intense attention, care, and love that are at once cognitive, affective, and spiritual. Archipelagic feeling, in this sense, is a practice of engaging with the bodies of water—yours, others', and nonhumans'—with utmost compassion.

For further clues, I (re)turn to my own subjectivities as mixed (Yamato) Japanese and Zainichi Korean. Whereas both Korean and Japanese hegemonic nationalisms would perhaps reject my impure Koreanness and Japaneseness (let alone my queerness!), only a few traces of my Koreanness appear intelligible in the eyes of the nation-state. Three generations into the diaspora, I have full Japanese citizenship and only a Japanese name. None of my family members and known relatives speak Korean, and many of them, including my father and older brother, are stateless "special permanent resident aliens," born and living in Japan. When I visit the ROK, some South Koreans would not think fondly of me for mixing with the colonizer and losing the ancestral language. Under the South Korean National Security Act, Zainichi Koreans were once prone to drawing suspicion as potential North Korean spies. I want to tell every shopkeeper and restaurant server I interact with that I am Korean, that I belong there, but I also excuse myself for not being fluent in our tongue. I want them to welcome me home, but I feel invisible and inconsistent. Instead, I let the smell and taste of the food suffice. I let the vibrant sounds of Korean as my third language constitute my skin. When I visit the DPRK, though, North Koreans make me feel somewhat more welcome as a compatriot. Between 1959 and 1984, some ninety thousand Zainichi Koreans repatriated to North Korea for the compelling promise of national liberation and material wealth, even though most Zainichi Koreans came from the southern parts of the Korean Peninsula (Morris-Suzuki 2009). For many Zainichi Koreans whose dream of national unification is written into the DPRK constitution, North Korea is the closest place to their imagined homeland. Pyongyang feels as foreign as the world could ever be, but I feel myself present at every moment. I have learned their songs, and I let my singing voice do the work. I am not entirely sure if I feel connected to anything or anyone, but I think of the Korea Strait. I feel myself flowing between Koreanness and Japaneseness to reach the open ocean, collecting all of my shipwrecked desires along the way.

I feel archipelagic when I situate my sense of diasporic longing in the history of the Japanese and US empires. Perhaps I would not have existed without these geopolitical forces, but I am as much a product of geophysical relationalities in the Northwest Pacific archipelagoes. Like Maximin's "fruit of the cyclone," my tectonic subjectivity is as "natural" as the earthquakes, volcanos, tsunamis, and typhoons, brimming with alternative possibilities.

From this location, I seek solidarity with the Ainu, Ryūkyūans, and Oceanians, among other indigenous peoples, in their struggle for self-determination and decolonization. I draw inspiration from the figure of Ryūkyūan kayak protestors, who have staged offshore direct actions against the planned construction of yet another military facility off the coast of rural Henoko district in Okinawa Island since 2015. The Japanese government originally proposed this construction in 1996 as the only solution to replace Marine Corps Air Station Futenma in the same island, a facility referred to as "the world's most dangerous base" by Donald Rumsfeld (Latman 2015). Equipped with life vests and paddles on vividly colored kayaks, these protestors occupied the sea surface of the Ōura Bay, whose marine biodiversity and fragile ecosystem are in danger. These protestors' maritime and land-based resistance sheds light on the dilemma of the relocation issue, as some Okinawans and Japanese have proposed Guam as an alternative, without regard to the Chamorros' resistance against the same geopolitical force of US imperialism. Ultimately, the only solution seems to be comprehensive and simultaneous demilitarization and pacifist diplomacy in the Trans-Pacific. With this awareness, the kayak protestors have cultivated solidarity with other base resistors in Jeju Island, where the South Korean government constructed a naval base that US naval ships would also use. The Japanese government continues to ignore the voices of the people of Okinawa and to push the construction further despite the prefecture-wide referendum expressing opposition. At the end of 2019, the government announced that the project would not be completed until 2032, adding ten more years to the five-year construction plan that was developed in 2013.

Across the Pacific, activists and community organizers of Korean, Ryūkyūan, Japanese, Filipino, Chamorro, and Hawai'ian descent in the United States are also connecting their homeland antimilitarization resistance to their diasporic struggles against police brutality and the prison-industrial complex in their local communities. Their analysis also connects the neoliberal Trans-Pacific Partnership Agreement with the ever-intensifying militarization of the Pacific. The geopolitics of dissidence against neoliberalism and securitization in the Trans-Pacific, therefore, brings together various diasporic and indigenous subjectivities to foster new forms of transnational solidarity. These mobilizations emerged in the context of the post–Cold War neoliberal global order, the shifting strategic focus of declining US hegemony, and the reemergence of Chinese and Russian economies. However, these new mobilizations also draw inspiration, imagination, and knowledge from the legacies of their ancestors who fought against the colluding and colliding imperial powers. I feel archipelagic when I practice solidarity with all these emergent formations to end militarization, exploitation, and the destruction of lives and livelihoods as well as land and water.

PRACTICAL AND METHODOLOGICAL IMPLICATIONS

Throughout this chapter, I sought to demonstrate how archipelagic thinking inspires embodied methodologies for decolonial solidarity, foregrounding the connectivities between indigenous and diasporic formations in the Trans-Pacific. Specifically, reading archipelagoes through a geopoetics of water allows me to analyze the entanglements of subject formation and territory formation without reinscribing geographic domination and methodological nationalism. My attempt to provide an archipelagic historiography of Japanese imperialism reveals that various imperial powers colluded with, as much as collided into, each other across the Pacific. In this historical process, mapmaking and place-naming practices helped consolidate Japanese national identity by naturalizing the dominant Yamato worldview as the single legitimate mode of geospatial knowledge. To disrupt this geopolitical hegemony in the cognitive and affective dimensions, I propose counter-mapping and archipelagic feeling as methodological tools for centering the bodies of water and eclipsing the neoliberal masculinist rhetoric of land ownership. My aim has been to offer a nuanced framework for political solidarity among ethnic minorities in Japan—starting with my situatedness as a Zainichi Korean/Japanese writer who came of age in Asian American communities and studies. Despite the preliminary scope of this chapter, I hope that my thinking and feeling contribute to the centuries-long struggles for cultivating sacred relationships in the Trans-Pacific and beyond.

REFERENCES

Akerman, James. 2008. *The Imperial Map: Cartography and the Mastery of Empire*. Chicago: University of Chicago Press.

Anderson, Benedict. 1983. *Imagined Communities: Reflections on the Origin and Spread of Nationalism*. New York: Verso Books.

Anthias, Floya. 1998. "Evaluating 'Diaspora': Beyond Ethnicity?" *Sociology* 32, no. 3: 557–580.

Balasopoulos, Antonis. 2008. "Nesologies: Island Form and Postcolonial Geopolitics." *Postcolonial Studies* 11, no. 1: 9–26. doi: 10.1080/13688790801971555.

Baldacchino, Godfrey. 2015. *Archipelago Tourism: Policies and Practices*. London: Routledge.

Bouvet, Rachel, and Stephanie Posthumus. 2016. "20 Eco- and Geo-approaches in French and Francophone Literary Studies." *Handbook of Ecocriticism and Cultural Ecology* 2: 385.

Caprio, Mark E. 2009. *Japanese Assimilation Policies in Colonial Korea, 1910–1945*. Seattle: University of Washington Press.

Chen, Mel Y. 2012. *Animacies: Biopolitics, Racial Mattering, and Queer Affect*. Durham, NC: Duke University Press.

Clifford, James. 2001. "Indigenous Articulations." *Contemporary Pacific* 13, no. 2: 468–90. doi: 10.1353/cp.2001.0046.
Cresswell, Tim. 2004. *Place: A Short Introduction*. Hoboken, NJ: Wiley.
Eda, Haruki. 2015. "Disaster Justice: Mobilizing Grassroots Knowledge against Disaster Nationalism in Japan." In *Research Justice: Methodologies for Social Change*, edited by Andrew Jolivette. Bristol, UK: Policy Press.
Fujikane, Candace, and Jonathan Y. Okamura. 2008. *Asian Settler Colonialism: From Local Governance to the Habits of Everyday Life in Hawai'i*. Honolulu: University of Hawai'i Press.
Hau'ofa, Epeli. 1994. "Our Sea of Islands." In *Oceania: Rediscovering Our Sea of Islands*, edited by Vijay Naidu, Eric Waddell, and Epeli Hau'ofa. Suva, Fiji: School of Social and Economic Development, University of the South Pacific.
Hodgson, Dorothy, and Richard Schroeder. 2002. "Dilemmas of Counter-mapping Community Resources in Tanzania." *Development and Change* 33: 79–100.
Hoskins, Janet, and Viet Thanh Nguyen, eds. 2014. *Trans-Pacific Studies: Framing an Emerging Field*. Honolulu: University of Hawai'i Press.
Italiano, Federico. 2008. "Defining Geopoetics." *TRANS-. Revue de Littérature Générale et Comparée* 6.
Jackson, Shona N. 2012. *Creole Indigeneity: Between Myth and Nation in the Caribbean*. Minneapolis: University of Minnesota Press.
Kim, Il Men. 1978. *Chōsenjin ga naze minzokumei wo nanorunoka*. Tokyo: Sanichi Shobo.
Last, Angela. 2015. "Fruit of the Cyclone: Undoing Geopolitics through Geopoetics." *Geoforum* 64 (August): 56–64. https://doi.org/10.1016/j.geoforum.2015.05.019.
Latman, Jon. 2015. "70 Years after the War, Okinawa Protests New U.S. Military Base." August 24. *Al Jazeera America*. http://america.aljazeera.com/articles/2015/8/24/70-years-after-the-war-okinawa-protests-new-us-military-base.html.
Lie, John. 2008. *Zainichi (Koreans in Japan): Diasporic Nationalism and Postcolonial Identity*. Berkeley: University of California Press.
Massey, Doreen. 1994. *Space, Place, and Gender*. Cambridge, UK: Polity Press.
Matsushima, Yasukatsu. 2014. *Ryūkyū dokuritsu-ron: Ryūkyūminzoku no manifesuto*. Tokyo: Basilico.
Maximin, Daniel. 2006. *Les fruits du cyclone: Une géopoétique de la Caraïbe*. Paris: Seuil.
McCormack, Gavan, and Satoko Oka Norimatsu. 2012. "Ryukyu/Okinawa, from Disposal to Resistance." *Asia-Pacific Journal: Japan Focus* 10, no 1 (38) (September).
McKittrick, Katherine. 2006. *Demonic Grounds: Black Women and the Cartographies of Struggle*. Minneapolis: University of Minnesota Press.
Mills, Amy. 2012. "Critical Place Studies and Middle East Histories: Power, Politics, and Social Change." *History Compass* 10, no. 10: 778–88. doi: 10.1111/j.1478-0542.2012.00870.x.
Mirsepassi, Ali, Amrita Basu, and Frederick Weaver. 2003. "Introduction: Knowledge, Power, and Culture." In *Localizing Knowledge in a Globalizing World: Recasting the Area Studies Debate*, edited by Ali Mirsepassi, Amrita Basu, and Frederick Weaver, 1–24. Syracuse, NY: Syracuse University Press.

Miyoshi, Masao, and Harry Harootunian. 2002. *Learning Places: The Afterlives of Area Studies*. Durham, NC: Duke University Press.
Morris-Suzuki, Tessa. 2009. "Freedom and Homecoming: Narratives of Migration in the Repatriation of *Zainichi* Koreans to North Korea." In *Diaspora without Homeland: Being Korean in Japan*, edited by Sonia Ryang and John Lie, 39–61. Berkeley: University of California Press.
Neimanis, Astrida. 2017. *Bodies of Water: Posthuman Feminist Phenomenology*. New York: Bloomsbury Publishing.
Okamoto, Masataka. 2014. *Minzoku no sōshutsu: Matsurowanu hitobito, kakusareta tayōsei*. Tokyo: Iwanami.
Okinawa Prefectural Government. 2016. "U.S. Military Base Issues in Okinawa." Okinawa Prefectural Government, Washington DC Office. http://dc-office.org/basedata.
Ryang, Sonia. 1997. *North Koreans in Japan: Language, Ideology, and Identity*. Boulder, CO: Westview Press.
———. 2013. "Reading Volcano Island: In the Sixty-Fifth Year of the Jeju 4.3 Uprising." *Asia-Pacific Journal: Japan Focus* 11, no. 2 (36) (September).
Sakamaki, Shunzo. 1964. "Ryukyu and Southeast Asia." *Journal of Asian Studies* 23, no. 3: 383–89. https://doi.org/10.2307/2050757.
Shinya, Gyo. 2015. *Ainu minzoku teikō-shi*. Tokyo: Kawade Shobo Shinsha.
Siddle, Richard. 1996. *Race, Resistance and the Ainu of Japan*. Abingdon, UK: Routledge.
Smith, Andrea. 2011. "Queer Theory and Native Studies: The Heteronormativity of Settler Colonialism." In *Queer Indigenous Studies: Critical Interventions in Theory, Politics, and Literature*, edited by Qwo-Li Driskill, Chris Finley, Brian Joseph Gilley, and Scott Lauria Morgensen, 43–65. Tucson: University of Arizona Press.
Soh, C. Sarah. 2008. *The Comfort Women: Sexual Violence and Postcolonial Memory in Korea and Japan*. Chicago: University of Chicago Press.
Stratford, Elaine, Godfrey Baldacchino, Elizabeth McMahon, Carol Farbotko, and Andrew Harwood. 2011. "Envisioning the Archipelago." *Island Studies Journal* 6, no. 2: 113–30.
Thompson, Lanny. 2010. *Imperial Archipelago: Representation and Rule in the Insular Territories under U.S. Dominion after 1898*. Honolulu: University of Hawai'i Press.
Vovin, Alexander. 2009. *Man'yōshū to Fudoki ni mirareru fushigina kotoba to jōdai Nihon rettō ni okeru Ainu go no bunpu*. Kyoto: Kokusai Nihon Bunka Kenkyū Sentā.
Walker, Brett L. 2001. *The Conquest of Ainu Lands: Ecology and Culture in Japanese Expansion, 1590–1800*. Berkeley: University of California Press.
White, Kenneth. 1992. "Elements of Geopoetics." *Edinburgh Review* 88: 163–78.
Wimmer, Andreas, and Nina Glick Schiller. 2002. "Methodological Nationalism and Beyond: Nation-state Building, Migration and the Social Sciences." *Global Networks* 2, no. 4: 301–34. doi: 10.1111/1471-0374.00043.
Yoneyama, Lisa. 2016. *Cold War Ruins: Transpacific Critiques of American Justice and Japanese War Crimes*. Durham, NC: Duke University Press.
Zerubavel, Eviatar. 2003. *Terra Cognita: The Mental Discovery of America*. Piscataway, NJ: Transaction Publishers.
———. 2012. *Time Maps: Collective Memory and the Social Shape of the Past*. Chicago: University of Chicago Press.

Part V

INTER-ISLAND DYNAMISMS

Small Islands/Big Worlds

Off-Island Chamorros

Craig Santos Perez

My family migrated to California when I was 15 years old. During the first day at my new high school, the homeroom teacher asked me where I was from. "The Mariana Islands," I answered. He replied: "I've never heard of that place. Prove it exists." When I stepped in front of the world map on the wall, it transformed into a mirror: the Pacific Ocean, like my body, was split in two and flayed to the margins. I found Australia, then the Philippines, then Japan. I pointed to an empty space between them and said: "I'm from this invisible archipelago." Everyone laughed. And even though I descend from oceanic navigators, I felt so lost, shipwrecked

on the coast of a strange continent. "Are you a citizen?" he probed. "Yes. My island, Guam, is a U.S. territory." We attend American schools, eat American food, listen to American music, watch American movies and television, play American sports, learn American history, and dream American dreams. "You speak English well," he proclaimed, "with almost no accent." And isn't that what it means to be a diasporic Chamorro: to feel *foreign in a domestic sense*.

Over the last 50 years, Chamorros have migrated to escape the violent memories of war; to seek jobs, schools, hospitals, adventure, and love; but most of all, we've migrated for military service, deployed and stationed to bases around the world. According to the 2010 census, 44,000 Chamorros live in California, 15,000 in Washington, 10,000 in Texas, 7,000 in Hawaii, and 70,000 more in every other state and even Puerto Rico. We are the most "geographically dispersed" Pacific Islander population within the United

States, and off-island Chamorros now outnumber
our on-island kin, with generations having been born
away from our ancestral homelands, including my daughter.

Some of us will be able to return home for holidays, weddings,
and funerals; others won't be able to afford the expensive plane
ticket to the Western Pacific. Years and even decades might pass
between trips, and each visit will feel too short. We'll lose contact
with family and friends, and the island will continue to change
until it becomes unfamiliar to us. And isn't that, too, what it means
to be a diasporic Chamorro: to feel foreign in your own homeland.

And there'll be times when we'll feel adrift, without itinerary
or destination. We'll wonder: What if we stayed? What if we return?
When the undertow of these questions begins pulling you
out to sea, remember: migration flows through our blood
like the aerial roots of i trongkon nunu. Remember: our ancestors
taught us how to carry our culture in the canoes of our bodies.
Remember: our people, scattered like stars, form new constellations
when we gather. Remember: home is not simply a house,
village, or island; home is an archipelago of belonging.

Chapter Seventeen

"Together, but Not Together, Together"

The Politics of Identity in Island Archipelagoes

Godfrey Baldacchino

I am writing while staying for a few weeks in Grenada, a "small island developing state" in the Caribbean, with a total population of around 105,000. A closer look reveals that Grenada—the country known as "the Island of Spice"—is more than just Grenada (the main island). As its last two suggestions in its "Top 10 Places to Visit in Pure Grenada," TripAdvisor recommends visiting two other islands within the country. The first is Carriacou (population: 4,600), "known as 'the isle of reefs,' with shallow clear waters: ideal for snorkelling, and surrounding offshore islands, perfect for a tranquil getaway." The other is Petite Martinique (population nine hundred): we are told that the "residents of Petite Martinique and C'cou are fiercely traditional, welcoming you to discover the traditional customs of African ancestry at annual events like the Carriacou Maroon and String Band Music Festival."[1]

The implications are self-evident: on a website like this, Grenada will showcase itself as a tourism destination, and Carriacou and Petite Martinique are not meant to miss out on the action. (There are various other islets in the country; all are uninhabited.)

Grenada welcomed some 528,000 visitors in 2018 (latest year for which complete statistics are available); 65 percent were cruise ship travelers, spending a few hours on the island and often venturing only as far as the nearest downtown shopping mall or the spectacular Grande Anse beach, a ten-minute taxi or bus ride from the harbor.[2] Carriacou is hardly touched by tourism, and Petite Martinique even less so: this may explain why the islands that are a ferry ride or short flight away from the "mainland" of Grenada are presented as unspoiled and traditional.

Not so in the official imaginary: the six stars in the red border on the country's flag, adopted upon independence from the United Kingdom in 1974, represent the six parishes that are found on the main island, while the

Figure 17.1. The flag of Grenada: designed by Anthony C. George and adopted on February 7, 1974. *Source*: http://www.gov.gd/our_nation/flag.html

larger middle star, encircled by a red disk, represents Carriacou and Petite Martinique (see Figure 17.1). In spite of its small resident population (in both relative and absolute terms), Carriacou has birthed two of Grenada's heads of state: Chief Minister Herbert Blaize and Prime Minister Sir Nicholas Brathwaite. Elvin Nimrod, of the New National Party, in power since 2013, is the representative for the Carriacou and Petite Martinique constituency and is the minister for Carriacou and Petite Martinique affairs.

The locals call their archipelago the "tri-islands." When still in early-childhood centers and kindergartens, Grenadian children are reminded that they live in an archipelago, where some islands are smaller or bigger than others. A short, cute poem is written on a wall poster in various early-childhood settings. It doubles as an English-language lesson in how to use the comparative and superlative:

> Somewhere in the blue Caribbean Sea, lies a chain of islands three.
> The first one said: I am big, you see.
> The second one said: I am bigger than you.
> The third one said: I am the biggest of the three.
> Grenada, Carriacou and Petite Martinique.
> They are three islands in the sea.

As with Grenada, the rest of the world follows, as this chapter sets out to argue. The "world of islands"[3] is the plurality for a complex and often unacknowledged interplay of centrality and peripherality, status upgrades and downgrades, visibility and negligence *between* islands, enacted by their politico-economic elites. In this endeavor, references to identity and identity politics are rife, possibly culminating in threats or initiatives in favor of that ultimate schism: secession. After all, an island is the quintessential single jurisdiction, thus preordained by nature. Its geography suggests a unitary polity: indeed, islands that are shared by or between more than one country tend to have uneasy relationships.

Although we discuss and refer to islands, as in "island tourism" or "island studies," the reality "on the ground" is often one of archipelagoes: of multiple islands in turbid relationship. This observation flies in the face of a widespread yet confusing practice that acts to camouflage the main islands of archipelagoes *as if they were single islands*, by officially or colloquially ascribing the whole archipelago with the same name as that of its largest component or the one that contains its main settlement. Åland, Chatham, Hawai'i, Iceland, Madagascar, Madeira, Montreal, Pitcairn, Socotra, Svalbard, Tasmania, and Zanzibar . . . the list goes on. Grenada undergoes a similar conflation: that is both the name of the country as well as the name of its main island.

A careful examination of the world's islands reveals that most are actually members of archipelagoes, assemblages that range from a minimum size of two (say, Samoa) to over fifty thousand in the island-dense archipelago sea of Southwest Finland, with the "mainland" of Åland—but itself an island—being its largest component.[4] There are very few island jurisdictions that really are nothing but single land blocks: the Pacific small island state of Nauru is one of these. Even places that might appear as unitary islands to a casual observer may have rocks offshore that are significant components of their respective tourism appeal—one should never miss an opportunity to frame and lock an offshore geological feature into the touristic imaginary. Hence the photogenic lure of Petra tou Romiou, alleged birthplace of the goddess Aphrodite, near Paphos, Cyprus,[5] and Bathsheba (Mushroom) Rock, at Bathsheba Beach, on the wild east coast of Barbados.[6] And what appears to us today as the single landmass of Bermuda is actually a series of discrete islands interconnected by bridges; there used to be as many as 123 islands at one point: some were destroyed to create UK and US military bases.[7]

Contemporary scholarship tends to settle upon two rather overworked topological relations of islands. The first presents a clear focus on an island's singularity and exotic signature: its unique history, culture, and biota, crafted by evolution, migration, and invasion and inscribed by the border between *land* and *sea*. The second is relational and distinguishes an island from a *mainland/continent*: it dwells on its differences from, tensions with, and dependencies on the larger player. The concept of the archipelago evokes a third topological relation that now has its own robust intellectual genealogy. It foregrounds interactions between and among islands themselves. The relation of "island-to-island" is characterized by repetition and assorted multiplicity,

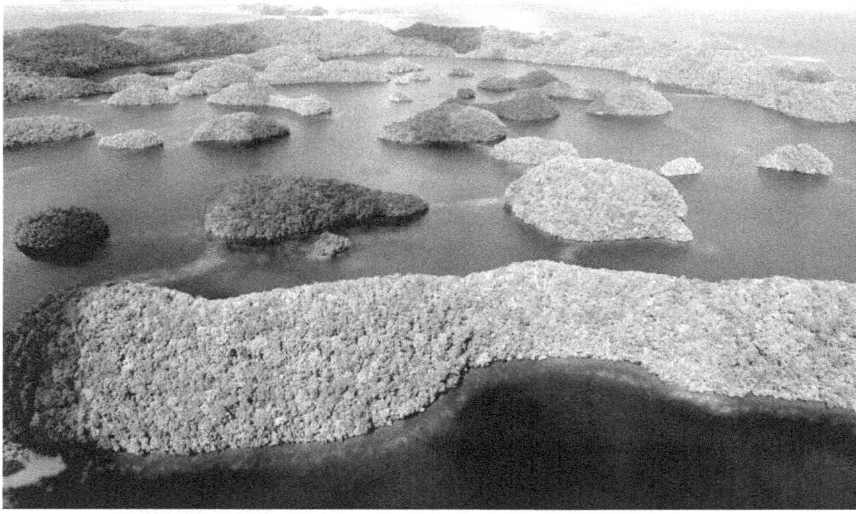

Figure 17.2. A detail of the archipelago of Palau. *Source*: Photograph by Lux Tonnerre, Wikimedia Commons, https://commons.wikimedia.org/wiki/File:Palau_archipelago.jpg

which intensify, amplify, and disrupt relations of both land and water, as well as island and continent/mainland.[8]

The archipelago is a component of the study of islands that is well established, especially within the natural sciences (see Figure 17.2). Thus, the juxtaposition of variation and similarity among the finches on the islands in the Galápagos archipelago played a key role in sharpening Charles Darwin's ideas on species evolution, speciation, and natural selection.[9] The Aru Islands of the Indonesian archipelago had a similar, inspirational effect on Alfred Russell Wallace.[10] Other examples have included the study of birds in Indonesia,[11] Hawaiian honeycreepers,[12] and Croatian lizards.[13]

Taking this powerful perspective systematically beyond fauna and flora and into social science is a recent undertaking[14] and indeed explains the timeliness of this volume. This is now a core concern of the international law of the sea, where the island members of an archipelago are located in such proximity to each other that they can naturally be seen as an integral unit, assigning them specific rights and obligations, such as designating sea lanes.[15] An archipelagic imagination has also been deployed in order to analyze the politics of the Azores, Portugal.[16] Still in political geography, an archipelagic lens adds fresh insights to such complex jurisdictions as Taiwan;[17] the United Kingdom;[18] Australia and its shifting identity as island, continent, nation, and archipelago;[19] the United States;[20] and Japan: the latter's identity as a *shima-*

guni (island nation) has a bearing on how it tackles geopolitical tensions in the East and South China Seas.[21]

THINGS ARE NOT ALWAYS WHAT THEY SEEM

If, then, what appear to be single islands are mostly and actually archipelagoes, then physical separation separates them not just from mainlands but also from each other. Thus, crossings between (especially inhabited) islands, often at levels of demand that preclude profitability, will be hot items on the local political agenda. Cultural differences between islanders within the same island cluster may be enhanced and exaggerated, even invented, in order to appeal to specific tourist niches or political status; a stretch of water can make a world of difference.

In an archipelago, the ability to attract tourists may need to stretch to include the ability to share and spread the tourist dollar to the far-flung corners of the island group. And an archipelago's political establishment would need to reconcile the need to offer a sharp, strong, consistent, and clearly identifiable brand as a single destination with the need to allow each of the islands within the archipelago to develop and express its own voice and identity. The articulation of a harmonious complementarity of a plurality of voices and interests, often in acute competition for scarce fiscal and spatial resources, will not align with notions of an earthy island paradise, but most tourists would be oblivious to this cacophonic and complex power play.

Concurrently, separate island identities can breed equally distinct political movements that will engage with "mainland" parties to extract concessions and more autonomy, but not necessarily to seek outright independence.[22] While often in a clear voter minority, secessionist or autonomist parties can still smartly and successfully shift mainstream politics to accommodate their aspirations, in part or in full.

The most common definition of an archipelago is simply as "a group of islands." This construct is simple, yet powerful. Archipelagic formations are common: planet Earth (with its two supercontinents of Eurasia-Africa and North-Central-South America) is one example.[23] The significance of the assemblage, however, goes beyond that of a simple gathering, collection, or composition of things that are believed to fit together. Much like the thinking behind systems theory, assemblages emphasize fluidity and exchange: they actively map out, select, piece together, and allow for the conception and conduct of individual units as members of a group.[24] An archipelago is similar: its framing as "such and such an assemblage" draws our attention to the ways in which practices, representations, experiences, and affects assemble and reassemble to take particular dynamic forms at specific historical moments. Archipelagos are fluid cultural constructs, sites of abstract and material relations of movement and rest, dependent on changing conditions of articulation or connection.[25]

CENTRIFUGAL TENDENCIES

The archipelago is referred to as a "societal subtype" in a short but seminal contribution,[26] where the author offers a more nuanced definition. He proposes "four major attributes" of the archipelago state. These involve (1) a "large number of islands," (2) small and underdeveloped component islands, (3) surrounding waters that are considered within the assemblage's boundaries and an integral part of its heritage and territory, and (4) a centrifugal tendency: "in an archipelago, the temptation is always great, at worst to secede and at best to disregard the political jurisdiction of the centre."[27]

This "inter-island and intra-archipelago" dynamic strikes right at the heart of invariably uneven "island-to-island" politics. How do multi-island jurisdictions, especially if run by democratically elected politicians, balance the wishes of their various island publics and constituencies with the rationale of hub-and-spoke transport logistics (versus costly repetitive infrastructure), tourism differentiation (versus repetition), and complementary (rather than similar) and cooperative (rather than competitive) economic development trajectories? Most islanders implicitly know that they experience tense relations with their island neighbors. Often made fun of in popular idiom, these tensions and rivalries may find expression in discriminatory practices of various kinds, official or otherwise. These strains become more likely if specific islands claim, or are represented as claiming, a linguistic, racial/ethnic, religious, historical, occupational, and/or economic status and identity distinct from those pertaining to other islands. Island nationalism knows no limits, as various attempts at secession demonstrate. A case in point is the Federated States of Micronesia (FSM): it may be the world's smallest federation and the nineteenth-smallest country in the world (land area: 702 square kilometers; population: 105,000), but this has not quashed secessionist aspirations on Chuuk, one of the FSM's four constituent "states," which the federate model was meant to appease.[28]

Perhaps the most dramatic of all archipelagic fissions is the unraveling of the West Indies Federation (WIF) (see Figure 17.3). This was a bold plan by London that would have seen its former Caribbean territories band together as one jurisdiction. But, at the end of the day, the centrifugal "Tuvalu effect" (see further below) could not be stopped: island nationalism trumped eco-

Figure 17.3. Flag of the short-lived West Indies Federation (1958-1962). *Source*: Wikipedia, https://en.wikipedia.org/wiki/Flag_of_the_West_Indies_Federation

nomic rationalism, also because the political class had island-specific (and not region-wide) power bases and constituencies. The largest member (Jamaica) absconded first; then the next largest (Trinidad and Tobago); then the next largest (Barbados): the separate island units opted for sovereignty rather than having to carry along a family of smaller and less developed nations.[29] As a regional institution, the University of the West Indies is one of the surviving vestiges of the WIF. But, even here, and in spite of diseconomies of scale, nationalist aspirations are ascendant: many Caribbean small states have their own national higher education institutions.

The act of naming a country bears witness to these tribulations. In poly-island jurisdictions, no island is *usually* large enough to expect any special status; so generic names usually serve their purpose. No better example here than the seventeen-thousand-island Indonesia (literally, Indian islands), the world's most populated archipelago state, or the seven-thousand-island Philippines (which, by the way, may eventually get to change its name from that of a former king of Spain).

However, the politics of identity can be fanned by fear of what the future might hold. The Polynesian citizens of the Ellice Islands were uncomfortable being shepherded to independence along with their Gilbert Island neighbors. A referendum cemented the divorce, and London was obliged to comply, though initially reticent, given the diminutive size of the wannabe country: Tuvalu—literally eight (islands) standing together—was born as an independent nation of twelve thousand citizens in 1978. Kiribati (pronounced Kiribass, from the English word "Gilberts") followed as its own independent small island state in 1979.[30]

Figure 17.4. A mechanical shovel at the Panguna copper mine. *Source*: Photo by Robert Owen Winkler, Wikimedia Commons, https://commons.wikimedia.org/wiki/File:Bougainville_Panguna_mine_shovel.jpg

Still in the Pacific lies the high-profile example of Bougainville, notionally part of Papua New Guinea, which has been ravaged by a violent civil war triggered by what was felt to be an unfair share of proceeds from the giant Panguna open-cut copper mine on the island (see Figure 17.4). A fragile peace has held, but a secession referendum is imminent.[31]

Not too far away, lies the eighty-three-island archipelago state known today as Vanuatu, which, until securing independence in 1980, was a British-French condominium known as the New Hebrides/Nouvelles Hebrides. In the late 1970s, as the islands moved toward independence, various secessionist movements emerged, mainly on the island of Espiritu Santo but also on Aoba, Malakula, and Tanna. They tended to find support among French-speaking residents, who feared becoming citizens of a new state led by Anglophones and so hoped for another dramatic, Tuvalu-style secession. In May 1980, members of the Nagriamel movement, led by a Jimmy Stevens, took control of Santo town and proclaimed a provisional government (see Figure 17.5). The insurrection was short-lived, however. Negotiations proved inconclusive, and two hundred British and French troops intervened and took over the town in July; these were eventually replaced by troops from Papua New Guinea.[32]

Figure 17.5. Jimmy Stevens and supporters, June 1980. *Source*: http://eyecontactsite.com/media/thumbs/uploads/2017_05/Jimmy_Stevens_June_1980_jpg_380x125_q85.jpg

BEWARE OF TWO, OR EVEN THREE

Expect more serious and more common problems to arise, however, when the country consists in just two or three populated islands: the smallest possible archipelago. The names of various island states bear testimony to attempts, many ongoing, at an awkward reconciliation of internal difference.

There is no better example than the intended tri-island country of St. Kitts-Nevis-Anguilla, until 1967 a British colony in the Caribbean (see Figure 17.6). Smelling then that plans for independence were threateningly in the air and preferring to remain a colony of London rather than to become a colony of Basse-Terre (the capital of St. Kitts), the Anguillans (some five thousand at the time) proceeded to disrupt the ongoing attempts at state building. The members of the St. Kitts police force were evicted from the island, a secession referendum was passed (1,813 votes in favor; 5 against), and a declaration of independence was drafted and read. The request was eventually granted, and Anguilla became a UK overseas territory, while St. Kitts-Nevis proceeded to become a two-island sovereign state.[33] Note that Anguilla itself is an archipelago: its largest subcomponent, Scrub Island (eight square kilometers), is privately owned.

Even more drama was in store here. Nevis joined St. Kitts in independence in 1983, but only after having secured an "exit clause" in the new country's constitution. Thus, it could secede, as long as such a momentous decision would be approved by a two-thirds majority of voters via referendum. The

Figure 17.6. A 1969 stamp from now defunct St Kitts-Nevis-Anguilla. *Source*: Cricket Philatelic Society, http://www.cricketstamp.net/associated2.htm

threshold looks unsurmountable; but Nevis (population twelve thousand) failed only by a whisker to pass a referendum in 1998 that would have seen it secede from St. Kitts (population thirty-two thousand).[34]

Next, consider Antigua and Barbuda. In 1981, Barbudan leaders, with an eye on the Anguilla episode, sought to convince the British government to sever their island from an independent Antigua and maintain its status as a British territory. A compromise was only secured by allocating significant autonomy to a local Barbuda Council.[35] A Barbuda People's Movement continues to advance the notion of sovereignty for the island. The recent repeal of the 2007 Barbuda Land Act has the local population (sixteen hundred) once again threatening secession and independence[36] (see Figure 17.7).

Figure 17.7. A 1968 stamp showing the island of Barbuda. *Source*: https://stampaday.wordpress.com/2016/08/31/barbuda-12-1968

Figure 17.8. The Tobago House of Assembly in session. *Source*: CNC3 TV, http://www.cnc3.co.tt/press-release/tobagonians-go-polls-january-23rd-tha-elections

There are also separatist tendencies in Tobago, which is part of the twin island state of Trinidad and Tobago. The quest for self-determination has been asserted at various times; it comes and goes as (mainly Afro-Caribbean) Tobagonians, who generally regard themselves as different from (mainly Indo-Caribbean) Trinidadians, react to events that impact adversely on their lives, often accusing Port of Spain with discrimination, neglect, and indifference.[37] Since 1996, the island of Tobago (population sixty thousand) has had its own twelve-member House of Assembly (see Figure 17.8), responsible for the management of various internal affairs, including sports and culture, fishery, state land, food production and agriculture, tourism, and health services.[38] These provisions may have been intended to assuage Tobago's quest for more autonomy. However, there has been a passionate call for internal self-government by the people of Tobago, who hold the view that "Tobago needs to stand side-by-side with Trinidad and not be an appendage or a ward of Trinidad."[39]

Beyond the Caribbean lies the West African bi-island microstate and former Portuguese colony of São Tomé y Príncipe (total population around 190,000). After securing a smooth transition to independence in 1975, the smaller island of Príncipe (population seven thousand) continued to feel the adverse consequences of its double insularity. In December 1981, a lack of food supplies provoked a revolt against the government in São Tomé, which was accompanied by secessionist slogans. In 1995, a few years after the country's democratic transition, Príncipe became an autonomous region: the island's first regional government was elected in the same year.[40]

The smallest member of the European Union is my home country of Malta (population: 420,000), which consists of three populated islands: the main island, also called Malta (no surprise here), Comino (now with just three year-round residents), and Gozo (around 30,000). The last of these is separated from the main island by an eight-kilometer channel. Gozo has no airport, and its plight as a victim of double insularity has the ear of the Maltese government (which has also installed a Ministry for Gozo, with a Gozitan as minister). Gozo enjoys, and has enjoyed intermittently throughout its history, a clutch of jurisdictional privileges that no other subterritory of the Maltese islands has ever entertained, let alone benefitted from. This is because the "mainland" Maltese have a love-hate relationship with Gozo's autonomy. The allegedly parochial and money-wise Gozitans are the regular butt of mainland Malta-driven jokes, commentaries, and proverbs, and most mainland Maltese cannot fathom why the residents of Gozo have their own special Gozo-Malta ferry fares and their own identity cards, not to mention their own minister[41] (see Figure 17.9). The proposed construction of a thirteen-kilometer Gozo-Malta undersea tunnel has reignited debates about Gozitan identity and the smaller island's development prospects.[42]

Politicians may pander to the presumed national distinctiveness of the Gozitans; however, they have been extra careful not to let the word "Gozo"

Figure 17.9. Ferries, fishing boats and pleasure craft at Mġarr, the main harbor of Gozo, with the main island of Malta in the background, 8km away. *Source*: Wikimedia Commons, https://commons.wikimedia.org/wiki/File:Malte,_Gozo,_port_de_Mgarr_%26_ferrys_%26_Comino.jpg

creep into the 1974 republican constitution, except in the definition of what territory constitutes "the Maltese state" and in finally recognizing Gozo (and Comino) as a single electoral district in a 2007 revision.[43] This is arguably because Gozo has serious claims to autonomy. During the Roman occupation, Gozo was its own *municipium*, autonomous from Malta, with a republican sort of government that minted its own coins.[44] From 1798 to 1800, the island enjoyed another brief period of formal autonomy, granted by French general Napoléon Bonaparte after his conquest of the Maltese Islands.[45] Ecclesiastically, it has been a diocese separate from Malta, with its own bishop, since 1864.[46] Gozo had its own Civic Council from 1961 to 1973, at a time when "mainland" Malta had no local government. It has claims to a distinct culture[47] and a distinct dialect.[48] Claims for the formal recognition of Gozo as a subnational jurisdiction have come and gone and are likely to persist.[49]

In archipelagoes, politicians and their publics often anchor their aspirations, affiliations, and identities to specific island geographies, culminating in expressions of statehood/nationhood that are ultimately sealed via independence. Thus, there may be a Caribbean region—as the world's best-known and -branded tourism playground, for example—but the region is articulated as comprising Antiguans, Barbadians, Dominicans, Trinidadians, and so many more, and this is just in the Anglo-speaking world (there are also US, Spanish, Dutch, and French island territories in the region). Serious attempts at federation have not waned: the Organization of Eastern Caribbean States (OECS) provides one successful example of how small island states combine resources and usurp the limitations of scale by crafting *confederal* institutions of shared governance, which therefore do not threaten the national basis of power politics.[50] The OECS slate of institutions includes a common currency, central bank, regional airspace, and court of highest appeal (see Figure 24.10). Such "bottom-up," transterritorial approaches to governance and policy—rather than federal, "top-down" ones, as in the case of the doomed West Indies Federation—are probably critical for long-term economic survival.

Figure 17.10. Logo of the Organisation of Eastern Caribbean States (OECS). *Source*: www.oecs.org

Island studies is part of the spatial turn in the social sciences that rejects the non- and antimaterialism that dominates the wave of postmodernity.[51] Geography still matters: ask Mexicans wanting to enter the United States or sub-Saharan migrants wanting to move to Europe. Of course, one should be careful not to essentialize islands, seeing only insular dynamics when internal island politics—like any other politics—are riddled with their own rich tensions and divisive practices. Yet archipelagic movements are a clear example of the richly tumultuous ways in which island units converge . . . and diverge too. As Errol Barrow, former prime minister of Barbados, once said, "We live together very well, but we don't like to live together together."[52]

PRACTICAL AND METHODOLOGICAL IMPLICATIONS

Taking its cue from the archipelagic state of Grenada, this chapter offers a global review of episodes of political tension that have arisen in archipelagic jurisdictions in recent decades. For all the hesitation to acknowledge the relevance of material geography among many social scientists, and in spite of the contrarian arguments of hard-nosed economic rationalists, the evidence on the ground is galling. The "archipelagic turn" helps to identify and assert the significance of inter-island dynamism and island-based nationalism, which continue to enjoy clean bills of health.

Deploying an archipelagic imagination illuminates and transforms identity theory beyond the nation-state framework. By definition, states will be keen to propose and present a unitary narrative, complicit in the social construction of nationalism. As Count Azeglio famously wrote in the 1860s, after the unification of Italy, "Italy is now made; we now need to produce Italians."[53] And yet, even after 160 years in the making, Italy still struggles with strong pockets and sentiments of regionalism, including autonomist movements in both its island provinces, Sicily and Sardinia.[54] This chapter reminds us, additionally, that both Sardinia and Sicily are themselves archipelagoes: they are the respective mainlands of even smaller islands where a distinctive type of politics prevails.[55]

In a landmark text, geographers Martin Lewis and Kären Wigen issued a powerful challenge, contesting and rebuking a dominant continent-oriented model of human geography and culture. Instead, they advocated the merits of "devising a *creative cartographic vision* capable of effectively grasping unconventional regional forms. . . . [A]nalyzing contemporary human geography requires a different vocabulary. Instead of assuming contiguity, we need a way to visualize discontinuous 'regions' that might take the spatial form of lattices, archipelagos, hollow rings, or patchworks" (emphasis in original).[56] The challenge is an ongoing one. Meanwhile, the archipelago is part of the set of mnemonic and diagnostic tools that lend themselves to this "creative cartographic vision."[57] Visualizing discontinuity sounds messy, even contradictory, but that is no excuse for not trying.

It would be fitting to end here by returning to the tri-island state of Grenada. The current New National Party government, elected in 2013, has proposed renaming the country as "Grenada, Carriacou and Petite Martinique." This was one of seven propositions that were put to a referendum vote in November 2016, the first in Grenada's history. This particular amendment was intended to respect the three-populated-island nature of the jurisdiction, while specifying constitutionally the areas that formed part of Grenada upon its independence in 1974—just in case someone had second thoughts.[58] The measure, along with the other six propositions, was summarily rejected by an apathetic and antagonistic public: only 43.7 percent voted in favor, with a voter turnout of just 32.5 percent. It appears that the Grenadian voting public is not entertaining a more explicit (and more unwieldy) recognition of its archipelagic jurisdiction just yet.

NOTES

1. "Top 10 Places to Visit in Pure Grenada," Trip Advisor, 2017, https://www.tripadvisor.com/Destinations-g147295-o52482-Grenada.html.

2. "Visitor Arrivals Increased by 12.7% in 2015," Pure Grenada, 2016, http://www.grenadagrenadines.com/news/visitor-arrivals-to-grenada-increased-by-12.7-in-2015.

3. G. Baldacchino, ed., *The Routledge International Handbook of Island Studies: A World of Islands* (London: Routledge, 2018).

4. C. Depraetere and A. L. Dahl, "Island Locations and Classifications," in *A World of Islands: An Island Studies Reader*, ed. G. Baldacchino, 57–105 (Charlottetown, Canada, and Luqa, Malta: Institute of Island Studies, University of Prince Edward Island and Agenda Academic, 2007).

5. "Aphrodite's Rock, Cyprus," Sacred Destinations, 2017, http://www.sacred-destinations.com/cyprus/aphrodites-rock-petra-tou-romiou.

6. "Mushroom Rock Barbados: Fairy Tales," video posted to YouTube by Victoria Fitton on May 13, 2013, https://www.youtube.com/watch?v=kxZJSBAYbXg.

7. "Welcome to Bermuda," Bermuda Online, 2017, http://www.bermuda-online.org/islands.htm.

8. G. Baldacchino, "More Than Island Tourism: Branding, Marketing and Logistics in Archipelago Tourist Destinations," in *Archipelago Tourism: Policies and Practices*, ed. G. Baldacchino, 1–18. London: Routledge, 2015.

9. J. Weiner, *The Beak of the Finch: A Story of Evolution in Our Time* (New York: Vintage, 2014); P. R. Grant and B. R. Grant. "Speciation and Hybridisation in Island Birds." *Philosophical Transactions of the Royal Society B* 351 (1996): 765–72.

10. D. Quammen, *The Song of the Dodo: Island Biogeography in an Age of Extinctions* (New York: Random House, 1996).

11. E. Mayr, *Systematics and the Origin of Species* (New York: Columbia University Press, 1942).

12. W. L. Wagner and V. A. Funk. *Hawaiian Biogeography: Evolution on a Hot Spot Archipelago* (Washington, DC: Smithsonian Institution Press, 1995).

13. "Lizards Rapidly Evolve after Introduction to Island," *National Geographic*, April 21, 2008, http://news.nationalgeographic.com/news/2008/04/080421-lizard-evolution.html.

14. J. Pugh, "Island Movements: Thinking with the Archipelago," *Island Studies Journal* 8, no. 1 (2013): 9–24; E. Stratford et al., "Envisioning the Archipelago," *Island Studies Journal* 6, no. 2 (2011): 113–30; E. Stratford, "The Idea of the Archipelago: Contemplating Island Relations," *Island Studies Journal* 8, no. 1 (2013): 3–8.

15. D. R. Rothwell, "UNCLOS Navigational Regimes and Their Significance for the South China Sea," in *UN Convention on the Law of the Sea and the South China Sea*, edited by S. Wu, M. Valencia, and N. Hong, 149–89 (Farnham, UK: Ashgate, 2015).

16. L. Marrou, "Quand l'île cache l'archipel: L'inscription des îles-escales dans l'archipel des Açores," in *Les dynamiques contemporaines des petits espaces insulaires: De l'île-relais aux réseaux insulaires*, ed. L. Marrou, N. Bernardie, and F. Taglioni, 181–97 (Paris: Karthala, 2005).

17. G. Baldacchino and H.-M. Tsai, "Contested Enclave Metageographies: The Offshore Islands of Taiwan," *Political Geography* 40, no. 1 (2014): 13–24.

18. J. G. A. Pocock, *The Discovery of Islands* (Cambridge: Cambridge University Press, 2005), 29.

19. S. Perera, *Australia and the Insular Imagination: Beaches, Borders, Boats and Bodies* (New York: Palgrave Macmillan, 2009).

20. B. R. Roberts and M. A. Stephens, eds., *Archipelagic American Studies* (Durham, NC: Duke University Press, 2017).

21. J. C. Suwa, "Shima and Aquapelagic Assemblages," *Shima: The International Journal of Research into Island Cultures* 6, no. 1 (2012): 12–16; G. Baldacchino et al., *Solution Protocols to Festering Island Disputes: "Win-Win" Solutions for the Diaoyu/Senkaku Islands* (London: Routledge, 2017).

22. E. Hepburn and G. Baldacchino, eds., *Independence Movements in Subnational Island Jurisdictions* (London: Routledge, 2016).

23. J. Diamond, Guns, Germs and Steel*: The Fates of Human Societies* (New York: W. W. Norton, 1997).

24. G. Deleuze and F. Guattari, *Mille plateaux* (Paris: Minuit, 1980). Translated by B. Massumi as *A Thousand Plateaus* (Minneapolis: University of Minnesota Press, 1987).

25. E. Stratford et al., "Envisioning the Archipelago," *Island Studies Journal* 6, no. 2 (2011): 113–30.

26. A. G. LaFlamme, "The Archipelago State as a Societal Subtype," *Current Anthropology* 24, no. 3 (1983): 361–62.

27. La Flamme 1983, 361.

28. C. Y. Mulalap, "Federated States of Micronesia," *Contemporary Pacific* 28, no. 1 (2016): 172–84.

29. E. Wallace, *The British Caribbean from the Decline of Colonialism to the End of Federation* (Toronto: University of Toronto Press, 1977); S. Mawby, *Ordering Independence: The End of Empire in the Anglophone Caribbean, 1947–69* (New York: Palgrave Macmillan, 2012).

30. B. K. Macdonald, "The Separation of the Gilbert and Ellice Islands," *Journal of Pacific History* 10, no. 4 (1975): 84–88; W. D. McIntyre, "The Partition of the Gilbert and Ellice Islands," *Island Studies Journal* 7, no. 1 (2012): 135–46.

31. J. Woodbury, "The Bougainville Independence Referendum: Assessing the Risks and Challenges before, during and after the Referendum," *Australian Defence College, Centre for Defence and Strategic Studies, Indo-Pacific Strategic Paper* 1 (2015).

32. J. Beasant, *The Santo Rebellion: An Empirical Reckoning* (Richmond, Australia: Heinemann, 1984).

33. D. Westlake, *Under an English Heaven* (New York: Simon & Schuster, 1972), 78–79; K. F. Olwig and B. Dyde, *Out of the Crowded Vagueness: A History of the Islands of St. Kitts, Nevis and Anguilla* (New York: Interlink Publishing Group, 2009).

34. R. Premdas, "Self-Determination and Secession in the Caribbean: The Case of Nevis," in *Identity, Ethnicity and Culture in the Caribbean*, ed. R. Premdas, 447–84 (Saint Augustine, Trinidad, and Tobago: School of Continuing Studies, University of the West Indies, 2000).

35. J. Minahan, *Encyclopaedia of the Stateless Nations: Ethnic and National Groups around the World* (Westport, CT: Greenwood Publishing Group, 2002), 2:268–69.

36. "Gov't Says an Independent Barbuda Is Not Feasible," *Daily Observer*, June 15, 2016, http://antiguaobserver.com/govt-says-an-independent-barbuda-is-not-feasible.

37. R. R. Premdas and H. Williams, "Self-Determination and Secession in the Caribbean: The Case of Tobago," *Canadian Review of Studies in Nationalism* 19, no. 1–2 (1992): 117–27.

38. "About the Assembly," Tobago House of Assembly, http://www.tha.gov.tt/about-the-assembly.

39. T. L. Roberts, "Tobago's Quest for Internal Self-Government," *Your Commonwealth: Youth Perspective*, December 16, 2016, http://www.yourcommonwealth.org/social-development/democracy-participation/tobagos-quest-for-internal-self-government.

40. G. Seibert, "São Tomé and Príncipe, 1975–2015: Politics and Economy in a Former Plantation Colony," *Estudos Ibero-Americanos* 42, no. 3 (2016): 888–912.

41. G. Baldacchino, ed., *A World of Islands: An Island Studies Reader* (Charlottetown, Canada, and Luqa, Malta: Institute of Island Studies, University of Prince Edward Island and Agenda Academic, 2007).

42. A. Attard, "The Social Impacts of a Fixed Link between Malta and Gozo" (unpublished BA thesis, University of Malta, 2018), https://www.um.edu.mt/library/oar//handle/123456789/36985.

43. C. Micallef-Borg, "Parliament: Constitution Amendment Bill," *Malta Independent*, September 27, 2007, http://www.independent.com.mt/articles/2007-09-27/news/parliament-constitution-amendment-bill-197362.

44. "Island of Gozo," Gozo Tourism Association, 2017, http://islandofgozo.org/about/history-of-gozo.

45. J. Bezzina, *Gozo's Government: The Autonomy of an Island through History* (Gozo: Gaulitana & Gozo Local Councils Association, 2005).

46. J. Bezzina, *Religion and Politics in a Crown Colony: The Gozo-Malta story, 1798–1864* (Malta: Bugelli Publications, 1985).

47. L. Briguglio and J. Bezzina, eds., *Gozo and Its Culture* (Malta: Formatek, 1995).

48. J. Xuereb, "A Sociolinguistic Analysis of Select Linguistic Features in Gozo" (unpublished MA thesis, University of Malta, 1996).

49. K. Sansone, "Airport and Elected Assembly for Gozo?," *Times of Malta*, March 29, 2017, https://www.timesofmalta.com/articles/view/20170329/local/airport-and-elected-assembly-for-gozo.643838.

50. S. Roberts, J. N. Telesford, and J. V. Barrow, "Navigating the Caribbean Archipelago: An Examination of Regional Transportation Issues," in *Archipelago Tourism: Policies and Practices*, ed. G. Baldacchino, 147–62 (Ashgate, UK: Farnham, 2015).

51. Pugh, "Island Movements.

52. E. Barrow, "A Role for Canada in the West Indies," *International Journal* 19, no. 2 (1964): 181.

53. S. Stewart-Steinberg, *The Pinocchio Effect: On Making Italians, 1860–1920* (Chicago: University of Chicago Press, 2007), 1.

54. N. Hansen, "The New Regionalism and European Economic Integration," in *Regional Development: Problems and Policies in Eastern and Western Europe*, edited by G. Demko, 57–82 (London: Routledge, 2017).

55. G. Baldacchino, "Lingering Colonial Outlier Yet Miniature Continent: Notes from the Sicilian Archipelago," *Shima: The International Journal of Research into Island Cultures* 9, no. 2 (2015): 89–102; R. Cannas and E. Giudici, "Tourism Relationships between Sardinia and Its Islands: Collaborative or Conflicting," in *Archipelago Tourism: Policies and Practices*, ed. G. Baldacchino, 67–81 (Aldershot, UK: Ashgate, 2016).

56. M. W. Lewis and K. Wigen, *The Myth of Continents: A Critique of Metageography* (San Francisco: University of California Press, 1997), 200.

57. Lewis and Wigen, *The Myth of Continents.*

58. "Grenada Constitutional Reform," Now Grenada, 2016, http://nowgrenada.com/2016/09/fact-sheet-grenada-constitution-reform.

Chapter Eighteen

Small Islands, Large Radio

Archipelagic Listening in the Caribbean

Jessica Swanston Baker

When I began fieldwork in St. Kitts-Nevis in the summer of 2010, many of my first interactions with interlocutors started with a lighthearted back-and-forth centered on a question about whether there was anything at all worthy of "serious" study on those islands. During one of the first interviews I conducted with the then director of the ministry of culture in St. Kitts, Creighton Pencheon, he noted, "Nothing is really from here. Everything we have here is really from somewhere else," after a brief discussion that traced the history of the Bull character (a staple of Kittitian-Nevisian Christmas sports) back to Jamaica and even further to West Africa more broadly. This apprehension about the validity of things being *from* St. Kitts-Nevis is continually echoed by Caribbeanist academic colleagues who have asked, candidly and in earnest, "Is there really music from St. Kitts-Nevis? From Anguilla? From Saba?" My initial goal in researching the popular music of St. Kitts and Nevis, then, was to disprove the conceit that the small islands of the Caribbean are historically and musically empty. After four years of fieldwork in and around St. Kitts-Nevis and its North American diasporic hubs—especially New York City and Toronto—I had amassed enough proof of a rich musical present and past by documenting the diverse styles of folkloric and popular music forms reverberating around the annual Christmas carnival. My work on wilders, a genre of carnival popular music that most closely resembles soca, demonstrates that contemporary music practices in St. Kitts-Nevis encapsulate the past and present of small island musical production. Having experienced wilders certainly offered a personal counter to popular understandings of origins and legitimacy as qualifiers elusive to small Caribbean islands. However, that my Caribbeanist and Caribbean interlocutors are unconvinced by the idea that something can be *from* St. Kitts-Nevis and other small islands points to larger questions about notions of legitimacy and origination as they intersect with geographical size and colonial history.

St. Kitts and Nevis, along with many of the other islands in the Leeward chain, are often unnamed or invisible altogether on world and regional maps.[1] Not surprisingly, there exists an internalized idea about the invalidity of such small places within the cultural landscape of the Caribbean and the wider world. For a genre of music or a discernible musical sound to be *from* somewhere there must be a substantiation of place and historical importance and recognizable influence that is not easily associated with small places. This is especially so in relation to small islands that are imagined as small within the context of the Caribbean, where the global and social economy render all Caribbean countries as *small* places.

In perhaps one of the most famous expositions about life on a small island, the 1998 essay "A Small Place," Jamaica Kincaid describes the deleterious effects of tourism and globalization on the eastern Caribbean island of Antigua. In an interesting deployment, particularly given the history between contrastive islands like Jamaica (population 2.7 million) and Antigua (population 90,000), Stephanie Black's 2001 documentary *Life and Debt* performatively deploys Kincaid's text to describe the negative effects of the International Monetary Fund and neoliberalism on Jamaica's economy, which sets the political scene for the emergence of dancehall music as the voice of the disenfranchised, black, urban Jamaican youth in the late 1980s. What these examples underscore is the more obvious fact that being "small" and being a "small island" are relative notions. However, the same types of large agencies that Stephanie Black rails against in *Life and Debt* are the entities that have historically been most interested in specifically pinpointing which islands should be defined as consistently small in a global frame.

Taking up these concerns as a main point of focus, in 1994 the inaugural Global Conference on the Sustainable Development of Small Island Developing States was convened through sponsorship of the United Nations; it was the first UN conference to focus exclusively on this group of island nations (Hein 2004, 1). Most of the issues attendant to the usefulness of the meeting and subsequent actions were based on a previous inability to adequately identify which islands qualified as small enough to receive additional economic and political support. In an effort to more succinctly define the intersection of environmental and geographical features that affected developing islands the most, researchers and policymakers suggested "vulnerability" as a guiding frame.

Relating primarily to "ecological fragility, proneness to natural disasters, and concentration of exports on limited ranges of products and markets," vulnerability defines small islands in terms of a transactionalism of possibilities (Hein 2004, 10). Small islands persistently face the possibility of damage, disaster, and death—the very foreclosure of certain kinds of possibilities for abundance and development. And still, the forever state of vulnerability precipitates what island studies scholar Godfrey Baldacchino calls "pseudo-development strategies," the "wily and adroit" complex of "commercialized resources" (2014, 3).

Much of the task of navigating the notion of small-island vulnerability is based on embracing the tropes that it conjures for international audiences. As Elizabeth DeLoughrey notes, "Islands are . . . outposts of empire that are deemed remote, exotic, and isolated by their continental visitors" (2010, 2). Many tourist boards and travel literatures emphasize remoteness as a key and endearing feature of temporary Caribbean inhabitance. The state of being a small island and, as an extension, I suggest, being *from* a small island within the Caribbean archipelago and worse, within the smaller archipelago of the Leeward Islands, is located in the interstices of these contradictory renderings of vulnerability, smallness, and islandness—between isolation and relation, dependence and self-sufficiency, and the openness of possibility and the probability of its foreclosure.

As I argue in this chapter, the contradistinction between islands and continents that informs our definition of small islands has, in an effort to practically represent the music of the entire Caribbean archipelago, centered on big islands. Lastly, I offer an ethnographic case study of what I term "archipelagic listening": an online streaming radio station and chatroom, largeradio.com, that has created a space and a mode of listening to and engaging with Caribbean music that counters the singularized generality of Caribbean music scholarship by broadening the horizons of the Caribbean imagination to include different kinds of *being from*. Through this virtual space and attendant listening practices, members of largeradio.com enact an archipelagic counter to singular originality through focus on a multicentric conception of Caribbean cultural production through archipelagic listening. To listen archipelagically is to be always aware of the catachresis of the term "origins" and to be equally wary of any form of status or legitimacy that misnaming accords. The largeradio.com members recognize mobile sounds to be perhaps the only surety of Caribbean music and use national belonging as a heuristic to assure inclusivity, not as a defining feature of the people or sounds themselves.

ETHNOMUSICOLOGY OF THE CARIBBEAN

Caribbean popular music histories have been especially silent when it comes to the contributions and developments of music in the Leeward archipelago.

Most scholarship and informal discourse that describes the trajectory of Caribbean popular music especially cites Jamaica as the birthplace of reggae and its extensions, points to Trinidad as the melting-pot homeland of calypso and soca, and suggests that other Anglophone islands have merely adopted and adapted these "big island" styles. The prevalence of singular island-nation origin stories of contemporary Caribbean music genres, particularly within former and current British colonies, still remains the norm despite the fact that Caribbeanists and Caribbean music scholars have long acknowledged the Caribbean as an area distinctively characterized by hybridity and mobility.[2] The notion of the Caribbean crucible, where the Caribbean functions as a container, a Dutch pot of pressure and violence that created new things out of the old from elsewhere, is made obvious in the long and meandering etymology of terms such as "creole" and "creolization" that have found new life in non-Caribbean spaces.[3] Trinidad, for example, has enjoyed much more historical attention from music scholars. With US occupation of Trinidad in the 1940s and the global popularity of calypso and calypso-derived acts as markers of black artistic expression, calypso came to be representative of the great majority of Caribbean music.

Islands have long held the attention of anthropologists. Many seminal cultural anthropology studies such as Bronisław Malinowski's *Argonauts of the Western Pacific* about the Trobriand islands and Margaret Mead's *Coming of Age in Samoa* have focused on island societies. However, notwithstanding their methodological indebtedness to anthropology, and despite the Caribbean's archipelagic geography, ethnomusicologists were disinterested in carrying out research in the Caribbean region. Early comparative musicological efforts were focused on "pure" and indigenous musical forms—antithetical to the necessarily creole nature of post-plantation-era Caribbean cultural expression (Guilbault 1991; see also Myers 1993, 461). Early social scientific interest in the Caribbean region described the Caribbean as a "laboratory," given its perceived geographical and cultural isolation from global centers such as the United States and Europe. North American anthropology turned its attention to the "New World Negro," and the West Indian petri dish took on greater significance as a space of the discovery of African pasts within the Americas.[4] American anthropologists Frances and Melville Herskovits carried out some of the most influential fieldwork within the circum-Caribbean and Africa, proving ultimately that vestiges of African culture were maintained through the Middle Passage and are evident in contemporary Afro-descendant cultures throughout the Americas.

The Herskovitses' research trips and recordings from Haiti and the Toco village in northern Trinidad in the 1930s, in particular, and the resulting monograph *Trinidad Village* (Herskovits and Herskovits 1947) are the anthropological precursors to extended musicological research in the Caribbean taken up by researchers such as Richard Waterman (1943), Alan Merriam (1951), and Daniel Crowley (1955, 1957). Within this intellectual genealogy, Trinidad, Haiti, and later Jamaica became the ethnomusicological locus of what David Scott calls an "authentic past" that is "anthropologically identifiable, ethnologically recoverable, and textually re-presentable" (1991, 263). Even more compelling, however, is the anthropological impulse to find such an authenticatable past on a singular island. If ethnology is dependent on the ability to categorize and compare groups, a bounded island has been a more naturalized (though not natural) site for study of a people and its culture. Singular Caribbean islands were the perfect bounded locales for ethnological study of New World African peoples.

In prefacing *Trinidad Village*, the Herskovitses suggest that their decision to undertake fieldwork in Trinidad was a matter of chance. They were waiting for transport back to the United States in Port of Spain via Dutch Guiana in 1929 and saw a letter published in the *Trinidad Guardian* lambasting local Shango worshippers. The Herskovitses resolved that they would return to Trinidad to investigate Shango worship, declaring, "We know it could derive only from the Yoruban peoples of Nigeria, and must represent an important body of direct African cultural retentions" (Herskovits and Herskovits 1947, i). The Herskovitses saw the immediate generalizable and broad import of this work as a "picture of Toco life" that could help those "charged with the solution of more immediate problems, both in the Caribbean and in the United States," which included "facilitating adjustments of an inter-racial character" and "ameliorating the position of underprivileged Negro groups" through investigation of "traditionally sanctioned modes of behavior" (vi–vii). It is fair to suggest that the immediate continued interest in that location was in part due to the Herskovitses' initial research. Many of the early discussions of Caribbean music in the US academy were undertaken by students of Melville Herskovits.

Take, for example, Daniel Crowley's 1959 *Ethnomusicology* journal article "Toward a Definition of Calypso (Part I)." As a starting place for pinpointing the singularly Trinidadian roots of the form, Crowley explains, "The popularity of such songs as 'Day-O' and 'Jamaica Farewell' sung in pseudo–West Indian dialect and called 'calypso' requires a fresh investigation into this traditional Trinidadian musical form." In a follow-up to this article, Crowley reiterates what he gathers is the common Trinidadian sentiment that

> the efforts of commerical-minded [*sic*] foreigners, small islanders, and what Trinidadians call "country bookies," have tended to obscure and distort the subtle art form which is contemporary calypso in Trinidad. The clarification attempted in this paper would not be important if it were not that serious artists

> are being robbed of their rightful rewards both in fame and in money, and the urban Creole community of Port-of-Spain denied the credit it has earned as a major source of creativity in the Caribbean. (Crowley 1959b, 120)

Crowley comes to this conclusion after quoting several earlier folklorists who, contrarily, assert a more inclusively Caribbean understanding of calypso's origin and cultural ownership in noting that the form "is no more the product of the island of Trinidad than it is of the islands of St. Thomas, St. John, and St. Croix" (Adams 1953, quoted in Crowley 1959b, 117). What makes these two standpoints most interesting is that they rehearse the long-standing tensions between big islands and small islands. They not only demonstrate the contrasting historical understandings of Caribbean creative output with regard to the geographical locus of authenticity via origination but also shed light on modes of imagining where Caribbean traditions can emerge. In Crowley's presentation, there is a juxtaposition of Trinidad (sans Tobago) and "the small islands," which in this case included the US Virgin Islands.

My tracing of Trinidadian music scholarship in relation to investigations of the small islands is not intended to suggest that the multiple and globally influential music traditions of Trinidad are not deserving of such enduring attention. Instead, I wish to offer some historically contextualized explanation for the typical exclusion of other *smaller* Caribbean islands from similar scholarly interest.[5] As Franklin Knight suggests regarding the small Leeward and Windward Islands, scholarly "neglect of those extremely interesting places of course, derives not from malice but from pragmatism. . . . [However] the small islands provide important and necessary qualifications for the generalizations about the region" (2011, xxvii). It is my suggestion here that the pragmatism that made study of Trinidad the main focus of early ethnomusicology of the Caribbean has morphed into a generalized understanding of Caribbean music based on largely singular fieldwork. Further, this generalization hinges on colonial approaches to regarding the islands of the Caribbean, approaches that are related to and emblematic of social and political tensions between big islands and small ones.

CARIBBEAN NATIONALISM, ISLAND SINGULARITY, AND MUSIC ETHNOGRAPHY

Despite a long-standing disciplinary aversion to the types of music styles that emerged in the Caribbean Basin, famed ethnomusicologist and folklorist Alan Lomax devoted considerable attention to the Caribbean region throughout the middle part of the twentieth century. After early visits to the Bahamas in 1935 and Haiti in 1937 that were driven by the anthropological notion of

African retentions among African descendants in the New World, Lomax turned his attention to the Lesser Antilles. Recalling his 1962 Caribbean fieldwork, Lomax writes,

> The Rockefeller Foundation sent me to the Lesser Antilles with a fascinating goal in mind. At that time there was a possibility that these various islands polities could be joined together with Jamaica to form a new democracy, the Federation of the West Indies. The enlightened presidents of both Jamaica and Trinidad were working hard to bring this idea to reality. My task as a folklorist was to look for the creative cultural commonalities among these many powers in support of their great dream of unity. . . . Back in Trinidad, the home port of the trip, I wrote my report for the Rockefeller Foundation, putting together the patterns that, as I saw them, provided a rich soil of unity for the future Federation. The most striking was this very unity. No matter what their language or dialect, the hundreds of Windward and Leeward Islanders we had met were stamped with a common Creole style. Whether their vocabularies are French-, Dutch-, or English-based, all are clearly related to the black transformation of West African linguistic structures. (2003, 337)

In many ways, the Lomax recording project unintentionally mirrored the social and economic relationship between larger islands such as Trinidad and Jamaica and the small islands that were the focus of the 1962 research. Gordon K. Lewis has noted that the failure of the West Indian Federation can be attributed to "British neglect and West Indian betrayal" (2004, 129). Less materially, however, Gordon suggests that a psychological "small island complex"—the imposed inferiority of small islands to their large island counterparts—also played a major role in the demise of the proposed political arrangement. Where "the Trinidadian attitude to the small islands has been one of genial contempt, [and] the Jamaican attitude one of sheer disinterest, compounded by gross ignorance," small islands and islanders have been imagined as economic "parasites" (Lewis 2004 129). Given the long and episodic nature of Caribbean negotiations surrounding federalization, incorporation, and regional nationalization, creative culture—broadly construed—appears to be the only area in which there is consensus surrounding the compatibility of Caribbean islands. Music, specifically, constitutes the most promising and convincing area of common Caribbean inhabitance. Cuban cultural critic Antonio Benítez-Rojo, in a reflection on the cultural similarities among people of different Caribbean islands—on the archipelagic possibilities of the region—reminds us of seventeenth-century ethnographer Pére Labat's analysis of the artificial separation of the Caribbean in stating,

> Nationality and race are not important, just small and insignificant labels compared to the message that the spirit brings me: and this is the place and predica-

> ment that history has imposed on you. . . . I saw it first in the dance. . . . The merengue in Haiti, the beguine in Martinique, and today I hear inside my old ear, echo of the calypsoes of Trinidad, Jamaica, Saint Lucia, Antigua, Dominica and the legendary Guyana. . . . It is not accidental that the sea which separates your lands makes no difference in the rhythm of your bodies.[6] (1997, 1)

Indeed, the notion of the Caribbean archipelago outside topographical inquiries is largely raised in the course of describing a Caribbean culturally unified through music and dance practices. Along with Benitez Rojo, key Caribbean thinkers such as Édouard Glissant and Kamau Brathwaite have considered the similar "tossings about" of all Caribbean peoples and the products of such a history as one of the main uniting features of Caribbeanity. In spite of these circumstantial similarities, however, the social relationships between Caribbean people as products of colonization and imperial expansion still bear the marks of colonial distribution. While the processes may have been the same in different islands under colonial rule, the overarching effect is a separation rooted in nationalism and insularity. Lewis traces the history of the contentious relationship between these islands back to the height of the sugar plantation period when "the planter class of one sugar island [feared] competition from its neighbors" (Lewis 2004, 129).

The breaking up of the Caribbean islands into competing nations is, as has been noted elsewhere, an outcome of imperial colonial rule. Even texts that seek to address the Caribbean in its entirety have found few ways of parsing the different areas of the circum-Caribbean. Dividing areas nationally, where possible, has been one mode. Other compendiums have opted for a linguistic scheme that, largely, defines micro areas within the circum-Caribbean by their relationship to a colonial legacy. The Eurocentric metageography that undergirds our common terrestrial and littoral understanding, the recognition of formal geographical areas, leads Caribbean music scholars to subsume smaller islands with a shared colonial history (e.g., the French Caribbean or the Dutch Caribbean) into historical narratives and analyses of larger places. These language blocs are, in themselves, a barrier to meaningful contemplation of the Caribbean region or Caribbeanity more broadly. However, further, within the linguistic silo of the Anglophone Caribbean, there is a practical base assumption that where small islands lack space and the capacity to participate politically in a global economy as insular island states, they are not sites capable of unique cultural emergences. Instead small islands are basins that catch the runoff of the musicality of larger islands.

This geographical and cultural hierarchy between big islands and small islands is an example of what Nelson Maldonado-Torres (2006), following Lisa Lowe (2001) and Walter Mignolo (2005), considers "continentality." Maldonado-Torres draws on postcontinental philosophy as a rejection of

European continental philosophy as well as of cultural geographers Martin W. Lewis and Kären Wigen's notion of metageography and the (il)logic of continents. Where metageography is "the set of spatial structures through which people order their knowledge of the world" through deployment of "often unconscious frameworks," continentality as an epistemology emerges as product (Lewis and Wigen 1997, 2). As Maldonado-Torres (2006) notes, continentality refers to "a deep affiliation to national or continental ontologies" that, he adds, "are not restricted to Europe." Continentality adheres to a metageography in which "the lived time and space of many peoples as well as of the liminal folk or condemned is subsumed into an organic and homogeneous temporality and spatiality" (ibid.). Lewis and Wigen have emphatically argued the contingency of continents as the characteristics and existences of what are naturalized concepts have a very human story of definition. Continentality, then, that is rooted in similarly contingent and humanistic endeavors of relation, negation, and comparison is a characteristic that can be given to or taken on through human action and epistemological renderings where an island such as Trinidad can be regarded continentally with relation to other places against which it is understood. The Leeward Islands, historically dwarfed and economically unstable, become the literal and metaphorical islands against which "big island" continentality is drawn. But we should not conflate the continent's failure to see beyond itself with a failure of the archipelago to sound.

ARCHIPELAGIC THINKING

If we consider continents and continentality in relation to the Caribbean specifically, poet and literary critic Édouard Glissant's reflections on the geography and cultural landscape offer a useful counter to continentality by positing continental ontology and epistemology as linked in his description of the difference between continents and archipelagoes. His notion of archipelagic thinking seeks to offer a more comprehensive vision of the Caribbean. Specifically, where Glissant (1997) describes continental thought as something that strives for fixity and singularity, he sees archipelagoes as connecting disparate parts. Further developments of the idea of the archipelago and archipelagic thinking have posited the archipelago as both a topological and theoretical frame for "unsettl[ing] certain tropes: singularity, isolation, dependency and peripherality; perhaps even islandness and insularity" that overbearingly characterize discussions of islands and island cultures (Stratford et al. 2011, 114). Music and the study of music through ethnographic practice sit at an interesting intersection between archipelagic thinking as a relatively new field of coalesced thought focusing on islands, oceans, and their relations and social and cultural commentary in and of the Caribbean.

Archipelagic thinking as an extension of island studies and ocean studies (among other fields dealing with issues of decolonization, postcoloniality, and intersections of space and culture) has focused on the archipelago as an understudied and undertheorized geographical form that opens up the possibility of refocusing our attention away from insularity and borders and toward "island-to-island" connections and assemblages (Stratford et al. 2011, 114). Within an archipelagic frame, we may escape narrow conceptions of extreme insularity or unfettered openness as main characteristics of islands. Further, by focusing on geographical imaginations at work, archipelagic thinking denatures the dichotomy of island versus continent such that the archipelago as "a sea of islands" refers not just to "islands of the world" but to "the world of islands"—expansively construed. Archipelagic thinking, as Jonathan Pugh (2016, 1042) has suggested, emerged as a framework attentive to that which is "relational, material and mobil[e]" in the humanities and social sciences, particularly in ocean and island studies.

Nevertheless, even archipelagic thinking, in many ways, has fallen short as a useful framework for recovering the invisible—the inaudible—small islands within Caribbean music scholarship. In the field-defining article "Envisioning the Archipelago," Elaine Stratford et al. suggest that "much has been done to delineate collective identity, shared sense of diaspora and history, settlement, mobility and change" through focus on the material and metaphorical aspects of the archipelago (2011, 123). Asserting "the archipelago's cultural relevance," they echo a popular trope of noting that the Caribbean's "most important exports—including various types of music and, of course, tourism—are likely more suggestive of the region than of its colonially fragmented political spaces" (123). But in the case of music of the Anglophone Caribbean, archipelagic thinking has not yet managed to escape the hierarchies of continental logic or decenter the spatial and temporal metageography of "big islands" and "small islands."

What has become clear is that this hierarchy is, and has been historically, more than an intellectual or institutional conception. It is my suggestion here that Caribbean music scholarship especially has been plagued by continentality in that the search for the origins and histories of Caribbean music genres and traditions has been delineated by national or continentalized borders. Consequently, this leads to the rejection of the legitimacy of certain islands and extra-national, extra-colonial archipelagoes as units of investigation and sites of valid musical production. Mobility, relation, and regional generalizability, while redemptive in many cases, have always been the tropes through which Caribbean music has been discussed and analyzed. That is to say, small islands and their interconnected inhabitances are imagined as locales where big sounds are received and localized but not as places of origin and exchange within the larger scope of the entire circum-Caribbean. Where Caribbean histories and narratives sprout from spatial ontologies, small islands come to be (un)known—undecipherable—as subsets of their larger counterparts. Further, the continentality of Caribbean music scholarship has created a methodology and analytical frame that while archipelagic in its insistence on the regional does not account for alternative—decontinental—metageographies. The question for investigations into small island music, then, is how are people responding to this historical narrative?

ARCHIPELAGIC LISTENING

If, as Michelle Stephens and Brian Roberts offer, engaging in archipelagic thinking "is to strive for different ways of seeing" (2017, 17), perhaps recovering the small islands within Caribbean music narratives requires a different sensorial approach. Archipelagic thinking has been, rightfully, concerned with compensatory epistemological and cosmological histories that attempt to recover and take seriously indigenous and otherwise suppressed frameworks through an investigation of the past (Hau'ofa 2008; Santos Perez 2017). Homing in on listening as a cultural practice may open up space to consider similar alternate models and modes contemporarily and ethnographically as opposed to exclusively historically. Most obviously, it is possible that archipelagic listening could serve as a more apt framework for recovering the silenced and forgotten small islands within the Caribbean music historical narrative.

Paul Carter (2004, 44) has defined listening as "engaged" and "intentional" hearing, which points to listening as a mode of communication and relation. In the remaining pages I offer a case study of archipelagic listening among members of an online radio station and website who have curated a virtual space in which to engage in listening practices that make audible all areas and epochs of Caribbean music in an effort to truly represent archipelagic totality. In the following analysis I am concerned with both the sounds the site's participants are listening to that emanate from within and around the circum-Caribbean and also with the listening practices that help to make the archipelagic worlds in which Caribbean people live. Archipelagic listening attunes us to certain listening practices that create the space out of which new genres emerge. It takes into consideration a contemporary Caribbeanity of interconnection that does not supersede older forms of mobility and interaction but, instead, augments, doubles, and crosses with other inter-island, island-continental relations.

SMALL ISLANDS, LARGE RADIO

Between 2000 and 2008, the origins of calypso and soca were hotly debated in the chat rooms and forums of one of the most popular trans-Caribbean internet rum shops called islandmix.com. A great deal of the previously untold history of small island music from the northern Caribbean islands was discussed as part of the feud between "big islands" and "small islands" of the West Indies. There currently exist hundreds of pages of archived forum discussions between two factions comically referred to as the "More Than Four Missionaries" and "Trickdadians."[7] The former refers to forum contributors who seek to spread the good word that the small islands contributed fast tempo, among other things, to the larger Caribbean musical landscape. Trickdadians, on the other hand, the totalizing force against which the missionaries are fighting, are forum members who have a tendency to offer Trinidadianized versions of music history in the forums. One such Trinidadian online historian known as "SocaPhD" participated heavily in the feuds on the site in the early 2000s and agitated nationalist tensions by signing each of his forum comments with a signature that read, "Hailing from Trinidad & Tobago and very proud of it!! Land of Calypso, Steelband, Limbo, Parang, Rapso, Chutney-Soca, Soca, Jamoo, Panjazz and the Biggest, Best & Most Influential Caribbean Carnival in the World with no apology!"[8]

Given that the small island Caribbean has not enjoyed a fraction of the archival and ethnographic attention that Trinidad has, much of what the "missionaries" offered in rebuttal to the grand claims of the pro-Trinidadians was based on firsthand experience and personal musical collections. It is beyond the scope and aim of this chapter to describe with much detail the rhythmic and technological specificities that bind these small island sounds together as a unique approach to Caribbean music according to forum members. But it is safe to say that the sentiments that fueled this communal analysis and research precipitated the creation of another type of virtual space that would account for the silenced histories within music scholarship and Caribbean imagination.

LARGERADIO.COM

Stany "DJ Stanman" Benoit is one of the original creators of the streaming radio station, website, chatroom, and forum largeradio.com. In describing the impetus behind the site's creation, Benoit noted, "In 2001 me and my partner from the Netherlands were in college in Miami and we thinking Miami has all these Caribbean people but we never heard any music on the radio or at parties except the standard stuff: Trinidad soca and some dancehall. We from St. Maarten are used to hearing music from everywhere: Spanish music, Haitian, Nevis, Aruba, Curacao. We were like, 'we have to do something!'"[9] During this time, Benoit created a DJ crew called "Small Island Massive" (DJs dedicated to playing songs from all over the Caribbean at parties and events) and a related site called "SXM-link" (an abbreviation for St. Marten) that was intended to keep the small community of St. Marteners updated on local events and happenings within the music world of that island. Stany Benoit noted, however, that very quickly they realized that the issue of being neglected and silenced by radio stations, DJs, and historians was relevant to all small islanders. The site was then renamed and revamped to be smallislandmassive.com with an official tagline reading "Small Island Massive, Taking Over to Unite Caribbean People, and to show the world what small island people can do. The Caribbean does not only consist of Four Islands. We are here to bring the REAL island Flavour to you."

"More Than Four" (Figure 18.1), according to Benoit, is the rallying cry and central concern of the work he sets out to do as a serious, professional

Figure 18.1. largeradio.com logo (October 25, 2017).

DJ in addition to being a graphic designer. He recounts, "People always ask which islands are the 4 in 'More than Four' and it has many meanings." The motto refers simultaneously to the various levels of forgetting and silencing small islands that occurs within the realm of contemporary popular music. On the broadest scale, this phrase forces the acknowledgment that there are more than four islands in the Caribbean and that these islands cannot represent the whole region. Moreover, Benoit stresses that the small repertoire of songs by small island artists that have made it into the canon of soca classics is still not representative of the true breadth of the Caribbean's musical offerings—"there are more than four small island songs and more than four small island artists."[10]

By 2005, internet innovations and the proliferation of web forums and instantaneous peer-to-peer messenger platforms such as ICQ and MSN Messenger created a new way for small islanders to "express themselves to one another" and to extend inter-island contacts and connections that had been a major part of the small island imaginary (particularly through music listening practices). During this time, in the early 2000s, traditional radio stations within the small islands were experimenting with a chatroom format that allowed listeners to engage with radio hosts and DJs directly and express their like or dislike of a selection.[11] Experimenting with this form, Benoit, a graphic web designer, created a new twenty-four-hour streaming radio station with an embedded chatroom script that allowed listeners to post flags and small graphics. This small act of "representation" was, according to Benoit, "addictive" for the small islanders who frequented the site—"they never got anything like that before."[12] While it is hard to say how many people were active on the site during this time, Benoit noted that the site was "shut down every month because it needed a bigger bandwidth" to accommodate all of the activity.

I stumbled upon largeradio.com after searching for online repositories for recordings of music from the Leeward Islands. Even in the era of digital music and streaming services, very little music from these islands is available on large streaming services. Instead, performances and recordings are uploaded onto sharing websites such as youtube.com or hulkshare.com.[13] The site listed a weekly schedule of DJs and the particular focus of their sets: a DJ in New York "representing" St. Kitts and the Virgin Islands; a DJ in Toronto representing Antigua. Benoit, who has been the most prominent DJ and the lifeblood of the site lives in Miami, and although he is listed as representing St. Maarten, he is staunchly dedicated to amassing, curating, and disseminating music from all islands of the Caribbean.

Offering his take on that long-standing argument about the origins of calypso and soca between big islands and small islands, Benoit remarked,

"The Caribbean is bouillon and pelau: mix up! Everybody 'from' your island is really from some other island, your grandmother, your father, everybody moving around from island to island. That's how life is for us . . . yet they [big islanders] believe they invented everything."[14] Benoit's sentiment about the movement between islands as undercutting any real argument about origins is what inspires his extremely time-consuming dedication to the site and to amassing, knowing, and broadcasting music from every Caribbean island, especially islands that have historically been understood as silent listeners to big styles and not producers of their own unique and legitimate sounds.

Benoit understands largeradio.com to be a space for a different kind of listening. With the "More Than Four" concept at its core, DJ Stanman's listeners tune in religiously and participate equally dutifully in the chatroom with a sense that there is a serious obligation as self-identified "Caribbean people" to listen actively to music from all around the region: participants listen, respond, ask questions, post examples of samples and possible precursors to songs, and deliberately place each sound in the context of the wider Caribbean. Benoit noted that "people modify their lives for the show. People have met one another in person—some people are still best of friends after meeting here 11 years ago. Long lost cousins on different islands have found each other there."[15] Here Benoit points to the power of this type of communal listening to make a space for such intimacy and relation. Furthermore, however, the site is imagined as strengthening primordial connections between small islanders anchored in the physical proximity of the islands and the long history of movement and even visibility between these Leeward Island spaces.

The format of the streaming radio station and chatroom that makes up the most popular portion of largeradio.com is what Henry Jenkins refers to as a "convergence" of "old and new media," noting that "audiences, empowered by these new media technologies, occupying a space at the intersection between old and new media, are demanding the right to participate within the culture" (2006, 24). "The culture" for largeradio.com participants refers to the Caribbean cultural imagination. Whereas a version of regional history has positioned small islands on the outside of notable history, they are demanding to be heard and recorded in the historical ledger and to hear themselves as participating in a larger soundtrack of the circum-Caribbean as it extends beyond the geographical arc of the archipelago.

Benedict Anderson's definition of the nation as an imagined community is instructive here. Anderson (2006) connects the emergence of "print capitalism," through the modern newspaper and the modern novel in the eighteenth century, to the burgeoning of the collective idea of belonging to a political community. An online streaming platform in the twenty-first century such as largeradio.com is a counterfigure to these text-based conduits of collective consciousness. Here the extra-national community is "conceived as a deep, horizontal comradeship" not through communal reading practices but, instead, through collective listening (37). More broadly, largeradio listeners are pushing for a reimagining of the boundaries of origination such that Caribbean cultural production can be regarded as a multisited process. The intentionally archipelagic listening practices of the site participants acknowledge the diversity, intertraversibility, and intervisibility of small island Caribbeanity and demand a fuller rendering of the Caribbean archipelago. In this way, largeradio.com members listen to Dominican Bouyon alongside Kittitian-Nevisian wilders and consider the relationship of each to Barbadian Bashment soca. Listeners are aware of the numerous carnival-like national celebrations and the associated sounds occurring throughout the calendar year such that Trinidad's carnival becomes one musical node amid others, such as Curaçao's similar pre-Lenten carnival and Houston, Texas's Caribbean festival.

The archipelago exists at the intersection of geomaterial reality and imaginative metaphoricity—between the tangible and predetermined and the culturally constituted. Largeradio.com and the mantra of "More Than Four" is archipelagic in its occupation of a more literal space of the dispersed Caribbean community over the virtual internet. Moreover, largeradio.com emerged as a response to continentalized history both externally and within the Caribbean. This response has been archipelagic in its emphasis on the musical interconnection and uniqueness of each island and nonisland Caribbean space.

In invoking music as proof of the archipelagic nature of the region, scholars of the Caribbean have slipped into continental ontologies they may otherwise seek to avoid by eliding the specificities of small sounds, small moves between small places, in the service of highlighting big trends that are traceable within a continental logic. Our listening practices, practically and as we deploy them in our work, must also be as expansive as the borderless region itself. I see largeradio.com as a public space of contemporary Caribbeanity and of archipelagic praxis. Changes in sea travel, national boundaries, economic stability, sovereignty, and neoliberal capitalism have changed the modes of mobility that made genres like calypso and reggae as mobile and ubiquitous as they currently are. Largeradio.com as an online radio station replete with instantaneous person-to-person communication—the creation of a virtual public—forces our attention to the entirety of the Caribbean, geographically, imaginatively, and sonically. The website's model of a borderless Caribbean collapses hierarchies based on size or location and reimagines the sea between islands and nonisland Caribbean spaces as equally fertile for cultural production. This is the stuff of archipelagic totality that we may be able to think about, with, and through if we learn to listen.

PRACTICAL AND METHODOLOGICAL IMPLICATIONS

Archipelagic listening is an attunement to a functional, geographic whole as seen, inhabited, experienced, and conceptualized by the people who imagine themselves as part of its dispersed community. A call for archipelagic thinking via sound reminds us to listen to and take seriously even the smallest satellite, ancillary, and seemingly derivative cultural products, particularly those emanating from seemingly small and vulnerable places. In this way, we may also be aware of the ways the primary modes of sonic investigation are largely prescribed by a global history of colonization. So, for example, while it is necessary and logical to look to the United States and Britain to understand the routes of Trinidadian popular music, continentality precludes, in many ways, an archipelagic frame that focuses attention on other islands in the Caribbean archipelago that make up an equally important, robust, and generative network of accumulated sounds. Ultimately, archipelagic listening forces us to think beyond the fiction of isolation within an archipelago and to consider listening practices, our own and those of our interlocutors, as indicative of a conceptualization of geography and history. Listening with the archipelagic spatiality of the Caribbean forces us to hear not variants on an original tune but rather strains of creative possibility that resound from many places and times.

NOTES

1. The Leeward island archipelago includes, from northwest to southwest, Vieques, Culebra, St. Thomas, St. John, St. Croix, Water Island, Jost Van Dyke, Tortola, Virgin Gorda, Anegada, Angulla, Saint Martin, Saint Barthelemy (St. Barts), Saba, Saint Eustasius (Statia), St. Kitts, Nevis, Barbuda, Antigua, Redonda, Montserrat, Guadeloupe, La Desirade, Iles des Saintes, Marie-Galante, and Dominica.
2. Jocelyn Guilbault's work on Zouk in the French Caribbean (Guilbault 1993) and Timothy Rommen's scholarship on popular music in the Bahamas (Rommen 2011) offer notable exceptions to this single-island rule.
3. See Palmié 2006 for a discussion of the etymology and use of "creolization."
4. It has been noted that while Melville and Frances Herskovits sought to address "the Negro Problem" in the United States by finding African cultural retentions among non-U.S. Afro-descendant communities, their desire to find resemblances led the Herskovitses to "find Africa" everywhere (Mintz and Price 1992; Price and Price 2003).
5. Daniel Crowley's work on calypso has tried to parse out the particular origins of calypso and in so doing offers a snapshot of both the historical understanding of the origins of the genre and a more social animosity between Trinidadians and other West Indians who utilize the term "calypso" to refer to local folk musics.

6. The author's translation. This quote has been used quite often by Caribbean scholars and politicians interested in forwarding a point about Caribbean unity and the artificiality of nationalist borders within the region. See, for example, Guyana Minister of State S. S. Rhampal in his 1971 address to the Ecumenical Consultation for Development in Trinidad, quoted in Jagan 1979. See also Jesús 2003.

7. Within the last decade, Islandmix.com has crashed and been rebuilt several times, and some aspects of these archives have been lost.

8. "Vincy Taught Trinidad, and They Plagiarized Our Music," Islandmix.com, http://www.islandmix.com/backchat/f16/vincy-taught-trinidad-they-plagiarized-our-music-260369 (accessed April 2015).

9. Benoit, interview with the author, April 2017.

10. Benoit, interview with the author, April 2017.

11. According to Benoit, Anguilla's KoolFm is notable in this regard in being one of the first convergences of new media within the Leeward Islands.

12. Benoit, interview with the author, April 2017.

13. There are some studio albums by bands such as Nu Vybes and Small Axe available on streaming music services such as Apple's iTunes. But by and large, other, less formal and more immediate services have been used to disseminate music since, at least, 2010.

14. Benoit, interview with the author, April 2017.

15. Benoit, interview with the author, April 2017.

REFERENCES

Adams, Alton A. 1953. "Whence Came the Calypso?" *The Caribbean* 7 (October): 218–20, 230, 235, reprinted in *Virgin Islands Magazine* 8.

Anderson, Benedict. 2006. *Imagined Communities: Reflections on the Origin and Spread of Nationalism*. New York: Verso Books.

Baldacchino, Godfrey. 2014. "Small Island States: Vulnerable, Resilient, Doggedly Perseverent or Cleverly Opportunistic?" *Études Caribéennes* 27–28.

Benitez Rojo, Antonio. 1997. "Nueva Atlántica: Reflexiones sobre un archipiélago posible." *Atlántica: Revista de Arte y Pensamiento*, no. 18.

Black, Stephanie, dir. 2001. *Life and Debt*. New York: New Yorker Films.

Carter, Paul. 2004. "Ambiguous Traces, Mishearing, and Auditory Space." In *Hearing Cultures: Essays on Sound, Listening and Modernity*, edited by Veit Erlmann, 43–63. Oxford, UK: Berg.

Crowley, Daniel J. 1955. "Festivals of the Calendar in St. Lucia." *Caribbean Quarterly* 4, no. 2: 99–121.

———. 1957. "Plural and Differential Acculturation in Trinidad." *American Anthropologist* 59, no. 5: 817–24.

———. 1959a. "Toward a Definition of Calypso (Part I)." *Ethnomusicology* 3, no. 2: 57–66.

———. 1959b. "Toward a Definition of Calypso (Part II)." *Ethnomusicology* 3, no. 3: 117–24.

DeLoughrey, Elizabeth M. 2007. *Routes and Roots: Navigating Caribbean and Pacific Island Literatures*. Honolulu: University of Hawai'i Press.
Glissant, Édouard. 1997. *Poetics of Relation*. Ann Arbor: University of Michigan Press.
Guilbault, Jocelyne. 1991. "Ethnomusicology and the Study of Music in the Caribbean." *Studies in Third World Societies* 45: 117–40.
———. 1993. *Zouk: World Music in the West Indies*. Chicago: University of Chicago Press.
Hau'ofa, Epeli. 2008. *We Are the Ocean: Selected Works*. Honolulu: University of Hawai'i Press.
Hein, David. 2004. "Is a Special Treatment of Small Island Developing States Possible?" United Nations Conference on Trade and Development. New York and Geneva.
Herskovits, Melville J., and Frances S. Herskovits. 1947. *Trinidad Village*. New York: Alfred A. Knopf.
Jagan, Cheddi. 1979. *The Caribbean Revolution*. Prague: Orbis Press Agency.
Jenkins, Henry. 2006. *Convergence Culture: Where Old and New Media Collide*. New York: NYU Press.
Jesús, Lara Ivette López de. 2003. *Encuentros sincopados: El caribe contemporáneo a través de sus prácticas musicales*. Mexico, DF: Siglo XXI.
Kincaid, Jamaica. 2000. *A Small Place*. 1st ed. New York: Farrar, Straus and Giroux.
Knight, Franklin W. 2011. "The Caribbean: The Genesis of a Fragmented Nationalism." Oxford: Oxford University Press.
Lewis, Gordon K. 2004. *The Growth of the Modern West Indies*. Kingston: Ian Randle Publishers.
Lewis, Martin W., and Kären Wigen. 1997. *The Myth of Continents: A Critique of Metageography*. Berkeley: University of California Press.
Lomax, Alan. 2003. *Alan Lomax: Selected Writings, 1934–1997*. Edited by Ronald D. Cohen. New York: Routledge.
Lowe, Lisa. 2001. "Epistemological Shifts: National Ontology and the New Asian Immigrant." In *Orientations: Mapping Studies in the Asian Diaspora*, edited by Kandice Chuh and Karen Shimakawa, 267–76. Durham, NC: Duke University Press.
Maldonado-Torres, Nelson. 2006. "Post-Continental Philosophy: Its Definition, Contours and Fundamental Sources." *Worlds and Knowledges Otherwise* 1–3 (fall): 1–29. https://globalstudies.trinity.duke.edu/sites/globalstudies.trinity.duke.edu/files/file-attachments/v1d3_NMaldonado-Torres.pdf.
Merriam, Alan Parkhurst. 1951. *Songs of the Afro-Bahian Cults: An Ethnomusicological Analysis*. Evanston, IL: Northwestern University.
Mignolo, Walter. 2005. *The Idea of Latin America*. Malden, MA: Blackwell.
Mintz, Sidney Wilfred, and Richard Price. 1976. *An Anthropological Approach to the Afro-American Past: A Caribbean Perspective*. Philadelphia: Institute for the Study of Humane Issues.
Myers, Helen. 1993. *Ethnomusicology: Historical and Regional Studies*. New York: W. W. Norton & Company.
Palmié, Stephan. 2006. "Creolization and Its Discontents." *Annual Review of Anthropology* 35, no. 1: 433–56.

Pearse, Andrew. 1956. "Mitto Sampson on Calypso Legends of the Nineteenth Century." *Caribbean Quarterly* 4, no. 3/4: 250–62.
Perez, Craig Santos. 2017. "Guam and Archipelagic American Studies." In *Archipelagic American Studies*, edited by Brian Russell Roberts and Michelle A. Stephens. Durham, NC: Duke University Press.
Price, Richard, and Sally Price. 2003. *The Root of Roots: Or, How Afro-American Anthropology Got Its Start*. Chicago: Prickly Paradigm.
Pugh, Jonathan. 2016. "The Relational Turn in Island Geographies: Bringing Together Island, Sea and Ship Relations and the Case of the Landship." *Social & Cultural Geography* 17, no. 8: 1040–59.
Roberts, Brian Russel, and Michelle Ann Stephens, eds. 2017. *Archipelagic American Studies*. Durham, NC: Duke University Press.
Scott, David. 1991. "That Event, This Memory: Notes on the Anthropology of African Diasporas in the New World." *Diaspora: A Journal of Transnational Studies* 1, no. 3: 261–84.
Stratford, E., G. Baldacchino, E. McMahon, C. Farbotko, and A. Harwood. 2011. "Envisioning the Archipelago." *Island Studies Journal* 6, no. 2: 113–30.
Torres-Saillant, Silvio. 2016. "The Hispanic Caribbean Question: On Geographies of Knowledge and Interlaced Human Landscapes." *Small Axe* 20, no. 3 (51): 32–48.
Waterman, Richard A. 1943. *African Patterns in Trinidad Negro Music*. Evanston, IL: Northwestern University.
———. 1948. "'Hot' Rhythm in Negro Music." *Journal of the American Musicological Society* 1, no. 1: 24–37.

Chapter Nineteen

The Insular and the Transnational Archipelagoes

The Indo-Caribbean in Samuel Selvon and Harold Sonny Ladoo

Anjali Nerlekar

The archipelagic imagination of the Indo-Caribbean manifests itself in varying ways in the early novels of two Trinidadian authors, Samuel Selvon and Harold Sonny Ladoo. Both authors question the role of the Trinidadian writer in the larger world of English letters and also the role of the Indo-Caribbean in the social world of Trinidad. In this chapter, I examine Selvon's *An Island Is a World* and Ladoo's *No Pain like This Body* to see how they employ the geography of the island and a cartographic imagination to layer Indo-Caribbean histories within their Trinidadian narratives. In 2006, Peter Hay asked for a closer look at the "sharpness" of the edge of the island, its "wave lined boundary," to decipher the sense of identity of the island (22). I take that provocation to read the novels of Samuel Selvon and Harold Sonny Ladoo, who negotiate the opposing ideas of transnational affiliations and insular boundedness in order to fashion a Trinidadian home with an Indo-Caribbean belonging, a maneuver between interspatiality and a sedimentary locality that highlights the fluctuating nature of archipelagic investments in the idea of the Indo-Caribbean.

Both Selvon and Ladoo emphasize the location of Trinidad in their novels, with each of them looking at the island nature of the place differently and examining the "edge" of this island location in his own way. Reading them together under the framework of archipelagic theory allows us to uncover the common project of local and translocal identities that haunt Indo-Caribbean narratives, when they try to weave a claim for the Trinidadian home along with a transnational connection to India/the United Kingdom/the United States. Brian Russell Roberts and Michelle Stephens ask, "Is the archipelago the sea, or is it the islands of the sea? This liminal and terraqueous mode of existence is produced by a history of being betwixt and between land and water, at one moment existing as water (studded with islands), and at the next moment existing as islands (surrounded by water)" (7). The idea of "the betwixt and the between" is what comes to the surface in this juxtaposition of two novels that interrogate as well as construct the Indo-Caribbean in Trinidadian literature. The Indo-Caribbean experience in Trinidad appears as part of a sedimentation of histories; it lives in insular space but also passes through it. This idea and identity lies between states, neither hardening into a continental mass nor completely aqueous and flowing, a middle state of solidifying that is the result of certain submerged narratives.

In *The Discovery of Islands: Essays in British History*, the antipodean historian J. G. A. Pocock calls for a study of "the world as an archipelago of histories rather than a tectonic of continents"; he sees "histories both transplanted by voyagings and generated by settlements and contacts, and consequently never at home" (2005, 3). This is the "archipelagraphy" Elizabeth DeLoughrey speaks of in her desire not to "romanticize indigeneity nor pathologize diaspora" (2010, xi). Her call represents an important methodological directive to heed the roots and routes of both land and water interconnections in any study of text, history, or culture.[1] Exploding the idea of the bounded nation through a study of the spatial model of the archipelago (Pugh 2013), archipelagic theory uses geography rather than the abstract concept of the nation and demands an examination of islands as transnationally connected spaces even as the theory also alerts us to a grounded view of the location of the island and its histories.

The island features in some typical ways in the literatures of the world (Brannigan 2015; Patke 2018). Owen Ronström gives a useful list of the features of the "island as model" when he names the representational lead weights that drag down the concept of the island: "homogeneity . . . boundedness . . . endemism—an emphasis on isolation . . . [and] the self-contained; . . . archaism—an emphasis on the old . . . and authentic; remoteness" (2013, 160). In 2011, Stratford et al. emphatically rejected the trope of island as small and isolated, reiterating the complicated formation of an island that includes three vectors of "land and water, island and continent/mainland, and island and island": the island is never singular and bounded (115). To similarly combat the static and fixed image of the island, Baldacchino and Clark call for a use of "island as noun" and "islanding" as verb: "If islandness is a particular state or condition of being, there is a corresponding action in islanding. . . . We need this verb to mediate and attenuate dizzying oscillations between paradise and prison, openness and closure, roots and routes, materiality and metaphor" (2013, 129). John Brannigan specifically uses the idea of the archipelago to debunk the imperial worldview of the British nation; through his study of Irish, English, and Caribbean writers' works, he shows how "the myth that islands could be 'places apart' proves unsustainable, and instead a new understanding of archipelagic commutuality begins to emerge in the wake of a dissolving Empire, and a weakening Union" (2015, 16). The island and the archipelago are joint concerns in all such theoretical frameworks.

Selvon and Ladoo employ many of these concepts of islandness—boundedness, isolated separateness, but also islands as points of departure and return and as hubs of globality. Each of their novels use and play with these contradictory significations of the topography as well as the metaphoricity of the island, while also revealing archipelagic connections across spaces and histories. Through their cartographic imaginations, these two novels represent the push and pull that characterizes many of the stories of the Indian in the Caribbean. On the one hand, there is the desire to reach across the oceans, across the *kala pani* journeys and the transnational affiliations with diasporic locations in the Western world.[2] On the other hand, there is the eagerness to lodge one's pasts in the ground of the Caribbean islands in order to belong to the history and geography that was foisted on one through the colonial trade. There is the urge to travel and leave, as well as the urge to stay and dig a foundation for the Indian in the Caribbean. The archipelagic framework—which allows for the focus to be both on the location as well as on the traveling itineraries that crisscross it—seems the most versatile theoretical approach to such flexing and tensile routes of identity. As Michel de Certeau states, "Every story is a travel story—a spatial practice" (1984, 115).[3] Taking that cue, I focus on the literary cartography of these two writers to show the fluid and ambiguous nature of the Indian experience that does and does not map easily onto Trinidadian space. The result is a mixed and uncertain belonging that is expressed by both novelists as it relates to the spaces and histories of the islands of Trinidad and Tobago. The in-betweenness of this Indo-Trinidadian experience—not just between India and Trinidad but also between rootedness and travel—begs for a different framework than most conventional historical frames applied so far. It is the roving cosmopolitan questionings of Selvon's characters and the dogged place-making in Trinidadian land of the characters of Ladoo's novel that need to be read in tandem in order to trace the complex delineations of the Indo-Caribbean.

The Indo-Caribbean harks back to 1838 when the British abolished slavery in theory and, in their effort to keep the plantations of the Caribbean functional and profitable, engineered another passage across the seas called the *kala pani* (the black waters). The British convinced the Indian colonial administration to allow the export of labor from India to the Caribbean. The official abolition of slavery in the colonies during emancipation by the British administration in the mid-nineteenth century marked the beginning of another crossing of the waters, when indentured laborers from India were chosen to replace the Afro-Caribbeans on the sugar plantations, starting with Guyana in 1838 and Trinidad in 1845. These were mostly poor, illiterate, and marginalized people from India who ended up living in practically segregated and insular communities because of the policies of the British administration that restricted them to the plantations.

The indentured Indians provided the needed labor on the sugar plantations during this time, but they also ended up occupying a distinct space between the emancipated Afro-Caribbeans and the British colonial administration. This contested history of indenture became the continual source of the divide between the Indo-Caribbeans and the Afro-Caribbeans even after independence (Munasinghe 2001; Puri 2004; Nerlekar 2011). The Indo-Caribbeans do not find much visibility in the historical narratives of the Caribbean. While the process of creolization is seen as the quintessential marker of Caribbean history, Viranjini Munasinghe notes the "tendency of analytic models of creolization to bypass Indo-Caribbean elements": "Although the concept of Creole society nominally includes Indo-Caribbeans as a part of the heterogeneity characterizing the region, attempts to theorize the nature of Creole societies in terms of a complex dialectical movement between contradictory principles structuring the creolization process have largely focused on the African and the European elements in the New World" (2001, 135).

Others, such as Lisa Outar, have documented this invisibility in various areas of thought and the arts. In the novels of the Windrush period, Outar notes that the Indians were consistently viewed as outsiders to a singularly imagined national project at that time. At the same time, Outar emphatically states that the Indo-Caribbean writers "were in the process of defining their place within Caribbean societies and within larger frameworks of British and colonial identity . . . as well as the region's non-Indians," and they actively sought "to figure out what role Indianness should play in the new national formations being imagined" (2015, 27). In the face of such struggle for the visibility of the Indo-Caribbean community, Outar's study of the print culture of this period demonstrates that Indo-Caribbeans were making complex claims to a Caribbean home and identity.

Similarly, in this chapter I demonstrate that studying the cartographic and spatial impetus of the narratives of Indo-Caribbean writers is an equally important method of documenting the archipelagic nature of such multiple affiliations of home and away. Samuel Selvon and Harold Sonny Ladoo present two separate cartographic visualizations of the "Indian" in Trinidad, as well as of Trinidad in the larger world. Read together, they reveal the many-layered presence of such attempts to include the Indian in the Caribbean—one asking for transnational solidarities, the other emphasizing local emplacement.

THE FICTION OF SAMUEL SELVON

Of part Indian ancestry, Sam Selvon did not necessarily see himself as Indo-Caribbean even though he featured characters with Indo-Caribbean background in much of his work. As he said in his interview with John Thieme, "You see I grew up in Trinidad as a Trinidadian and my mother's father was a Scotsman and my father was an Indian. So I'm an offspring of that and I grew up in Trinidad completely Westernized, completely Creolized, not following any harsh, strict religious or racial idea at all" (1990, 72).

Born in Trinidad in 1923, he spent his first twenty-seven years in the Caribbean and moved to England in 1950, still during the colonial years of the nation. Selvon claims that it was only after leaving Trinidad that he developed a sense of Caribbeanness. Together with V. S. Naipaul and George Lamming, among others, he was part of the group of writers who established the idea of the West Indian novel in the Western and Caribbean imaginary.

Selvon's two books from 1956 provide complementary visions of the Caribbean: *The Lonely Londoners* features a group of memorable Caribbean characters from different islands, who migrate to London and make it a brown place of immigrants rather than a white space of the colonial British. These Caribbean characters restructure the life and streets of London through their memories of home in the Caribbean and thus destabilize this colonial center. The novel *An Island Is a World* looks at these mobilities from the other side, from that of the Caribbean islands of Trinidad and Tobago, with travels starting in the islands to the metropolitan centers in the Anglo-American world. In this novel, there are Trinidadian characters who come in and go out of Trinidad, making the island a movable node of transnational routes; Trinidad gets constructed as a buzzing space of belonging and exile.

An Island Is a World features the Trinidadian common man as a thinking man, with intellectual quests and learned desires, something akin to the existentialist hero of European novels at the time, thus countering the colonial racialized stereotype of the Caribbean man as mere bodily presence, fit only for indentured and slave labor on plantations. In writing this intellectual Caribbean in the English genre of the novel, Selvon was staking a claim to the space of Trinidad as a global home and also to the English literary culture and its genres of writing, creolizing the novel form to insert his Trinidadian hero. This Trinidadian, with his archipelagic home, is at the center of the narrative project, and Trinidad itself is at the center of multiple travel itineraries while Selvon questions the idea of home and belonging in the context of a large global, unheeding world outside the Caribbean.

Selvon's early writing, especially his poems, give us a clue to his thematic preoccupations in this novel. In a long poem aired on the BBC on *Caribbean*

Voices in 1950, Selvon expresses a persistent theme of his writings—how to reclaim the humanity of the Caribbean experience:

> Sometimes in a pensive mood
> I go to the harbor's edge at eve
> And watch the world sail in and out
> And say: "Little Trinidadian, bound
> Your talks to humming birds
> And West Indianisms. Do not
> Aspire to globe-girdling ideas,
> Symphonic music or epics.
> No reception for you abroad
> Unless you play in steel bands,
> Chant calypsos or plant cane."
> To lie spreadeagled, roofless
> And project thought beyond the sky
> In spaces bare and clean
> As a shiver caused by cold.
> To heighten consciousness, philosophise,
> Devise morals, solve mysteries,
> Spiral through heavy riddles
> Is not a Colonial's lot.
>
> —Selvon and Salick 2012, 41–42

The colonial imagination of the world considers the Trinidadian incapable of producing deep thought or great art. This perspective is reflected in an image from 1938, an advertisement for a cruise of the Amazon and the Antilles, where the very imagination of the Caribbean man is annexed by the colonial world (see Figure 19.1).[4]

The advertisement is dominated by the face and head of a somewhat minacious Caribbean man, wearing a large straw hat signifying sunny locales, smoking a cigarette. The baleful attitude along with his black skin reinforces clichéd narratives of rebellious black bodies that are nevertheless available to the tourist's gaze. In fact, the Caribbean islands emanating from his head in this image, with the joyous red line of pleasure-seeking travel that prances from one Caribbean location to another, embodies the takeover of his very mind and imagination, replacing it with the colonial fantasy of the modern-day tourist. In the contrast between the resistant Caribbean native and the festive red travel line of the cruise ship (the hop, skip, and jump from one island to another) lies the proposed exotic thrill that is for sale for privileged tourists and the untroubled joy of such travel. Juxtapose this postindenture cruise ship with the idea of the older colonial/slave ship of the *kala pani*, and it is clear that the colonial world of plantations is still available under a modern guise and for the modern gaze. This advertisement is an image of an

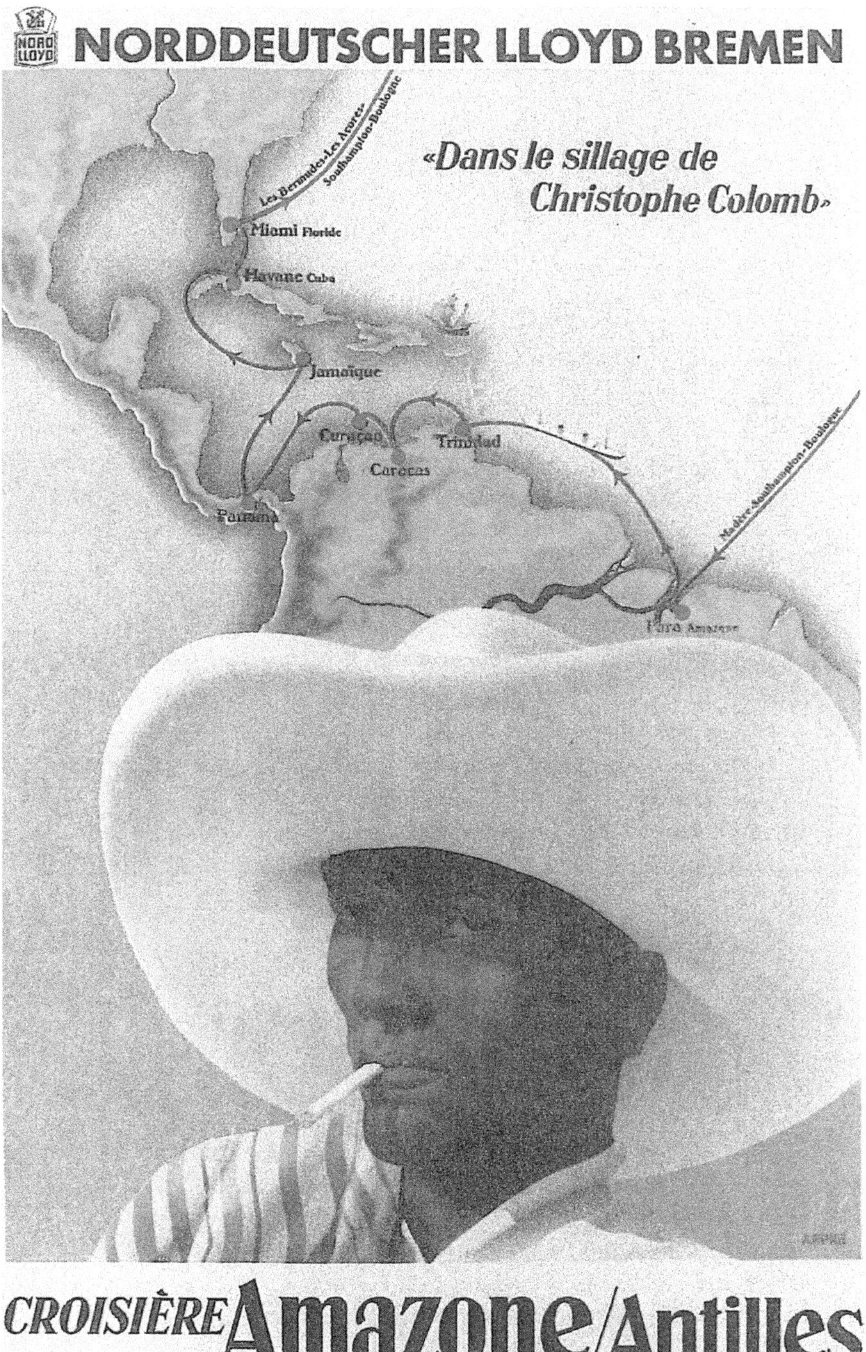

Figure 19.1. The advertisement for a Norddeustcher Lloyd Bremen Caribbean cruise, thanks to the exhibition and the catalog of the "Antillean Visions or, Maps and the making of the Caribbean" hosted by the Lowe Art Museum, University of Miami.

unconflicted world that uses the Caribbean man's body and mind for play as much as his island home.

An Island Is a World responds to such global stereotyping of the Caribbean body and history. Selvon frames the narrative of his multiple characters with the opening musings of his protagonist, Foster, doing nothing significant, thinking and pondering on national and world affairs, his mind exploding with the large view of the entire globe and then focusing the perspective smaller to see life at home in Trinidad. In bringing the world to Trinidad, Selvon uses a similar image as the advertisement but turns the structure upside down—instead of the Caribbean being fixed in Foster's gaze, it is the larger globe that spins away from him: "Looking through the windscreen, he tried to shake the despondency he felt. He could see the world spinning ahead of them. It was as if they were going towards it, but it kept its distance, they were never nearer. Somehow it didn't seem to matter anymore" (237). The Caribbean man's imagination is colonized by the tourist's agenda in the advertisement discussed above, but the colonial and colonizing world spins away from Foster's gaze, leaving him located in Trinidad and, somewhat reluctantly, the master of his own narrative.

Foster *does* nothing in the prologue of the novel: wake up, get breakfast, take a shower, and ride the bus. A first reading of the novel might mystify the reader about the relevance of this prologue to the narrative. In terms of the structure of Selvon's novel, however, if we were to list the main themes of the work, the italicized words on the left (in the sentences below) would build a striated statement of intertwined invisibilities that are juxtaposed to the form, characters, and location of the writing to create a new kind of writing for the Trinidadian novelist:

Nothing can be the stuff of a novel.

The colonial dismissal of Caribbean lives as of no value—they are "nothing"—is challenged here by putting them at the center of this colonial literary form.

A Trinidadian can be an intellectual.

As Selvon bemoans in his poem, "To heighten consciousness, philosophise, / Devise morals, solve mysteries, / Spiral through heavy riddles / Is not a Colonial's lot." Here he challenges this frontally in his novel, where it opens with the Trinidadian Foster's mental wanderings.

Trinidad can be a place to live in and thrive.

The novel shows this neglected Caribbean location as a complex, civilized place of multiple histories.

The dismissal of Trinidad as space, of the Trinidadians as protagonists, meant that neither could be taken seriously as the subject of the literary novel. Selvon's work challenges such assumptions with a novel ostensibly about "nothing," or rather that which is deemed as "nothing" by the metropolitan cultures: "It didn't seem fair that West Indians had to know all about England in the schools, and on the other hand, Englishmen didn't know anything about the West Indies" (50).

Selvon's novel, then, is engaged in a reclamation of the marginality of home. The "nothing" of the prologue—Foster merely wakes up and rides the bus—is deliberate when Selvon ventriloquizes through Foster: "He stopped suddenly in front of a grocery and watched his dim reflection in the glass. 'My God,' he whispered to himself . . . 'is it possible for a man to think as I have been thinking since I got up this morning?'" (5). Selvon tries to validate the multiple aspects of his nationality, vocation, and ethnicity in one stroke when he juxtaposes the colonial past and the geography of the islands against the idea of the globe, through the image of Foster ruminating on this image of the spinning Earth.

An Island Is a World is a story of travels—to the United States, to the United Kingdom and back, and to India, but it is mainly a novel of Trinidad, with the protagonist, Foster (and also Andrew and Father Hope), settling down in Trinidad. Alison Donnell notes how this novel of the early 1950s "can be read as a dense, almost compressed, representation of its own recent historical moment . . . [of the] era of decolonization and the promise of Federation" in its multiple travel stories each with a different destiny and destination (119–120).

While many of the characters in the novel are Indo-Trinidadian, Selvon makes a conscious effort to include Creole characters of diverse and mixed ancestry. As Kenneth Ramchand notes in his introduction to the novel, the character of Andrew in the book is modeled on Selvon's own brother, but Andrew is not of Indian ancestry. Moreover, Johnny, the inveterate drunk, and his submissive wife, Mary, represent the dark side of the legacy of Indian indentured history (with its violence toward women) even as these characters also reveal themselves as people who seem to have something to cling to, an imaginary Indian home to which they try to return.[5]

Selvon was invested in the idea of a West Indian Federation that would secure the freedom of the entire region from the British. Therefore, even as he showcased Indo-Trinidadians in his writing, his work shows a conscious move away from creating a binary between the Indian and the Creole: "I feel that in the Third World, I could think of myself as being Indian, but outside of the Third World, I have to think of myself as being black, because in my experience all the people of the Third World are considered black" (Nazareth 1988, 93). The sense of the Indian in Selvon's work therefore fluctuates between the poles of black history, Trinidad, India, and England and locates itself on the island that lies somewhere in the middle of these trajectories.

THE FICTION OF HAROLD SONNY LADOO

Harold Sonny Ladoo is a younger writer with a terrible biography: born in poverty in Trinidad, he migrated to Canada with his family in 1968. His first novel, *No Pain like This Body*, was published by House of Anansi press in Toronto in 1972. Sadly, he was found murdered by the roadside in Trinidad in 1973. A second novel, *Yesterdays*, was published posthumously in 1974.

No Pain like This Body is a fictional narrative about rural Trinidad (called "Carib island" in the novel), in a location by the edge of a scorpion- and snake-infested rice land. The story is narrated through the eyes of the four children, Sunaree (twelve), Balraj (ten), and twin boys Rama and Pandey (eight), who flee from a brutally abusive father as the novel opens. They live in a seemingly self-enclosed community and in unrelenting poverty and degradation, facing the wrath of nature outside (lightning, snakes, scorpions, torrential rain) and of the brutal father inside the home. This is compounded by the disconnect between the Hindu religious beliefs of the community and the brutal and unforgiving harshness of their existence. One of the children, Rama, gets a high fever and dies in the rice land, an event that is beyond the comprehension of his siblings, who make up childlike stories about what must have happened to him. The mother loses her mind after the son's wake and runs off into the Tola forest, while the grandparents are left behind with the children. The novel ends with the continuous beating of the Indian drum by the grieving grandmother, a sign of mourning but also an indecipherable statement of survival and persistence against all odds.

The language of the novel is remarkably spare, replete with repetitions and with a distinct lack of smooth transitions across sentences. It is notable for its use of onomatopoeic constructions and elemental similes that help to construct an invisible narrator as young and clueless as the protagonist of the novel, Rama, and his three siblings. Here is Balraj running away from his abusive father: "The sky was black as Sunaree's hair, and Pa was watching Balraj. Balraj was almost out of the water. Pa leaned over the edge of the riceland and tried to hold his hand. Balraj ran *splash splash*. Pa ran eastwards along the rice. His feet went *tats tats tats*. There were many snakes in the riceland; they lived inside the deep holes near the barahar tree" (10).

The novel also employs a recognizable Hindi-laced dialect that marks the people and their lives as Indian:

> Rama and Panday walked up to Sunaree. They were not walking easy as a fly walks; they were walking like mules; feet went *splunk splunk splunk* in the water.
>
> "Rama and Panday all you walk easier dan dat," Balraj told them. "Dem crappo fish smart as hell. Wen dey hear all you walkin hard hard dey go run away." (6)

"Dey go run away" could easily be a translation of the Hindi phrase "wo bhaag jaate hain" (literally, "they run go away").[6]

But if the language harks back to Indian roots, the novel is also a pitiless rejection of the Hindu rituals and beliefs that are followed by the larger community but feel alien and inadequate to speak to the harsh lives they lead in this indentured world. Shalini Khan has noted the various references to the disregarding God, the inversion of accepted mythical stories from the Ramayana, and the final death of the child named Rama, thus sealing the inefficacy of religion in providing succor in this unforgiving world.

The spatialities of the novel seem claustrophobic, mimicking the small world of the children but also the bounded world of the Indo-Caribbean community here, with very little of the outside to counter the rain-soaked, swampy rice land that becomes the desperate refuge of the fleeing children. This is a land that is soaked with water, a land that is shown early in the novel as a place where there are small fish that the children play with and try to catch, part sea and part land. It is a liminal place between land and sea, and the hand-drawn map at the beginning of the novel indeed shows the Gulf of Tola on the West insistently pushing into the map and threatening to encroach into the ground (see how the neatline of the map on the east is broken by the Gulf of Tola) (Figure 19.2).

In this way, through the watery land of rice fields, Ladoo borrows from the sea journeys of the past that are now trying to ground themselves in new geographies. The novel retains transnational belongings through language and custom, but it also digs itself into the very soil of the land in the depiction of the rice land, claiming a Trinidadian belonging by rejecting many

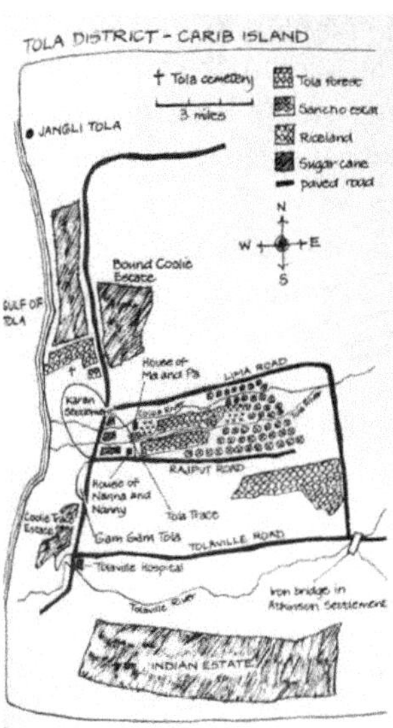

Figure 19.2. A hand-drawn map that opens the novel, *No Pain Like This Body*. It is accompanied by a page that makes it clear that the work equates the imaginary Carib Island with Trinidad.

of the rigidities of Hindu religious and ritualistic beliefs. Or as Sten Pultz Moslund puts it, "The transcultural centripetalizes or is *gathered in* by the novel's locality" (2015, 199; emphasis in the original). Ladoo's novel tries to lay down Indian stilts in the new land, making a claim that originates in a transnational journey, but it also attempts to ground itself in the marshy fields of the Tola District.

THE CARTOGRAPHIC IMPULSE IN SELVON AND LADOO

> Give me a map; then let me see how much
> Is left for me to conquer all the world"
>
> —Christopher Marlowe, *Tamburlaine the Great*

> If you must go abroad on continents
> You must travel with maps
> Prepared to argue longitude and latitude
> Or where is your country?
>
> —Samuel Selvon (Selvon and Salick 2012)

> Could it be that what cartographers do, albeit unwittingly, is to transform by mapping the subject they seek to mirror so as to create not an image of reality, but a simulacrum that redescribes the world?
>
> —J. B. Harley

Christopher Marlowe's character speaks those lines in a play written near the end of the sixteenth century; J. B. Harley states his suspicion about cartographers in 1990; and in between is Selvon's poem aired on the BBC on *Caribbean Voices* in 1950. *Tamburlaine*'s poetic statement reveals a knowledge of cartographic theory that became the stuff of debates only in the second half of the twentieth century. Marlowe presents the imperial gaze of the map in the era of exploration and shows its instrumentality as a tool of colonialism. Harley asks for a more self-critical study and practice of cartography where one focuses on not just what the cartographer does but also the politics of the practice. And then there is Sam Selvon, who bemoans the fact that while continents can be understood as spaces with histories and identities, his travel abroad necessitates the use of maps where other people can never find his home, which is a small island in the Caribbean.

Michel de Certeau takes the map as an example of the institutional gaze and notes the erasure of the earlier maps that included "art" (indicating the artifice or the craftedness of the map) by the later map with the unmarked view from the distant above. Today, the map seems to declare no origins or provenance, no markers of its begetters or its context, and de Certeau finds the claims of these maps to veracity and fact misleading: "The map, a totalizing stage on which elements of diverse origin are brought together to form the tableau of a 'state' of geographical knowledge, pushes away into its prehistory or into its posterity, as if into the wings, the operations of which it is the result or the necessary condition. It remains alone on the stage. The tour describers have disappeared" (1984, 94).

The "scientific" map places an invisible eye on the top at a distance, and the space becomes homogenized and cleared of its histories. Such a "clean" space that is geometrically arranged can also be carved up, broken up, and rejoined piecemeal according to the requirements of the imperial eye: "The tour describers have disappeared," as de Certeau says, and the spatial knowledge is just another consumable *product* of this exercise. In this "scientific" perspective, globalization is seen as the only way of looking at international networks. The viewpoint is distant, disembodied, and all-encompassing. Donna Haraway describes this as the "god trick" of universal and unmarked vision and says that instead "we need to learn in our bodies . . . how to attach the objective to our theoretical and political scanners in order to name where we are and where we are not" (1988, 581). As Yolanda Martinez-San Miguel states, "The Caribbean is on one hand an imperial/colonial archipelago that is imagined from the top by European cartographers . . . and on the other hand it is a decolonial archipelago, reimagined from below as a cultural and historical constellation or network of communities and individuals" (2017, 169).

Through their individual cartographic initiatives, both Selvon and Ladoo offer a more embodied, located, and partial perspective of this crushing global imposition from above. Selvon's novel takes the insularity of his home/place as the theme and shows transnationalism from below, a provincializing of the colonial and the global cartographic imagination. *An Island Is a World* opens with Foster imagining a spiraling globe and a distant dot that then come into focus as his own Trinidadian location:

> The world spun in his brain, and he imagined the island of Trinidad, a half degrees north of the Equator. He saw it on a globe, with the Americas strong like giant shadows above and below, and the endless Atlantic lapping the coastlines of the continents and the green islands of the Caribbean. The globe spun and he saw Great Britain and Europe, and Africa, the Eastern countries, Australia. Foster imagined Trinidad as it was, a mere dot on the globe. But he saw himself in the dot. (1)

In claiming a place on the world map for Trinidad, he also claims a place for the Indo-Caribbeans, flawed as they are in his novel, in the location of

Trinidad. The focus of the novel is on Trinidad and from there, out into the globe: "He got himself interested in the news by glancing at the newspaper a man was reading in the seat before him. New York, London, the Middle East. Things happening all over the world. And he, in a dilapidated bus traveling from Petite Bourg to Port of Spain—a city with emotion and life and radios and the modern sewerage system, but unknown to the billions of people on the earth—with the world spinning in his mind" (4). In the prologue and in the ending of the novel, the Indian self is subsumed under that of the Trinidadian identity. In keeping with his desire for West Indian Federation, Selvon emphasizes a unity of diverse communities under the national and regional umbrella. The Indian becomes part of the multiverse of Trinidad, and even while a dot on the world map, Trinidad joins the visuality of the globe in this novel. What Jonathan Pugh says of the archipelagic imagination coincides here with Selvon's vision of the intermixed and interrelated stories and spaces: "The notion of the archipelago unsettles static tropes of singularity, isolation, dependency and peripherally that presently dominate how islands are conceptualized in the literature" (2013, 11).

However, Ladoo's novel, *No Pain like This Body*, seemingly goes against this idea of the transnational and interconnected histories in the closed space of Indo-Caribbean misery and history in the rice plantations of Trinidad. Here the space on the map (see Figure 19.2) is squared off and surrounded by the white nothingness of the outside world, of the globe. Nevertheless, Ladoo's novel attempts to carve out that space and that history as peculiarly Indian and Trinidadian at the same time and thus effectively to reinsert the Indian into the Caribbean and the Trinidadian.

No Pain like This Body opens with an actual map of the relevant location, the Tola District, on the fictional "Carib Island." The verso of the text gives the reader an appendix for

> "Carib Island" (the map is titled "Tola District—Carib Island"):
> Area: 1000 square miles
> History: Discovered by Columbus in 1498. Taken over by the British in 1797. East Indians came to Carib Island to work on the sugar plantations from 1845 to 1917.
>
> Chief Exports: Sugar, petroleum, rum, cocoa, coffee, citrus fruits and asphalt.

The information clearly marks this island as Trinidad, and "Tola District," in its location on the map, fits the coastal region of the predominantly Indian-inhabited central Trinidadian space. The place "Tola District," while fictional, references the Bhojpuri/Hindi word *tola*, meaning "village" or "district." The locations marked on the map are predominantly of Indian origin: Jangli Tola (*jungli* is "wild" in Hindi); Bound Coolie Estate (noting the indentured laborers), Rajput Road (referring to a community in western India). But there is also Sancho estate, which references the Spanish past of Trinidad, and Atkinson settlement, which marks the British presence, something that does not immediately register on the reader's mind at first sight.

Ladoo's novel places the narrative in 1905 Trinidad—indenture would be over in a decade or so, and the laborers had received a piece of land to till from the planters in the settlement, but much of this land was unarable. This novel presents the reader with an extremely Indocentric story; in fact, there is little *overt* outside to the Indian story, no stated "other" here (even though, as I showed earlier, there are "other" histories and presences marked on the map less prominently). Similarly, the map has no delineated outside and takes the coast of central Trinidad as its focus. The outside is present in its unmarked state instead. The larger world is there, but the map does not purport to tell the whole story, only that of this isolated, enclosed, desperately poor Indian community. By overlaying this Indian story on the map of "Carib island," which is Trinidad, Ladoo makes a textual and visual claim to national identity and to a Caribbeanness that has not been readily ascribed to the Indian community in Trinidad, certainly not in the pre-independence setting of the novel.

Ladoo rejects traditional practices of Hindu religion in the narrative through repeated references to the Indian religious epic *Ramayana*. Like this textual reference to the Indian epic, the novel also transfers Indian geography to Trinidadian land and history. The map shows the "Tola District" (the name deriving from Bhojpuri and Hindi) superimposed upon the places on the Trinidadian map, just as the indentured laborers were transferred from northern India to this island. Such transpositions of geographical markers are not uncommon in the cultural life of the Indo-Trinidadians as a whole. Sharda Patesar writes about the public performance of *Ganesh utsav*, or Ganesh festival, in Suchit Trace in southern Trinidad: every year, the Indo-Trinidadian devotees perform the ritual immersion of the Ganesh idol at the meeting of three waterways or "the three river trails—the Dodge, Bhagmania (as it is locally named), and the Cut Channel—[where the waters] make a confluence" (2016, 191). This is a transposition of the sacredness of the space of the *prayag* (which is the meeting of the three rivers of Ganga, Yamuna, and Saraswati in India) to the hybridized creation of a similar confluence in Trinidad, and it reveals the palimpsestic layers of history that lie beneath contemporary practices and namings.

The language of the novel (lots of Hindi linguistic structures used in the Creole and English phrases of the narrative), as well as the references to the beliefs and the cultural practices of Hinduism and to the gods of the religion, all create a network across the Tola District in the novel and beyond Trinidad, across the Indian Ocean to reach India and the history of indenture that started there. In that sense, the map is not really self-enclosed or bounded, only seemingly so.

Note that Ladoo's map has no intact east and west neatlines—the innermost line near the edge of a printed map that separates it from the margin—and on the west the Gulf of Tola (which represents the Gulf of Paria between Venezuela and Trinidad) breaks the neatline of the map and strongly threatens to flow in. On the east, the district disappears into the white margin, which one assumes is the rest of Trinidad/Carib island. Thus, pressed by the waters of the sea, the map posits an openness in the direction of the interior of Trinidad but chooses to concentrate on a small section of Indo-Caribbean geography and history. The focus on the hyperlocal is a way of denying the global gaze and adopting the partial standpoint of embodied perspective. This map fades out not only the rest of the world—we only see the insistent waters threatening to flood the Tola District—but also the country as a whole: one zoomed-in piece of the land, the Tola District, is available for the reader, while the acknowledgment of the island-nation is in the accompanying appendix and in the open borders of the map. If Selvon tried to chart the transnational network where the Indo-Caribbeans are embedded within a global network, Ladoo also gestures toward this transnationalism but keeps the reader's gaze firmly fixed on the hyperlocal Tola District. In the map, the waters of the Gulf of Tola come washing into the space that is the Tola District; the world is out there and present, but the author and his map pull our eyes toward the tiny settlement, its forest and the rice land and its bordering houses.

The two novels, Samuel Selvon's *An Island Is a World* and Harold Sonny Ladoo's *No Pain like This Body*, manipulate the notion of boundedness and travel and keep both ideas in play in novels that map the travels as well as the travails of the Trinidadian with Indian belongings. They allow us to see the multiple ways in which the spatial model of the island and the archipelago get energized in the project of establishing an Indo-Caribbean claim to Trinidadian histories. It is a complicated sense of belonging to an Indian past that does not claim India, that lives in a transversal network of sea and land between Trinidad and India and the West. Ladoo's novel projects a middle stage between the "tectonic of continents" and these voyagings and contacts—an emerging sedimentation of histories, a layering of partially obscured and ambiguous Indo-Trinidadian literature and histories, a swamp of material belongings that is neither land nor sea and which makes claim to a native self, based on a fictional sense of boundedness in this swamp. The acknowledged transnational trajectories attempt to find ground and material space here. Selvon's novel, on the other hand, takes the global turn and places Trinidad on the world map through a network of travels, including those of Indo-Caribbean colonial history, and attempts to destabilize established notions of Trinidad in the world and of the Indian in Trinidad by highlighting these exchanges.[7] In this way, these two novels instantiate the multiple ways in which Indo-Caribbean authors tried to map their journeys from past to present, to create a home that is also an away-from-home, a layering of travel, dispersal, arrival, and struggle onto the island that they want to call home.

In the introduction to *The Cartography of Exile*, Karen Bishop makes a necessary connection between the act of mapmaking and the exilic imagination of the writer/mapmaker: "[M]apping the world around her provides for a new world in which she might live and how she draws that world determines and how and what she will know of that world" (2016, 250). Together, Selvon's roving imagination starting from Trinidad and encompassing the globe is a world-forming device, a way of creating the world according to his perspective. Ladoo's map similarly tries to do away with the rest of the nation and the world and concentrates on the small rice-growing community of Indian laborers—in doing so, Ladoo tries to unsettle the solidities of linear history and insert an Indian claim of belonging to Trinidadian history and land through the depiction of an unnostalgic picture of Indian heritage and a terrible but real connection to Trinidadian space.[8] The two novels present a holographic perspective on Indo-Caribbean history as well as narrative—one shows the travels that lead to and from the Caribbean and make Trinidad the nub of the story of the self; the other takes a hyperclose look at the small, beleaguered community eking out a survivalist mode of living and yet having nowhere else to go or call home. Together they alert us to the idea that both the stay on landmasses (islands, continents) and the journey on sea passages that lead to them need to be taken into account in any complete story of the Indian in the Caribbean.

In discussing how the Caribbean is constructed in multiple registers, Mimi Sheller uses the image of "moving and maneuvering, ducking and weaving" to describe a process of globalization from below: "Moving and maneuvering, ducking and weaving in the narratives and debates on globalization are, and have been, actual or potential migrants, flipping through the one-way traffic of globalization by the cultural back roads, absorbing and transforming the global agenda into that of their own, at the same time transforming the cultures and societies into which they enter, momentarily or forever" (2003, 179). I would add "expanding and contracting" as the verbs of movement that describe the cartographic impulses of Sam Selvon and Harold Sonny Ladoo in their efforts to map the multiple natures of their Indo-Trinidadian experience into the literature of the region and the world.

PRACTICAL AND METHODOLOGICAL IMPLICATIONS

The Indo-Caribbean is frequently seen as an ethnically bounded concept, as an idea that limits the subject positions to a nostalgic return to the Indian motherland. However, the notion of the Indo-Caribbean includes a complex allegiance to both here and away, the local and the translocal, and it codes the essence of what are seen as essentially Caribbean contradictions in a renewed formulation.

Khal Torabully invented the idea of "coolitude" to connect the different articulations of the Caribbean to the Indian presence in the region. Brinda Mehta uses the idea of *kala pani* (the dark waters) to describe this world, while Shalini Puri uses the Caribbean notion of *dougla*, or bastardization, to explain the radical practices of this community. In this chapter I add to those discussions by examining the possibilities of applying archipelagic thinking to the writings of Indians in Trinidad in order to explore how the immured sense of being Indo-Caribbean in two novels can be read in connection with the hemispheric travels of the authors themselves and the fraught connection of the Indo-Caribbean subject to archipelagic Caribbean cultures. Reading the texts of the Indo-Caribbean allows us to note the Indo-Caribbean subject as an authorial position that takes the Caribbean/the island/the archipelagic frame of the region as a starting point in order to interrogate colonial/global networks of power. The archipelagic framework, one that insists on the intersections of geography and history, allows for both a politics of place and the politics of translocal travel—that is, an insistence on acknowledging the grounded realities of place but also the fluid changes associated with transnational affiliations.

NOTES

1. See Walcott's Nobel Prize speech (1992) where he connects Indian traditions to African heritages in the Caribbean context. Also see Glissant (1997), Hay (2006), Baldacchino (2008), Stratford et al. (2011), and Pugh (2013).

2. *Kala Pani* refers to the belief among caste Hindus that crossing of the seas would result in the loss of one's caste and exile from one's caste community. This belief among the Indians also made the contractual journey of the laborers (supposedly to end after a period of time) into a permanent move to settle—after all, many of these Indians could not return to India after the expiration of their contracts even if they somehow managed to scrape together the money because Indian society would not accept them back into the fold.

3. Also see Pugh 2013, where Pugh states, "Archipelagic thinking denaturalizes the conceptual basis of space and place, and therefore engages 'the spatial turn' currently sweeping the social sciences and humanities" (9).

4. A Norddeutscher Lloyd Bremen poster, exhibited in the *Antillean Visions* exhibition at the University of Miami in 2018. William Pestle has a commentary on it in the exhibition catalog (44).

5. Many writers, such as Naipaul, Shani Mootoo, and Gaiutra Bahadur, among others, have depicted this in their works. The Indo-Caribbean community is marked

by patriarchal violence against women, which has historical roots in the British policies during the years of indenture.

6. For more on the complex language in the novel, see Gonzalez 2018.

7. Also see Selvon's keynote, "Three in One Can't Go," delivered to the "East Indies in the Caribbean" conference held at the University of the West Indies, Trinidad, in 1979.

8. For more on place in Ladoo's novel, see Moslund 2015.

REFERENCES

Baldacchino, G. 2008. "Studying Islands: On Whose Terms? Some Epistemological and Methodological Challenges to the Pursuit of Island Studies." *Island Studies Journal* 3, no. 1: 37–56.

Baldacchino, G., and Eric Clark. 2013. "Introduction: Islanding Cultural Geographies." *Cultural Geographies* 20, no. 2: 129–34.

Bishop, Karen. 2016. "Introduction." In *Cartographies of Exile: A New Spatial Literacy*, edited by Karen Elizabeth Bishop, 1–22. New York: Routledge.

Brannigan, John. 2015. *Archipelagic Modernism: Literature in the Irish and British Isles, 1890–1970*. Edinburgh: Edinburgh University Press.

De Certeau, Michel. 1984. *The Practice of Everyday Life*. Berkeley: University of California Press.

Deloughrey, Elizabeth. 2010. *Routes and Roots: Navigating Caribbean and Pacific Island Literatures*. Honolulu: University of Hawai'i Press.

Donnell, Alison. 2012. "The Island and the World: Kinship, Friendship and Living Together in Selected Writings of Sam Selvon." *Journal of West Indian Literature* 20, no. 2: 54–69.

Glissant, E. 1997. *Poetics of Relation*. Translated by Betsy Wing. Ann Arbor: University of Michigan Press.

Gonzalez, Shawn. 2018. "Ethics of Opacity in Harold Sonny Ladoo's *No Pain like This Body*." *CLR Journal* 4, nos. 1–2 (fall): 215–37.

Haraway, Donna. 1988. "Situated Knowledges: The Science Question in Feminism and the Privilege of Partial Perspective." *Feminist Studies* 14, no. 3 (autumn): 575–99.

Hay, P. A. 2006. "A Phenomenology of Islands." *Island Studies Journal* 1, no. 1: 19–42.

Khan, Shalini. 2013. "Harold Sonny Ladoo's *No Pain like This Body*: The Ramayana and Indo-Caribbean Experience." *Wasafiri* 28, no. 2: 21–27.

Ladoo, Harold Sonny. 1987. *No Pain like This Body*. London: Heinemann.

Martinez-San Miguel, Yolanda. 2017. "Colonial and Mexican Archipelagoes: Reimagining Colonial Caribbean Studies." In *Archipelagic American Studies*, edited by Brian Russell Roberts and Michelle Ann Stephens, 155–73. Durham, NC: Duke University Press.

Moslund, Sten Pultz. 2015. *Literature's Sensuous Geographies: Postcolonial Matters of Place*. New York: Palgrave Macmillan.

Munasinghe, Viranjini. 2001. *Callaloo or Tossed Salad? East Indians and the Cultural Politics of Identity in Trinidad.* Ithaca, NY: Cornell University Press.
Nazareth, Peter. 1988. "Interview with Sam Selvo." In *Critical Perspectives on Sam Selvon*, edited by Susheila Nasta, 77–94. Washington, DC: Three Continents Press.
Nerlekar, Anjali. 2011. "Living Beadless in a Foreign Land: David Dabydeen's Poetry of Disappearance." In *Talking Words: New Essays on the Work of David Dabydeen*, edited by Lynn Macedo, 15–29. Kingston: University of West Indies Press.
Outar, Lisa. 2015. "Indianness and Nationalism in the Windrush Era." In *Beyond Windrush: Rethinking Postwar Anglophone Caribbean Literature*, edited by J. Dillon Brown and Leah Reade Rosenberg, 27–40. Jackson: University Press of Mississippi.
Patesar, Sharda. 2016. "The Ritual Art of the Ganesh Utsav in Trinidad." *Asian Diasporic Visual Cultures and the Americas* 2: 181–99.
Pestle, William. 2018. "Tourism." In the catalog for *Antillean Visions: Maps and the Making of the Caribbean: An Exhibition of Cartographic Art* at the Lowe Art Museum, University of Miami, 44.
Patke, Rajeev, S. 2018. *Poetry and Islands: Materiality and the Creative Imagination.* Lanham, MD: Rowman & Littlefield International.
Pocock, J. G. A. 2005. *The Discovery of Islands: Essays in British History*. Cambridge: Cambridge University Press.
Pugh, Jonathan. 2013. "Island Movements: Thinking with the Archipelago." *Island Studies Journal* 8, no. 1 (May): 9–24.
Puri, Shalini. 2004. *The Caribbean Postcolonial: Social Equality, Post-nationalism, and Cultural Hybridity*. New York: Palgrave Macmillan.
Ramchand, Kenneth. 1993. "Introduction." In *An Island Is a World*, by Samuel Selvon, v–xxv. Toronto: TSAR.
Ronström, Owen. 2013. "Finding Their Place: Islands as Locus and Focus." *Cultural Geographies* 20, no. 2: 153–65.
Selvon, Samuel. 1986. "Three into One Can't Go—East Indian, Trinidadian or West Indian? Samuel Selvon Discusses the Question of an East Indian Identity," *Wasafiri,* 3, no. 5: 8–11.
Selvon, Samuel. 1993. *An Island Is a World.* Toronto: TSAR.
Selvon, Samuel, and Roydon Salick. 2012. *The Poems of Sam Selvon*. Royston, UK: Cane Arrow Press.
Sheller, Mimi. 2003. *Consuming the Caribbean: From Arawaks to Zombies*. London: Routledge.
Stratford, Elaine, Godfrey Baldacchino, Elizabeth McMahon, Carol Farbotko, and Andrew Harwood. 2011. "Envisioning the Archipelago." *Island Studies Journal* 6, no. 2: 113–30.
Thieme, John. 1990. "Old Talk: Interview with Sam Selvon." *Caribana* 1: 71–76.
Walcott, D. 1992. "Derek Walcott: Nobel Lecture." The Nobel Prize. December 7. https://www.nobelprize.org/prizes/literature/1992/walcott/lecture.

Chapter Twenty

On Archipelagic Beings

Gitanjali Pyndiah

CREOLIZATION AND ARCHIPELAGIC THINKING

In 1981, Franco-Martinican novelist and essayist Édouard Glissant wrote in *Le discours antillais* that "creolization is, first, the unknown awareness of the creolized" (1996, 3). His early reflections (1960s–1990s) focused on the specificity and singularity of Caribbean creolization in relation to the history of European slavery and the displacement of the African peoples. Glissant pinpointed the racialized hierarchies of power, which continued in the present, and described the assimilation of French cultural values as a form of paralysis. The French Caribbean was "paralyzed by being scattered geographically and also by one of the most pernicious forms of colonization: the one by means of which a community becomes assimilated" (5). Glissant also highlighted the colonial erasure of the culture of the enslaved people and their descendants: the "communities [were] condemned as such to painless oblivion" and the "definitive muteness of those peoples [were] physically undermined" (2).

However, in the author's later theoretical work (1990s–2000), such as *Traité du tout-monde* (1997b), the discourse on creolization moved away from its Caribbean singularity to mark globally occurring processes of mixture (Stewart 2007, 3). Glissant viewed the Caribbean space as a philosophical geography, a space from which the philosopher could theorize processes of cultural mixture. In so doing, Glissant universalized the notion of creolization as a phenomenon of "archipelagic thinking" and cultural hybridization—one that occurs across the world—where languages and other practices like music or literature emerge creatively from the interaction of cultural difference. In the evocative style of his later writings, Glissant proposed to view the world as an archipelago witnessing a process of creolization.

The term "creolization," within the Euro-American (white) academic fields of linguistics, anthropology, and sociology, has mostly been used to define linguistic or cultural mixture (Stewart 2007; Cohen 2007a, 2007b; Hannerz 1987). On the other hand, the term "archipelagic thinking," in island studies, geography, or philosophy, focuses on topology and physical geography (island, continent) to formulate modes of thinking about relations, in the same way that Glissant theorizes from the standpoint of creolization. This chapter examines how such concepts, "creolization" or "archipelagic thinking," attempt to disrupt the topological relations associated with islands, predominantly articulated in the humanities and social sciences as land/sea, island/continent, island/island (Baldacchino et al. 2011), but omit to reveal the "coloniality of power" (Dussel, Jáuregui, and Moraña 2008) that informs the human histories of archipelagoes.

The critical argument here is a caution to scholars who use archipelagic thinking as an all-embracing framework. This chapter brings the Jamaican philosopher Sylvia Wynter into conversation with Glissant to examine how the creolization in the Caribbean/Indian Ocean (drawing from the Mauritian context) is embedded in the creative practices of the Black/Creole people, a historical occurrence that is well explored in Caribbean scholarship. It is thus important to be attentive to the ways in which such semantic terms, which center "culture," "language," "society," or "topology," can undermine the "cartographies of struggle" (McKittrick 2006) that are the originating point from which those processes of archipelagic thought emanate.

THE PHYSICAL ARCHIPELAGO AND GEOPOLITICS

To ground the comparison of the approaches of the two Caribbean theorists, Glissant and Wynter, this chapter draws from the archive of a different archipelagic site—the history of the Indian Ocean archipelagoes such as Mauritius, which consists of the mainland and the islands of Rodrigues, Agaléga, Tromelin, St. Brandon, and the Chagos Archipelago. Together with Réunion island, an overseas department of France, mainland Mauritius and Rodrigues form part of what has been described as the Mascarene Islands, and they share a common history of European colonization, slavery, and indenture with the Seychelles (another archipelagic country in the Indian Ocean). These archipelagoes were uninhabited before the monopolization of trade by Europeans in the Indian Ocean region in the seventeenth century. Within a few centuries of colonialism, the islands were reduced to sugar plantation economies under Dutch exploitation colonialism (1638–1710), French settler colonialism (1710–1810), and English administration (1810–1968). This established a plantation structure informed by two exploitative systems of labor, slavery and indenture, which displaced people from Africa, Madagascar, India, and China for over three hundred years.

The Indian Ocean islands offer a productive space to reflect on how creolization processes intersect with the physical archipelago. During the French and English period, "Creole" was used as a noun to designate descendants of settlers who were born in the colony. This is similar to the ways in which *criollo*, meaning Creole, was used to refer to settlers in the Latin American context (Mazzotti 2016). In the Latin American context, white settler Creoles have preeminence over Amerindians, the African diaspora, and peoples of mixed descent (Mazzotti 2016). The term Creole further undermines the experience of Black people and the creolization process that took place under the violence of colonization. In French Mauritius, the term "Creole" also appears in marriage legislation to categorize enslaved people of mixed races (mulattoes or *gens de couleur*) or to denote people of African and Indian origins (TJC 2011, 107). As a marker of inferiority, the use of the term "Creole" came to reflect the racialized structures of imperial power that justified enslaved and indentured labor from Africa, Madagascar, Southeast Asia, and India on the islands.

The term "Creole" was also used to describe the language created by the enslaved Malagasy and African peoples in the eighteenth century after the epistemic erasure of their mother tongues (Pyndiah, 2018a). Furthermore, the colonial terms "creole" and "ti kreol" in Mauritius are still used to refer to Black/Creole people whose mother tongue is the Creole language (Boswell 2006, 47). In the context of the Indian Ocean islands, the Black/Creole people have less political power and are subjected to epistemic and structural marginalization, stigmatization and ghettoization (46–55). Racism and antiblackness in mainland Mauritius are discussed lengthily in the reports of the Mauritian Truth and Justice Commission (TJC 2011).[1]

The notion of creolization used in the 1960s and 1970s during the Mauritian independence period (1968) was deployed to build a nationalist narrative based on the Creole language as a marker of national unity. I argue here that a Glissantian notion of archipelagic thinking was promoted that focused on prioritizing cultural mixture over and above ethnic, racial, and religious identities. However, in contemporary Mauritius, it is common to hear different semantic usages of the term "Creole": *mezon kreol* (colonial house), *kuizinn kreol* (cuisine of blended influences), *ti kreol* (a Black/Creole person, often of the rural areas), *langaz Kreol* (Creole language), *nou tou kreol* (we are all Creole, denoting those born on the island and culturally mixed, that is, from a process of creolization). As some of these usages reveal, and as explained by the report of the Truth and Justice Commission, racism is well engrained on mainland Mauritius. Furthermore, exiled Chagossians have been paradoxically called *zilwa* (from the French word *illois*), meaning "islanders," used as a derogatory term by mainland Mauritians who have assimilated with a settler conception of the (is)land that they colonized and on which they were colonized.[2] The binary imaginaries of island/sea and land/continent are reproduced within the island itself under the nationalist narrative of the mainland versus the other island(s). In this sense, the physical geographical entity of the island connected to other islands is rendered void through the hierarchies of power that govern the archipelago.

It is in this sense that this chapter discusses how a nationalist narrative of archipelagic thinking can undermine the afterlives of a history of enslavement, exile, and the state oppression of the Black/Creole people. While archipelagic thinking valorizes "the power of cross-currents and connections, of complex assemblages of humans and other living things, technologies, artefacts and the physical scapes they inhabit," it does not critically engage with the coloniality of power within the politics governing the physical archipelago itself. This chapter raises the question of whether archipelagic thinking is engaged critically enough with the reality of the geopolitics of the archipelago, in its myriad global settings, before it is abstracted to a conceptual framework. As Baldacchino et al. caution, "The archipelagic relation has, of course, been used cynically and opportunistically in the processes of colonial acquisition: island constellations have been convenient stepping stones of dominion; those shadows should not be forgotten in further analysis of the relationality that archipelagic thinking engenders" (2011, 125). This danger, I argue here, is very real in Mauritius, where a settler/plantation mentality of those in power determines a national narrative that homogenizes the experiences of different cultural/ethnic/racial groups under the banner of the "Creole island."

The appropriation of a Glissantian discourse that universalizes or romanticizes creolization and archipelagic thinking has been problematized across Caribbean literature and scholarship between 1970 and 1990, as well as by contemporary Caribbeanist social scientists in relation to the plantation societies of the post-1492 New World (Britton 1999; Gallagher 2007). Following this logic, this chapter also seeks to problematize the universalization of the terminology of "creole/ization" as a process of cultural mixture/*métissage* and to challenge the "spatial turn" of archipelagic thinking. Glissant's reflection resonates here: "If we posit *métissage* as, generally speaking, the meeting and synthesis of two differences, creolization seems to be a limitless *métissage*, its elements diffracted and its consequences unforeseeable. . . . Here it is devoted to what has burst forth from lands that are no longer islands" (Glissant 1997a, 34).

EX-SLAVE LABOR ARCHIPELAGO

Both Katherine McKittrick, whose work is central to the scholarship on Black geographies, and the philosopher Sylvia Wynter refer to the "ex-slave labor archipelago of the post-1492 Caribbean and the Americas" as "global archipelagos of poverty" and "Black cartographies of struggle" (McKittrick and Wynter 2015). They reiterate that across the post-1492 New World, colonial organization followed a logic of racialized segregation between white elite geographies and Black cartographies of struggle. Wynter refers to the symbolical year of 1492 as representing the beginning of "a founding politico-statal mercantilist economic system" based on racialized power structures of labor (McKittrick and Wynter 2015, 46). By 1900, more than half of Asia, 98 percent of Africa, and most of the former slave plantation archipelago islands of the Caribbean (and the Indian Ocean) were under direct colonial rule. Furthermore, while islands, as geographic and geological formations, are physically separated by water and have been idealized as pristine, virginal, and isolated in the colonial imaginary, they were discovered, inhabited, or visited by indigenous people whose worlds were brutally displaced by the post-1492 New World order.

While archipelagic thinking may be a useful framework for breaking binary topological relations between islands and continents, the analysis of the binaries and networks of power within the archipelago as part of a postcolonial geography of the present, based on class, race, gender, sex, and ability, exclusions operative on both continents and on islands, remains undertheorized. Thinking about the archipelago using a topographical lens promotes an eagle-eye view of the physical geography, which encourages distance and abstraction from the human geographies of both islands and continents. Wynter's reflections on the "post-aesthetic creative practices" developed from cartographies of struggle (for example, cultural films from and about the underclass of Kingston, Jamaica), what Wynter calls "Third World shantytown archipelagos," offers insights from the ground of intra-island and intra-archipelago cultural politics (1992, 240–241). For Wynter, archipelagoes represent first and foremost islands of social exclusion that are connected to a historical process of coloniality. However, this chapter also draws from another aspect of Wynter's philosophical configurations, what Caribbean political theorist Anthony Bogues (2010, xii) theorizes as her "politics of being," to make two observations. First, Wynter's notion of spatiality is intertwined with an idea of archipelagic being; second, her notion of indigenization represents the process by which archipelagic beings—for example, the Black/Creole people in the context of the Indian Ocean islands of Mauritius—resist, survive, rehumanize the landscape and create new modes of existence (xxiii).

In contemporary discussions of indigeneity, Indigenous studies scholars (Jackson 2012; Newton 2013) have raised probing questions problematizing current usage of the term "indigenization." Melanie Newton is particularly attentive to how the "new natives," predominantly Africans and their descendants, replaced the original Antilleans and *became* indigenous to the Caribbean. In this historical scenario, processes of indigenization undermine the history of the struggle of the indigenous people of the Americas who continue to face violence by the state. In the context of mainland Mauritius, a notion of indigeneity has also been appropriated by former colonized (enslaved and indentured) people who then paired it with a settler mentality after independence. Shona Jackson describes a similar phenomenon in Guyana, where the Black/Creole subjects' labor in the colonies becomes their justification for the appropriation of the land on which they worked. Critiques against the appropriation of the notion of indigeneity are particularly attentive to Caribbean writers' adoption of an imperial narrative that makes the struggle of the Indigenous peoples of the Americas invisible (Newton 2013).

However, what this chapter draws from Wynter's discussion is how she establishes a correlation between indigenization and the (re)humanization of land. Wynter's work links to and illuminates McKittrick's notion of geography as mapping stories. For McKittrick, "a flesh-and-body worldview [is] implicit to the production of space" and allows us "to consider the ways in which space, place, and poetics are expressing and mapping an ongoing human geography story" (2006, 122). Within this view, the archipelago takes the form of a *tropological* human relation to the topology of the material geography—that is, the "ongoing" human story mapped by the history as well as the geography of the space. In the same vein, creolization as cultural hybridity or universalized as global phenomenon examines the process from the distant position of the linguist or the anthropologist. Creolization, as lived materially and geographically in the Caribbean or the Mascarene Islands, is historically situated in the creative process of the humanization of the enslaved.

The structure of the plantation island was informed by the racialized cartography that still defines the postcolonial space of the islands. For Françoise Vergès, expert scholar of Indian Ocean colonial histories, creolization "is not a harmonious process; it involves exclusion and discrimination" and was determined by "conditions of the plantation by which the process of creolisation emerged" (2007, 148). Rinaldo Walcott, theorist in Black Diaspora Cultural Studies, gender and sexuality, explains that creolization "takes place in the context of unequal and brutal power arrangements alongside forms of severe cultural dominance. . . . I thus do not read creolization as simply assimilation/integration; nor do I understand it as a process through which hybridity emerges" (2015, 187).

The late Caribbean poet, Kamau Brathwaite, who coins the term "creolization" in 1974, explained in the context of Jamaica that "Creolization began with 'seasoning'—a period of one to three years, when the slaves were branded, given a new name and put under apprenticeship to creolized slaves. During this period the slave would learn the rudiments of his new language and be initiated into the work routines that awaited him" (1995, 203).

It is in this sense that Wynter's notion of the spatiality of the ex-slave archipelagoes, across the Atlantic and Indian Oceans, is intertwined with the realities of the creolized/archipelagic human beings who recreated lives under systems of oppression and, as Wynter articulates, indigenized and rehumanized colonized lands with their creative practices. Archipelagic beings, in the context of the Indian Ocean, are the Black/Creole people who still live in the most precarious conditions in contrast to settler islanders (who are of colonial descent). In my research, I look at the creative practices in the Creole language that developed during slavery (Pyndiah 2016, 2018a, 2018b). It is in this creative space that the people of the Indian Ocean islands find a shared oral, aural, and sonic Creole culture, where the sega, a polymorphous performative art form embodied in dance, music, and song in the Creole mother tongue, emerged from the context of slavery. While Creole and the creative practices in the language are consumed as part of the discourse of national culture, the creative labor of the Black/Creole people is not recognized in the Mauritian narrative of creolization.

BEING HUMAN AS PRAXIS: INDIGENIZATION AND HUMANIZATION

Wynter's *being human as praxis* project and the difference between creolization and indigenization provide alternate lines of thought in thinking about agency within ex-slave archipelagoes and how that, in turn, can inform archipelagic thinking. For Wynter, "while the 'creolization' process represents . . . a more or less 'false assimilation' in which the dominated people adopt elements from the dominant . . . in order to obtain prestige or status, the 'indigenization' process represents the more secretive process by which the dominated culture survives; and resists" (Bogues 2010, xiii).

Wynter (1970) explores the Jonkonnu folkdance, a creative practice brought by Africans to the Caribbean islands, and elaborates on the cultural process of "indigenization" that the dance practice went through, where being human is central to the cultural and aesthetic production of the dance (Walcott 2015). For Wynter, this cultural history is to be found not in "writing" (referring to the culture of literature brought by colonization versus the generational knowledges ingrained in the orality of indigenous people) but in "those 'homunculi' who humanize the landscape by peopling it . . . with all the rich panoply of man's imagination" (Bogues 2010, xii). The process of indigenization, according to Wynter, is interrelated with a process of rehumanization that allowed enslaved Africans to create possibilities of life and culture under the violence of the system. These practices allowed the enslaved to resist the negation of their being and to foster processes of affirmation (xxiv).

In the case of Mauritius, the Creole language brought forward new autopoetically instituted cultural roots, traditions, and knowledges within the colony that were birthed out of a process of indigenization (Wynter 2007, 13). While there were no indigenous people on the islands of Mauritius, I would argue that the enslaved people were compelled to "autopoetically institute" a Creole culture in order to rehumanize the colonized landscape and create possibilities of life and culture under regimes of oppression. Mauritian Creole originated from the new languages, living systems, and modes of existence in the colony and carries the oral histories of resistance, resilience, and survival through the oral, aural, scribal, sonic, performative, and embodied aspects of the language.

The first people who were brought under Dutch slavery indigenized the island (Malagasy, Africans, and South/Southeast Asians in the Indian Ocean contexts), similar to the history of the Caribbean islands, which for Wynter is, in large part, the history of the indigenization of the Black person. Wynter looks at "cultural forms of subaltern creative invention [that] were constitutive of a drive to find new modes of existence," and in her formulation it is the Black people who have created life practices that are not part of the dominant (colonial) regime since slavery (Bogues 2010, xii). Hence the process of indigenization is directly related to a process of humanization, and *being human* is at the center of its epistemology and aesthetics.

Wynter's critical observation of creative practices in Jamaican culture, her theoretical framework on indigenization, and her intellectual approach to unsettle coloniality—by positioning *being human* as central to the imagining of an ongoing human geography story—is productive here for rethinking the archipelagic. We may ask, Where are the archipelagic beings within both islands and continents who are indigenizing and rehumanizing the present landscape, outside the geographies of land extraction, exploitation, and policed migration? How can we rethink coloniality or the present systems of control and discipline that regulate the precarious crossing of oceans as well as continental frontiers? Can we imagine archipelagic thinking as a reconfiguration of knowledges that does not abstract human stories of geography from the eagle-eye topography of islands and landmasses?

PRACTICAL AND METHODOLOGICAL IMPLICATIONS

The concepts of "coloniality of power" and "decoloniality of knowledge" emerged from the work of Latin American scholars Aníbal Quijano, Enrique Dussel, and Nelson Maldonado-Torres (Dussel, Jáuregui, and Moraña 2008; Quijano 2008; Maldonado-Torres 2007) and was promoted in Anglo-American academia by the Argentinian semiotician Walter Mignolo (2012). They have influenced the present momentum toward "decolonising knowledge" in British universities as well as cultural institutions across various continents. "Coloniality" "encompasses the transhistoric expansion of colonial domination and the reproduction of its effects in contemporary times" (Dussel, Jáuregui, and Moraña 2008, 2). It defines the practices and structures put in place during the European colonization of the "global South"—regrouping historically, more than geographically, the Caribbean, the Americas, Africa, South/Southeast Asia, the Pacific Islands, and the Indian Ocean region—and the residues of colonization in the reproduction of racialized structures of power, knowledge, and hierarchies of labor.

In this sense, this discursive reflection on archipelagic beings takes the Mascarene Islands in the Indian Ocean, namely Mauritius, as a case study to reflect on the relevance of archipelagic thinking in challenging coloniality. Scholarship from Caribbean and Black studies provides critical insights in rethinking binary formations in island studies around the archipelago—land/sea and island/continent, for example—which exclude the geopolitical analysis of islands as spaces of exile, quarantine, militarization, or consumption as tourist destinations.[3] This reflection follows a Glissantian genealogy of archipelagic thinking and creolization to discuss how a universalized notion of archipelagic thinking can render invisible the human geography and histories of coloniality that bind both islands with continents and archipelagoes from one ocean to another (Martinez-San Miguel 2017). Yolanda Martinez-San Miguel discusses the need to bring into conversation the colonial/imperial contexts of the Caribbean with other regions that share a similar set of conditions (155). It in this sense that this chapter brings the Indian Ocean context into conversation with the Caribbean.

The chapter briefly explored the genealogy of Caribbean archipelagic thinking and the reproduction of the concept of creolization in European scholarship. In contrast, it drew from Sylvia Wynter's concept of indigenization and the work of Katherine McKittrick on global archipelagoes of poverty to resituate an archipelagic analysis within the hierarchies of power that inform the precarity of Black/Creole people. Reflecting on the binaries of colonial power that still inform the dynamics within archipelagic and island states as well as on the continents, the chapter gestured toward theories that aim at the collective rehumanization of spaces.

The conceptual framework of archipelagic thinking has the potential to examine the human relation to material geographies that exist outside Euro-American scholarship, in a way that excludes the racialized cartography of modernity/coloniality that informs the present. A philosophy of land as epistemology borrowed from Indigenous scholarship ultimately shapes this chapter's investigations. The land (continents, oceans, islands, rivers, air, soil) does not belong to us; we belong to the land. This allows us to reimagine how we can break the colonial epistemology of private ownership of land as capital, the sea as site of exploration of islands, and islands as isolated specks of land to be consumed as tourist destinations by people from the continent.

NOTES

1. The Mauritian Truth and Justice Commission (TJC) was established in 2009. Two of the aims of the TJC were to make an assessment of the consequences of slavery and indentured labor during the colonial period up to the present and to make recommendations on reparation for slavery.

2. The Chagos Archipelago was excised from Mauritius before independence, contrary to the United Nations resolution, which specifically banned the breakup of colonies prior to independence, and one of the main islands, Diego Garcia, was leased to the United States, by the United Kingdom, for a military base. About two thousand Chagossians were exiled in Mauritius and Seychelles with no resettlement programs despite compensation negotiated between the British and Mauritian governments. While the lease ended in 2016, the United Kingdom has refused to transfer sovereignty over the island to Mauritius on the basis that one of the clauses stipulates that the lease be renewed for another twenty years if the military base is still considered relevant. The Mauritian government received, in 2017, an unprecedented majority of votes from a UN committee (ninety-four countries voted for Mauritius compared to fifteen that supported the United Kingdom and sixty-five that abstained) to support an advisory report from the International Court of Justice (ICJ) on the issue of sovereignty of the Chagos. On February 25, 2019, the advisory committee announced that the process of decolonization of Mauritius was not lawfully completed when that country acceded to independence and that the United Kingdom is under an obligation to bring to an end its administration of the Chagos Archipelago as rapidly as possible. The United Kingdom replied, on May 6, that it refuses to accept the ruling of ICJ.

3. These islands have been strategically used as spaces of quarantine, refuge, and exile. Rodrigues, which has hoped for independence from Mauritius, was initially

considered for a leprosy center and the Chagos for a space for incarceration. Diego Garcia has a specific history of "illegal" occupation and militarization.

REFERENCES

Baldacchino, Godfrey, E. McMahon, and Elaine Stratford. 2011. "Envisioning the Archipelago." *Island Studies Journal* 6, no. 2: 113–30.

Bogues, Anthony. 2010. "Introduction." In *The Hills of Hebron*, by S. Wynter. 2nd ed. Kingston: Ian Randle Publishers.

Boswell, Rosabelle. 2006. *Le Malaise Creole: Ethnic Identity in Mauritius*. New York: Berghahn Books.

Brathwaite, Kamau. 1995. "Creolization in Jamaica." In *The Post-colonial Studies Reader*, edited by B. Ashroft, G. Griffiths, and H. Tiffin. London: Routledge.

Britton, Célia. 1999. *Edouard Glissant and Postcolonial Theory: Strategies of Language and Resistance*. New World Studies. Charlottesville: University Press of Virginia.

Cohen, Robin. 2007a. "Creolization and Cultural Globalization: The Soft Sounds of Fugitive Power." *Globalizations* 4, no. 3: 369–84.

———. 2007b. "Creolization and Diaspora—the Cultural Politics of Divergence and Some Convergence." In *Opportunity Structures in Diaspora Relations: Comparisons in Contemporary Multi-level Politics of Diaspora and Transnational Identity*, edited by G. Totoricagüena. Reno: University of Nevada Press.

Dussel, Enrique, Carlos Jáuregui, and Mabel Moraña. 2008. "Colonialism and Its Replicants." In *Coloniality at Large: Latin America and the Post-colonial Debate*, edited by Enrique Dussel, Carlos Jáuregui, and Mabel Moraña, 1–22. Durham, NC: Duke University Press.

Gallagher, Mary. 2007. "The *Créolité* Movement: Paradoxes of a French Caribbean Orthodoxy." In *Creolization: History, Ethnography, Theory*, edited by Charles Stewart, 220–37. Walnut Creek, CA: Left Coast Press.

Glissant, Édouard. 1996. *Caribbean Discourse: Selected Essays*. Charlottesville: University Press of Virginia.

———. 1997a. *Poetics of Relation*. Ann Arbor: University of Michigan Press.

———. 1997b. *Traité du tout-monde: Poétique IV*. Paris: Gallimard.

Grosfoguel, Ramon. 2007. "The Epistemic Decolonial Turn: Beyond Political-Economy Paradigms." *Cultural Studies* 2, no. 2–3: 211–23.

Hannerz, Ulf. 1987. "The World in Creolisation." *Journal of the International African Institute* 57, no. 4: 546–59.

Jackson, Shona N. 2012. *Creole Indigeneity: Between Myth and Nation in the Caribbean. First Peoples: New Directions in Indigenous Studies*. Minneapolis: University of Minnesota Press.

Maldonado-Torres, N. 2007. On the Coloniality of Being: Contributions to the Development of a Concept. *Cultural Studies* 21, nos. 2–3: 240–70.

Martínez-San Miguel. 2017. "Colonial and Mexican Archipelagoes: Reimagining Colonial Caribbean Studies." In *Archipelagic American Studies*, edited by Brian Russell Roberts and Michelle Ann Stephens. Durham, NC: Duke University Press.

Mazzotti, José Antonio. 2016. "*Criollismo*, Creole and *Créolité*." In *Critical Terms in Caribbean and Latin American Thought: Historical and Institutional Trajectories*, edited by Yolanda Martínez-San Miguel, Ben. Sifuentes-Jáuregui, and Marisa Belausteguigoitia. New York: Palgrave.

McKittrick, Katherine. 2006. *Demonic Grounds: Black Women and the Cartographies of Struggle*. Minneapolis: University of Minnesota Press.

McKittrick, Katherine, and Sylvia Wynter. 2015. *Sylvia Wynter: On Being Human as Praxis*. Durham, NC: Duke University Press.

Mignolo, Walter. 2012. *Local Histories/Global Designs: Coloniality, Subaltern Knowledges, and Border Thinking*. Princeton, NJ: Princeton University Press.

———. 2015. "Sylvia Wynter: What Does It Mean to Be Human?" In *Sylvia Wynter: On Being Human as Praxis*, edited by Katherine McKittrick. Durham, NC: Duke University Press.

Newton, Melanie J. 2013. "Returns to a Native Land: Indigeneity and Decolonization in the Anglophone Caribbean." *Small Axe* 17, no. 2: 108–22.

Pugh, Jonathan. 2013. "Island Movements: Thinking with the Archipelago." *Island Studies Journal* 8, no. 1: 9–24.

Pyndiah, Gitanjali. 2016. "Decolonizing Creole on the Mauritius Islands from Creative Practices in the Language." *Island Studies Journal* 11, no. 2: 485–504.

———. 2018a. "Performative Historiography of the Mothertongue: Reading 'Kreol' outside a Colonial and Nationalist Approach." In *The Mauritian Paradox: Fifty Years of Development, Diversity and Democracy*, edited by Thomas Hylland Eriksen and Ramola Ramtohul, 213–30. Réduit: University of Mauritius Press.

———. 2018b. "Sonic Landscape of Seggae: Mauritian Sega Rhythm Meets Jamaican Roots Reggae." *Interactions: Studies in Communication and Culture* 9. 119–36.

Quijano, Aníbal. 2008. "Coloniality of Power, Eurocentrism, and Latin America." In *Coloniality at Large: Latin America and the Post-colonial Debate*, edited by E. Dussel, M. Moraña, and C. Jáuregui, 181–224. Durham, NC: Duke University Press.

Stewart, Charles. 2007. *Creolization: History, Ethnography, Theory*. Walnut Creek, CA: Left Coast Press.

Truth and Justice Commission (TJC). 2011. *Truth and Justice Commission*. Vols. 1–4. Mauritius: Government Printing.

Vergès, Françoise. 2007. "Indian-Oceanic Creolizations: Processes and Practices of Creolization on Réunion Island." In *Creolization: History, Ethnography, Theory*, edited by Charles Stewart, 133–53. Walnut Creek, CA: Left Coast Press.

Walcott, Rinaldo. 2015. "Genres of Human: Multiculturalism, Cosmo-politics, and the Caribbean Basin." In *Sylvia Wynter: On Being Human as Praxis*, edited by Katherine McKittrick and Sylvia Wynter. Durham, NC: Duke University Press.

Wynter, Sylvia. 1970. "Jonkonnu in Jamaica: Towards the Interpretation of Folk Dance as a Cultural Process." *Jamaica Journal*, no. 2: 34–72.

———. 1992. "Rethinking 'Aesthetics': Notes towards a Deciphering Practice." In *Ex-iles: Essays on Caribbean Cinema*, edited by M. B. Cham, 237–79. Trenton, NJ: African World Press.

———. 2007. "Human Being as Noun? Or Being Human as Praxis? Towards the Autopoetic Turn/Overturn: A Manifesto." Unpublished essay.

Thanksgiving in the Anthropocene, 2015

Craig Santos Perez

Thank you, instant mashed potatoes, your bland taste
makes me feel like an average American. Thank you,

incarcerated Americans, for filling the labor shortage
and packing potatoes in Idaho. Thank you, canned

cranberry sauce, for your gelatinous curves. Thank you,
Ojibwe tribe in Wisconsin, your lake is now polluted

with phosphate-laden discharge from nearby cranberry
bogs. Thank you, crisp green beans, you are my excuse

for eating apple pie à la mode later. Thank you, indigenous
migrant workers, for picking the beans in Mexico's farm belt,

may your children survive the season. Thank you, NAFTA,
for making life dirt cheap. Thank you, Butterball Turkey,

for the word, *butterball*, which I repeat all day *butterball,*
butterball, butterball because it helps me swallow the bones

of genocide. Thank you, dark meat for being so juicy
(no offense, dry and fragile white meat, you matter too).

Thank you, 90 million factory farmed turkeys, for giving
your lives during the holidays. Thank you, factory farm

workers, for clipping turkey toes and beaks so they don't scratch
and peck each other in overcrowded, dark sheds. Thank you,

genetic engineering and antibiotics, for accelerating
their growth. Thank you, stunning tank, for immobilizing

most of the turkeys hanging upside down by crippled legs.
Thank you, stainless steel knives, for your sharpened

edge and thirst for throat. Thank you, de-feathering
tank, for your scalding-hot water, for finally killing the last

still conscious turkeys. Thank you, turkey tails, for feeding
Pacific Islanders all year round. Thank you, empire of

slaughter, for never wasting your fatty leftovers. Thank you,
tryptophan, for the promise of an afternoon nap;

I really need it. Thank you, store bought stuffing,
for your ambiguously ethnic flavor, you remind me

that I'm not an average American. Thank you, gravy,
for being hot-off-the-boat and the most beautiful

brown. Thank you, dear readers, for joining me at the table
of this poem. Please join hands, bow your heads, and repeat

after me: "Let us bless the hands that harvest and butcher
our food, bless the hands that drive delivery trucks

and stock grocery shelves, bless the hands that cooked
and paid for this meal, bless the hands that bind

our hands and force feed our endless mouth.
May we forgive each other and be forgiven."

Planetary Constellations

An Afterword

Susan Stanford Friedman

Stunning fragments of aquamarine colors, scattered at random, making a pattern. Poems as islands in a sea of interdisciplinary scholarship. *Contemporary Archipelagic Thinking* uses the arts to suggest the double function of the volume: first, to insist on the distinctive materiality of archipelagoes, including their histories, cultures, peoples, languages, geographies, and ecologies; second, to suggest that this materiality introduces a new way of thinking relationally, a hermeneutic about the interconnected, interlapping world. The book moves back and forth from the specificity of archipelagic times and places to the epistemology of *thinking with* islands, oceans, and archipelagoes as a broadly applicable mode of perception.[1] *Contemporary Archipelagic Thinking* at its most ambitious offers a new framework for understanding space and time, land and water, and human histories and diversities on planet Earth. But it does so while also insisting that we not forget what has been largely sidelined: the specificity of the peoples and the land/sea histories of islands and archipelagoes, especially their relationship to colonialism, slavery, indentured labor, environmental crisis, and militarization.

CONSTELLATIONS

> The archipelagoes resemble constellations.
>
> —Craig Santos Perez, from "The Fifth Map"[2]

Archipelagoes, like archipelagic thinking, are constellations—that is, figurations, patterns, and cartographies of relationship and interconnection. The word "constellation," often used in Betsy Wing's translation of Édouard

Glissant's *Poetics of Relation*, originally referred to a grouping of stars perceived and named by people from the Earth and associated with astrological signs and stories from Greek mythology. As the term broadened its meaning, "constellation" came to signify a "pattern" or "arrangement," an "assemblage, collection, or group of usually related persons, qualities or things."[3] Thus, the leap from material stars to a human perception of patterns in interstellar space references the connection as well as disconnection between material things and perceptions of relationships among them: from, in other words, materiality to epistemology, just the connection that *Contemporary Archipelagic Thinking* makes. Moreover, the chapters in the volume are themselves disciplinary islands in an archipelagic sea. Historians, geographers, literary critics, art historians, and environmental scientists and humanists constitute an archipelago of humanities approaches. While Caribbean and Pacific Island studies have often been at the foreground of the new interlocking fields of oceanic/island/archipelagic studies, this volume extends well beyond these regions to encompass essays on the Anglo-Saxon, British-Irish, Mediterranean, Japanese-Korean, Indonesian, and Indian Ocean seas and archipelagoes. The geographical reach of the volume also insists upon a temporal engagement with history, and to reference Lanny Thompson's chapter in the volume, archipelagic thinking invokes the Bakhtinian *chronotope*, the imperative to interconnect space and time in narrative—geography and history together.

> . . . I examine the map closely,
> navigating beyond the violent divisions
> of national and maritime borders, beyond
> the scarred latitudes and longitudes
> of empire, to navigate the cartography
> of our most expansive legends
> and deepest routes.
>
> —Craig Santos Perez, from "The Fifth Map"

THE COUNTERDISCOURSES OF ARCHIPELAGIC THINKING

In its insistence on materiality, archipelagic thinking is a counterdiscourse. It goes against how islands have been desired, imagined, overpowered, exploited, and militarized by hegemonic forces located on continents, in metropolitan cultural capitals, and through powerful navies and air forces. It insists that islands have what Craig Santos Perez, in the foreword to this volume, calls "positive insularity," that is, a grounded uniqueness, a specificity of location, language, culture, people, geography, and history. As opposed to imaginary ones, real islands have a singularity of material location but often a plurality of peoples, languages, cultures, and histories. As a counterdiscourse, archipelagic thinking locates a Real that challenges a continentalist imaginary: the desire, within a binary logic of civilized/savage, for the remote, the isolated, the exotic, the primitive. As Elizabeth DeLoughrey writes in "The Litany of Islands, the Rosary of Archipelagoes," "In the Western imagination, 'island' and 'islandness' have metaphorical nuances which are contingent upon the repercussions of European colonialism and continental migration towards islands spaces" (2001, 26). Islands in the literatures and arts of the West, she continues, are often tropical, associated with southern warmth, adventure, and journeys far from metropolitan centers of the north (22). Archipelagic thinking unmasks the pervasive island tropes that accompanied conquest and colonialism.[4]

> thanks Bougainville
> for desiring 'em young
> so guys like Gauguin could dream
> and dream
> then take his syphilitic body
> downstream to the tropics
> to test his artistic hypothesis
> about how the uncivilized
> open like pawpaw
> are best slightly raw
> delectably firm
> dangling like golden prepubescent buds
> seeding nymphomania
> for guys like Gauguin.
>
> —Selina Tusitala Marsh,
> from "Guys like Gauguin,"
> in *Fast Talking PI* (2003, 36)[5]

The counterdiscourse of archipelagic thinking about real islands participates in a particular form of what W. E. B. Du Bois called "double consciousness," that is, the difficult combination of internalized hegemonic representations of an oppressed people with those people's sense of a different identity. In his influential essay "Our Sea of Islands" (2003, 30), Epeli Hau'ofa narrates an epiphany so paradigm shifting that he likens it to Paul's revelation on the road to Damascus. As a social scientist trained in development theory, he suddenly realized that he was deflating his students at the University of the South Pacific in Fiji in teaching them that "the small island states and territories of the Pacific" were "too small, too poorly endowed with resources, and too isolated from the centres of economic growth for their inhabitants to rise above their present condition of dependence on the largesse of wealthy nations" (29). No, he thought, "the world of Oceania is not small; it is huge and growing bigger every day" (30). His recognition that Oceania is not "islands in the sea" but rather "a sea of islands" represents archipelagic counterdiscourse. Real archipelagoes—that is, the material places and peoples in them—are locations of land and sea, encompassing vast spaces that archipelagic peoples have traversed for millennia in widespread networks of contact, trade, exchange, conquest, migration, and diaspora.

Brian Russell Roberts' chapter in this volume narrates how Indonesia convinced the United Nations to recognize the territory of nation-state archipelagoes as the perimeter around the outermost islands, thereby embedding in its declaration on the Law of the Sea the Malay concept of archipelagoes as *tanah air*—that is, land-and-sea. This history shows the counterdiscourse of archipelagic thinking at work in the geopolitical world represented by the United Nations. While such declarations do not in themselves create transformative change, they do enter the realm of international law, and even without mechanisms for enforcement, they can affect the way archipelagic states claim their territorial waters.

Equally material, however, are the dire effects of climate change, rising seas, militarization, plastic, and overfishing on many of the planet's archipelagoes, as this volume's various engagements with the Anthropocene demonstrate.[6] The counterdiscourse of archipelagic thinking suggests that we see the conditions of archipelagoes, including the oceans within and surrounding them, as the canary in the coal mine of planetary crisis. In "Praise Song for Oceania," Perez presents a utopian prayer that the ocean will somehow survive, although his litany of dire human effects suggests otherwise.

> . . . praise
>
> your capacity
> to survive / our trawling
> boats / breeching /
> your open body /
> & taking from your
> collapsing depths / praise
>
> your capacity
> to dilute / our sewage
> & radioactive waste /
> our pollutants & plastics /
> our heavy metals
> & greenhouse gases / praise
>
> —Craig Santos Perez, from "Praise Song for Oceania"

In "Girl from Tuvalu," Selina Tusitala Marsh paints a spare portrait with mythological overtones to invoke the threat of rising seas, imaged here as *moana* (the Maori and Hawaiian word for ocean):

girl sits on porch
back of house
feet kicking
salt water skimming
like her nation
running fast
nowhere to go
held up by
Kyoto Protocol
An Inconvenient Truth

this week her name is Siligia
next week her name will be
Girl from Tuvalu: Environmental Refugee

her face is 10,000
her land is 10 square miles
she is a dot
below someone's accidental finger
pointing outwards

the bare-chested boys
bravado un sea spray
running on tar-seal
they are cars
they are bikes
they are fish out of water
moana waves a hand
swallows
a yellow median strip

—Selina Tusitala Marsh, "Girl from
Tuvalu" from *Dark Sparring*[7]

ARCHIPELAGIC MODERNITIES

The modernities of real archipelagoes and archipelagic epistemology blend, reinforcing each other and calling into question the center/periphery mapping of global relations in Eurocentric thought. As I have discussed at length in *Planetary Modernisms: Provocations on Modernity across Time*, the dominant narrative of modernity asserts that modernity was invented in Europe after 1500 and spread to other parts of the globe along with the so-called rise of Europe and "the West's" subsequent colonization of "the Rest," doomed to a belated and imitative modernization.[8] As geographer J. M. Blaut summarizes this ideological narrative of European exceptionalism, "Europe eternally advances, progresses, modernizes. The rest of the world advances more sluggishly or stagnates; it is 'traditional society.' Therefore, the world has a permanent geographical center and a permanent periphery: an Inside and an Outside. Inside leads. Outside lags. Inside innovates; Outside imitates" (1993, 1). As the major world-system theorizer of modernity, Immanuel Wallerstein began asserting in 1974 that Europe developed capitalism starting around 1500, thereby initiating modernity with a distribution of power in terms of itself as the central core and the rest of the world as semiperiphery or periphery. In *A Singular Modernity*, Fredric Jameson (2002) even more firmly defines modernity as capitalism, invented in Europe and spreading as a singular formation throughout the world. Sociologist Anthony Giddens and cultural studies scholar Stuart Hall broadened the definition of modernity beyond capitalism to include such things as the nation-state, industrialism, military superiority, surveillance, secularism, and individualism, but for them, Europe nonetheless initiated these expanded characteristics and spread them elsewhere with its superior military and economic power.

Along with Blaut, many world historians have challenged this center/periphery ideology of the West and the Rest. But they have done so from a profoundly continentalist perspective, as I myself did in *Planetary Modernisms*, in which I used the modernities of the sedentary Tang/Song dynasties in China and the nomadic Mongol Empire of Central Asia as examples of pre-1500 modernities that greatly influenced the later rise of European modernity.[9] What I didn't realize until I began reading in archipelagic, island, and oceanic studies in preparation for a conference on oceanic modernism in Fiji in 2015 is just how much archipelagic thinkers had permeated my analysis of continental modernities. Paul Gilroy (1993), in *The Black Atlantic*, opened my mind to thinking about modernity in the plural with his assertion that the first modern subjects were not the Enlightenment philosophers in France but the people stolen from Africa who experienced the massive dislocations and alienations of the Middle Passage and slavery in the Americas. Simon Gikandi's *Writing in Limbo* (1992) about the Caribbean and *Maps of Englishness* (1996) about how colonialism in Africa was constitutive of British modernism, not peripheral to it, taught me to think relationally and transnationally about different formations of interlinked modernities. Aimé Césaire's *Notebook of a Return to the Native Land* (1939) led me to think about a diasporic modernism based in return to the home, not exile from it. And above all, Glissant's *Poetics of Relation* theorized global relations—including different sites of modernity—as part of mobile networks. His critique of both center/periphery and "rooted" filiation and his lyric theorizations of "worldness" and the "planetary" as a rhizomatic *créolité* were the single most influential ideas that led me to hypothesize modernity as multiple, polycentric, and recurrent periods of rapid, transformative, dystopic/utopic change across time and space. Archipelagic thinking as an epistemology is what shaped my revisionist approach to the modernities of sedentary and nomadic continental histories. As for me, Glissant is a major theorist for *Contemporary Archipelagic Thinking*, which in its epistemological project is a volume promoting a new hermeneutic for understanding the world.

Archipelagic thinking about modernity is more than an epistemology, however. It is also an important approach to the material specifics of how various archipelagoes experienced and constituted their own modernities. These histories are not the same, even as they share certain structural characteristics. The modernities of the Caribbean include the near erasure of indigenous populations, the institutions of slavery and indenture, the palimpsestic layerings of different colonial histories, economies based on tourism and money laundering, the creative explosions of global musics, and so forth. The modernities of Austronesia, Melanesia, Micronesia, and Polynesia,[10] in contrast, have deep time histories of migration, complex interactions in Oceania of conquest and trade, the more recent histories of European and American colonialism, including the missionary cultures of conversion and attack on indigenous cultures, and periods of Japanese imperialism, partial liberation, and militarization of the Pacific islands in the twentieth century. Both the Caribbean and Oceania, in other words, have experienced waves of recurrent periods of rapid change and transformation as they have interacted within and outside their archipelagoes.[11] One aspect of recent island modernity is tourism.

> the man in the blue mother hubbard
> languors in the doorway
>
> in Suva heat on Ellery Street
> his face a frayed Punja's Flour sack,
>
> unrolls as he greets me
> with mad familiarity
>
> mouth wide open
> he is tasting the souls of passersby
>
> with hungry eyes and mad mad skin
> jaina, balgani, painapiu
>
> the children with red balloons
> rush at the taxi
>
> bags-ing it
> for their nana
>
> —Selina Tusitala Marsh, "Bound for Sigatora," from *Dark Sparring*[12]

ARCHIPELAGIC AGENCY

Archipelagic modernities share a dialogic relation with the mainlands and metropoles of continental modernities, each taking shape through interaction with the other, in creative as well as oppressive ways. What archipelagic thinking importantly accomplishes is a recognition of the agency of archipelagic cultures, whether it takes the form of resistance, hybridic transculturation, or the creative and performing arts, especially as these blend orature and cultural traditions with new conditions of modernity. By *agency*, I do not mean *autonomy*, because complete freedom to act is clearly impossible with slavery, indenture, colonialism, military power, and climate change. (Autonomy is, in my view, incompatible with living in any human society.) One form of archipelagic agency is the counterdiscourse I discussed above. But even more dialogic is the archipelagic engagement with dominant cultural and institutional forms, the ability to wrestle with them, transplant them, transform them in a process of indigenization, a kind of translational process that Edward Said describes in "Traveling Theory" and "Traveling Theory Reconsidered." It's worth noting that colonial centers like London and Paris were themselves indigenizing aspects of cultures they colonized—from the African and Oceanian masks in Picasso's cubist breakthrough in *Les demoiselles d'Avignon* to the *tiki masala* of British cuisine. But for archipelagic thinking about real archipelagoes, I am interested in the dialogic agency of archipelagic cultures to engage with continental powers in creative ways, translating what comes from outside into their own cultures and circumstances. Matthew Hayward and Maebh Long's *New Oceania: Modernisms and Modernities in the Pacific* explores ways in which Oceanian writers and performers revision such early twentieth-century Euro-American modernist writers as T. S. Eliot and James Joyce in ways that address the modernities of the post-1960s boom in Pacific writing.

The agency to "make it new" in Oceanian writing, as with the creative arts more generally, is not always an interactive relationship with the arts of the West, however. As Yolanda Martínez-San Miguel and Michelle Stephens argue in their introduction to this volume, archipelagic thinking about material archipelagoes fosters comparison across archipelagoes: not just "North-South" and "South-North" but also "South-South" comparisons. In "Relational Comparison," Shu-mei Shih (2013) draws on Glissant's poetics of relation to propose a relational model of comparison that focuses on the long arc of common formations across different parts of the colonized world. Building on Glissant's work, she uses as her examples literatures articulating the plantation system in the Caribbean, the American South, and Malaysia, where Chinese settler colonialism breaks the model of Euro-American colonialism to demonstrate how Asian powers like China have engaged in imperial patterns. Françoise Lionnet and Shu-mei Shih's *Minor Transnationalism* makes a similar point about "South-South" comparisons, breaking the hegemony of center/periphery thinking that always discusses the colonized in relation to the metropolitan/imperial center. This volume, while it does not limit its discussions to islands and archipelagoes, exhibits the kind of comparative archipelagic thinking that Martínez-San Miguel and Stephens promote. So does the Bandung Conference held in Indonesia in 1955, which both Christopher J. Lee and Brian Russell Roberts discuss as instances of relational political activism that actively excluded the post–World War II Euro-American and Soviet powers during the Cold War.

In solidarity with the Standing Rock Sioux tribe and all peoples protecting the sacred waters of the earth

. . .
water is life becuz we proclaim water a human right
becuz we grant bodies of water rights to personhood
becuz some countries signed the UN Convention on the Law of the Sea
becuz my wife says the Hawaiian word for wealth, waiwai, comes from
  their word for water, wai
water is life becuz corporations steal, privatize, dam & bottle our waters
. . .
water is life becuz we say stop, you are hurting our ancestors
. . .
becuz they say we thought you were vanishing
becuz we are water warriors & peaceful protectors
becuz they call us savage & primitive & riot
becuz we bring our feathers & lei & sage & shells & canoes & drones &
  hashtags & totems
becuz they bring their bulldozers & drills & permits & surveillance drones
  & helicopters
becuz we bring our treaties & the UN Declaration on the Rights of
  Indigenous Peoples
becuz they bring their banks & politicians & police & private militia &
  national guard & lawyers
becuz we bring our songs & schools & lawyers & prayers & chants &
  ceremonies
becuz they bring their barking dogs & paychecks & pepper spray & rubber
  bullets
becuz we bring all our relations & all our generations & all our
  livestreams
. . .
Water is life becuz my daughter loves playing in the ocean

—Craig Santos Perez, from "Chanting the Waters"

NEW RESEARCH QUESTIONS

Martinez-San Miguel and Stephens affirm in their introduction that archipelagic thinking is a lens that functions "beyond the geographical or geopolitical. Instead, it is another framework to conceive human social formations and historical experiences." As such, "the archipelagic makes possible a whole new series of research questions that interrogate the primacy of continental frameworks in several disciplines of study." While the interdisciplinary range of *Contemporary Archipelagic Thinking* charts an impressive array of "new research questions," I find myself stimulated by the volume to ask even more questions, some of which arise out of the locational specificity of material archipelagoes, others that emerge from the volume's richly suggestive arguments about archipelagic epistemology, and still others that are a combination of both.

THE PLANET

Environmental crisis, climate change and rising seas, plastic clogging the oceans and coastlines: the Anthropocene appears as a vital archipelagic lens in the volume. But I want to step back just a moment to think about the planet, specifically about how archipelagic *thinking with* and *about* islands, archipelagoes, and oceans opens up new approaches to *thinking with* and *about* the planet Earth. Most people live on land, whether coastal or interior, whether continents or islands; this fundamental reliance on land leads understandably to what we might call a land-based centrism in thinking about the planet, a land-centrism that encourages a focus on the formation of cities as constitutive of civilizations in human history, fed by the land, the domestication of animals, and the development of agriculture. As Margaret Cohen writes in "Literary Studies on the Terraqueous Globe" for *PMLA*'s special section on oceanic studies, "Despite the preeminence of maritime transport in making the modern world, literary scholars across the twentieth century passed over its impact with their gazes fixed on land" (2010, 657).

The Earth, however, is about 70 percent water, with 97 percent of that being ocean. Oceans are supplemented by seas, lakes, rivers, straits, and canals. Seas like the Mediterranean have been the condition of civilizational relations and the rise of what world historians call fulcral cities (Constantinople/Istanbul, Alexandria, Venice, Tangiers). Rivers like the Nile, Tigris and Euphrates, Yangtze, Yellow, Congo, five Indus rivers, Mekong, Amazon, Danube, and Mississippi—to name a few—connected regions and peoples. Straits like the Malacca, Hormuz, and Gibraltar have been integral to human history.

The completion of the Grand Canal linking the east-west Yangtze and Yellow Rivers in seventh-century China unleashed an explosion of population, agricultural advancement, and cultural renaissance reinforced by the nomadic incursions from central Asia in the formation of the Tang dynasty. How might our conceptions of human and nonhuman histories change if water is at the center instead of the periphery of our thinking?

> *On June 8, 2016, World Oceans Day*
>
> . . .
> praise
> your capacity
> for birth / your fluid
> currents and trenchant
> darkness / praise your contracting
> waves & dilating
> horizons / praise our briny
> beginning, the source
> of every breath / praise
> your endless bio-
> diversity / praise
>
> —Craig Santos Perez, from "Praise Song for Oceania"

ARCHIPELAGOES, ISLANDS, OCEANS, AND CONTINENTS

"What is an archipelago?" Roberts asks in his chapter in this volume. I want to add, What is a continent? An island? A sea? An ocean?[13] What are the differences between them? What are the differences within each formation? How are they related? Archipelagic thinking begins with the geoformations of the planet Earth. But as an epistemology, it is necessarily conceptual, a product of the human mind and thereby fluid, mobile, historically produced, and potentially ideological, critical, or some mixture of both. As an epistemology of relationality, archipelagic thinking may begin with material islands and archipelagoes, but it asks us to understand not only how these are connected to ocean and continent but what in fact we mean by these terms and how the distinctive fields of oceanic studies, island studies, archipelagic studies, and continental studies are both distinct and linked.

Archipelagoes are not continents, but what is a continent? Standard notions, for example, identify seven continents, defined as large and distinctive landmasses: Africa, Antarctica, Asia, Australia, Europe, North America, and South America. But as Martin W. Lewis and Kären Wigen (1997, 1–46) point out in *The Myth of Continents: A Critique of Metageography*, this designation of seven continents is a product of Eurocentric colonial ideology.[14] Much like the still-prevailing sixteenth-century Mercator projection governing two-dimensional maps of the world and the prime meridian based in Greenwich determining global longitude, the notion of seven continents has an imperial history. A different lens might challenge Europe's status as a continent, since physically speaking, it appears more like a large peninsula of western Asia than its own continent. The designation of Australia as a continent has also been fluid; as the smallest continent, it is unambiguously surrounded by ocean and is sometimes regarded as a large island rather than a continent. As an ideological concept, continent reinforces the center/periphery mode of thinking that marginalizes the land-sea life of archipelagoes and islands. With continents as the primary geoformation, the significance of oceans, islands, and archipelagoes recedes. How might archipelagic thinking change the way we understand continents?

In "Envisioning the Archipelago," Elaine Stratford (2011) and her coauthors suggest that "thinking with the archipelago" changes fundamentally how we think of large landmasses, including continents. Canada is a case in point, with "the world's largest shoreline," which "straddles three oceans," and a north/south movement from Arctic islands, to extensive lake regions, to borders east, west, and south that are islands and archipelagoes. "Thus, it is timely," they write, "to consider Canada not as a unitary land mass but as a series of multiple assemblages of coast, oceanic and insular identities, even as its centre of politico-economic gravity remains stuck in the Alberta tar sands" (121). Canada doesn't have the greatest number of islands on the planet, a distinction held by Sweden. But if Canada is more than a vast landmass, then just what is the cogency of the category "continent"?

Archipelagoes are land-sea combinations, but what is an ocean? "The sea is not a metaphor" (2010, 670), writes Hester Blum in "The Prospect of Oceanic Studies." Like the word "continent," the word "ocean" designates a real phenomenon, but the definition is fluid, conceptual, and potentially political. The phrase "seven seas" goes back at least to the Sumerian hymn to the goddess Inanna in 2,300 BCE and down through the centuries has designated different bodies of water. More recently, the Earth is said to have five oceans, typically listed in descending order of size: the Pacific, Atlantic, Indian, Southern (Antarctic), and Arctic. But the National Oceanic and Atmospheric Administration (NOAA) declares on its National Ocean Service website, "There is only one global ocean," because all the Earth's oceans exchange water.[15] Yet, the NOAA retains the old notion of the seven seas by dividing the Global Ocean into the Artic, North Atlantic, South Atlantic, Indian, North Pacific, South Pacific, and Southern (Antarctic) seas.

The spatial turn in cultural studies starting in the late twentieth century began focusing on oceans as continental and metropolitan connectors, with the notions of the Atlantic world, the Pacific Rim, and Indian Ocean cultures from Africa to Southeast Asia. But interdisciplinary oceanic studies has more recently shifted the emphasis to the ocean itself, insisting for starters that the ocean is mobile, with both horizontal and vertical dimensions, from the surface to the ocean floor, and existing in close symbiosis with the atmosphere. In this context, archipelagoes are mountain chains, beginning on the sea floor with their peaks above water. The tallest mountain on earth is not Mt. Everest (29,035 feet above sea level) but Mauna Kea on Hawaii (33,000 feet from the sea floor).[16] As Blum writes in her introduction to a special issue of *Atlantic Studies* on Oceanic Studies, the ocean is "not just a theme or organizing metaphor, with which to widen a landlocked critical prospect: in its geophysical, historical, and imaginative properties, the sea instead provides a new epistemology—a new dimension—for thinking about surfaces, depths, and the extra-terrestrial dimensions of planetary resources and relations" (2013, 151).[17]

Islands and archipelagoes are unthinkable without a concept of the ocean. But what is an island? What is an archipelago? Islands and archipelagoes are the front lines of contact with the Global Ocean, a contact that can breed disaster in the weather chaos of climate change, from Hurricane Maria in Puerto Rico to Cyclone Winston in Fiji.

> *For Fiji, after Cyclone Winston, the most powerful storm ever recorded in the Southern Hemisphere*
>
> This is when the warm ocean gives birth to a cyclone—
> This is when we give a human name to a thing we can't control—
> This is when wind triggers warning system & rain echoes violent refrain,
> This is when evacuation is another word for shelter & home for debris
> and broken sugar cane—
> This is when we hold our children close and whisper: *don't be scared, we won't let go*—
> This is when flood waters whisper: *let go*—
>
> —Craig Santos Perez, from "Storm Tracking, 2016"

Roberts reminds us in "What Is an Archipelago?," however, that the word "archipelago" has an etymology; it's a linguistic entity with a history that attests to the role of language in our polylingual planet. The word has Greek roots but was first used in an Italian source written in Latin in the twelfth century to allude specifically to the chain of islands in the Aegean Sea. As such, the word "archipelago" embodies a kind of medieval Eurocentrism in its application to island chains outside ancient Greece, which later played such an ideological role in the formation of the concept of "the West." This linguistic history does not mean, as Roberts explains, that we should abandon the term in reference to other parts of the world. But it does suggest that we should be attuned to when and how and in what language(s) the concept of archipelago is applied to a group of islands. Indonesia, for example, *became* an archipelago in its postcolonial formation as a nation-state, one covering a vast latitudinal distance that reaches from Myanmar to east of the Philippines. Forming these diverse islands into a single archipelagic state with its perimeter around its outermost islands resulted from a lengthy and complex political process. So too was the constitution of the much smaller state of Tuvalu as an archipelagic state, whose name (as Roberts explains) alludes to its union of eight islands. Geoformations of the Earth are also political and cultural formations, blending the material and epistemological dimensions of archipelagoes together.

The case of Indonesia's postcolonial emergence as an archipelagic state raises the question of size and relation to power as well. Britain and Fiji, for example, are both island archipelagoes.[18] Fiji's "mainland" has its outlying islands, while Britain is an archipelago that has smaller archipelagoes offshore—the Hebrides, the Orkneys, the Shetlands, and so forth. Do the differences in size, history, and power among the planet's archipelagoes challenge the very coherence and usefulness of the category? "Archipelago" is an umbrella term that encompasses very different kinds of island chains—from the relatively large imperial British and Japanese archipelagoes to the much smaller colonized archipelagoes of the South Pacific, Caribbean, and Aleutian Islands. Other large archipelagoes, like the Philippines and Indonesia, have complex histories, both colonizing and colonized, and the divisions among the islands within each archipelago are significant for the current politics of each country. Further layering these histories are the migrations of poor, merchant, and landowning Chinese into these archipelagoes. In short, the category "archipelago," significant for the development of an alternative epistemology, encompasses vast differences of size, power, and cultures. Analysis of these variations is critical to both the material and the epistemological aspects of archipelagic thinking.

The imperative to consider the differences among archipelagoes raises yet another question. Given the distinctions made between archipelagoes and continents, how, then, are we to speak conceptually of their interconnections, even their similarities? Can we go beyond tracking mainland fantasies and exploitation of islands to see how they might "interlap," to invoke Roberts's term? Urban studies is developing just such a line of inquiry, as Elaine Stratford's chapter in this volume notes briefly. Long-standing concepts and methodologies developed in urban studies to examine continental cities are being adapted for studying cities on islands and archipelagoes. For all that the sea, land, flora, and fauna of islands are at times more fully present in writings about them—think of the rich physical presence of the natural world in Derek Walcott's *Omeros*, for example—it is important to remember that islands big and small often contain cities that invite comparison with their continental cousins. The journal *Urban Island Studies*, published since 2015 out of Prince Edward Island, features articles on coastal cities built on islands (Venice, Manhattan, Mombasa, Singapore, Hong Kong, Mumbai), Caribbean cities like Kingston and San Juan, cities on large archipelagoes like the Philippines, and new island cities such as "eco-cities" built as environmental models hoping to affect urban planning on the mainland.

Archipelagic thinking has also been variously applied to reconceiving continents in which cities have archipelagic enclaves within them or are themselves archipelagoes across the large landmass. Ethel Baraono Pohl and César Reyes Nájera's "A Tale of Two Cities: The Archipelago and the Enclave" (2015) contrasts the planned 1977 project "Berlin as a Green Archipelago," with the unplanned border disputes about 160 Indian enclaves within Bangladeshi territory unresolved until 2015. The *Journal of Urban Ecology* features an article titled "Urban Forests Form Isolated Archipelagoes" (Olejniczak et al. 2018). Maps of continental countries sometimes identify urban areas as islands within a sea of rurality, like the interactive red/blue precinct map of the United States during the 2016 election featured in the *New York Times*. Instead of dividing the United States into red and blue states based on the electoral college, the map shows each state as a patchwork of Republican red and Democratic blue. The map looks almost entirely red, except for islands of blue in metropolitan areas *in every state*. The politically "blue" archipelago of the United States shows the divisions in the country as metropolitan and rural rather than regional ("An Extremely Detailed Map" 2018).

The international journal *continent* (2019), produced in Denmark since 2011, "maps a topology of unstable confluences and ranges across new thinking, traversing interstices and alternate directions in culture, theory, biopolitics and art." Not only do its issues undermine the geographical and cultural fixities of continentalist thinking, but its special issues also revisit the planetary in provocative ways. One special issue, titled *drift* (2013), imaged as a ship at sea, experiments with how essays can be experimentally written, not as isolated islands but as archipelagoes. The three geographically scattered guest editors—April Vannini (Gabriola Island, British Columbia), Jeremy Fernando (Singapore), and Berit Soli-Holt (Denver, Colorado)—sent a brief call on space/place issues "adrift" over the internet and asked potential contributors to respond to these "drifts" and then set their own essays adrift to other potential contributors. Based on images of mobility, the formation of this special issue of *continent* conceives of its contributors as islands linked relationally by their responses to each other.

Selina Tusitala Marsh's experiments with an archipelagic use of page space in a number of her poems, including "Chant from Matiatia to Orapiu," in which the swaying placement of lines mimics both waves and interconnected island mobilities, while the images feature urban life on the touristy island of Waiheke, not far from Auckland, New Zealand:

surveyors plotting
land-lubbers allotting

supermarket squatting
parked cars rashing
transfer station trashing
Levi Hawken's grinding at the skateboard park

headless tiger flexing
fuming motors vexing
one tree mountain standing
dozer teeth man-handling
agents picket signing
cyclists white lining

—Selina Tusitala Marsh, from "Chant from Matiatia to Orapiu," from *Dark Sparring*[19]

INTERSECTIONALITY

Elaine Stratford's "Contemporary Archipelagic Thinking" in this volume proposes that we not only *think with* archipelagoes but also *think within* and *think about* them. Islands differ from each other and from the mainlands for sure, but there is also difference *within* the islands. How might the concept of intersectionality, rooted in but not limited to feminist theory, generate new research questions about cultural, linguistic, political, historical, and geographical diversities *within* archipelagoes? By intersectionality, I mean especially the lines of power and hierarchy that crisscross *within* any given island or archipelago, factors such as class, race, ethnicity, lineage, religion, gender, sexuality, national origin, migration, age, and so forth.[20] Intersectional methodology provides strategies for seeing how these different systems of power, these different elements of identity, mediate each other and complicate simple binaries of self/other, oppressor/oppressed. Intersectionality allows for more complex understandings of situational and contradictory identities—not all privileged, not all oppressed. Hybridities of culture, lineage, and individual identities come into sharp focus with an intersectional lens. Intersectionality also encourages attention to agency—both individual and communal—as a form of negotiating and navigating whatever ideological and institutional forces pressure life on an archipelago. Postcolonial studies and environmental studies have been defining elements in volumes like *Contemporary Archipelagic Thinking*; the intersectional approaches in these fields can also illuminate new directions in archipelagic studies, particularly those that address how (post)coloniality and environmental crisis can impact sectors of island peoples differently depending on such categories as ethnicity, race, class, gender, and so forth.

By definition, archipelagoes big and small contain islands that differ from each other. How significant are these differences—say, between Jamaica and Barbados? Do archipelagoes have centers and peripheries? Do nation-states within an archipelago set up barriers to migration from other islands within the archipelago? Is power, privilege, and cultural capital distributed differently among them? What are the struggles *within* an archipelago, however much different islanders unite in the face of land-centrism and continental dominance? Elaine Yee Win Ho and Shirley Geok-Lin Lee's special issue of *ARIEL* on microstates emphasizes what small island states share, but one of their contributors, Curwen Best (2001), argues for the distinctiveness of different island cultures in the Caribbean and the importance of island literary criticism to recognize the different aesthetics characterizing islands in the region.

Indigeneity is a vital and volatile issue in many of the world's archipelagoes, perhaps most intensely in Oceania because of the relatively recent waves of colonizing and labor migrations moving into the area. Whereas most of the indigenous peoples were wiped out in the Caribbean, sizeable populations remain in North and South Pacific archipelagoes, as well as in diasporic communities in the Americas. In Hawaii, migrant Japanese plantation workers and Euro-American settlers and tourists have greatly outnumbered native peoples; an indigenous movement asserting political, land, and cultural rights has brought about some limited improvements. But in Fiji, for example, the Melanesian population is in the majority, with the Indian Fijian population decreasing, in large part due to the nativist politics on the island, which has led many to migrate to Australia and New Zealand. The history of original and subsequent migrations into the Pacific Islands is complex and different for each Pacific archipelago. How do these archipelagoes negotiate racial, ethnic, and cultural differences, especially in the context of Euro-American colonialism, the missionary cultures that denigrated traditional cultures, the arrival of Japanese imperialism, and the growing militarization of the islands in the postwar period?

The status of "tradition" in archipelagoes with substantial indigenous populations is another divisive issue for which an intersectional approach is useful. Does tradition mandate hierarchical or relatively equitable structures?

Do these traditions continue or disintegrate in the face of outside influences? What are the traditional forms of gender difference and especially the status of women? Conflict between urban and rural communities, old and young, men and women, different religions and ethnic groups, and "pure" and mixed-race people often permeates debates about the importance of "tradition" in the face of "modernity," particularly when modernity is associated with the ways of the colonizers, the powerful, the outsiders. Whether understood as fixed or fluid, unchanging or hybridic, tradition remains central for indigenous people on archipelagoes that have been colonized by outsiders. Intersectionality offers a flexible framework for understanding divisions within and intersections among indigenous groups and with nonindigenous peoples as well.

> *Written for the 2016 Guam Educators Symposium*
> *on Soil and Water Conservation*
>
> Before we enter the jungle, my dad
> asks permission of the spirits who dwell
> within. He walks slowly, with care,
> to teach me, like his father taught him,
> how to show respect. Then he stops
> and closes his eyes to teach me
> how to *listen. Ekungok*, as the winds
> exhale and billow the canopy, tremble
> the understory, and conduct the wild
> orchestra of all breathing things.
>
> —Craig Santos Perez, from "Family Trees"

A paternal lineage is one thing, but what about literary father to poet daughter? Selina Tusitala Marsh (2018) ponders these questions intersectionally in her essay "Make It *Niu*: Blacking Out Albert Wendt's *Pouliuli* the *Tusitala* Way." "Make It *Niu*" puns on Ezra Pound's dictum for modernism, "Make It New," by using the word *niu* and its rich cultural meanings for coconut, known in Oceania, as she writes, as "the tree of life," a "powerful symbol for island sustenance and identity" (71). The essay defines a poetic she calls "the *Tusitala* way" as a combination of drawing on tradition and engaging creatively with it. As example, she uses her own volume, *Black Out Poetry*, a long poem that embeds an intersectional understanding of tradition/modernity, past/future, and male/female. The poem takes a copy of Albert Wendt's seminal novel *Pouliuli* ([1977] 1990), adapts a contemporary American poetic form called blackout poetry, and literally blacks out most of Wendt's words, leaving visible (on what looks like a backdrop of traditional *tapa* cloth) only a few scattered words here and there that assert a strong woman's voice.

in darkness

Women

speak

a loud silence

—Selina Tusitala Marsh, "Black Out # 33"[21]

Why Albert Wendt? Why *Pouliuli*, whose title means "darkness" and alludes to the common way islanders refer to the period before conversion to Christianity? Wendt, whom Marsh describes as her much-valued mentor, was born in 1939 on Samoa and is a leading novelist, poet, and critic of Oceania. *Pouliuli* is well known for its powerful attack on missionary colonialism as well as for its misogyny. To "make it *niu*" the Tusitala way is to both honor and challenge the novel, to value tradition and also enable tradition to change its silencing of women. Like *niu* itself, indigenous forms of knowing and histories are incorporated into Marsh's Oceanian modernity, one that allows women to "speak / a loud silence."

DIASPORA

Diaspora is one dimension of communal identities that intersectional approaches can illuminate. Island peoples are and have always been on the move—from one island to another, one archipelago to another, from islands to continents and back. The "land-sea" that the archipelago conjoins in its very etymology suggests mobility. Diaspora, too, is about mobility; it implies old and new homelands, what James Clifford (1997, 254) calls "dwelling-in-displacement." In *Cartographies of Diaspora: Contesting Identities*, Avtar Brah writes that all diasporas are "lived and re-lived through multiple modalities: modalities, for example, of gender, 'race,' class, religion, language and generation. As such, all diasporas are differentiated, heterogeneous, contested spaces, even as they are implicated in the construction of a common 'we'" (1996, 184). How, she asks, is the "collective 'we' constituted"? "Who is empowered and who is disempowered in a specific construction of the 'we'?" (184).

The "we" of all archipelagoes is similarly heterogeneous. Archipelagic thinking in its dual attention to materiality and epistemology needs to remain attuned to both the diversities and potential for collectivities around the planet: land and sea, oceans, archipelagoes, and continents. "Care," one of the poems in this volume by Craig Santos Perez (born in Guam, migrant to California at fifteen), illuminates these planetary mobilities as the lines move from the nurturing father on O'ahu, to Syria and the refugee boats on the Mediterranean, to those seeking asylum over the hot desserts, to those in buses and trains ending in detention centers, and finally to the utopian image of a "horizon of care."

> My 16-month old daughter wakes from her nap
> and cries. I pick her up, press her against my chest
>
> and rub her back until my palm warms
> like an old family quilt, "Daddy's here, daddy's here,"
>
> I whisper. Here is the island of O'ahu, 8,500 miles
> from Syria. But what if Pacific trade winds suddenly
>
> became helicopters? Flames, nails, and shrapnel
> indiscriminately barreling towards us? What if shadows
>
> cast against our windows aren't plumeria
> tree branches, but soldiers and terrorists marching
>
> in heat. Would we reach the desperate boats of
> the Mediterranean in time? If we did, could I straighten
>
> my legs into a mast, balanced against the pull and drift
> of the current? . . .
> . . .
> To all parents who brave the crossing:
> you and your children matter. I hope
>
> your love will teach the nations that emit
> the most carbon and violence
>
> that they should, instead, remit the most
> compassion. I hope, soon, the only difference
>
> between a legal refugee and an illegal migrant
> will be how willing we are to open our homes,
>
> offer refuge, and carry each other
> towards the horizon of care.
>
> —Craig Santos Perez, from "Care"

CONCLUSION

Scattered fragments big and small of Kathryn Chan's *Árchipelago* form a fitting visual metaphor for *Contemporary Archipelagic Thinking* and the multidirectional, multidimensional epistemology it generates. The blue-green of its aquamarine pieces suggest how the sea is in the land, the land is in the sea. Each piece has its distinctive shape, edges, and distance from the others. We see the pieces at a slant, from above, as if from the window of an airplane, not from the ground. And it is from this distance that patterns emerge. The image perfectly captures the double focus of this volume: the real geographies, geopolitics, and histories of these land-sea geoformations and the networking patterns of these planetary constellations as a mode of thinking, a lens on human and nonhuman interlappings.

NOTES

1. The volume's emphasis on epistemology builds on the concept of *thinking with* islands and archipelagoes in the special section of the 2013 issue of *Island Studies Journal* titled "Reframing Islandness: Thinking with the Archipelago." See especially Stratford's "The Idea of the Archipelago," "Imagining the Archipelago," and chapter in this volume; Pugh 2013.

2. All quotations from Perez's poems are taken from *Contemporary Archipelagic Thinking*. For a discussion of Perez, see Rowe 2017.

3. "Constellation," Merriam-Webster.com, https://www.merriam-webster.com/dictionary/constellation.

4. See also Gillis 2004 and Shell 204.

5. March 2003, 36. Selina Tusitala Marsh was the poet laureate of New Zealand (2017–2019) and a scholar and poet who teaches at the University of Auckland. She describes herself as of Samoan, Tuvalu, English, and French descent, all heritages that make their way into her work. Bougainville Island is currently an autonomous region of Papua New Guinea; this largest of the north Solomon Islands was named by French explorer Louis Antoine de Bougainville and was successively ruled by France, Germany, Australia, Japan, and Australia until 1975. *Fast Talking Pi* includes a section of poems titled "Talkback" (36–56) that perform satiric sass on the Western imaginary of the Pacific islands. In "Guys like Gauguin," Marsh also "thanks" Balboa, Vasco, and the "North," thus extending her "gratitude" to the so-called Age of Discoveries and its impact on "the Africas/the Orient, the Americas/and now us" (36–37).

6. In addition to essays in this volume, see DeLoughrey 2010, 2015, 2017.

7. Marsh 2013, 23. The Kyoto Protocol, which the United States did not sign, is the UN treaty signed by 192 nations to limit greenhouse gases. *An Inconvenient Truth* (2006) is the documentary film directed by Davis Guggenheim featuring Al Gore's campaign to educate people about the looming environmental crisis. Siligia is a well-known Samoan kickboxer. Marsh's use of "moana" in the poem's code-switching predates the Disney production of *Moana* (2016), in which the heroic girl protagonist is named Moana.

8. See especially Friedman 2015, 83–142. For discussions of modernity centered on oceans, see "Theories and Methodologies: Oceanic Studies," 2010.

9. For other world historians challenging Eurocentrism, see, for example, Abu-Lughod 1989, Frank 1995, and Hobson 1983.

10. I am aware that these categories have been critiqued as colonial constructions, but the deep-time studies of migration through genome data have also established the complex array of cultural and biological distinction and interaction. The sheer number of languages spoken in Oceania attest to cultural diversity. Like Oceania, Europe is studied in its diversities of genome, culture, and history—with names for different regions of Europe that are often weighted ideologically.

11. For comparisons of Caribbean and Pacific Island archipelagoes, see especially DeLoughrey 2007 and Roberts and Stephens 2017.

12. Marsh 2013, 13. Suva is the capital city on the east side of Fiji, far away from the beaches and tourist islands. Sigatoka is the town on Viti Leva, the island center of Fiji's coastal tourism. The "man in the blue mother hubbard" evokes contemporary urban poverty and the colonial tunic that missionaries imposed on PI women. Punja's Flour is a company in Fiji that makes plastic packaging and containers, with a name that recalls the large migration of Indians into Fiji starting in the nineteenth century, as do "jaina, balgaigani, painapiu," most likely the foods and people associated with Indian Fijians.

13. With a focus particularly on islands, Marc Shell's transhistorical and transglobal *Islandology: Geography, Rhetoric, Politics* asks these questions as well. He writes in "Planet," for example, "It takes little imagination nowadays to consider that North America and the other continents are islands—along with the other hundred thousand 'open-sea islands' on planet Earth. By almost inevitable logical extension, Earth itself is an island" (38). With his interrogation of the rhetoric of islands, the materiality of islands recedes, marking a considerable difference from *Contemporary Archipelagic Thinking*.

14. In their introduction to *Archipelagic American Studies* (6–11), Roberts and Stephens reference *The Myth of Continents* in developing their critique of continentalism as a distorting lens in American studies.

15. "How Many Oceans Are There?," National Ocean Service, https://oceanservice.noaa.gov/facts/howmanyoceans.html.

16. "Highest Mountain in the World," Geology.com, https://geology.com/records/highest-mountain-in-the-world.shtml.

17. See especially Steinberg 2013 in this issue; see also the introduction in Bystrom and Hofmeyr 2017 and DeLoughrey 2017.

18. For productive uses of archipelagic categories for the study of Britain and Ireland, see Esty 2004 and Brannigan 2015.

19. Marsh 2013, 6. Matiatia and Orapiu are bays located on Waiheke Island. See also "Airport Road to Apia" and "Afkasi Archipelago" for an archipelagic use of poetic page space (Marsh 2003, 8–9, 18–19).

20. Although the concept has a long history in feminist theory, the term "intersectionality" entered feminist discourse with Kimberlé Crenshaw's 1989 law review article, spread rapidly to become a central and much debated concept and methodology in gender and women's studies (Lykke 2009; Collins and Bilge 2016) and has been widely adapted to social and cultural theory and methodology more generally.

21. "Black Out # 33" is a page from Marsh's unpublished long poem *Blacking Out* Pouliuli and appears on the cover of *Contemporary Revolutions* (Friedman 2019), which includes her essay and her reference to Wendt as her mentor. Marsh published a number of additional pages from *Blacking Out* Pouliuli in *Tightrope* (2017). Wendt's novel *Ola* (1991) appears to answer the feminist criticism of *Pouliuli* with its focus on a complexly drawn female protagonist.

REFERENCES

Abu-Lughod, Janet. 1989. *Beyond European Hegemony: The World System, AD 1250–1350*. Oxford: Oxford University Press.

Best, Curwen. 2001. "Barbadian Aesthetics: Toward a Conceptualization." *ARIEL* 32, no. 1: 197–215.

Blaut, J. M. 1993. *The Colonizer's Model of the World: Geographical Difusionism and Eurocentric History*. New York: Guilford.

Blum, Hester. 2010. "The Prospect of Oceanic Studies." *PMLA* 125, no. 3: 670–77.

———. 2013. "Introduction: Oceanic Studies." *Atlantic Studies* 10, no. 2: 151–55.

Brah, Avtar. 1996. *Cartographies of Diaspora: Contesting Identities*. London: Routledge.

Brannigan, John. 2015. *Archipelagic Modernism: Literature in the Irish and British Isles, 1890–1970*. Edinburgh: Edinburgh University Press.

Bystrom, Kerry, and Isabel Hofmeyr. 2017. "Oceanic Routes: (Post-it) Notes on Hydro-colonialism." *Atlantic Studies* 69, no. 1: 11–16.

Césaire, Aimé. [1939] 1982. *Cahier d'un retour au pay natal / Notebook for a Return to the Native Land*. Trans. Clayton Eshleman and Annette Smith, 34–85. Berkeley: University of California Press.

Clifford, James. 1997. "Diasporas." In *Roots and Routes: Travel and Translation in the Late Twentieth Century*, 244–78. Cambridge, MA: Harvard University Press.

Cohen, Margaret. 2010. "Literary Studies on the Terraqueous Globe." *PMLA* 125, no. 3: 657–62.

Collins, Patricia Hill, and Silma Bilge. 2016. *Intersectionality*. Cambridge, MA: Polity Press.

continent. 2019. 8.1-2. www.continentcontinent.cc/index.php/continent.

Crenshaw, Kimberlé. 1989. "Demarginalizing the Intersection of Race and Sex: A Black Feminist Critique of Antidiscrimination Doctrine, Feminist Theory and Antiracist Politics." *University of Chicago Legal Forum* 4: 139–67.

DeLoughrey, Elizabeth M. 2001. "'The Litany of Islands, the Rosary of Archipelagoes': Caribbean and Pacific Archipelagraphy." *ARIEL* 32, no. 1: 21–51.

———. 2007. *Routes and Roots: Navigating Caribbean and Pacific Island Literatures*. Honolulu: University of Hawai'i Press.

———. 2010. "Heavy Waters: Waste and Atlantic Modernity." *PMLA* 125, no. 3: 703–12.

———. 2015. "Ordinary Futures: Interspecies Worldings in the Anthropocene." In *Global Ecologies and the Environmental Humanities: Postcolonial Approaches*, edited by Elizabeth DeLoughrey, Jill Didur, and Anthony Carrigan, 352–72. London: Routledge.

———. 2017. "Submarine Futures of the Anthropocene." *Comparative Literature* 69, no. 1: 32–44.

Du Bois, W. E. B. [1904] 1994. *The Souls of Black Folk*. New York: Dover Publications.

Esty, Jed. 2004. *Shrinking Island: Modernism and National Culture in England*. Princeton, NJ: Princeton University Press.

"An Extremely Detailed Map of the 2016 Election." 2018. *New York Times*. July 25. https://www.nytimes.com/interactive/2018/upshot/election-2016-voting-precinct-maps.html.

Frank, André Gunder. 1995. "The Modern World System Revisited: Rereading Braudel and Wallerstein." In *Civilizations and World Systems*, edited by Stephen K. Sanderson, 149–80. Walnut Creek, CA: Altamira.

Friedman, Susan Stanford. 2015. *Planetary Modernisms: Provocations on Modernity across Time*. New York: Columbia University Press.

———, ed. 2019. *Contemporary Revolutions: Turning Back to the Future in 21st Century Literature and Art*. London: Bloomsbury Academic Press.

Giddens, Anthony. 1990. *The Consequences of Modernity*. Stanford, CA: Stanford University Press.

Gikandi, Simon. 1992. *Writing in Limbo: Modernism and Caribbean Literature*. Ithaca, NY: Cornell University Press.

———. 1996. *Maps of Englishness: Writing Identity in the Culture of Colonialism*. New York: Columbia University Press.

Gillis, Joh R. 2004. *Islands of the Mind: How the Human Imagination Created the Atlantic World*. New York: Palgrave.

Gilroy, Paul. 1993. *The Black Atlantic: Modernity and Double Consciousness*. Cambridge, MA: Harvard University Press.

Glissant, Eduard. 1997. *Poetics of Relation*. Translated by Betsy Wing. Ann Arbor: University of Michigan Press.

Hall, Stuart, and Bram Gieben, eds. 1992. *Formations of Modernity*. Cambridge, UK: Polity Press.

Hau'ofa, Epeli. 2003. "Our Sea of Islands." In *We Are the Ocean: Selected Works*, 27–40. Honolulu: University of Hawai'i Press.

Haywood, Matthew, and Maebh Long, eds. 2020. *New Oceania: Modernisms and Modernities in the Pacific*. London: Routledge.

Ho, Elaine, Yee Lin, and Chirley Geok-Lin Lim, eds. 2001. "The Cultures and Literatures of Micro-states." Special issue, *ARIEL* 32, no. 1.

Hobson, John H. 1983. *The Eastern Origins of Western Civilization*. Cambridge: Cambridge University Press.

Jameson, Fredric. 2002. *A Singular Modernity: Essay on the Ontology of the Present*. London: Verso.

Lewis, Martin W., and Kären Wigen. 1997. *The Myth of Continents: A Critique of Metageography*. Berkeley: University of California Press.

Lionnet, Françoise, and Shu-mei Shih, eds. 2005. *Minor Transnationalism*. Durham, NC: Duke University Press.

Lykke, Nina. 2009. *Feminist Studies: A Guide to Intersectional Theory, Methodology and Writing*. London: Routledge.

Marsh, Selina Tusitala. N.d. *Blacking Out* Pouliuli. Unpublished long poem.

———. 2003. *Fast Talking PI*. Auckland: Auckland University Press.

———. 2013. *Dark Sparring*. Auckland: Auckland University Press.

———. 2017. *Tightrope*. Auckland: Auckland University Press.

———. 2019. "Make It *Niu*: Blacking Out Albert Wendt's *Pouliuli* the *Tusitala* Way." In *Contemporary Revolutions: Turning Back to the Future in 21st-Century*

Literature and Art, edited by Susan Stanford Friedman, 71–99. London: Bloomsbury Academic Press.

Olejniczak, M. J., D. J. Spiering, D. L. Potts, and R. J. Warren. 2018. "Urban Forests Form Isolated Archipelagoes." *Journal of Urban Ecology* 4, no. 1. doi .org./10.1093/jue/juy007.

Pohl, Ethel Baraona, and César Reyes Nájera. 2015. "A Tale of Two Cities: The Archipelago and the Enclave." *continent* 4, no. 3: 3–8.

Pound, Ezra. 1934. *Make It New*. London: Faber and Faber.

Pugh, Jonathan. 2013. "Island Movements: Thinking with the Archipelago." *Island Studies Journal* 8, no. 1: 9–24.

Roberts, Brian Russell, and Michelle Ann Stephens, eds. 2017. *Archipelagic American Studies*. Durham, NC: Duke University Press.

Rowe, John Carlos. 2017. "'Shades of Paradise': Craig Santos Perez's Transpacific Voyages." In *Archipelagic American Studies*, edited by Brian Russell Roberts and Michelle Ann Stephens, 213–31. Durham, NC: Duke University Press.

Said, Edward W. 1983. "Traveling Theory." In *The World, the Text, and the Critic*, 236–47. Cambridge, MA: Harvard University Press.

———. [1994] 2002. "Traveling Theory Reconsidered." In *Reflections on Exile and Other Essays*, 436–52. Cambridge, MA: Harvard University Press.

Shell, Marc. 2014. *Islandology: Geography, Rhetoric, Politics*. Stanford, CA: Stanford University Press.

Shih, Shu-mei. 2013. "Comparison as Relation." In *Comparison: Theories, Approaches, Uses*, edited by Rita Felski and Susan Stanford Friedman, 78–98. Baltimore, MD: Johns Hopkins University Press.

Steinberg, Philip E. 2013. "Of Other Seas: Metaphors and Materialities in Maritime Regions." *Atlantic Studies* 10, no. 2: 156–69.

Stratford, Elaine. 2013. "The Idea of the Archipelago: Contemplating Island Relations." *Island Studies Journal* 8, no. 1: 3–8.

———. 2017. "Imagining the Archipelago." In *Archipelagic American Studies*, edited by Brian Russell Roberts and Michelle Ann Stephens, 74–94. Durham, NC: Duke University Press.

Stratford, Elaine, Godfrey Baldacchino, Elizabeth McMahon, Carol Farbotko, and Andrew Harwood. 2011. "Envisioning the Archipelago." *Island Studies Journal* 6, no. 2: 113–30.

"Theories and Methodologies: Oceanic Studies." 2010. *PMLA* 125, no. 3: 657–763.

Vannini, April, Jeremy Fernando, and Berit Soli-Holt, eds. 2013. "Drift." Special issue, *continent* 3, no. 2.

Walcott, Derek. 1990. *Omeros*. New York: Farrar, Straus and Giroux.

Wallerstein, Immanuel. 1974. *The Modern World-System: Capitalist Agriculture and the European World Economy in the Sixteenth Century*. New York: Academic Press.

Wendt, Albert. [1977] 1990. *Pouliuli*. Honolulu: University of Hawai'i Press.

———. 1991. *Ola*. London: Penguin.

Yaeger, Patsy. 2010. "Editor's Column: Sea Trash, Dark Pools, and the Tragedy of the Commons." *PMLA* 123, no. 3: 523–45.

Index

Page references for figures are italicized.

About the Contributors

Jessica Swanston Baker is a Caribbeanist ethnomusicologist and assistant professor of ethnomusicology and the humanities at the University of Chicago. Her first book project, *Island Time: Speed, Music, and Modernity in St Kitts and Nevis*, focuses on tempo analysis as it relates to issues of postcoloniality and modernity within the small-island Caribbean. She holds a PhD in ethnomusicology from the University of Pennsylvania and was a postdoctoral fellow in critical Caribbean studies at Rutgers University.

Godfrey Baldacchino is prorector (international development and quality assurance) and professor of sociology at the University of Malta. At the University of Prince Edward Island, Canada, he also served as Canada Research Chair in Island Studies (2003–2013) and UNESCO Co-chair in Island Studies and Sustainability (2016–2020). He set up *Island Studies Journal* and served as its executive editor from 2006 to 2016. He has written and edited several books, including, for example, *Archipelago Tourism: Problems and Prospects* (2015).

Jeremy DeAngelo received his doctorate in medieval studies from the University of Connecticut and has held teaching and research positions at Rutgers University, Carleton College, and North Central University. He has published in *Scandinavian Studies*, *Peritia*, and *Anglo-Saxon England* and has a book published by Amsterdam University Press, *Outlawry, Liminality, and Sanctity in the Literature of the Early Medieval North Atlantic*.

Sarah DeMott is a Middle East research librarian at Harvard University. Her forthcoming book, *Mediterranean Nomads: Mobility between Tunisia and Sicily*, focuses on defamiliarizing cartographies of power through networks of gender, migration, and mobility. DeMott's recent scholarship draws on archipelagic connections across the Arab world from North Africa and the Sahel to the Levant and Indochina. Her current manuscript project, *Tropical by Design: French Empire and Afro-Asian Circulations across the Tropical World, c. 1880–1980*, mines colonial archives to understand the relationships between empire, technology, and the environment.

Haruki Eda is completing his PhD in sociology at Rutgers University, where he teaches sociology, Asian American studies, East Asian studies, and English composition. His first book project, *Queer Unification: Community and Healing in the Korean Diaspora*, examines ethnicity, sexuality, and place in US-based Korean community organizing. He is the 2014 winner of the Philips G. Davies Graduate Student Paper Award by the National Association for Ethnic Studies, and his entry was published in *Research Justice: Methodologies for Social Change* (2015). His work has also appeared in *Social Text Periscope* (2018) and anthologies *Gender and Love: Interdisciplinary Perspectives* (2013) and *Converging Identities: Blackness in the Contemporary African Diaspora* (2013).

Mary Eyring is associate professor of English and American studies at Brigham Young University. She has recently published a book, *Captains of Charity: The Writing and Wages of Postrevolutionary Atlantic Benevolence*, and with Matthew Mason and Christopher Hodson coedited a special issue of *Early American Studies* titled "The Global Turn in Early American Studies." She is writing a book titled *Saltwater: Globalizing Early American Grief.*

Susan Stanford Friedman is Hilldale Professor in the Humanities and Virginia Woolf Professor of English and Women's Studies Emerita at the University of Wisconsin, Madison. Recent books include *Planetary Modernisms: Provocations on Modernity across Time* (2015) and the edited volumes *Comparison: Theories, Approaches, Uses* (2013, with Rita Felski) and *Contemporary Revolutions: Turning Back to the Future in 21st-Century Literature and Art* (2018). Publishing widely on modernism, feminist theory and women's writing, migration, and religion, she is the author of *Mappings: Feminism and the Geographies of Encounter* (1998), among other books. Her work has been translated into eleven languages, and she is at work on *Sisters of Scheherazade: Religion, Diaspora, Muslim Women's Writing.*

Jenny R. Isaacs is a PhD candidate in geography at Rutgers University, where she is a part-time lecturer. Her critical environment research in conservation applies an archipelagic perspective within feminist decolonial more-than-human geography, critical animal studies, science and technology studies, and political ecology. Her work has been published in several disciplinary collections, and she recently served as guest editor for a special theme section on "more-than-human contact zones" for *Environment and Planning E: Nature and Space* (2019).

Christopher J. Lee is associate professor of history and Africana studies at Lafayette College in Easton, Pennsylvania. He has published five books, including *Making a World after Empire: The Bandung Moment and Its Political Afterlives* (2010, 2019). In 2016 he was a research associate with the Center for Cultural Analysis at Rutgers University as part of its seminar on archipelagoes.

Thomas P. Leppard is an assistant professor in the Department of Anthropology at Florida State University and codirects the Landscape Archaeology of Southwest Sardinia project. His research is primarily concerned with the transition to complex social organization and has appeared in *Current Anthropology*, *Human Ecology*, *Environmental Conservation*, *Antiquity*, *Archaeological and Anthropological Sciences*, *World Archaeology*, *Nature + Culture*, and other journals.

Pippa Marland is a Leverhulme Early Career Research Fellow based at the University of Leeds, where she is a member of the Environmental Humanities Research Group. Her research project is a study of the representation of farming in modern British nature writing. She has published widely on ecocriticism, ecopoetry, and nature writing and is coeditor of the new Routledge collection *Walking, Landscape and Environment*. She is author of a monograph titled *Ecocriticism and the Island: Readings from the British-Irish Archipelago*, forthcoming in 2021 as part of the Rowman & Littlefield Rethinking the Island series.

Yolanda Martínez-San Miguel is the Marta S. Weeks Chair in Latin American Studies at the University of Miami. She specializes in colonial and postcolonial Latin American and Caribbean literatures. She is author of four books: *Saberes americanos: Subalternidad y epistemología en los escritos de Sor Juana* (1999); *Caribe Two-Ways? Cultura de la migración en el Caribe insular hispánico* (2003); *From Lack to Excess: "Minor" Readings of Latin American Colonial Discourse* (2008); and *Coloniality of Diasporas: Rethinking Intra-colonial Migrations in a Pan-Caribbean Context* (2014). She has coedited two anthologies: *Critical Terms in Caribbean and Latin American Thought* (2016, with Ben. Sifuentes-Jáuregui and Marisa Belausteguigoitia) and *Trans Studies: The Challenge to Hetero/Homo Normativities* (2016, with Sarah Tobias).

Elizabeth A. Murphy is assistant professor of Roman archaeology in the Department of Classics at Florida State University and codirects the Landscape

Archaeology of Southwest Sardinia project. She specializes in the archaeological investigation of crafts production, technology, ancient work and labor, and economic history. Her work has appeared in the *American Journal of Archaeology*, *Oxford Journal of Archaeology*, *Journal of Mediterranean Archaeology*, and *Anatolian Studies*.

Anjali Nerlekar is associate professor in the Department of Africa, the Middle East, and South Asia at Rutgers University. She teaches courses on Indian and Indo-Caribbean literature at Rutgers, and her research interests include multilingual Indian modernisms, Indo-Caribbean literature and the Indian diaspora, spatial and cartographic studies, translation and multilingualism, postcolonial archives, and Indian print cultures. Her first book is on the post-1960 bilingual poetry from Bombay, *Bombay Modern: Arun Kolatkar and Bilingual Literary Culture* (2016). She has also coedited a special double issue of *Journal of Postcolonial Writing* (2017) titled "The Worlds of Bombay Poetry." Along with Bronwen Bledsoe, she has started the ongoing Bombay Poets' Archive at Cornell University.

Lizabeth Paravisini-Gebert is professor of Caribbean culture and literature in the Department of Hispanic Studies and the Environmental Studies Program at Vassar College, where she holds the Sarah Tod Fitz Randolph Distinguished Professor Chair. She is author of a number of books, among them *Phyllis Shand Allfrey: A Caribbean Life* (1996); *Jamaica Kincaid: A Critical Companion* (1999); *Creole Religions of the Caribbean* (2003, with Margarite Fernández Olmos; 2nd ed., 2011); *Literatures of the Caribbean* (2008); and *Extinctions: The Ecological Cost of Colonization in the Caribbean* (forthcoming). She coedits *Repeating Islands*, a popular blog on Caribbean culture, with Ivette Romero-Cesareo.

Jonathan Pugh, reader in island studies, Department of Geography, Newcastle University, is associated with the "relational turn" in island studies, exploring the relational characteristics disrupting insular island geographies. He has given a range of keynote addresses on this theme, authored more than sixty publications, and lectured at universities including Taipei, West Indies, Zurich, London, Rutgers, California, Virginia Tech, Cornell, Harvard, and Princeton. He is on the executive board of the International Small Island Studies Association (ISISA), the editorial boards of *Islands Studies Journal* (editor for island theory and philosophy) and *Shima*, and the steering committee of the International Geographical Union on Islands, and he is moderator of the "Islands Philosophy" email list islands-philosophy@newcastle.ac.uk (to join, email Jonathan.Pugh@ncl.ac.uk). He is presently working on the monograph "Anthropocene Islands: A Critical Agenda for Island Studies in the Anthropocene" as part of the Rowman & Littlefield Rethinking the Island series.

Gitanjali Pyndiah is a London-based Mauritian writer and researcher with a PhD in cultural studies from Goldsmiths, University of London. She explores language genealogies and extinctions, decolonial historiography, and creative practices (art, poetry, and music in Creole) in the Caribbean and the Indian Ocean regions. Her journal articles, book chapters, and encyclopedia entry feature in *Children's Geographies*, *Island Studies Journal, Interactions*, and publications by Bloomsbury Academic and the University of Mauritius Press. Her nonfiction and poetry appear in *Parentheses Journal, POETRY* magazine, *ADDA*, Amberflora, and the anthology *We Mark Your Memory* by Commonwealth Writers.

Brian Russell Roberts is associate professor of English and director of American studies at Brigham Young University. He has been a Fulbright Senior Scholar at Universitas Sebelas Maret in Solo, Indonesia, and has recently published two edited collections: *Indonesian Notebook: A Sourcebook on Richard Wright and the Bandung Conference* (with Keith Foulcher) and *Archipelagic American Studies* (with Michelle Ann Stephens). His book titled *Borderwaters: Amid the Archipelagic States of America* is forthcoming from Duke University Press in 2021.

Andrea Roppa is a STARS Research Fellow at the Università degli Studi di Padova and codirects the Landscape Archaeology of Southwest Sardinia project. His research interests lie in the study of the materiality, connectivity, and cultural identity of urban and rural communities of the Iron Age to Hellenistic western Mediterranean, with a focus on the island of Sardinia.

Craig Santos Perez is an indigenous Chamorro poet and scholar from the Pacific island of Guam. He is author of four books of poetry and coeditor of four anthologies. He is associate professor in the English Department at the University of Hawai'i, Mānoa.

Mimi Sheller is professor of sociology and founding director of the Center for Mobilities Research and Policy at Drexel University in Philadelphia. She is founding coeditor of the journal *Mobilities* and past president of the International Association for the History of Transport, Traffic & Mobility. She is author or coeditor of ten books, including *Island Futures: Caribbean Survival in the Anthropocene* (2020); *Mobility Justice: The Politics of*

Movement in an Age of Extremes (2018); *Aluminum Dreams: The Making of Light Modernity* (2014); *Citizenship from Below: Erotic Agency and Caribbean Freedom* (2012); *Consuming the Caribbean: From Arawaks to Zombies* (2003); and *Democracy after Slavery: Black Publics and Peasant Radicalism in Haiti and Jamaica* (2000).

Michelle Stephens is a licensed psychoanalyst and dean of the humanities at Rutgers University, New Brunswick, New Jersey. She is the author of *Skin Acts: Race, Psychoanalysis and the Black Male Performer* (2014) and co-editor of *Archipelagic American Studies* (2017, with Brian Russell Roberts); *Relational Undercurrents: Contemporary Art of the Caribbean Archipelago* (2017, with Tatiana Flores); and *Contemporary Archipelagic Thinking: Toward New Comparative Methodologies and Disciplinary Formations* (2020, with Yolanda Martínez-San Miguel). She writes regularly on Caribbean art, the intersections of race and psychoanalysis, and the emerging field of archipelagic American studies.

Elaine Stratford is professor of geography at the University of Tasmania. Her key focus is on fundamental, strategic, and applied research into how people flourish in place, in everyday and unexpected movements and migrations, in daily life, and over the life-course. Islands and archipelagoes have formed an important part of that work over time. Her most recent major works include *Geographies, Mobilities, and Rhythms over the Life-course* (2016) and *Home, Nature, and the Feminine Ideal* (2019).

Lisa Swanstrom is author of *Animal, Vegetable, Digital: Experiments in New Media Aesthetics and Environmental Poetics* (2016) and coeditor of *Science Fiction Studies* and the *electronic book review*. Her research and teaching interests include science fiction, natural history, media theory, and the digital humanities. She is associate professor of English at the University of Utah.

Lanny Thompson is professor in the Department of Sociology and Anthropology, University of Puerto Rico, Río Piedras. His work focuses on the comparative analysis of visual representations and aesthetics, education and governmentality, politics and governance, and military cartography among the islands that constitute the United States' overseas empire. He is author of *Imperial Archipelago* (2010) and *Nuestra isla y su gente* (2007). His articles have appeared in journals in the Philippines (*American Studies Asia*), Spain (*Culture and History Digital Journal*), Puerto Rico (*Op.Cit.*), and the United States (*Pacific Historical Review*). His contributions are included in the recent anthologies *Formation of U.S. Colonialism* (2014) and *Archipelagic American Studies* (2017).

www.ingramcontent.com/pod-product-compliance
Ingram Content Group UK Ltd.
Pitfield, Milton Keynes, MK11 3LW, UK
UKHW020128290926
15471UKWH00003B/352

9 781538 149911